DIGITAL EVIDENCE AND COMPUTER CRIME

SECOND EDITION

COMPUTER FORENSIC AND COMPUTER SECURITY RELATED BOOK TITLES:

- Casey, *Handbook of Computer Crime Investigation,* ISBN 0-12-163103-6, 448pp, 2002.
- Kovacich, *The Information Systems Security Officer's Guide,* ISBN 0-7506-7656-6, 361pp, 2003.
- Boyce and Jennings, *Information Assurance,* ISBN 0-7506-7327-3, 261pp, 2002.
- Stefanek, *Information Security Best Practices: 205 Basic Rules,* ISBN 0-878707-96-5, 194pp, 2002.
- De Clercq, *Windows Server 2003 Security Infrastructures: Core Security Features,* ISBN 1-55558-283-4, 752pp, 2004.
- Rittinghouse, *Wireless Operational Security,* ISBN 1-55558-317-2, 496pp, 2004.
- Rittinghouse & Hancock, *Cybersecurity Operations Handbook,* ISBN 1-55558-306-7, 1336pp, 2003.
- Speed & Ellis, *Internet Security,* ISBN 1-55558-298-2, 398pp, 2003.
- Erbschloe, *Implementing Homeland Security for Enterprise IT,* ISBN 1-55558-312-1, 320pp, 2003.
- XYPRO, *HP NonStop Server Security,* ISBN 1-55558-314-8, 618pp, 2003.

For more information, visit us on the web at **http://books.elsevier.com**

COMPUTER FORENSIC AND COMPUTER SECURITY RELATED PRODUCTS FROM ELSEVIER:

Compsec Newsletters and Journals:

- Biometric Technology Today
- Card Technology Today
- Computer Fraud & Security
- Computer Law and Security Report
- Computers & Security
- Information Security Technical Report
- Network Security
- Digital Investigation

Compsec Market Reports:

- Biometric Industry Report
- Smart Card Report

Compsec Conferences:

- Compsec2004
- Biometrics2004
- IDSmart

For more information, visit us on the web at **http://www.compseconline.com**

DIGITAL EVIDENCE AND COMPUTER CRIME

FORENSIC SCIENCE, COMPUTERS AND THE INTERNET

Second Edition

by Eoghan Casey

with contributions from

Robert Dunne
Monique Mattei Ferraro
Troy Larson
Michael McGrath
Gary Palmer
Tessa Robinson
Brent Turvey

ELSEVIER
ACADEMIC
PRESS

Amsterdam • Boston • Heidelberg • London • New York • Oxford
Paris • San Diego • San Francisco • Singapore • Sydney • Tokyo

First published 2000
Reprinted 2001, 2003
Second edition 2004

Academic Press
An Imprint of Elsevier
84 Theobald's Road, London WC1X 8RR, UK
http://www.academicpress.com

Academic Press
An Imprint of Elsevier
525 B Suite, Suite 1900, San Diego, California 92101-4495, USA
http://www.academicpress.com

ISBN 0-12-163104-4

A catalogue record for this book is available from the British Library

Library of Congress Cataloging-in-Publication Data

Casey, Eoghan
Digital evidence and computer crime: forensic science, computers, and the Internet/Eoghan Casey.—2nd ed.
p. cm.
Includes bibliographical references.
ISBN 0–12–163104–4 (alk.paper)
1. Computer crimes. 2. Evidence, Criminal. I. Title

HV6773.C35.C35 2004
363.25′968—dc22 2003063576

Typeset by Newgen Imaging Systems (P) Ltd, Chennai, India
Printed and bound in Great Britain

CONTENTS

ACKNOWLEDGMENTS

The substance and structure of this book are the result of several years of intensive case work, research, and teaching. Many colleagues, students, and my family and friends assisted me during this period. I am deeply grateful to each of you for your support and I would like to give special thanks to the following.

The contributors Robert Dunne, Monique Mattei Ferraro, Troy Larson, Mike McGrath, Gary Palmer, Tessa Robinson, and Brent Turvey for your inspiration, dedication, and for accepting the ambitious schedule. Barbara Troyer for your assistance with the figures in the text and your friendship over the years.

Colin Harris and Stephen Douglas for your direction and calming influence during the rough patches. Clare O'Connor for your lifelong encouragement and guidance. Jim Casey for your sage advice. Ita O'Connor for your clarity of thought and for making this all possible. Genevieve Gessert for your boundless love, friendship, and support.

H. Morrow Long, Andrew Newman, Shawn Bayern, and everyone else at Yale University for the supportive and challenging work–learning environment.

Bruce Patterson, Andy Russell, Jim Smith, Joe Sudol, Ken Gray, John Blawie, Mike O'Connor, Mark Califano and everyone in the Connecticut State Crime Laboratory, FBI, and State's and US Attorney's Offices for your dedication and camaraderie.

Fred Cotton, Todd Colvin, Jim Jolley, Keith Daniels, Glenn Lewis, and everyone at SEARCH for your continued support.

Tony Noble, Javier Torner, Larry Amos, Don Allison, Harlan Carvey, Paul Gillen, Harold Jones, Gary Gordon, Sarah Mocas, Warren Harrison, Mark Morrisey, Mark Bowser, Warren Kruse, and Carrie Whitcomb for your personal encouragement and contributions.

Brian Carrier for your technical review of Chapters 10–12 and E. Larry Lidz for your technical review of Chapters 16–18.

Brian Carrier, Joe Grand, Dan Mares, John Patzakis, Amber Schroader, Eric Thompson, Bob Weitershausen, and Walker Whitehouse for assistance with your digital evidence examination tools.

Mark Listewnik, Linda Beattie, Jennifer Rhuda, and the others at Academic Press who fostered this project over the years.

Eoghan Casey is a founding member of Knowledge Solutions LLC, a partnership of practicing forensic professionals who have made a commitment to providing quality training, information resources, and case consultations. He investigates network intrusions, intellectual property theft, and other computer-related crimes, and has extensive experience analyzing digital evidence. He has assisted law enforcement in a wide range of criminal investigations including homicide, child exploitation, cyber'stalking, and larceny. Eoghan also has extensive information security experience. As an Information Security Officer at Yale University and in subsequent consulting work, he has performed vulnerability assessments, deployed and maintained intrusion detection systems, firewalls and public key infrastructures, and developed policies, procedures, and educational programs. Eoghan holds a B.S. in Mechanical Engineering from the University of California at Berkeley, an M.A. in Educational Communication and Technology from New York University, and is currently working towards a Ph.D. in Computer Science at University College Dublin. Eoghan also brought together forensic experts to create the Handbook of Computer Crime Investigation: Forensic Tools and Technology. He can be contacted at eco@corpus-delicti.com.

Robert Dunne is an attorney and member of the faculty in the Department of Computer Science at Yale University, where he teaches "Computers and the Law," "Legal Implications of Computing Technology," and "Intellectual Property in the Digital Age." He has written on alternative paradigms for behavioral control in cyberspace, the impact of cyberspace on the legal profession, and Internet crime. Robert is Co-Director of Yale's Center for Internet Studies, an interdisciplinary enterprise whose goal is to explore the Internet's effect on society, and vice versa, from technological, legal, political, economic, cultural, and educational perspectives.

Monique Mattei Ferraro is an attorney with the Connecticut Department of Public Safety Computer Crimes and Electronic Evidence Unit and a Certified Information Systems Security Professional. She has been with the Department of Public Safety since 1987. She advises the Computer Crimes Unit and the Internet Crimes Against Children Task Force, develops training curricula for law enforcement, prosecutors and the public regarding Computer Crime Investigation and Internet Safety. Monique is co-author of Connecticut's Law Enforcement Guidelines for Computer and Electronic Evidence Search and Seizure, and is currently coauthoring a book on Investigating Child Exploitation with Eoghan. She holds a Master's degree from Northeastern University and a Law Degree from the University of Connecticut Law School.

Troy Larson is president of Digital Evidence Solutions, Inc., based in Seattle, Washington. Mr. Larson specializes in assisting attorneys with electronic evidence throughout all facets of litigation, particularly discovery and expert testimony. He is a member of the Washington State Bar and received both his undergraduate and law degrees from the University of California at Berkeley. He can be contacted at ntevidence@comcast.net.

Dr. Michael McGrath divides his time between clinical, administrative, teaching and research activities. His areas of special expertise include forensic psychiatry and criminal profiling. He has lectured on three continents and is a founding member of the Academy of Behavioral Profiling. He has published articles and/or chapters related to criminal profiling, sexual predators and the Internet, false allegations of sexual assault, and sexual asphyxia.

Gary Palmer is an INFOSEC Research Scientist for the MITRE Corporation, Bedford, MA in the Security and Information Operations Group (G021). He currently supports the Digital Forensic Research programs at the Air Force Research Laboratory's (AFRL) Rome Research Site in Rome, New York where he is focusing efforts on forensic identification, recovery and analysis of database systems as well as the forensic implications of wireless technology. Gary is also co-founder and a lead organizer of the Digital Forensic Research Workshop (DFRWS), sponsored by AFRL, which provides a forum for dialog between academic research and practice in the field. He attained a BS in 1979 from The Virginia Polytechnic Institute and State University (VATech). He has been active in the field of computer, network and information security since 1981 and wrote his first macro assembler program on a paper tape attached to a DEC PDP/11-44 running RSX11/M with 64K overlays. He lives in Sanford, FL, where he rides his motorcycle, plays the guitar and is currently enrolled in a Computer Forensic Graduate Certificate Program at the University of Central Florida.

Tessa Robinson B.L. studied at Trinity College Dublin and the Kings Inns. She is a practising barrister, called to the Irish bar in 1998. Her areas of practice include criminal, commercial, administrative and family law. Prior to commencing at the bar in Ireland she worked in New York with the Lawyers Committee for Human Rights, in Brussels with White & Case, in San Francisco with Morrison Foerster and in Washington D.C. with Hogan Hartson.

Brent Turvey received his Masters of Science in Forensic Science after studying at the University of New Haven, in West Haven, Connecticut. He also holds a Bachelor of Science degree from Portland State University in Psychology, with an emphasis on Forensic Psychology, and an additional Bachelor of Science degree in History. He has been studying violent sex offenders since 1990. He has consulted with law enforcement, attorneys, and private agencies in the United States, New Zealand, Canada, Australia, Korea and China on a range of serial rapes, homicides, staged crime scenes, and multiple death cases, as a forensic scientist and criminal profiler. He is author of the textbook Criminal Profiling: An Introduction to Behavioral Evidence Analysis, 2nd Ed., which is used in colleges and universities all over the world. He is currently a full partner, Criminal Profiler, and Instructor with Knowledge Solutions LLC.

INTRODUCTION

In the years since the first edition of this book, there has been an explosion of interest in digital evidence. This growth has sparked heated debates about tools, terminology, definitions, standards, ethics, and many other fundamental aspects of this developing field. It should come as no surprise that this book reflects my positions in these debates. Most notably, this text reflects my firm belief that this field must become more scientific in its approach. The primary aim of this work is to help the reader tackle the challenging process of seeking scientific truth through objective and thorough analysis of digital evidence. A desired outcome of this work is to encourage the reader to advance this field as a forensic science discipline.

AREAS OF SPECIALIZATION

Currently, there is little clarity in this field regarding areas of specialization and who should receive what training. For instance, there is no clear distinction between digital crime scene technicians (a.k.a. first responders) and digital evidence examiners, despite the fact that data recovery requires more knowledge than basic evidence documentation, collection, and preservation. The investigative process detailed in Chapter 4 suggests three distinct groups with different levels of knowledge and training.

- *Digital Crime Scene Technicians*: Individuals responsible for gathering data at a crime scene should have basic training in evidence handling and documentation as well as in basic crime reconstruction to help them locate all available sources of evidence on a network.
- *Digital Evidence Examiners*: Individuals responsible for processing particular kinds of digital evidence require specialized training and certification in their area.
- *Digital Investigators*: Individuals responsible for the overall investigation should receive a general training but do not need very specialized training or certification. Investigators are also responsible for reconstructing the actions relating to a crime using information from first responders and forensic examiners to create a more complete picture for investigators and attorneys.

Training and certification programs in this field should take into account these different areas of expertise.

For the purposes of this text, the more general term "digital investigator" is used to refer to individuals who play a key role in digital investigations, including computer security professionals, attorneys, law enforcement officers and forensic examiners.

RELIABILITY OF DIGITAL EVIDENCE

Digital investigators do not currently have a systematic method for stating the certainty they are placing in the digital evidence they are using to reach their conclusions. This lack of formalization makes it more difficult for courts and other decision makers to assess the reliability of digital evidence and the strength of digital investigators' conclusions. The Certainty Scale presented in Chapter 7 provides a consistent method of referring to the relative certainty of different types of digital evidence. The immediate aim of the Certainty Scale is to improve our ability to assess the reliability of digital evidence.

Ultimately, it is hoped that this Certainty Scale will point to areas that require additional attention in digital evidence research. Debate over C-values in specific cases may reveal that certain types of evidence are less reliable than was initially assumed. For some types of digital evidence, it may be possible to identify the main sources of error or uncertainty and develop analysis techniques for evaluating or reducing these influences. For other types of digital evidence, it may be possible to identify all potential sources of error or uncertainty and develop a more formal model for calculating the level of certainty for this type of evidence.

THE NEED FOR STANDARDIZATION

> Digital evidence is just another form of "latent" evidence that must be handed with scientific principles and legal boundaries. There is an investigative component for electronic crimes and a laboratory component for the digital evidence associated with those crimes. (Carrie Whitcomb, 2001, "*A Forensic Science Perspective on Digital Evidence Training, Education, and Certification,*" National Center of Forensic Science)

In 1994, the O.J. Simpson trial exposed many of the weaknesses of criminal investigation and forensic science. The investigation was hampered from the start with incomplete evidence collection, documentation and preservation at the crime scenes. Arguably, as a result of these initial errors, experienced forensic scientists were confused by and incorrectly interpreted important exhibits, introducing sufficient doubt for the jurors. The controversy surrounding this case made it clear that investigators and forensic scientists were not as reliable as was previously believed, undermining not just their credibility but also that of their profession. This crisis motivated many crime laboratories and investigative agencies to revise their procedures, improve training, and make other changes to avoid similar problems in the future. More recently flaws have been found in the fingerprint and DNA analysis performed by some crime laboratories, calling many convictions into questions and creating doubts about the analytical techniques themselves.

A similar crisis is looming in the area of digital evidence. The lack of generally required standards of practice and training allows weaknesses to persist, resulting in incomplete evidence collection, documentation and preservation as well as errors in analysis and interpretation of digital evidence. Innocent individuals may be in jail as a result of improper digital evidence handling and interpretation allowing the guilty to remain free. Failures to collect digital evidence have undermined investigations, preventing the apprehension or prosecution of offenders and wasting valuable resources on cases abandoned due to faulty evidence. If this situation is not corrected, the field will not develop to its full potential, justice will not be served, and we risk a crisis that could discredit the field. The only reason we have not already encountered such as crisis is that our mistakes have been masked by obscurity. As more cases become reliant on digital evidence and more attention is focused on it, we must take steps to establish standards of practice and compel practitioners to conform to them.

There have been several noteworthy developments toward standardization in this field. The International Organization of Computer Evidence (www.ioce.org) was established in the mid-1990s "to ensure the harmonization of methods and practices among nations and guarantee the ability to use digital evidence collected by one state in the courts of another state." In 1998, the Scientific Working Group on Digital Evidence (www.swgde.org) was established to "promulgate accepted forensic guidelines and definitions for the handling of digital evidence." In 2001, the first Digital Forensics Research Work Shop (www.dfrws.org) was held, bringing together knowledgeable individuals from academia, military and the private sector to discuss the main challenges and research needs in the field. This workshop also gave new life to an idea proposed several years earlier – a peer-reviewed journal – leading to the creation of the *International Journal of Digital Evidence* (www.ijde.org). In 2003, the American Society of Crime Laboratory Directors/Laboratory Accreditation Board (ASCLD/LAB) updated its accreditation manual to include standards and criteria for digital evidence examiners in US crime laboratories. In 2004 the UK Forensic Science Service plans to develop a registry of qualified experts, and several European organizations, including the European Network of Forensic Science Institutes (ENFSI) will publish examination and report writing guidelines for digital investigators. Also, Elsevier will begin publishing Digital Investigation: The International Journal of Digital Forensics and Incident Response (http://www.compseconline.com/digitalinvestigation/).

Historically, Forensic Science disciplines have used certification to oversee standards of practice and training. Certification provides a standard that individuals need to reach to qualify in a profession and provides an incentive to reach a certain level of knowledge. Without certification, the target and rewards of extra effort are unclear. This is not to say that everyone who handles digital evidence requires the same level of skill or training. A strong certification program needs to have tiered levels of certification facilitating progression upwards, setting basic requirements for crime scene technicians, and setting higher standards for specialists in a laboratory and for investigators who are responsible for analyzing evidence.

Although there are a growing number of certification programs for digital investigators, many are only available to law enforcement personnel and none are internationally accepted. In 2004, representatives from around the world convened to discuss the feasibility of an internationally accepted certification for digital investigators. The outcome is not decided and there are obstacles to such a certification. Some feel that proposed training requirements are too high while others fear that certification will enable anyone to enter the field and obtain specialized knowledge, even individuals who work for the defense on criminal cases. There is also the fear that setting standards and placing additional requirements on practitioners will make it more difficult to get digital evidence admitted in court.

Paradoxically, some of those concerned that training requirements will exclude them also want to exclude individuals who perform criminal defense work. In addition to being unethical, any attempt to withhold knowledge from criminal defense attorneys and experts stifles improvement and progress in the field by allowing misunderstandings and poor practices to persist. If we cannot work together despite our differences to improve the field, the only winners will be the criminals and the losers will be the innocents. The aim of everyone in this field should be to ensure the best reasonable standards and quality. In the long run, digital evidence processed properly by certified professionals is less likely to be impeached or cause an injustice.

The investigation into the Starnet Internet gambling company provides a good example of the successes of proper training and preparation. The August 1999 raid of Starnet's offices in Vancouver, BC, was the culmination of more then a year's worth of investigative effort and preparation by the Royal Canadian Mounted Police. Over 100 personnel from all over Canada were brought together to search and seize Starnet's systems. Search teams were trained to implement standard operating procedures to ensure consistency and were given sufficient equipment to store the large amounts of data that were anticipated. As a result of this planning, Starnet's office building and the network it contained were secured in a few minutes. Although it took several days, digital evidence from more than 80 computers was preserved. In 2001, Starnet pled guilty to violating Section 202 (1) b of the Canadian criminal code by having a machine in Canada for gambling or betting.

Although professionalization may not be desirable for some, it is necessary for all. Without generally accepted standards, there is no basis to judge work. Without certification, there is no basis upon which to assess qualifications. Our community has a duty to agree upon standards of practice and training, and to require practitioners to meet these standards through certification.

> This duty exists because in the forensic disciplines our opinions and interpretations are allowed to impact whether people are deprived of their liberties, and potentially whether they live or die. (Turvey, B., 2000, "*The Professionalization of Criminal Profiling*" in *Criminal Profiling*, Academic Press)

ROADMAP TO THE BOOK

This book draws from four fields: Law, Computer Science, Forensic Science, and Behavioral Evidence Analysis. The Law provides the framework within which all of the concepts of this book fit. Computer Science provides the technical details that are necessary to understand specific aspects of digital evidence. Forensic Science provides a general approach to analyzing any form of digital evidence. Behavioral Evidence Analysis provides a systematized method of synthesizing the specific technical knowledge and general scientific methods to gain a better understanding of criminal behavior and motivation.

This book is divided into five parts, beginning with a presentation of relevant legal issues and investigative methods in Part 1 (Chapters 1–7). Chapter 1 provides an overview. Chapter 2 (History and Terminology) provides relevant background, history, and terminology. Chapter 3 (Technology and Law) discusses legal issues that arise in computer related investigations, comparing US and European law. Chapter 4 (Investigative Process) discusses a systematic approach to investigating a crime based on the scientific method, providing a context for the remainder of this book. Chapter 5 (Investigative Reconstruction) describes how to use digital evidence to reconstruct events and learn more about the victim and the offender in a crime. Chapter 6 (Technology, MO, and Motive) is a discussion of the relationship between technology and the people who use it to commit crime. Understanding criminal motivation and behavior is key to assessing risks (will criminal activity escalate?), developing and interviewing suspects (who to look for and what to say to them), and focusing investigations (where to look and what to look for). Chapter 7 (Digital Evidence in Court) provides an overview of issues that arise in court relating to digital evidence.

Part 2 of this book (Chapters 8–13) begins by introducing basic Forensic Science concepts in the context of a single computer. Learning how to deal with individual computers is crucial because even when networks are involved, it is usually necessary to collect digital evidence stored on computers. Case examples and guidelines are provided to help apply the knowledge in this text to investigations. The remainder of Part 2 deals with specific kinds of computers and ends with a discussion of overcoming password protection and encryption on these systems.

Part 3 (Chapters 14–18) covers computer networks, focusing specifically on the Internet. A bottom-up approach is used to describe computer networks, starting with the raw data transmitted on networks and progressively building up to the types of data that can be found on networked systems and the Internet. The "top" of a computer network is comprised of the software that people use, like e-mail and the Web. This upper region hides the underlying complexity of computer networks and it is, therefore, necessary to examine and understand the underlying complexity of computer networks to appreciate fully the information found at the top of the network. Understanding the "bottom" of networks – the physical media (e.g. copper and fiber optic cables) that carry data between computers is also necessary to collect and analyze raw network traffic.

Part 4 of this book (Chapters 19–22) focuses on specific types of investigations starting with Computer Intrusions in Chapter 19. Tools and techniques specific to this type of investigation are presented and detailed case examples are used to demonstrate key points. Chapter 20 covers investigations of Cyberstalking. Chapter 21 details Sexual Predators on the Internet and Chapter 22 discusses computers as alibi.

Part 5 is a short segment that provides guidelines for handling and processing digital evidence. This text does not cover forensic image, video and audio analysis. For information about image/video/audio enhancement and other aspects of this kind of analysis, see Electronic Evidence by Gruber (Gruber 1995).

The Forensic Science concepts described early on in relation to a single computer are carried through to each layer of the Internet. Seeing concepts from Forensic Science applied in a variety of contexts will help the reader generalize the systematic approach to processing and analyzing digital evidence. Once generalized, this systematic approach can be applied to situations not specifically discussed in this text. In place of the CD-ROM in the first edition of this book, an interactive Web site (www.disclosedigital.com) provides practical exercises based on actual cases to demonstrate key aspects of investigating computer related crimes and to help the reader apply the concepts in this book to his/her own investigations. This Web site epitomizes a general educational model that others can replicate or borrow from to create inexpensive, educational resources to assist investigators.

DISCLAIMER

Tools are mentioned in this book to illustrate concepts and techniques, not to indicate that a particular tool is best suited to a particular purpose. Digital investigators must take responsibility to select and evaluate their tools.

Any legal issues covered in this text are provided to improve understanding only, and are not intended as legal advice. Competent legal advice should be sought to address the specifics of a case and to ensure that nuances of the law are considered.

PART 1

DIGITAL INVESTIGATION

CHAPTER 1

DIGITAL EVIDENCE AND COMPUTER CRIME

Within the past few years a new class of crime scenes has become more prevalent, that is, crimes committed within electronic or digital domains, particularly within cyberspace. Criminal justice agencies throughout the world are being confronted with an increased need to investigate crimes perpetrated partially or entirely over the Internet or other electronic media. Resources and procedures are needed to effectively search for, locate, and preserve all types of electronic evidence. This evidence ranges from images of child pornography to encrypted data used to further a variety of criminal activities. Even in investigations that are not primarily electronic in nature, at some point in the investigation computer files or data may be discovered and further analysis required.

(Lee *et al.* 2001).

Increasingly, criminals are using technology to facilitate their offenses and avoid apprehension, creating new challenges for attorneys, judges, law enforcement agents, forensic examiners, and corporate security professionals. Organized criminals around the globe are using technology to maintain records, communicate, and commit crimes. Offenders have obtained computer information about a police officer and his family to intimidate and discourage him from confronting them. As a result of the large amounts of drugs, child pornography, and other illegal materials being trafficked on the Internet, the US Customs Cybersmuggling Center has come to view every computer on the Internet in the United States as a port of entry. Felons have even broken into court systems to change their records and monitor internal communications.

CASE EXAMPLE (CALIFORNIA 2003):
William Grace and 22-year-old Brandon Wilson were sentenced to 9 years in jail after pleading guilty to breaking into court systems in Riverside, California, to alter records. Wilson altered court records relating to previous charges filed against him (illegal drugs, weapons, and driving under the influence of alcohol) to indicate that the charges had been dismissed. Wilson also altered court

Digital Evidence and Computer Crime Second Edition
ISBN: 0-12-163104-4

documents relating to several friends and family members. The network intrusion began when Grace obtained a system password while working as an outside consultant to a local police department. By the time they were apprehended, they had gained unauthorized access to thousands of computers and had the ability to recall warrants, change court records, dismiss cases, and read e-mail of all county employees in most departments, including the Board of Supervisors, Sheriff, and Superior Court judges. Investigators estimate that they seized and examined a total of 400 Gbytes of digital evidence (Sullivan 2003).

As more medical machinery, office equipment, home computers and appliances, and handheld devices are networked, there is greater exposure to abuse that could disrupt health care, office, and home life work. Network-based attacks targeting critical infrastructure such as power, health, communications, financial, and emergency response services are becoming a greater concern as terrorists become more technologically proficient.

CASE EXAMPLE (COWEN 2003):
Michael McKevitt was charged with directing terrorist activities. In addition to being accused of involvement in a bombing in Northern Ireland, McKevitt allegedly contacted an FBI informant on behalf of the Real IRA to obtain laptops for bomb detonation, encryption software, and personal digital assistants. McKevitt apparently saw cyberterrorism – the use of the networks to cause panic and loss of life – as the future over bombing and was taking steps to expand his terrorist organization's capabilities in this area. The evidence in the case includes laptops, e-mail messages, and mobile telephone records.

There is a positive aspect to the increasing use of technology by criminals – the involvement of computers in crime has resulted in an abundance of digital evidence that can be used to apprehend and prosecute offenders. For instance, computers played a role in the planning and subsequent investigations of both World Trade Center bombings. Ramsey Yousef's laptop contained plans for the first bombing and, during the investigation into Zacarias Moussaoui's role in the second attack, over 100 hard drives were examined (United States v. Moussaoui; United States v. Salameh *et al.*; United States v. Ramsey Yousef). Realizing the increasing use of high technology by terrorists compelled the United States to enact the USA Patriot Act and motivated the European Union to recommend related measures. E-mail ransom notes sent by Islamists who kidnapped and murdered journalist Daniel Pearl were instrumental in identifying the responsible individuals in Pakistan. In this case, the "threat to life and limb" provision in the USA Patriot Act enabled Internet Service Providers (ISPs) to provide law enforcement with information quickly, without waiting for search warrants.

While paper documents relating to Enron's misdeeds were shredded, digital records persisted that helped investigators build a case. Subsequent investigations of financial firms and stock analysts have utilized e-mail and other digital evidence to build a case. Realizing the value of digital evidence in such investigations, the Securities and Exchange Commission set an example in December 2002 by fining five brokerage houses a total of $8.25 million for failing to retain e-mail and other data as required by the Securities and Exchange Act of 1934 (SEC 2002).

Digital evidence can be useful in a wide range of criminal investigations including homicides, sex offenses, missing persons, child abuse, drug dealing, and harassment. Also, civil cases can hinge on digital evidence, and digital discovery is becoming a routine part of civil disputes. Computerized records can help establish when events occurred, where victims and suspects were, whom they communicated with, and may even show their intent to commit a crime. Robert Durall's Web browser history showed that he had searched for terms such as "kill + spouse," "accident + deaths," and "smothering" and "murder" prior to killing his wife (Johnson 2000). These searches were used to demonstrate premeditation and increase the charge to first-degree murder. Sometimes information stored on a computer is the only clue in an investigation. In one case, e-mail messages were the only investigative link between a murderer and his victim.

CASE EXAMPLE (MARYLAND 1996):
A Maryland woman named Sharon Lopatka told her husband that she was leaving to visit friends. However, she left a chilling note that caused her husband to inform police that she was missing. During their investigation, the police found hundreds of e-mail messages between Lopatka and a man named Robert Glass about their torture and death fantasies. The contents of the e-mail led investigators to Glass's trailer in North Carolina and they found Lopatka's shallow grave nearby. Her hands and feet had been tied and she had been strangled. Glass pled guilty, claiming that he killed Lopatka accidentally during sex.

Digital data are all around us and should be collected in any investigation routinely. More likely than not, someone involved in the crime used a computer, personal digital assistant, mobile telephone, or accessed the Internet. Therefore, every corporate investigation should consider relevant information stored on computer systems used by their employees both at work and home. Every search warrant should include digital evidence to avoid the need for a second warrant and the associated lost time and evidence. Even if digital data do not provide a link between a crime and its victim or a crime and its perpetrator, they can be useful in an investigation. Digital evidence can reveal how a crime was committed, provide investigative leads, disprove or support witness statements, and identify likely suspects.

This book provides the knowledge necessary to handle digital evidence in its many forms, to use this evidence to build a case, and to deal with the challenges associated with this type of evidence. This text presents approaches to handling digital evidence stored and transmitted using networks in a way that is most likely to be accepted in court. However, what is illegal, how evidence is handled, received, rejected, and how searches are authorized and conducted varies from country to country. Therefore, it is important to seek legal advice from a competent attorney, particularly since the law is changing to adapt to rapid technological developments.

1.1 DIGITAL EVIDENCE

For the purposes of this text, digital evidence is defined as *any data stored or transmitted using a computer that support or refute a theory of how an offense occurred or that address critical elements of the offense such as intent or alibi* (adapted from Chisum 1999).

> Digital evidence has been previously defined as any data that can establish that a crime has been committed or can provide a link between a crime and its victim or a crime and its perpetrator (Casey 2000). The definition proposed by the Standard Working Group on Digital Evidence (SWGDE) is any information of probative value that is either stored or transmitted in a digital form. Another definition proposed by the International Organization of Computer Evidence (IOCE) is information stored or transmitted in binary form that may be relied upon in court. However, these definitions focus too heavily on proof and neglect data that simply further an investigation. Additionally, the term *binary* in the later definition is inexact, describing just one of many common representations of computerized data.

The data referred to in this definition are essentially a combination of numbers that represent information of various kinds, including text, images, audio, and video. Take a moment to consider the types of digital data that exist and how they might be useful in an investigation. Computers are ubiquitous and digital data are being transmitted through the air around us and through wires in the ground beneath our feet.

The terms digital evidence and electronic evidence are sometimes used interchangeably. However, an effort should be made to distinguish between electronic devices such as mobile telephones and the digital data that they contain. Although this text necessarily covers certain aspects of electronic devices, the focus is on the digital evidence they contain. When considering the many sources of digital evidence, it is useful to categorize computer systems into three groups (Henseler 2000).

> *Open computer systems*: Open computer systems are what most people think of as computers – systems comprised of hard drives, keyboards, and monitors such as laptops, desktops, and servers that obey standards. These systems, with their ever increasing amounts of storage space, can be rich sources of digital evidence. A simple file can contain incriminating information and can have associated properties that are useful in an investigation. For example, details such as when a file was created, who created it, or that it was created on another computer can all be important.
>
> *Communication systems*: Traditional telephone systems, wireless telecommunication systems, the Internet, and networks in general can be a source of digital evidence. For instance, the Internet carries e-mail messages around the world. The time a message was sent, who sent it, or what the message contained can all be important in an investigation. To verify when a message was sent, it may be necessary to examine log files from intermediate servers and routers that handled a given message. To verify the contents of a message, it may be necessary to eavesdrop on the communication as it occurs.

Embedded computer systems: Mobile telephones, personal digital assistants, smart cards, and many other systems with embedded computers may contain digital evidence. For example, navigation systems can be used to determine where a vehicle has been and Sensing and Diagnostic Modules in many vehicles hold data that can be useful for understanding accidents, including the vehicle speed, brake status, and throttle position during the last five seconds before impact. Microwave ovens are now available with embedded computers that can download information from the Internet and some home appliances allow users to program them remotely via a wireless network or the Internet. In an arson investigation, data recovered from a microwave can indicate that it was programmed to trigger a fire at a specific time.

Given the ubiquity of digital evidence it is the rare crime that does not have some associated data stored and transmitted using computer systems. A trained eye can use these data to glean a great deal about an individual, providing such insight that it is like looking through a stained glass window into the individual's personal life and thoughts. An individual's personal computer and their use of network services are effectively behavioral archives, potentially retaining more information about an individual's activities and desires than even his/her family and closest friends. E-commerce sites use some of this information for direct marketing and a skilled digital investigator can delve into these behavioral archives and gain deep insight into a victim or offender (Casey 2002).

Despite its prevalence, few people are well versed in the evidentiary, technical, and legal issues related to digital evidence and as a result, digital evidence is often overlooked, collected incorrectly, or analyzed ineffectively. The goal of this text is to equip the reader with the necessary knowledge and skills effectively to use digital evidence in any kind of investigation. This text illuminates the technical, investigative, and legal facets of handling and utilizing digital evidence.

1.2 INCREASING AWARENESS OF DIGITAL EVIDENCE

By now it is well known that attorneys and police are encountering progressively more digital evidence in their work. Less obviously, computer security professionals and military decision makers are concerned with digital evidence. An increasing number of organizations are faced with the necessity of collecting evidence on their networks in response to incidents such as computer intrusions, fraud, intellectual property theft, child pornography, stalking, sexual harassment, and even violent crimes.

More organizations are considering legal remedies when criminals target them and are giving more attention to handling digital evidence in a way that will hold up in court. Also, by processing digital evidence properly, organizations are protecting themselves against liabilities such as invasion of privacy and unfair dismissal claims. As a result, there are rising expectations that computer security professionals have training and knowledge related to digital evidence handling.

In addition to handling evidence properly, corporations and military operations need to respond to and recover from incidents rapidly to minimize the losses caused by an incident. Many computer security professionals deal with hundreds of petty crimes each month and there is not enough time or resources to open a full investigation for each incident. Therefore, computer security professionals attempt to limit the damage and close each investigation as quickly as possible. There are three significant drawbacks to this approach. First, each unreported incident robs attorneys and law enforcement personnel of an opportunity to learn about the basics of computer-related crime. Instead, they are only involved when the stakes are high and the cases are complicated. Second, computer security professionals develop loose evidence processing habits that can make it more difficult for law enforcement personnel and attorneys to prosecute an offender. Third, this approach results in underreporting of criminal activity, deflating statistics that are used to allocate corporate and government spending on combating computer-related crime.

System administrators who find child pornography on computers in their workplace are in a perilous position. Simply deleting the contraband material and not reporting the problem may be viewed as criminally negligent. A system administrator who did not muster his employer's support before calling the police to report child pornography placed on a server by another employee was disavowed by his employer, had to hire his own lawyer, testify in his own time, and ultimately find a new job. Well meaning attempts to investigate child pornography complaints have resulted in the system administrator being prosecuted for downloading and possessing illegal materials themselves. Therefore, in addition to being technically prepared for such incidents, it is important for organizations and system administrators to have clear policies and procedures for responding to these problems.

Balancing thoroughness with haste is a demanding challenge. Tools that are designed for detecting malicious activity on computer networks are rarely designed with evidence collection in mind. Some organizations are attempting to address this disparity by retrofitting their existing systems to address authentication issues that arise in court. Other organizations are implementing additional systems specifically designed to secure digital evidence, popularly called Network Forensic Analysis Tools (NFATs). Both approaches have shortcomings that will be addressed gradually as software designers become more familiar with issues relating to digital evidence.

Government agencies are also interested in using digital evidence to detect terrorist activities and prevent future attacks. As a result, data mining technologies that were previously used to detect and investigate criminal activity that occurred in the past are now being adapted to identify suspicious, but not necessarily criminal, activities. Understandably, the possibility of the government freely sifting through every citizen's personal data for anything that looks suspicious is a privacy advocate's worst nightmare. There is certainly a risk that these pre-crime systems will do more harm than the problems they aim to address.

Ultimately, these systems will not achieve their intended goal because of inadequate training data sets, inaccurate data, high numbers of false positives, and information overload. With detailed knowledge of only several thousand known terrorists and ignoring the fact that terrorists regularly change their behavior to evade detection, it is statistically impossible to develop data mining methods that can reliably distinguish between normal and suspicious activity. The resulting inaccurate data mining methods would result in false positives that could ruin the lives of thousands, perhaps millions, of innocent individuals. Considering the amount of junk mail that is incorrectly addressed to Mr Eogliam Casey, Mr Bogan Caseui, and Ms Eileen Casey, it is likely that erroneous data in the underlying databases will increase the number of false positives in data mining. Even if data mining stumbled upon one actual terrorist, this lead would probably be lost among the false positives and bureaucracy created by the data mining process. Let us just hope that careless efforts to utilize these powerful data mining technologies do not cause too much damage and inhibit our ability to use them to investigate crimes.

Keep in mind that criminals are also concerned with digital evidence and will attempt to manipulate computer systems to avoid apprehension. Therefore, digital investigators cannot simply rely on what is written in this book to process digital evidence and must extend the lessons to new situations. With this in mind, in addition to presenting specific techniques and examples, this text provides general concepts and methodologies that can be applied to new situations with some thought and research on the part of the reader.

1.3 CHALLENGING ASPECTS OF DIGITAL EVIDENCE

Digital evidence as a form of physical evidence creates several challenges for forensic examiners. First, it is a messy, slippery form of evidence that can be very difficult to handle. For instance, a hard drive platter contains a messy amalgam of data – pieces of information mixed together and layered on top of each other over time. Only a small portion of this amalgam might be relevant to a case, making it necessary to extract useful pieces, fit them together, and translate them into a form that can be interpreted.

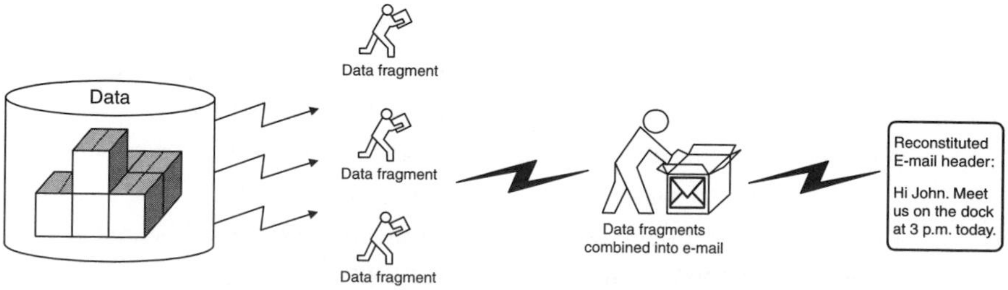

Figure 1.1

Conceptual depiction of data fragments being extracted from a hard drive platter, combined, and translated into an e-mail message.

Similarly, radio waves and microwaves traveling through the air contain a tangle of data, making it necessary to find the desired signal amongst the noise and translate it into the data that can be understood (Figure 1.1). This is conceptually similar to DNA analysis – the relevant information must be extracted from human fluid/tissue, processed, and translated into a form that we understand.

Second, digital evidence is generally an abstraction of some event or digital object. When a person instructs a computer to perform a task such as sending an e-mail, the resulting activities generate data remnants that give only a partial view of what occurred (Venema, Farmer 2000). Unless someone has installed surveillance equipment, individual mouse clicks, keystrokes, internal system commands, and other minutiae are not retained. Only certain results of the activity such as the e-mail message and server logs remain to give us a partial view of what occurred. Even when such minutiae are recorded, the electrical impulses of our mouse button clicks and keyboard depressions must be translated into data before they have any meaning. Similarly, an e-mail message and server log stored on a disk are the result of several layers of abstraction from magnetic fields on the disk to the letters and numbers that we see on the screen. Therefore, we never see the actual data but only a representation, and each layer of abstraction can introduce errors (Carrier 2003).

This situation is similar to that of the traditional crime scene investigation. In a homicide case, there may be clues that can be used to reconstruct events like putting a puzzle together. However, all of the puzzle pieces are never available, making it impossible to create a complete reconstruction of the crime. This book describes various sources of digital evidence and how these multiple, independent sources of corroborating information can be used to develop a more complete picture of the associated crime.

Third, the fact that digital evidence can be manipulated so easily raises new challenges for digital investigators. Digital evidence can be altered either maliciously by offenders or accidentally during collection without leaving any obvious signs of distortion. Fortunately, digital evidence has several features that mitigate this problem.

- Digital evidence can be duplicated exactly and a copy can be examined as if it were the original. It is common practice when dealing with digital evidence to examine a copy, thus avoiding the risk of damaging the original.
- With the right tools it is very easy to determine if digital evidence has been modified or tampered with by comparing it with an original copy.
- Digital evidence is difficult to destroy. Even when a file is "deleted" or a hard drive is formatted, digital evidence can be recovered.
- When criminals attempt to destroy digital evidence, copies and associated remnants can remain in places that they were not aware of.

CASE EXAMPLE (BLANTON 1995):
When Colonel Oliver North was under investigation during the Iran Contra affair in 1986, he was careful to shred documents and delete incriminating e-mails from his computer. However, unbeknown to him, electronic messages sent using the IBM Professional Office System (PROFS) were being regularly backed up and were later retrieved from backup tapes.

Fourth, digital evidence is usually circumstantial making it difficult to attribute computer activity to an individual. Therefore, digital evidence can only be one component of a solid investigation. If a case hinges upon a single form or source of digital evidence such as date–time stamps on computer files, then the case is unacceptably weak. Without additional information, it could be reasonably argued that someone else used the computer at the time. For instance, authentication mechanisms on more secure computers can be bypassed and many computers do not require a password, allowing anyone to use them. Similarly, if a defendant argues that some exonerating digital evidence was not collected from one system, this would only impact a weak case that does not have supporting evidence of guilt from other sources.

CASE EXAMPLE (UNITED STATES v. GRANT 2000):
In an investigation into the notorious online Wonderland Club, Grant argued that all evidence found in his home should be suppressed because investigators had failed to prove that he was the person associated with the illegal online activities in question. However, the prosecution presented enough corroborating evidence to prove their case.

1.4 FOLLOWING THE CYBERTRAIL

Many people think of the Internet as separate from the physical world. This is simply not the case – crime on the Internet mirrors crime in the physical world. There are several reasons for this cautionary note. First, a crime on the Internet usually reflects a crime in the physical world, with human perpetrators and victims and should be treated with the same gravity. To neglect the very real and direct link between people and the online activities that involve them limits one's ability to investigate and understand crimes with an online component. Auction fraud provides a simple demonstration of how a combination of evidence from the virtual and physical worlds is used to apprehend a criminal.

CASE EXAMPLE (AUCTION FRAUD 2000):
A buyer on E-bay complained to police that he sent a cashier's check to that seller but received no merchandise. Over a period of weeks, several dozen similar reports were made to the Internet Fraud Complaint Center against the same seller.

To hide his identity, the seller used a Hotmail account for online communications and several mail drops to receive checks. Logs obtained from Hotmail revealed that the seller was accessing the Internet through a subsidiary of Uunet. When served with a subpoena, Uunet disclosed the suspect's MSN account and associated address, credit card and telephone numbers. Investigators also obtained information from the suspect's bank with a subpoena to determine that the cashier's checks from the buyers had been deposited into the suspect's bank account. A subpoena to E-bay for auction history and complaints and supporting evidence from each of the buyers helped corroborate the connections between the suspect and the fraudulent activities. Employees at each mail drop recognized a photograph of the suspect obtained from the Department of Motor Vehicles. A subpoena to the credit card company revealed the suspect's Social Security Number and a search of real estate property in the suspect's name turned up an alternate residence where he conducted most of his fraud.

Second, while criminals feel safe on the Internet, they are observable and thus vulnerable. We can take this opportunity to uncover crimes in the physical world that would not be visible without the Internet. Murders have been identified as a result of their online actions, child pornography discovered on the Internet has exposed child abusers in the physical world, and local drug deals are being made online. By observing the online activities of offenders in our neighborhoods, jurisdictions, and companies, we can learn more about the criminal activities that exist around us in the physical world. Third, when a crime is committed in the physical world, the Internet often contains related digital evidence and should be considered as an extension of the crime scene. For instance, a program like Chat Monitor can be used to find individuals from a specific geographical region who are using Internet Relay Chat (IRC) networks to exchange child pornography.

The crimes of today and the future require us to become skilled at finding connections between crimes on the Internet and in the physical world, following the cybertrail if you will. By following the cybertrail, investigators of physical world crime can find related evidence on the Internet and investigators of crime on the Internet find related evidence in the physical world. The cybertrail should be considered even when there is no obvious sign of Internet activity. Criminals are learning to conceal their Internet activities and even the most obvious indication that a computer is used to access the Internet is disappearing: a cable connecting the computer to a jack in the wall. With the rise in wireless networks fewer computers have network cables.

The Internet may contain evidence of the crime even when it was not directly involved. There are a growing number of sensors on the Internet such as cameras showing live highway traffic on the Web as shown in Figure 1.2. These sensors may inadvertently capture evidence relating to a crime. In one investigation of reckless driving that resulted in a fatal crash, the position of the victim's car and average speed was determined using

Figure 1.2

Web camera of live traffic from www.marylandroads.com.

position data relating to a mobile telephone in the car, enabling investigators to locate a surveillance camera at a gas station along the route. The surveillance videotape showed the offender's car tailgating the victim at high speed, supporting the theory that the offender had driven the victim off the road. Conversely, a cyberstalker can access sensors over the Internet, such as a camera and microphone on a victim's home computer, to monitor her activities.

In addition to the Internet, digital evidence may exist on commercial systems (e.g. ATMs, credit cards, debit cards) and privately owned networks. These privately owned networks can be a richer source of information than the public Internet. In addition to having internal e-mail, chat, newsgroup, and Web servers, these networks can have databases, document management systems, time clock systems, and other networked systems that contain information about the individuals who use them. Also, private organizations often configure their networks to monitor individuals' activities more than the public Internet. Some organizations monitor which Web pages were accessed from computers on their networks. Other organizations even go so far as to analyze the raw traffic flowing through their network for signs of suspicious activity.

Furthermore, these smaller networks usually contain a higher concentration of digital information about the individuals who use them (more bits per square foot) making it easier to find and collect relevant digital data than on the global Internet. It is conceivable that a digital investigator could determine where an individual was and what he/she was doing throughout a given day, especially if the individual is an employee of an organization that makes heavy use of their network. The time an individual first logged into the network (and from where) would be recorded. E-mail sent and received by an individual throughout the day would be retrievable. The times an individual accessed certain files, databases, documents, and other shared resources might be available. The time an individual logged out of the network would be recorded. If the individual dialed in from home that evening, that would also be recorded and any e-mail sent or received may be retrievable.

1.5 CHALLENGING ASPECTS OF THE CYBERTRAIL

The dynamic and distributed nature of networks makes it difficult to find and collect all relevant digital evidence. Data can be spread over a group of adjacent buildings, several cities, states, or even countries. For all but the smallest networks, it is not feasible to take a snapshot of an entire network at a given instant. Also, network traffic is transient and must be captured while it is in transit. Once network traffic is captured, only copies remain and the original data are not available for comparison. The amount of data lost during the collection process can be documented but the lost evidence cannot be retrieved.

Also, networks contain large amounts of data and sifting through them for useful information can be like looking for a needle in a haystack and can stymie an investigation. Even when the vital digital evidence is obtained, networks provide a degree of anonymity making it difficult to attribute online activities to an individual. This text provides methods of addressing these obstacles.

1.6 FORENSIC SCIENCE AND DIGITAL EVIDENCE

The ultimate aim of this text is to demonstrate how digital evidence can be used to reconstruct a crime or incident, identify suspects, apprehend the guilty, defend the innocent, and understand criminal motivations. Forensic Science provides a large body of proven investigative techniques and methods for achieving these ends that are referenced extensively in this text. By *forensic* we mean a characteristic of evidence that satisfies its suitability for admission as fact and its ability to persuade based upon proof (or high statistical confidence). Strictly speaking, Forensic Science is the application of science to law and is ultimately defined by use in court. For instance, the scientific study of insects has many applications including the study of insects on a decaying corpse – *forensic entomology*. Entomological evidence has been

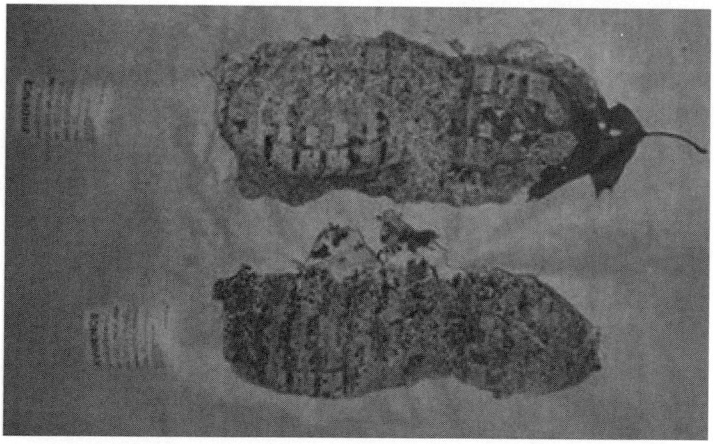

Figure 1.3

Shoe prints preserved using dental cement.

accepted in courts to help determine how long a body has been exposed to fauna in a specific area. Another example of forensic science is using shoe prints left at a crime scene to locate the source of the impressions. Different kinds of shoe prints on flat surfaces can be preserved using chemicals or electrostatic lifting devices, and impressions made in soil can be preserved using dental stone as shown in Figure 1.3. Forensic examiners use characteristics of these shoe prints to determine the type of shoe and ultimately to associate the impressions with the shoes that made them.

Similarly, the scientific study of digital data becomes a forensic discipline when it relates to the investigation and prosecution of a crime.

In Forensic Science, certainty is a word that is used with great care. We cannot be certain of what occurred at a crime scene when we only have a limited amount of information. Therefore, we can generally only present possibilities based on the limited amount of information.

This strict definition will disappoint readers whose work does not usually lead to prosecution, such as implementing a company's sexual harassment policy or investigating security incidents. Although prosecution is not always the goal of such investigations, they sometimes result in legal action. Therefore, it is important to handle digital evidence in such cases as if it were going to be used in a court of law. Furthermore, any investigation can benefit from the influence of Forensic Science. In addition to providing scientific techniques and theories for processing individual pieces of digital evidence, Forensic Science can help reconstruct crimes and generate leads. Using the scientific method to analyze available evidence, reconstruct the crime, and test their hypotheses, digital investigators can generate strong possibilities about what occurred.

To encourage corporate digital investigators to apply the principles of Forensic Science presented in this text, a broader definition of Forensic Science will be adopted. For the purpose of this text, *Forensic Science is the application of science to investigation and prosecution of crime, or the just resolution of conflict.*

1.7 SUMMARY

Digital evidence exists in abundance on open computer systems, communication systems, and embedded computer systems. A hard drive can store a small library, digital cameras can store hundreds of high-resolution photographs, and a computer network can contain a vast amount of information about people and their behavior. At any given moment, private telephone conversations, financial transactions, confidential documents, and many other kinds of information are transmitted in digital form through the air and wires around us – all potential sources of digital evidence. Even crimes that were not committed with the assistance of computers can have related digital evidence, including homicide, arson, suicide, abduction, torture, and rape.

Given the widespread use of computers and the wide use of networks, it would be a grave error to overlook them as a source of evidence in *any* crime. Computers should be collected in all criminal, civil, and corporate internal investigations and the cybertrail should be followed routinely. Also recall that privately owned networks may have more sources of digital evidence than the global Internet, detailed monitoring of individuals' activities, and a higher concentration of digital data per unit area.

Although there are many challenges to dealing with evidence stored on and transmitted using computers, we must not be daunted. Criminals will be especially eager to use computers and networks if they know that attorneys, forensic examiners, or computer security professionals are ill equipped to deal with digital evidence. Therefore, anyone who is involved with criminal investigation, prosecution, or defense work should be comfortable with personal computers and networks as a source of evidence. One of the major aims of this work is to educate students and professionals in the computer security, criminal justice, and forensic science communities about computers and networks as a source of digital evidence.

Education can only bring us so far. Ultimately, all of these groups must work together to build a case and bring offenders to justice. In addition to learning how to handle digital evidence, law enforcement officers must know when to seek expert assistance. Similarly, computer security professionals must know when to call law enforcement for assistance. Attorneys (both prosecution and defense) must also learn to discover digital evidence, defend it against common arguments, and determine whether it is admissible. Forensic computer examiners must continually update their skills effectively to support investigators, attorneys, and corporate security professionals in an investigation.

REFERENCES

Blanton T. (1995) "The Top-Secret Computer Messages the Reagan/Bush White House Tried to Destroy", National Security Archive (Available online at http://www.gwu.edu/~nsarchiv/white_house_email/).

Carrier B. (2003) "Defining Digital Forensic Examination and Analysis Tool Using Abstraction Layers", *International Journal of Digital Evidence*, Volume 1, Issue 4, Syracuse, NY (Available online at http://www.ijde.org/docs/02_winter_art2.pdf).

Casey E. (2000) *Digital Evidence and Computer Crime*, 1st Ed., London: Academic Press.

Casey E. (2002) "Cyberpatterns: Criminal Behavior on the Internet" in Turvey B. Criminal Profiling: An Introduction to Behavioral Evidence Analysis 2nd Ed., London: Academic Press.

Chisum J. W. (1999) "Crime Reconstruction and Evidence Dynamics", Presented at the Academy of Behavioral Profiling Annual Meeting, Monterey, CA.

Cowen R. (2003) "Real IRA trial told of £750,000 spy payment", *The Guardian*, June 19, 2003.

Henseler J. (2000) "Computer Crime and Computer Forensics" in the Encyclopedia of Forensic Science, London: Academic Press.

Johnson T. (2000) "Man searched web for way to kill wife, lawyers say", June 21, 2000, Seattle Post-Intelligencer (Available online at http://seattlepi.nwsource.com/local/mu rd21.shtml).

Lee H., Palmbach T. and Miller M. (2001) Henry Lee's Crime Scene Handbook, London: Academic Press.

Seaton D. (2002) "Hack attacks spook county", The Press Enterprise, 06/14/2002.

Securities and Exchange Commission (2002) "Order instituting proceedings pursuant to Section 15(b)(4) and Section 21c of the Securities Exchange Act of 1934, making findings and imposing cease-and-desist orders, penalties, and other relief: Deutsche Bank Securities, Inc., Goldman, Sachs & Co., Morgan Stanley & Co. Incorporated, Salomon Smith Barney Inc., and U.S. Bancorp Piper Jaffray Inc.," Administrative Proceeding, File No. 3-10957 (Available online at http://www.sec.gov/litigation/admin/34-46937.htm)

Sullivan B. (2003) "Pair who hacked court get 9 years" MSNBC 02/07/03.

Venema W. and Farmer D. (2000) Forensic Computer Analysis: an Introduction, Doctor Dobb's Journal (Available online at http://www.ddj.com/documents/s=881/ddj 0009f/0009f.htm).

CASES

United States v. Grant (2000) Case No. 99-2332, US District Court, District of Maine (Available online at http://laws.lp.findlaw.com/1st/992332.html).

United States v. Ramzi Yousef, Eyad Ismoil (2003) (Available online at http://caselaw.findlaw.com/data/circs/2nd/98104IP.pdf).

United States v. Mohammad Salameh (1993) S12 93 CR. 180, US District Court, Southern District of New York (Available online at http://laws.findlaw.com/2nd/941312v2.html).

United States v. Zacarias Moussaoui (2001) US District Court, Eastern District of Virginia (Available online at http://notablecases.vaed.uscourts.gov/1:01-cr-00455/).

## CHAPTER 2

HISTORY AND TERMINOLOGY OF COMPUTER CRIME INVESTIGATION

Seizing, preserving, and analyzing evidence stored on a computer is the greatest forensic challenge facing law enforcement in the 1990s. Although most forensic tests, such as fingerprinting and DNA testing, are performed by specially trained experts, the task of collecting and analyzing computer evidence is often assigned to untrained patrol officers and detectives. While most forensic tests are performed in the analyst's own laboratory, investigators are required to search and seize computers at unfamiliar and potentially hostile sites, such as drug labs, residences, "boiler rooms", small business offices, and warehouse-sized computer centers.

(Rosenblatt 1995)

Besides component theft, some of the earliest recorded computer crimes occurred in 1969 and 1970 when student protestors burned computers at various universities. At about the same time, individuals were discovering methods for gaining unauthorized access to large time-shared computers (essentially stealing time on the computers), an act that was not illegal at the time. In the 1970s, many crimes involving computers and networks were dealt with using existing laws. However, there were some legal struggles because digital property was seen as intangible and therefore outside of the laws protecting physical property. Since then, the distinction between digital and physical property has become less pronounced and the same laws are often used to protect both.

Computer intrusion and fraud committed with the help of computers were the first crimes to be widely recognized as a new type of crime. The first computer crime law to address computer fraud and intrusion, the Florida Computer Crimes Act, was enacted in Florida in 1978 after a highly publicized incident at the Flagler Dog Track. Employees at the track used a computer to print fraudulent winning tickets. The Florida Computer Crimes Act also defined all unauthorized access to a computer as a crime, even if there was no maliciousness in the act. This stringent view

Digital Evidence and Computer Crime Second Edition
ISBN: 0-12-163104-4

of computer intrusion was radical at the time but has since been adopted by every US state except Vermont. This change of heart about computer intrusions was largely in reaction to the growing publicity received by computer intruders in the early 1980s. It was during this time that governments around the world started enacting similar laws. Canada was the first country, to enact a federal law to address computer crime specifically in amending their Criminal Code in 1983. The US Federal Computer Fraud and Abuse Act was passed in 1984 and amended in 1986, 1988, 1989, and 1990. The Australian Crimes Act was amended in 1989 to include Offenses Relating to Computers (Section 76) and the Australian states enacted similar laws at around the same time. In Britain, the Computer Abuse Act was passed in 1990 to criminalize computer intrusions specifically as discussed in Chapter 3.

Hollinger's Crime, Deviance and the Computer consists of a collection of articles from various authors and is separated into four sections: The Discovery of Computer Abuse (1946–76), The Criminalization of Computer Crime (1977–87), The Demonization of Hackers (1988–92), and The Censorship Period (1993–present).

In the 1990s, the commercialization of the Internet and the development of the World Wide Web (WWW) popularized the Internet, making it accessible to millions. Crime on the global network diversified and the focus expanded beyond computer intrusions. One of the earliest large-scale efforts to address the problem of child pornography on the Internet was Operation Long Arm in 1992, involving individuals in the United States who were obtaining child pornography from a Danish bulletin board system. A more detailed view of the history of computer crime can be found in Hollinger (1997). As the range of crimes being committed with the assistance of computers increased, new laws to deal with copyright, child pornography, and privacy were enacted as discussed in Chapter 3.

2.1 BRIEF HISTORY OF COMPUTER CRIME INVESTIGATION

Responding to the growth in computer-related crime, in the late 1980s and early 1990s law enforcement agencies in the United States began to work together to develop training and build their capacity to deal with the problem. These initiatives led to law enforcement training programs at centers such as SEARCH, Federal Law Enforcement Center (FLETC), and National White Collar Crime Center (NW3C).

Subsequently, the United States and other countries established specialized groups to investigate computer-related crime on a national level. However, the demands on these groups quickly exhausted their resources and regional centers for processing digital evidence were developed. These regional centers also became overloaded, causing many local law enforcement agencies to develop their own units for handling digital evidence. Additionally, some countries have updated the training programs in their academies, realizing that the pervasiveness of computers requires every agent of law enforcement to have basic awareness of digital evidence. This rapid development has resulted in a pyramid structure of first responders with basic collection and examination skills to handle the majority of cases, supported by regional laboratories to handle more advanced cases, and national centers that assist with the most challenging cases, perform research, and develop tools that can be used at the regional and local levels.

The rapid developments in technology and computer-related crime have created a need for specialization: digital crime scene technicians who collect digital evidence, examiners who process the acquired evidence, and digital investigators who analyze all available evidence to build a case. These specializations are not limited to law enforcement and have developed in the corporate world as well. In addition to recovering from a security incident, it is often necessary to collect digital evidence to determine what occurred and help decision makers assess the problem. This is conceptually similar to the situation in violent crime when paramedics tend to the injured person's needs while crime scene experts process the evidence. Since paramedics are often the first people on the scene, investigators depend on them for information about the crime scene and victims in their original state. If paramedics changed anything at a crime scene, investigators need to know this before reconstructing the crime.

Even when a single individual is responsible for collecting, processing, and analyzing digital evidence, it is useful to consider these tasks separately. Each area of specialization requires different skills and procedures – dealing with them separately makes it easier to define training and standards in each area. Realizing the need for standardization in training and best practices, in 2002, the Scientific Working Group for Digital Evidence (SWGDE)[1] published guidelines for training and best practices. As a result of these efforts, the American Society of Crime Laboratory Directors (ASCLD) proposed requirements for digital evidence examiners in forensic laboratories (ASCLD 2003). There are similar efforts to develop digital evidence examination into an accredited discipline under international standards (ISO 17025; ENFSI 2003).

[1] *http:// www.swgde.org*

The development of these standards has created a need for standards of practice for individuals in the field. To answer this need, certification and training programs are being developed to ensure that digital evidence examiners have the necessary skills to perform their work competently and to follow approved procedures. The aim is to create several tiers of certification, starting with a general knowledge exam that everyone must pass, including first responders who handle digital evidence, and then more specialized certifications for individuals who handle more complex cases in a laboratory setting.

2.2 EVOLUTION OF INVESTIGATIVE TOOLS

In the early days of computer crime investigation, it was common for digital investigators to use the evidentiary computer itself to obtain evidence. One risk of this approach was that operating the evidentiary computer could alter the evidence in a way that is undetectable. Although programs such as dd on UNIX existed in the 1980s and could be used to capture deleted data stored on a hard drive, these tools were not widely used and most digital evidence examinations at that time were performed at the file system level, neglecting deleted data.

It was not until the early 1990s, that tools like SafeBack and DIBS were developed to enable digital investigators to collect all data on a computer disk, without altering important details. At around the same time, tools such as those still available from Maresware and NTI were developed by individuals from the US Internal Revenue Service (IRS) to help digital investigators process data on a computer disk. The Royal Canadian Mounted Police (RCMP) also developed specialized tools for examining computers. As more people became aware of the evidentiary value of computers, the need for more advanced tools grew. To address this need, integrated tools like Encase and FTK were developed to make the digital investigator's job easier. These tools enable more efficient examination, by automating routine tasks and display data in a graphical user interface to help the user locate important details. Recently, there has been renewed interest in Linux as a digital evidence examination platform and tools such as The Sleuthkit and SMART have been developed to provide a user-friendly interface. More sophisticated tools utilizing powerful microscopes are available to recover overwritten data from hard drives, but these are prohibitively expensive for most purposes.

Unfortunately, many individuals are still unaware of the need for these tools. Although courts have been lenient on investigators who mishandle digital evidence, this is changing as awareness of the associated issues grows. Gates Rubber Co. v. Bando Chemical Indus. Ltd. provides an example of one court that criticized an investigator for improper digital evidence handling. Instead of using specialized digital evidence processing tools, the investigator copied individual files from the computer and was criticized by the court for not using "the method which would yield the most complete and accurate results."

There has been a similar progression in the evolution of tools for collecting evidence on communication systems. In the late 1980s, Clifford Stoll described how he made paper printouts of network traffic in an effort to preserve it as evidence (Stoll 1989). Network monitoring tools like tcpdump and Ethereal can be used to capture network traffic but they are not specifically designed for collecting digital evidence. Commercial tools such as Carnivore, NetIntercept, NFR Security, NetWitness, and SilentRunner have been developed with integrated search, visualization, and analysis features to help digital investigators extract information from network traffic. As described in Part 3 of this book, there are other forms of evidence on computer networks, many of which do not have associated evidence collection tools, making this a very challenging area for digital investigators. Rather than relying on tools, networks often require an individual's ingenuity to collect and analyze evidence.

There has been a similar progression in the evolution of tools for collecting evidence on embedded computer systems. It is common for digital investigators to read data from pagers, mobile phones, and personal digital assistants directly from the devices. However, this approach does not provide access to deleted data and may not be possible if the device is password protected or does not have a way to display the data it contains. Therefore, tools such as ZERT, TULP, and Cards4Labs have been developed to access password protected and deleted data (van der Knijff 2001). More sophisticated techniques involving electron microscopes are available to recover encrypted data from embedded systems but these are prohibitively expensive for most purposes.

Over the years, bugs have been found in various digital evidence processing tools, potentially causing evidence to be missed or misinterpreted. To avoid the resulting miscarriages of justice that may result from such errors, it is desirable to assess the reliability of commonly used tools. The National Institute of Standards and Testing are making an effort to test some digital evidence processing tools.[2] However, testing even the most basic functionality of tools is a time intensive process making it difficult to keep up with changes in the tools. Also, it is unlikely that a single group can test every tool including those used to collect evidence from networks and embedded systems. Additionally, in some instances, it may not be possible to create standard tests for the advanced features of various tools, because each tool has different features.

[2] *http://www.cftt.nist.gov/*

Another approach that has been suggested to reduce the complexity of tool testing is to allow people to see the source code for critical components of the software (Carrier 2002). Providing programmers around the world with source code allows tool testers to gain a better understanding of the program and increases the chances that bugs will be found. It is acknowledged that commercial tool developers will want to keep some portions of their programs private to protect their competitive advantage. However, certain operations, such as copying data from a hard drive, are sufficiently common and critical to require an open standard. Ultimately, given the complexity of computer systems and the tools used to examine them, it is not possible to eliminate or even quantify the errors, uncertainties, and losses and digital investigators must validate their own results using multiple tools.

2.3 LANGUAGE OF COMPUTER CRIME INVESTIGATION

The movement towards standardization in this area is made more difficult by a lack of agreement on basic terminology. Several attempts have been made to develop a standard language to describe the various aspects of computer crime investigation. Despite decades of discussion, no general agreement has been reached on the meaning of even the most basic term, *computer crime.*

> There has been a great deal of debate among experts on just what constitutes a computer crime or a computer-related crime. Even after several years, there is no internationally recognized definition of those terms. Indeed, throughout this Manual the terms computer crime and computer-related crime will be used interchangeably. There is no doubt among the authors and experts who have attempted to arrive at definitions of computer crime that the phenomenon exists. However, the definitions that have been produced tend to relate to the study for which they were written. The intent of authors to be precise about the scope and use of particular definitions means, however, that using these definitions out of their intended context often creates inaccuracies. A global definition of computer crime has not been achieved; rather, functional definitions have been the norm.

Although there is no agreed upon definition of computer crime, the meaning of the term has become more specific over time. Computer crime mainly refers to a limited set of offenses that are specifically defined in laws such as the US Computer Fraud and Abuse Act and the UK Computer Abuse Act. These crimes include theft of computer services; unauthorized access to protected computers; software piracy and the alteration or theft of electronically stored information; extortion committed with the assistance of computers; obtaining unauthorized access to records from banks, credit card issuers, or customer reporting agencies; traffic in stolen passwords and transmission of destructive viruses or commands.

One of the main difficulties in defining computer crime is that situations arise where a computer or network was not directly involved in a crime but still contains digital evidence related to the crime. As an extreme example, take a suspect who claims that she was using the Internet at the time of a crime. Although the computer played no role in the crime, it contains digital evidence relevant to the investigation. To accommodate this type of situation, the more general term *computer-related* is used to refer to any crime that involves computers and networks, including crimes that do not rely heavily on computers. Notably some organizations such as the US Department of Justice and the Council of Europe use the term *cybercrime* to refer to a wide range of crimes that involve computers and networks.

The term *computer forensics* also means different things to different people. Computer forensics usually refers to the forensic examination of computer components and their contents such as hard drives, compact disks, and printers. However, the term is sometimes used to describe the forensic examination of all forms of digital evidence, including data traveling over networks (a.k.a. network forensics). To confuse matters, the term computer forensics has been adopted by the information security community to describe a wide range of activities that have more to do with protecting computer systems than gathering evidence.

In fact, computer forensics (and by extension, network forensics) is a syntactical mess that uses the noun *computer* as an adjective and the adjective *forensic* as a noun, resulting in an imprecise term. Imagine referring to forensic entomology as "bug forensics" – this lacks clear meaning and sounds unprofessional. Also, referring only to computers limits the scope of the term, neglecting important aspects of the field such as communication systems, embedded systems, and digital image, audio, and video analysis. In 2001, the first annual Digital Forensic Research Workshop (DFRWS)[3] recognized the need for a revision in terminology and proposed *digital forensic science* to describe the field as a whole. The terms *forensic computer analysis* and *forensic computing* have also become widely used.

[3] http://www.dfrws.org

Given these disagreements regarding terminology, such terms will be avoided in this book. Instead, more descriptive language, such as *digital evidence examination*, will be used. This term is specific enough to be clear in the context of digital forensic science, computer forensics, incident response, or any other situation that involves the examination of digital evidence. Additionally, there is room in this terminology to include digital evidence in a legal context as well as the process of persuading decision-makers in civilian or military operations.

2.3.1 THE ROLE OF COMPUTERS IN CRIME

In addition to clarifying the general terms describing this field, it is productive to develop terminology describing the role of computers in crime. More specific language is crucial for developing a deeper understanding of how computers can be involved in crime and more refined approaches to investigating different kinds of crime. For example, investigating a computer intrusion requires one approach, while investigating a homicide with related digital evidence requires a completely different procedure.

The specific role that a computer plays in a crime also determines how it can be used as evidence. When a computer contains only a few pieces of digital evidence, investigators might not be authorized to collect the entire computer. However, when a computer is the key piece of evidence in an investigation and contains a large amount of digital evidence, it is often necessary to collect the entire computer and its contents. Additionally, when a computer plays a significant role in a crime, it is easier to obtain a warrant to search and seize the entire computer.

Several attempts have been made to develop a language, in the form of categories, to help describe the role of computers in crime. Categories are necessarily limiting, ignoring details for the sake of providing general terms, but they can be useful provided they are used with an awareness of their limitations. The strengths and weaknesses of three sets of categories are discussed in this section in an effort to improve understanding of the role of computers in crime.

Donn Parker was one of the first individuals to perceive the development of computer-related crime as a serious problem back in the 1970s and played a major role in enacting Florida's Computer Crime Act of 1978. Parker studied the evolution of computer-related crime for more than two decades and wrote several books on the subject (Parker 1976, 1983, 1998). He proposed the following four categories – while reading through these categories, notice the lack of reference to digital evidence.

1 A computer can be the *object* of a crime. When a computer is affected by the criminal act, it is the object of the crime (e.g. when a computer is stolen or destroyed).
2 A computer can be the *subject* of a crime. When a computer is the environment in which the crime is committed, it is the subject of the crime (e.g. when a computer is infected by a virus or impaired in some other way to inconvenience the individuals who use it).
3 The computer can be used as the *tool* for conducting or planning a crime. For example, when a computer is used to forge documents or break into other computers, it is the instrument of the crime.
4 The *symbol* of the computer itself can be used to intimidate or deceive. An example given is of a stockbroker, who told his clients that he was able to make huge profits on rapid stock option trading, by using a secret computer program in a giant computer in a Wall Street brokerage firm. Although he had no such programs nor access to the computer in question, hundreds of clients were convinced enough to invest a minimum of $100,000 each.

The distinction between a computer as the object and subject of a crime is useful from an investigative standpoint because it relates to the intent of the offender. However, additional terminology is needed to clarify this distinction. For the purposes of this text, a *target* is defined as the object of an attack from the offender's point of view, and may include computers or information they contain. The *intended victim* is the term for the person, group, or institution that was meant to suffer loss or harm. The intended victim and the target may be one and the same. There may also be more than one intended victim. Because of the closely linked nature of computer networks, there may also be *collateral victims.* This term refers to victims that an offender causes to suffer loss or harm in the pursuit of another victim (usually because of proximity). When an arsonist burns down a building to victimize an individual or a group, innocent individuals can get hurt. Similarly, when an intruder destroys a computer system to victimize an individual or a group, unconnected individuals can lose data.

Considering the computer as a tool that was used to plan or commit a crime is also useful. If a computer is used like a weapon in a criminal act, much like a gun or a knife, this could lead to additional charges or a heightened degree of punishment. As stated, the symbolic aspect of computers may seem irrelevant because no actual computers are involved and, therefore, none can be collected as evidence. The symbolic aspect of computers comes up more frequently when they are the targets of an attack and can be useful for understanding an offender's motivations. In this context, a *symbol* is any person or thing that represents an idea, a belief, a group, or even another person. For example, computers can symbolize authority to a particular offender, an organization can symbolize failure to an ex-employee, and a CEO can symbolize an organization. Therefore, a computer, organization, or individual may become a victim or target because of what they symbolize. Identifying the targets, intended victims, collateral victims, and symbols of a crime is one of the issues that an investigation is intended to resolve as discussed in Chapter 5.

The most significant omission in Parker's categories, is computers as sources of digital evidence. In many cases, computers did not play a role in a crime but they contained evidence that proves a crime occurred. For example, a revealing e-mail between US President Clinton and intern Monica Lewinsky could indicate that they had an affair, but the e-mail itself played no role in Clinton's alleged act of perjury. Similarly, a few of the millions of e-mail messages that were examined during the Microsoft anti-trust case contained incriminating information, yet the e-mail message did not play an active role in the crime – they were simply evidence of a crime.

In 1995, Professor David L. Carter used his knowledge of Criminal Justice to improve upon Parker's categorization of computer-related crime (Carter 1995). Instead of describing a computer as an *object* or *tool* of crime as Parker did, Carter used the more direct and legally oriented terms *target* and *instrumentality,* respectively. Although Carter did not address the subtleties of target/victim/symbol, he corrected Parker's main omission, describing scenarios in which computers are incidental to other crimes but hold related digital evidence. However, Carter did not distinguish between physical evidence (computer components) and digital evidence (the contents of the computer components). Very different procedures are required when dealing with physical and digital evidence, as described in Chapter 9.

In 1994, the US Department of Justice (USDOJ) created a set of categories and an associated set of search and seizure guidelines (USDOJ 1994, 1998). These categories made the necessary distinction between hardware (electronic evidence) and information (digital evidence), which is useful when developing procedures and from a probative standpoint. For instance, developing a parallel process for physical crime scene investigation and digital crime scene investigation (Carrier and Spafford 2003). In this context, *hardware* refers to all of the physical components of a computer, and *information* refers to the data and programs that are stored on and transmitted using a computer. The final three categories that refer to information all fall under the guise of digital evidence:

1 Hardware as Contraband or Fruits of Crime.
2 Hardware as an Instrumentality.
3 Hardware as Evidence.
4 Information as Contraband or Fruits of Crime.
5 Information as an Instrumentality.
6 Information as Evidence.

These categories are not intended to be mutually exclusive. A single crime can fall into more than one category. For example, when a computer is instrumental in committing a crime, it usually contains evidence of the offense. The details of collecting hardware and processing digital evidence are introduced in Chapter 9 and developed in the context of computer networks throughout the remainder of the text. Conspicuously absent from these categories is the computer as target, possibly because this distinction is more useful from an investigative standpoint than an evidence collection standpoint, as discussed in Chapters 5 and 19.

In 2002, this USDOJ document was updated to keep up with changes in technology and law and developed into a manual (as opposed to guidelines) for "Searching and Seizing Computers and Obtaining Electronic Evidence in Criminal Investigations" (USDOJ 2002). While the guidelines gave hardware and information equal weight, the manual takes the position that, unless hardware itself is contraband, evidence, an instrumentality, or a fruit of crime, it is merely a container for evidence. Thus, there is a realization that the content of computers and networks is usually the target of the search rather than the hardware. However, the manual points out that even when information is the target, it may be necessary to collect the hardware for a variety of reasons.

> In light of these uncertainties, agents often plan to try to search on-site, with the understanding that they will seize the equipment if circumstances discovered on-site

> make an on-site search infeasible. Once on-site to execute the search, the agents will assess the hardware, software, and resources available to determine whether an on-site search is possible. In many cases, the search strategy will depend on the sensitivity of the environment in which the search occurs. For example, agents seeking to obtain information stored on the computer network of a functioning business will in most circumstances want to make every effort to obtain the information without seizing the business's computers, if possible. In such situations, a tiered search strategy designed to use the least intrusive approach that will recover the information is generally appropriate.

Although the manual does not explicitly categorize information as contraband, a fruit of crime, or an instrumentality, it makes occasional reference to child pornography as contraband. These distinctions can be useful as discussed later in this section.

Because each of these categories has unique legal procedures that must be followed, this manual has become required reading among investigators, prosecutors and defense attorneys.

> [Defense] counsel should carefully review the Manual in cases where clients' computers are searched, because in almost every case there will be deviations from the Manual's recommended procedures. Whether those deviations are the result of casual adherence to the Manual or utter ignorance of it, this is a fertile area for suppression practice (Hoover 2002).

Significantly, the manual takes a more network-centric approach than its predecessor, taking into account more of the real world complexities of collecting digital evidence. In addition to general discussions about dealing with networks as a source of evidence, the manual mentions the possibility of a network being an instrumentality of a crime, and provides a section "Working with Network Providers" and a lengthy chapter titled "Electronic Surveillance in Communications Networks" with updated information regarding the USA PATRIOT Act. These sections are of interest to both law enforcement and computer security professionals who may be required to respond to requests for data on their networks.

2.3.1.1 HARDWARE AS CONTRABAND OR FRUITS OF CRIME

Contraband is property that the private citizen is not permitted to possess. For example, under certain circumstances, it is illegal for an individual in the United States, to possess hardware that is used to intercept electronic communications (18 USCS 2512). The concern is that these devices enable individuals to obtain confidential information, violate other people's privacy, and commit a wide range of other crimes using intercepted data. Cloned cellular phones and the equipment that is used to clone them are other examples of hardware as contraband.

The fruits of crime include property that was obtained by criminal activity such as computer equipment that was stolen, or purchased using stolen credit card numbers. Also, microprocessors are regularly stolen because they are very valuable, they are in high demand, and they are easy to transport.

The main reason for seizing contraband or fruits of crime is to prevent and deter future crimes. When law enforcement officers decide to seize evidence in this category, a court will examine whether the circumstances would have led a reasonably cautious agent to believe that the object was contraband or a fruit of crime.

2.3.1.2 HARDWARE AS AN INSTRUMENTALITY

A sniffer is not always a piece of specialized hardware. With the right software, a regular computer that is connected directly to a network can be used as a sniffer, in which case the software might be considered the instrumentality of the crime. Specialized hardware and software can also be installed in standard handheld devices enabling them to monitor wireless networks, in which case both the hardware and software can be viewed as instrumentalities.

When computer hardware has played a significant role in a crime, it is considered an instrumentality. This distinction is useful because, if a computer is used like a weapon in a criminal act, much like a gun or a knife, this could lead to additional charges or a heightened degree of punishment. The clearest example of hardware as the instrumentality of crime is a computer that is specially manufactured, equipped and/or configured to commit a specific crime. For instance, sniffers are pieces of hardware that are specifically designed to eavesdrop on a network. Computer intruders often use sniffers to collect passwords that can then be used to gain unauthorized access to computers.

The primary reason for authorizing law enforcement to seize an instrumentality of crime is to prevent future crimes. When deciding whether or not a piece of hardware can be seized as an instrumentality of crime, it is important to remember that "significant" is the operative word in the definition of instrumentality. Unless a plausible argument can be made that the hardware played a significant role in the crime, it probably should not be seized as an instrumentality of the crime.

It is ultimately up to the courts to decide whether or not an item played a significant role in a given crime. So far, the courts have been quite liberal on this issue. For example, in a New York child pornography case the court ruled that a computer was the instrumentality of the offense because the computer hardware *might* have facilitated the sending and receiving of the images (United States v. Lamb 1996). Even more liberal, was the Eastern District Court of Virginia decision that a computer with related accessories was an instrumentality because it contained a file that detailed the growing characteristics of marijuana plants (United States v. Real Property 1991).

2.3.1.3 HARDWARE AS EVIDENCE

Before 1972, "mere evidence" of a crime could not be seized. However, this restriction was removed and it is now acceptable to "search for and seize any property that constitutes evidence of the commission of a criminal offense" (Federal Rule of Criminal Procedure 41[b]). This separate category of *hardware as evidence* is necessary to cover computer hardware that is neither contraband nor the instrumentality of a crime. For instance, if a scanner that is used to digitize child pornography has unique scanning characteristics that link the hardware to the digitized images, it could be seized as evidence.

2.3.1.4 INFORMATION AS CONTRABAND OR FRUITS OF CRIME

As previously mentioned, contraband information is information that the private citizen is not permitted to possess. A common form of information as contraband is encryption software. In some countries, it is illegal for an individual to possess a computer program that can encode data using strong encryption algorithms because it gives criminals too much privacy. If a criminal is caught but all of the incriminating digital evidence is encrypted, it might not be possible to decode the evidence and prosecute the criminal. Another form of contraband is child pornography. Information as fruits of crime include illegal copies of computer programs, stolen trade secrets and passwords, and any other information that was obtained by criminal activity.

2.3.1.5 INFORMATION AS AN INSTRUMENTALITY

Information can be the instrumentality of a crime if it was designed or intended for use or has been used as a means of committing a criminal offense. Programs that computer intruders use to break into computer systems are the instrumentality of a crime. These programs, commonly known as *exploits*, enable computer intruders to gain unauthorized access to computers with a specific vulnerability. Also, computer programs that record people's passwords when they log into a computer can be an instrumentality, and computer programs that crack passwords often play a significant role in a crime. As with hardware, the significance of the information's role is paramount to determining if it is the instrumentality of crime. Unless a plausible argument can be made that the information played a significant role in the crime, it probably should not be seized as an instrumentality of the crime.

2.3.1.6 INFORMATION AS EVIDENCE

This is the richest category of all. Many of our daily actions leave a trail of digits. All service providers (e.g. telephone companies, ISPs, banks, credit institutions) keep some information about their customers. These records can reveal the location and time of an individual's activities, such as items purchased in a supermarket, car rentals and gasoline purchases, automated toll payment, mobile telephone calls, Internet access, online banking and shopping, and withdrawals from automated teller systems (with accompanying digital photographs). Although telephone companies and ISPs try to limit the amount of information that they keep on customer activities, to limit their storage and retrieval costs and their liability, law makers in some countries are starting to compel some communications service providers to keep more complete logs. For instance, the US Computer Assistance Law Enforcement Act (CALEA) that took effect in 2000, compels telephone companies to keep detailed records of their customers' calls for an indefinite period of time. The European Union has created log retention guidelines for its member states. In Japan, there is an ongoing debate about whether ISPs should be compelled to keep more complete logs.

For fun, take a single day in a life as an example. After breakfast, Jane Doe reads and responds to her e-mail. Copies of this e-mail remain in various places so Jane takes care to encrypt private messages. However, even if her encrypted e-mail is never opened, it shows that she sent a message to a specific person at a specific time. This simple link between two people can be important in certain circumstances. Encrypted e-mail can be even more revealing in bulk. If Jane sends a large number of e-mails to a newspaper reporter just before publication of a story about a confidential case she is working on, a digital investigator would not have to decrypt and read the e-mails to draw some daring inferences. Similarly, if a suspect used encrypted e-mail to communicate with another individual around the time a crime was committed, this might be considered sufficient probable cause to obtain a warrant to examine the e-mail or even search the second person's computer or residence.

After checking her e-mail, Jane opens her schedule in her computerized planner. Jane's small planner contains vast amounts of information about her family, friends, acquaintances, interests, and activities. Next, on the way to the bank, Jane makes a few quick calls on her mobile telephone, propelling her voice through the air for anyone to listen to. At the bank, she withdraws some cash, creating a record of her whereabouts at a specific time. Not only is her transaction recorded in a computer, her face is captured by the camera built into the automated teller machine.

Although she pays for her lunch in cash, Jane puts the receipt in her wallet, thus keeping a record of one of the few transactions that might have escaped the permanent record. After lunch, Jane decides to page her husband John. From her computer she accesses a Web page that allows her to send John a short message on his pager. This small act creates a cascade of digits in Jane's computer, on the Web, and ultimately on John's pager. Unfortunately, the battery on Jane's telephone is low so when John tries to call, he gets Jane's voice mail and leaves a message. Then it occurs to him that Jane was probably at her computer when she sent him the short alphanumeric message, so he connects to the Internet and uses one of the many computer programs that allow live communication over the global network. These few minutes of digital tag create many records in many different places and though some of this information might dissolve in a matter of hours, some of it will linger indefinitely on backup tapes and in little-used crannies on Jane's hard drive.

As an exercise, think back on some recent days and try to imagine the trail of digits left by your activities on various computers at banks, telephone companies, work, home, and on the Internet.

2.4 SUMMARY

Despite being a new field, great advancements have been made in computer crime investigation. Powerful evidence processing tools have been developed and there is a move towards standardization.

One of the fundamental purposes of categories described in this chapter is to emphasize the role of computers in crime and to give guidance for dealing with computers in that role. These categories can be used to develop procedures for dealing with digital evidence and investigating crimes involving computers. Early categories were necessarily general and as the categories were refined, guidelines were developed to help investigators deal with electronic and digital evidence. These guidelines are still in their early stages, especially with regard to digital evidence. More detailed guidelines for dealing with information as evidence, also known as digital evidence, are presented throughout this book.

The language described in this chapter both enables and limits our ability to describe and interpret digital evidence. This language is useful for developing investigative and evidence-processing procedures but does not include other important aspects of investigating this type of crime. Concepts and techniques that are helpful for interpreting digital evidence, discerning patterns of behavior, understanding motives, generating investigative leads, linking cases, and develop trial strategies are presented in Chapters 4 and 5.

REFERENCES

American Society of Crime Laboratory Directors (2003) "Proposed Revisions to 2001 Accreditation Manual", (Available online at http://www.ascld-lab.org/pdf/aslabrevisions.pdf).

Carrier B. and Spafford E. H. (2003) "Getting Physical with the Digital Investigation Process", International Journal of Digital Evidence, Volume 2, Issue 2 (Available online at http://www.ijde.org/docs/03_fall_carrier_Spa.pdf)

Carter D. L. (1995) "Computer Crime Categories, How Techno-Criminals Operate", FBI Law Enforcement Bulletin (July).

ENFSI Forensic Information Technology Working Group (2003) Draft Guidelines for Best Practice in the Forensic Examination of Digital Technology (Available online at http://www.enfsi.org/docs/FITWG-BPM-001-003.pdf)

Hollinger R. C. (1997) Crime, Deviance and the Computer, Brookfield, VT: Dartmouth Publishing Company.

Hoover T. W. (2002) "An Introduction to the DoJ's Manual on Searching and Seizing Computers", Federal Public Defender Report, Volume 11, No 1, March, 2002.

Parker D. (1976) Crime by Computer, New York, NY: Charles Scribners' and Sons.

Parker D. (1983) Fighting Computer Crime, New York, NY: Charles Scribners' and Sons.

Parker D. (1998) Fighting Computer Crime: A new Framework for Protecting Information, New York, NY: John Wiley & Sons.

Rosenblatt K. S. (1995) High-Technology Crime: Investigating Cases Involving Computers, San Jose, CA: KSK Publications.

Stoll C. (1989) The Cuckoo's Egg: Tracking a Spy Through the Maze of Computer Espionage, New York, NY: Pocket Books.

United Nations (1995) international Review of Criminal Policy No. 43 and 44, United Nations Manual on the Prevention and Control of Computer-Related Crime, (available at http://www.uncjin.org/Documents/irpc4344.pdf).

US Department of Justice (1994) "Federal Guidelines for Searching and Seizing Computers", (Available online at http://www.usdoj.gov/criminal/cybercrime/search_docs/toc.htm).

US Department of Justice (1998) "Supplement to Federal Guidelines for Searching and Seizing Computers", (Available online at http://www.usdoj.gov/criminal/cybercrime/supplement/ssgsup.htm).

US Department of Justice (2002) "Searching and Seizing Computers and Obtaining Electronic Evidence in Criminal Investigations", (Available online at http://www.cybercrime.gov/searchmanual.htm).

van der Knijff R. (2001) "Embedded System Analysis" in Handbook of Computer Crime Investigation: Forensic Tools and Technology Ed. by Eoghan Casey, Academic Press: London.

CASES

Gates Rubber Co. v. Bando Chemical Indus., Ltd. (1996) 167 F.R.D. 90, 112 (D.C. Col.)

United States v. Lamb (1996) 945 F. Supp. 441, 462 (N.D.N.Y).

United States v. Real Property & Premises Known as 5528 Belle Pond Drive (1991) 783 F. Supp. 253 (E.D. Va.).

## CHAPTER 3

TECHNOLOGY AND LAW

Eoghan Casey, Robert Dunne, and Tessa Robinson

Given the intangible nature of "electronic property" and the legal ambiguity surrounding malicious intent, we should not be surprised to find states amending extant criminal law to cover abuse by computer … Most jurisdictions, however, adopted a very different tactic. They defined computer crime as a unique legal problem and thereby created separate computer crime chapters in their criminal codes.

(Hollinger and Lanza-Kaduce 1988)

Many cybercrimes can be addressed using existing laws. After all, cybercrime is just a new manifestation of age-old crimes – the primary difference is that a new technology is involved. However, the Internet creates new challenges that require legal issues to be rethought and legislation to be amended. For instance, laws prohibiting the creation and distribution of child pornography have been amended to include the use of computers and networks.

This chapter provides an overview of legal issues relating to technology from two perspectives – United States and Europe. By presenting legal issues from both sides of the Atlantic side-by-side, similarities and differences become evident.

PART A TECHNOLOGY AND LAW – A US PERSPECTIVE

Robert Dunne

In 1997, in Bensusan Restaurant Corporation v. King, a trademark infringement dispute, the plaintiff, owner of the famous Blue Note jazz club in New York City, argued that New York jurisdiction over the Missouri defendant was appropriate based on the existence of a Web site owned and maintained by the defendant and accessible in New York. The United States Court of Appeals for the Second Circuit in affirming the lower court's decision in favor of the defendant (on grounds unrelated to the Web site), opined that, "attempting to apply established trademark law in the fast-developing world of the Internet is somewhat like trying to board a moving bus."

Digital Evidence and Computer Crime Second Edition
ISBN: 0-12-163104-4

The analogy continues to be true, and not just of attempts to apply trademark law in the context of the Internet, but, in fact, to the general relationship between traditional law and its application in cyberspace and in the general context of other new technologies. Cyberspace is still largely undefined and, from the law's perspective, often frustratingly intangible. When dealing with activities in cyberspace, courts frequently struggle with issues that are glaringly obvious in the physical world, such as where the activity in question occurred. Where, after all, "are" we when we connect to the Internet? When involved in a chat room discussion, where is it happening? In the Bensusan case, the lower court actually said that if a computer user viewing the Missouri Blue Note club's Web site were confused about the club's relationship to the New York club, the confusion would be occurring in Missouri, not wherever the user was sitting at the time! This is, at best, counterintuitive thinking about a question with important legal implications, and it illustrates how confused the courts themselves are about these basic matters.

It is not just computing technology that has generated new and important legal issues. Developments in various surveillance and search technologies, for example, have raised fundamental questions about the extent to which people may reasonably expect to be free from intrusions on their privacy. From the perspective of US law, there is a direct correlation between what technology makes possible and what our privacy expectations are. Thus, the definition of "privacy" is continually evolving. What is "private" today, and therefore subject to protection against unreasonable search and seizure via the Fourth Amendment to the US Constitution, may no longer be "private," and thus exempt from such protection, tomorrow.

This part of the chapter is intended to provide an overview of a number of areas of the law in the United States and how they are affected by technology. These areas include jurisdiction; distribution of pornography and obscenity; child pornography and online solicitation of minors for sexual activity; privacy; and copyright and "theft" of digital intellectual property. Some basic legal principles are reviewed to enhance understanding of how the relevant laws have been interpreted and applied in the context of the Internet and other technologies. Specific legal procedural matters, such as those related to authentication and admissibility of evidence, and search warrant requirements and considerations, are addressed in later chapters.

3A.1 JURISDICTION

Jurisdiction is often the first question raised both in civil litigation and criminal prosecutions. Where should the suit be heard? Where should the defendant be tried? US law requires that a court have two types of jurisdiction in order to hear any case: subject matter jurisdiction and personal jurisdiction.

Subject matter jurisdiction is the power to hear the particular type of dispute being brought before the court. Certain types of disputes, such as civil suits between citizens of the same state, can only be brought in state court. Every state, however, has at least one court of *general* jurisdiction, that is, a court that can hear any type of dispute between any parties. Federal courts, on the other hand, are all courts of *limited* jurisdiction, meaning that they only have subject matter jurisdiction over types of disputes specifically defined by the statutes that created the courts. Primarily, subject matter jurisdiction for federal courts is defined by the Constitution (Article 3, Section 2). For example, federal courts have subject matter jurisdiction when the dispute is between citizens of different states (referred to as *diversity* jurisdiction), or when the dispute is one which arises from invocation of a federal law.

Personal jurisdiction is the power to enforce a judgment over a defendant. In civil suits, this is a two-part question that is often difficult to answer. Defendants may always be sued in the state in which they reside, but when is it permissible to sue them in other states? The first question is whether jurisdiction in a proposed state is appropriate under that state's "long arm statute." Every state has one of these. The long arm statute declares how far the state believes it can reasonably reach to assert personal jurisdiction over an out-of-state resident. Some states have long arm statutes that specify certain activities, such as intentionally causing harm within the state, which the state claims give rise to personal jurisdiction. Other states use long arm statutes that simply claim personal jurisdiction to the extent the Constitution permits.

This question of the appropriateness of personal jurisdiction under the Constitution is the second piece of the analysis and it is dependent on whether the court believes the defendant had sufficient "minimum contacts" with the proposed forum state, such that imposing personal jurisdiction over the defendant would not offend "traditional notions of fair play and substantial justice" – in other words, the Due Process clause of the Constitution. Certain contacts with the proposed forum state have become well understood as satisfying the "minimum contacts" standard. These contacts include things such as doing business in a state and forming contracts with citizens of a state. The less tangible the contacts are, however, the more difficult it becomes to assess whether or not they make imposition of personal jurisdiction constitutional. When the contacts are in cyberspace, this can become an extremely contentious question.

The US District Court for the District of Connecticut decided in the 1996 case Inset Systems, Inc. v. Instruction Set, Inc. that by creating a Web site, the defendant had "directed its advertising activities via the Internet" to all states, and that therefore Connecticut had personal jurisdiction over the Massachusetts defendant. The Minnesota Court of Appeals endorsed the Inset Systems decision a year later in its decision in the case of Minnesota v. Granite Gate Resorts, Inc. The trend, though, has been towards decisions that say the mere accessibility of a so-called "passive Web site" is not enough to satisfy the "minimum contacts" requirement – perhaps because the idea that a person or corporation should be subject to personal jurisdiction in any state from which it is accessible is unpalatable from a policy perspective. There are also, of course, important implications in terms of international jurisdiction. However, until the question is definitively addressed by the US Supreme Court, the issue is unsettled.

In criminal prosecutions in the physical world, the question of where jurisdiction is appropriate has traditionally been much clearer than in civil suits. Jurisdiction is appropriate if a state was either the site of the conduct or the site of its result. For many crimes committed in cyberspace, application of the traditional rule is easy. For example, if a hacker located in New York uses a computer there to break into a bank's computer in California the hacker can be charged with violations of both state and federal laws and tried in either of the two states. However, for certain other types of crimes, it can be unclear "where" something occurs in cyberspace.

Robert and Carleen Thomas, residents of Milpitas, California, ran an online Bulletin Board Service (BBS), called "Amateur Action," which offered members materials the Thomases assumed qualified as "pornography" but not "obscenity." The former is protected speech under the Constitution, the latter is not and its interstate distribution is prohibited by a federal criminal statute (18 U.S.C. 1465). In any given prosecution under the statute, the question of whether the materials in question are "obscene" and not merely pornographic is decided by a jury applying "contemporary community standards." In other words, the standard is the standard of the community from which the jury was chosen and in which the trial takes place.

After receiving a complaint about the Thomases' BBS from a resident of Tennessee, a US Postal Inspector in Tennessee signed up as a member, accessed the BBS from Tennessee, downloaded digitized photographs (GIF files), and ordered video tapes, which were sent to him. The Thomases were charged with interstate distribution of obscene materials and were tried and convicted in Memphis, Tennessee. Robert and Carleen were sentenced to 37 and 30 months in prison, respectively.

On appeal, the Thomases argued that, among other things, the statute applied only to distribution of *tangible* materials, that the GIF files were not transmitted to Tennessee by them, but rather by the Postal Inspector, and that the community standards of Memphis were the wrong ones to use in assessing the question of obscenity because their actions took place in cyberspace, a new "community" which has its own standards. The novel argument that activities in cyberspace should be judged by its own community standards received support from the American Civil Liberties Union, the Interactive Services Association, the Society for Electronic Access, and the Electronic Frontier Foundation. However, the US Court of Appeals for the Sixth Circuit affirmed the lower court's decision that the GIF files were tangible, not just a diaphanous collection of ones and zeros: they began as photographs on a computer in California and ended as photographs on a computer in Tennessee. The appellate court also managed to sidestep the troublesome questions of whether the GIF files were "sent" or "pulled" to Tennessee and whether special cyberspace community standards should apply by pointing to the videotapes that were shipped to the Postal Inspector as the sort of traditional physical world activity that made the Tennessee venue and community standards proper.

3A.2 PORNOGRAPHY AND OBSCENITY

Pornography is big business in cyberspace. In an October 23, 2000 article, *The New York Times* provided some useful data:

1. Two Web ratings services estimated that about one in four Internet users, or 21 million Americans, visited one of the more than 60,000 sex sites on the Web once a month.
2. Analysts from Forrester Research say that sex sites on the Web were generating at least $1 billion a year in revenue.
3. A single successful sex-related Web site was bringing in between $5 and $10 million a year – about the same amount a federal study thirty years previously estimated to be the total retail value of hard core pornography in the entire country.

Of course, pornography on the Web is only the latest step in the interplay between technology and pornography. The video cassette recorder began a trend towards using technology to provide increasingly anonymous – and thus less inhibited – access to pornographic materials. Viewing pornographic films no longer required a very public visit to a particular type of cinema, merely a brief stop at one of many "adult" video rental stores. Cable television and its pay per view options soon made even visits to video stores unnecessary. Now, pornography is not only *available* on the Web, it is often inescapable as pornographers take advantage of Internet technology to bombard users with unsolicited e-mail (spam) and unwanted pop-up advertisements.

As noted above, pornography is protected as speech under the First Amendment to the Constitution. Obscenity is not – for purposes of First

Amendment analysis, "speech" includes all sorts of expression including, obviously, photographs. Pornography may be freely distributed and possessed. Obscenity may be legally possessed in the privacy of a person's home, but its distribution is illegal. The Miller test, the source of the "community standards" issues raised by Robert and Carleen Thomas, is used to determine whether expression has crossed the line from pornography to obscenity. The test, as articulated by the Supreme Court in the 1973 case of Miller v. California, is as follows:

1 Would the average person, applying contemporary community standards, find that the work, taken as a whole, appeals to the prurient interest?

2 Does the work depict or describe, in a patently offensive way, sexual conduct specifically defined by applicable state law?

3 Does the work, taken as a whole, lack serious literary, artistic, political, or scientific value?

Justice William Douglas, in his dissenting opinion in the Miller decision, pointed out a serious flaw and danger with any test to determine obscenity. Douglas asked how could the Court sustain a conviction for distribution of materials prior to the time when those materials were deemed obscene? In other words, a person cannot know he or she is distributing obscene materials until convicted – at which point, obviously, it is too late. Nonetheless, the Miller test is still the standard used to differentiate unprotected obscenity from protected pornography.

As we have seen, Robert and Carleen Thomas discovered the dangers of national distribution of material that, probably, would have been considered pornographic, but not obscene, by their local community standards, but the aspect of Internet access to pornography that has most troubled legislators is the ease of access by children. Congress first tried to address this problem in the Telecommunications Act of 1996. Title V of the Act, known as the Communications Decency Act (CDA), contained provisions intended to regulate the dissemination on the Internet of material that was felt to be inappropriate for minors. Section 223(a) prohibited the "knowing transmission of obscene or indecent messages to any recipient under 18 years of age." Section 223(d) prohibited the "knowing sending or displaying of patently offensive messages in a manner that is available to a person under 18 years of age."

The CDA was almost immediately challenged in court as being unconstitutional. In 1997, in the first true landmark decision involving cyberspace, Reno v. American Civil Liberties Union *et al.*, the Supreme Court struck down Sections 223(a) and (d), primarily because the lack of clear definition of the terms "indecent" and "patently offensive" would make speakers uncertain about what sort of speech might violate the statute, thus chilling free expression. Would speakers, for example, know if discussions about birth control or homosexuality violate the CDA? Adult speech, the court found, could not be restricted to a level of discourse suitable for children.

In the course of the Reno decision, the Court took into account specific attributes of Internet technology and communication, comparing these with attributes of other communications technology and media, such as broadcasting. In particular, the Court revisited its 1978 decision, FCC v. Pacifica Foundation, in which it held that the FCC did have the power to regulate a broadcast that was "indecent" but not obscene. The Reno Court emphasized differences between broadcasting and Internet communication. Broadcasting has traditionally been regulated because of the scarcity of broadcast frequencies. Licenses are required for broadcasters. There is no scarcity of frequencies for communication on the Internet. Furthermore, a key factor in the Pacifica decision was that court's perception that radio was a "pervasive" medium, easily accessible to children, and one in which a listener might easily tune in unexpectedly to material the listener might find offensive. The Reno Court saw the Internet as less pervasive, requiring affirmative steps for access to pornography or other material that might be inappropriate for children. Ironically, Internet technology has now been successfully exploited by pornographers so that it is actually difficult to use the Internet without encountering potentially objectionable material.

Congress returned to the issue of Internet access by children to inappropriate material in 1998 by passing the Child Online Protection Act (COPA). COPA's language was intended to rectify the vagueness of the CDA's "indecent" and "patently offensive" terms by creating a "test" relying very closely on the language of the Miller test for obscenity. COPA prohibited persons from knowingly making a communication for commercial purposes that is available to any minors and that includes material that is "harmful to minors." COPA defines "material that is harmful to minors" as

> any communication, picture, image, graphic image file, article, recording, writing, or other matter that is obscene or that
>
> 1 the average person, applying contemporary community standards, would find, taking the material as whole and with respect to minors, is designed to appeal or pander to the prurient interest;
>
> 2 depicts, describes, or represents, in a manner patently offensive with respect to minors, an actual or simulated sexual act or sexual contact, an actual or simulated normal or perverted sexual act, or a lewd exhibition of the genitals of post-pubescent female breast; and
>
> 3 taken as a whole, lacks serious literary, artistic, political, or scientific value for minors.

A month before COPA was to go into effect, its constitutionality was challenged in the federal district court by the American Civil Liberties Union and others, arguing that some material contained on Web sites, although valuable for adults, might be construed as "harmful to minors" by some community standards, thus leading to the same burden on protected adult speech that was found unacceptably broad by the Supreme Court in its Reno decision regarding the CDA. The District Court, focusing on the plaintiffs' argument that the statute was overbroad and not the least restrictive means of preventing minors from accessing material that was "harmful to minors," enjoined the government from enforcement of COPA until the merits of the constitutional challenge had been adjudicated.

The government appealed to the Court of Appeals for the Third Circuit, which affirmed the lower court. However, the Court of Appeals completely ignored the District Court's rationale for the injunction, basing its analysis instead on the ground that COPA's use of "contemporary community standards" likely rendered the statute unconstitutionally overbroad. Web publishers, the court reasoned, are unable to limit access to their sites based on the geographic location of the users accessing them. Thus, COPA would require that any material potentially "harmful to minors" would be subject to the standards of the most puritan of communities in the nation and would have to be shielded behind an age or credit card verification system.

The case made its way to the Supreme Court which in May 2002, in Ashcroft v. American Civil Liberties Union *et al.*, held that the use of contemporary community standards in COPA's definition of "material harmful to minors" did not by itself make the statute unconstitutionally overbroad. The court carefully limited its holding and deferred any consideration of the statute's potential unconstitutionality for other reasons. The court left the injunction against enforcement of COPA in place, but remanded the case for further action by the District Court and the Court of Appeals.

The COPA decision makes it clear that the Supreme Court is evolving in its attitude towards the Internet and the potential for use of "community standards" to regulate conduct in cyberspace. The Miller court flatly rejected the idea that there might be some sort of national standard to determine obscenity, stating that it is "neither realistic nor constitutionally sound to read the First Amendment as requiring that the people of Maine or Mississippi accept public depiction of conduct found tolerable in Las Vegas, or New York City." The Reno court suggested that use of "community standards" to regulate Internet conduct would be unwise, since it would impose the most restrictive standards on communities everywhere. But in Ashcroft, it is possible to see some shift towards the notion that use of community standards to regulate conduct on the Internet might be a difficult, but possibly not insurmountable, problem. As with so many questions in cyberlaw, the answer to whether pornography and obscenity can be controlled on the Internet and, if so, how is a work in progress.

3A.2.1 CHILD PORNOGRAPHY

Like obscenity, child pornography is not expression protected by the First Amendment. Like obscenity, its distribution is illegal. However, unlike obscenity, its mere possession is criminal.

Historically, courts have given the states and the federal government more leeway in regulating child pornography. Attempts at the federal level to deal with the problem of child pornography have an extensive history. The original federal legislation on this topic was the Protection of Children Against Sexual Exploitation Act of 1977. In the following 14 years, four other Acts of Congress were targeted at the problem of child pornography. These included the Child Protection Act, the Child Sexual Abuse and Pornography Act, the Child Protection and Obscenity Enforcement Act, and the Child Protection Restoration and Penalties Enforcement Act.

The "compelling interest" that made all of these statutes constitutional was preventing the exploitation and abuse of the children used in making the pornography. Congress has always defined the problem of child pornography in terms of real children. In the case of New York v. Ferber, in 1982, the Supreme Court limited criminalization of child pornography to works that "visually depict explicit sexual conduct by children below a specified age." The Ferber court specifically stated that depictions of sexual conduct "which do not involve live performance or photographic or other visual reproduction of live performances, retain[s] First Amendment protection." Sketches from the imagination or literary descriptions of children engaged in sexual activities remained protected. Even the use of persons who looked younger than their actual age would be permissible.

However, digital technology now makes it possible to create "virtual" child pornography that is essentially indistinguishable from photographs of actual children engaged in sexual activities. Congress sought to prohibit the distribution of digitally created child pornography by passing the Child Pornography Prevention Act (CPPA) in 1996. The focus of the regulation shifted at this point radically from harm to real children to a determination that child pornography was evil in itself. The basis of the new law was the asserted impact of such images on children who view them and the notion that child pornography, real or virtual, increases the activities of pedophiles and child molesters. The law criminalized depictions that "appear to be" of a minor, or that "convey[s] the impression that the material is or contains a visual depiction of a minor" engaged in sexual activities.

Decisions by various federal appellate courts regarding the constitutionality of the CPPA differed. The Court of Appeals for the First Circuit decided in 1999 in United States v. Hilton that the statute was not unconstitutionally vague. In the same year, the Ninth Circuit Court of Appeals, in Free Speech

Coalition v. Ashcroft, held that "the First Amendment prohibits Congress from enacting a statute that makes criminal the generation of images of fictitious children engaged in imaginary but explicit conduct."

The Supreme Court reviewed the Ninth Circuit's decision to clarify the legal situation. It agreed with the Ninth Circuit that the assertion that virtual child pornography makes children who may view such images more susceptible to engaging in such acts, and that virtual as well as real child pornography increases the activities of child molesters and pedophiles were not adequately substantiated and did not meet the required standard of a compelling government interest. Therefore, the CPPA was unconstitutional. The Court was careful to point out that distributors of virtual child pornography might still be prosecuted successfully under laws prohibiting distribution of obscenity, emphasizing that the apparent age of the participants in the allegedly obscene materials is an acceptable factor to be included in a jury's application of contemporary community standards. Interestingly, the possibility of obscenity prosecution of those trafficking in virtual child pornography does not satisfy those wishing to define it as child pornography. Quite probably, this is because, if merely obscene, virtual child pornography, unlike actual child pornography, could be legally possessed in the privacy of a person's home. This suggests that the existence of such materials, not the harm being done to children used in creating it, has been the true target of child pornography laws. Technology has once again significantly changed the legal landscape.

As discussed in Chapter 19, the Internet makes solicitation of children by pedophiles and child molesters significantly easier than it is in the physical world. The Internet gives these offenders access to a greater number of victims in chat rooms and other "public" online venues. The anonymity possible on the Internet emboldens potential offenders who might never have approached a child in a physical world playground. On the Internet it becomes clear that other people share the same obsession and this tends to "validate" it, lowering their inhibitions. Furthermore, cyberspace's intangible nature removes the sort of physical indications children are told to be wary of when dealing with strangers. It is impossible to tell what percentage of Internet users might be engaging in the solicitation of children for sexual activity, but the fact that successful sting operations by police are extremely common suggests that another very serious problem has been exacerbated by new technology.

3A.3 PRIVACY

Most of us take for granted today that we have some sort of "right of privacy," but what is that "right" and how did we get it?

Privacy can be defined in different ways. Justice Cooley in 1888 defined it as a right to be left alone. Others define privacy as a right to be anonymous. These are very different definitions with significantly different implications, particularly in the context of technology.

In legal terms, privacy is a dual right: the right to be free from government intrusion in certain areas of our lives and the right to be free from intrusion by other individuals into our "private" lives. The former right is largely protected via constitutional interpretation and assorted statutes. The latter, the right to be free from intrusion by other individuals, is protected largely via common law, that is, the law of courts and precedent – sometimes referred to as "judge-made law" or as a tort. A tort, as defined by the great legal commentator William Prosser, is a civil, as opposed to criminal, wrong, other than breach of contract, for which the law will provide a remedy in the form of an action for damages.

Before 1890, no English or American court had ever recognized a "right of privacy." However, in 1890, a Harvard Law Review article by Samuel Warren and Louis Brandeis examined a number of cases ostensibly decided on other grounds, and concluded that these decisions were actually based on a broader principle: a right of privacy. By the 1930s, almost all jurisdictions had recognized the Right of Privacy, either by statute or common law.

The Common Law "Right of Privacy" actually is four different rights protected by four different tort causes of action:

1 appropriation of a person's name or likeness for the defendant's benefit,
2 unreasonable intrusion, defined as intentional interference with another person's interest in solitude and seclusion,
3 public disclosure of private facts,
4 false light, that is, publicity which presents a person to the public in a false light.

In addition to common law protections, there are a number of federal statutes specifically aimed at preserving privacy in particular circumstances. These include:

- the Electronic Communications Privacy Act, which regulates interception of electronic communications by both the government and private individuals,
- the Privacy Act of 1974, which imposes limits on the collection and use of personal information by federal agencies,
- the Family Educational Rights & Privacy Act, which permits students (and parents of minor students) to examine and challenge the accuracy of school records,
- the Fair Credit Reporting Act, which regulates collection and use of personal data by credit-reporting agencies,
- the Equal Credit Opportunity Act, which prohibits creditors from gathering certain types of data from applicants such as gender, race, religion, national origin, birth control practices, or child-bearing plans,

- the Federal Right to Financial Privacy Act of 1978, which limits the ability of financial institutions to disclose customer information to agencies of the federal government,
- the Federal Cable Communications Policy Act, which prohibits cable television companies from using cable systems to collect personal information about subscribers without their consent,
- the Video Privacy Protection Act, which prohibits video tape sale or rental companies from disclosing customer names and addresses and the subject matter of their purchases or rentals for direct marketing use.

While it is important to understand that the "right of privacy" is protected by common law and statutes, for the purposes of the criminal law, and this book, the focus is on our privacy protection as it is embodied in the Constitution. The word "privacy" does not appear in the Constitution and the right of privacy in this context is largely a separate body of law developed over many years through interpretations and analysis of the Fourth Amendment, which prohibits "unreasonable searches and seizures." It turns out that our right of privacy has a lot to do with our expectations and how reasonable they are. Consider some seminal Supreme Court criminal cases regarding our "right to privacy" under the Fourth Amendment.

The Fourth Amendment's prohibition of "unreasonable searches and seizures" applies to searches and seizures made by government without a warrant, unless either a warrant is unnecessary, or the search or seizure falls under one of the exceptions to the warrant requirement. It is illustrative to consider several key cases to gain a better understanding of when a warrant is unnecessary and when it is not, that is, when does a warrantless search or seizure not violate our "right of privacy?"

3A.3.1 KATZ V. UNITED STATES

In 1967, the Supreme Court considered the case of Katz v. United States. Katz had been convicted of transmitting wagering information from Los Angeles to Miami and Boston. The government introduced evidence of Katz's end of telephone conversations, which had been obtained without a warrant by placing electronic listening devices on the *outside* of the phone booth from which Katz made the calls. The government argued that a warrant was unnecessary in this case because there was no search or seizure since there had been no physical entrance into the area occupied by Katz – the phone booth.

The Katz court defined the parameters of "privacy" by saying that what a person exposes to the public, even in his own home, is *not* private but that, similarly, what a person seeks to preserve as private, even in an areas accessible to the public, *may* be constitutionally protected. Here, the Court said,

Katz was in a phone booth (the old-fashioned kind, which was actually an enclosed space with a door that closed). He was entitled to aural privacy. "One who occupies it, shuts the door behind him, and pays the toll is *surely* entitled to assume that the words he utters will not be broadcast to the world."

Katz is one of the earliest cases in which the Supreme Court addressed the question of privacy rights in *intangible* property, that is, information, and how strongly information is protected against searches or seizures made possible by technology. The court's decision makes it clear that such property is as highly protected as tangible property, the object of more traditional searches and seizures.

3A.3.2 CALIFORNIA V. GREENWOOD

Twenty-one years later, in California v. Greenwood, the Supreme Court explained the relationship between privacy expectations and the extent to which a right of privacy is safe from government intrusion. Acting on a tip from an informant and without a warrant, the police asked the garbage collectors in Greenwood's community to give them Greenwood's garbage. The police searched the garbage and found evidence of drug use. Based on this evidence, the police obtained a warrant, searched Greenwood's home, and found drugs. Greenwood was arrested on felony narcotics charges. Finding that the warrant to search Greenwood's house could not have been obtained without the evidence obtained from the warrantless trash searches, the California Superior Court dismissed the charges under a previous California decision holding that warrantless trash searches violated the Fourth Amendment. On appeal, both the California Court of Appeals and the California Supreme Court affirmed the dismissal of the charges against Greenwood. The state took its appeal to the United States Supreme Court.

The Supreme Court stated that the warrrantless search of the trash bags left outside Greenwood's home would violate the Fourth Amendment only if Greenwood manifested a "subjective expectation of privacy in [his] garbage that society accepts as objectively reasonable." Whether something is constitutionally protected as "private" (in cyberspace as well as the physical world) is therefore determined by a two-prong test. Did the individual do something to demonstrate that he or she personally had an expectation of privacy (the subjective prong), and is that person's expectation of privacy one that society believes is reasonable (the objective prong).

In Greenwood's case, the court said he may indeed have had a subjective expectation of privacy, but it was not an objectively reasonable one. By putting out his trash he exposed it to the public (i.e. made it accessible to snoops, scavengers, animals, etc.). In fact, by leaving the trash at the curb he expressly intended to convey it to a third party. Greenwood therefore had no constitutionally protected expectation of privacy in the contents of his garbage and no warrant was necessary for the police to search it.

3A.3.3 KYLLO V. UNITED STATES

Perhaps no Supreme Court decision illustrates the tension between technology and privacy law better than its 2001 decision in Kyllo v. United States. Kyllo was suspected of growing marijuana in his home. Without obtaining a warrant, Department of the Interior agents used a thermal imager to scan Kyllo's triplex apartment from the passenger seat of a car. The imager showed that the garage roof and sidewall of the home were relatively hot compared to the rest of the structure and neighboring homes. The agents concluded that Kyllo was using halide lights to grow marijuana. Based on the results of the thermal imaging, as well as tips from an informant and Kyllo's high electricity bills, a warrant was issued and Kyllo's home was searched. An indoor marijuana growing operation was found. Kyllo was indicted on one count of "manufacturing marijuana." Kyllo moved to suppress the evidence obtained during the search, arguing that a warrant was required for the thermal imaging of his home. The Ninth Circuit Court of Appeals held that no warrant was needed for the thermal imaging, reasoning that Kyllo had exhibited no subjective expectation of privacy because he had not attempted to conceal the heat escaping from his home, and that even if he had exhibited a subjective expectation of privacy, it was not one that society would consider objectively reasonable because the thermal imager did not expose intimate details of his life, merely amorphous "hot spots" on the exterior of his home.

The Supreme Court reversed. The Court noted that it is true that warrantless surveillance is generally legal and that previous holdings say that visual observation is simply not a "search," and thus, not subject to Fourth Amendment prohibition. The Court emphasized the critical issue in this case: "The question we confront today is what limits there are upon the power of technology to shrink the realm of guaranteed privacy." The Court found that using sense-enhancing technology to obtain information about the interior of a home that could not otherwise have been obtained without physical intrusion constituted a search and was presumptively unreasonable without a warrant. Interestingly, however, the court added that this was true "at least where the technology in question is not in general public use."[1] In the Kyllo case, the Court reasoned that the thermal imager was a device not in general public use and that it exposed details of activities within the home. The case was remanded to the District Court to determine whether there was sufficient evidence other than the thermal imaging results to support the issuance of the search warrant.

The dissenting opinion stressed the more traditional analysis that what a person knowingly exposes to the public is not a subject of Fourth Amendment

[1]The Court also noted its concern with other technology under development by the Department of Justice, such as a "Radar-based Through-the-Wall Surveillance System" and a "Radar Flashlight" that will enable law officers to detect individuals through interior building walls.

protection and that searches and seizures of property in plain view are presumptively reasonable. It characterized the thermal imaging as "off the wall" surveillance, as opposed to "through the wall." No details of the interior of the home were revealed, said the dissent, and the conclusions drawn by the officers from the results of the thermal imaging were simply inferences, not a search.

The dissent also criticized the majority's opinion because of the difficulty of defining "in general public use." The dissent noted that 12,000 thermal imagers had already been manufactured and are readily available to the public (they can even be rented by calling an 800 number).

When is a technology "in general public use", and therefore, one from whose use as a search tool individuals are no longer constitutionally protected? What about technologies developed to replace non-mechanical law enforcement techniques that have previously been held not to be "searches?" For example, the use of dogs to sniff out narcotics has not been considered a search. Would a device to detect the same odors be treated differently?

In many ways, the Kyllo decision raises more questions than it answers. However, it vividly illustrates how tightly connected technology and its use are with the reasonableness of our expectations about what is private, and therefore protected by the Fourth Amendment from warrantless search or seizure.

The ECPA prohibits anyone, not just the government, from unlawfully accessing (18 USC 2511-2521) or intercepting (18 USC 2701-2709) electronic communications. Rather than detail every aspect of these complex sections, several interesting aspects are highlighted here. A more detailed discussion of the ECPA can be found in (USDOJ 2002) and (Rosenblatt 1995). Rosenblatt interprets the ECPA twice. One interpretation is aimed at law enforcement and the other is directed at corporate investigators.

The ECPA stipulates that, to obtain authorization to intercept transmissions, law enforcement must follow a specific procedure and obtain a court order (or another certification in writing) that satisfies a given list of requirements. These rigid requirements make it more difficult to obtain authorization to intercept electronic communications. Notably, when dealing with intercepted transmissions, a search warrant will not satisfy the ECPA's court order requirement. There is more flexibility when it comes to stored electronic communications:

> 2703(1) A governmental entity may require a provider of remote computing service to disclose the contents of any electronic communication to which this paragraph is made applicable by paragraph (2) of this section –
>
> (i) without required notice to the subscriber or customer, if the governmental entity obtains a search warrant issued under the Federal Rules of Criminal Procedure or equivalent State warrant; or

(ii) with prior notice from the governmental entity to the subscriber or customer if the government entity –

(i) uses an administrative subpoena authorized by a Federal or State statute or a Federal or State grand jury subpoena; or

(ii) obtains a court order for such disclosure under subsection (d) of this section;

except that delayed notice may be given pursuant to section 2705 of this title

This distinction between stored and transmitted communications was made because intercepting transmissions is potentially a greater invasion of privacy than collecting stored communications. When intercepting communications, there is a high chance that unrelated, private information will also be intercepted, whereas stored communications are more discrete and the chance of collecting unrelated, private information is limited.

An interesting distinction between intercepted and stored communications arose during the Steve Jackson games case (detailed in Chapter 9), in which the Secret Service violated the ECPA by reading and deleting e-mail that had never reached the intended recipients. Steve Jackson Games argued that the Secret Service had intercepted the e-mail because it had not been delivered to the intended recipients. However, the court argued that the e-mail had been delivered to the recipient's mailboxes and the ECPA made a clear distinction between storage and transmission so there was no way that the deleted e-mail fell into both categories. The e-mail would have to have been actively traveling through a wire or computer to qualify for the transmission clause.

There is one aspect of the ECPA that is still hotly debated. It is argued that under certain conditions (e.g. prior consent by one of the participants in a communication) an organization can search employees' communications. Therefore, many organizations have policies that allow them to monitor communications and all employees are required to agree to the policy by signature before gaining access to e-mail or a network. However, some people feel that any random monitoring of communications is not in the spirit of the law and that an employee's consent should be obtained each time the employer needs to access or intercept communications.

The USA Patriot Act, enacted after the terrorist attacks of September 11, 2001, greatly expands the government's ability to use technology as a surveillance and data collection tool – both in the physical world and cyberspace. As mentioned in Chapter 1, the Terrorist Information Awareness (formerly Total Information Awareness) program seeks to exploit technology to merge information contained in thousands of different databases and use novel data mining techniques to create profiles of individuals in an attempt to identify suspected terrorists and potential terrorist activity. It is too soon to assess the impact of these and other efforts to monitor activities and behavior, or to know whether they will withstand constitutional scrutiny by the courts, but civil libertarians are concerned that the traditional privacy protections of the Fourth Amendment are rapidly being eroded.

3A.4 COPYRIGHTS AND THE "THEFT" OF DIGITAL INTELLECTUAL PROPERTY

3A.4.1 COPYRIGHT BACKGROUND

Copyright law – like patent law – involves a balancing act between providing an incentive for creators of work and generating "progress," that is, permitting others to build upon earlier work by others. Article I, Section 8 of the Constitution is the source of American copyright law, giving Congress the power "to promote the Progress of Science and useful Arts, by securing for limited Times to Authors and Inventors the exclusive Right to their respective Writings and Discoveries."

A copyright in a work is created automatically if the work is original expression that is fixed in a tangible form. So, for example, original expression written or drawn on paper, recorded on tape, photographed on film, or saved to a computer disk is automatically copyrighted. Computer software code is generally copyrightable as a "literary work." The requirement that the work be "original expression" is one that is easily met. Even the slightest spark of creativity is enough. Facts are not copyrightable, but compilations of facts may be copyrightable in themselves if they organize the facts in some creative way. Ideas are also not copyrightable, only the expression of ideas. So, for example, Peter Benchley does not have a copyright on the idea of a large shark terrorizing a summer resort community, merely on his particular expression of that idea in the novel *Jaws.*

No notice of copyright is required on a work in order to secure a copyright, nor is registration of the copyright required – although both are good ideas if the work is potentially valuable. Copyrights held by individuals last for the life of the author plus 70 years (recently increased from 50 by the Copyright Term Extension Act). For corporately owned copyrights, the term is 95 years.

A copyright holder gets more than merely the exclusive right to control reproduction of the work – a bundle of rights come with the copyright. These rights include the rights to create derivative works (e.g. to make a movie based on a book), the right to public performance or display of the work, the right to claim the work as one's own, and the right to prevent others from use of the author's name as creator of a distorted version of the work.

Remedies for copyright infringement include injunction prohibiting continued infringement, impoundment of allegedly infringing copies, and damages, either "actual" or "statutory." Actual damages are the copyright owner's actual losses plus the infringer's profits (to the extent that they differ). At any point during the course of an infringement suit, a copyright owner who has registered the copyright prior to the alleged infringing act may opt to pursue statutory damages instead of actual damages. The Copyright Act defines statutory damages and no proof of actual loss is required.

The Copyright Act also allows for criminal penalties in some cases. If the copying by the infringer is "willful," that is, an intentional violation of a known legal duty, penalties can be up to 5 years in prison and a fine of up to $250,000. Previously, criminal liability required a profit motive. For instance, in 1994 an indictment against David LaMacchia, an MIT student who was accused of running a bulletin board for use in copying popular software valued at over $1,000,000, was dropped because he did not charge anyone to use the bulletin board, and therefore had no profit motive. Congress responded in 1997 by passing the No Electronic Theft Act, which removed the requirement of a profit motive for criminal liability.

3A.4.2 FAIR USE

The Copyright Act contains language intended to address the need to allow some use of copyrighted material other than by the copyright owner. This is an attempt to create an appropriate balance between the need for both incentives for authors and progress in the arts and sciences.

The Act specifies that use of copyrighted material may be considered a "fair use" under certain circumstances. Fair use is a defense to copyright infringement and is analyzed on a case-by-case basis. Four factors are used in determining whether or not a particular use qualifies as a fair use:

1. The purpose and character of the use. Is it a commercial use, and presumptively unfair, or a non-profit use?
2. The nature of the copyrighted work. Where does the work fall on the spectrum of factual (lightly protected) to creative (highly protected) work?
3. The amount and substantiality of the portion of the copyrighted work used. "Substantiality" is an important consideration. A three-hundred-word quotation can be the "heart" of a long book.
4. Effect on the potential market for the work. Will the copyright owner incur losses as a result of the use?

No single factor is dispositive, although many courts view the fourth as the most important.

The 1993 case of Playboy Enterprises, Inc. v. Frena, provides an illustration of the application of the fair use doctrine to online activities. Frena operated a computer bulletin board system (BBS) that provided members with pictures from Playboy and Playmate magazines for download. One of the arguments he made in his defense was that his use of the photographs was a "fair use." A US District Court in Florida applied the four factors and found:

1. Frena's use of the copyrighted material was commercial since he charged users to become members of the BBS.
2. The copyrighted material was "fantasy and entertainment" and therefore towards the more protected end of the factual – creative spectrum.
3. Frena's argument that he had taken only a small portion of Playboy Enterprises' vast collection of photographs was incorrect because each photograph is individually protected by copyright. Thus, Frena had taken 100% of each copyrighted work.
4. If unchecked and engaged in by others, the activities in which Frena had engaged could ultimately have a serious effect on Playboy Enterprises' market.

Therefore, Frena's use of Playboy Enterprise's copyrighted work was not a fair use permitted by the Copyright Act.

In an earlier, 1984 landmark case in which technology and copyright law collided, Sony Corporation of America v. Universal City Studios, Inc., the US Supreme Court applied fair use analysis in assessing Sony's liability for the taping of copyrighted television shows by users of Sony's video cassette recorders (VCR). The Court ultimately found, in a close 5–4 decision, that recording a movie for viewing at later time ("time-shifting") was a fair use. The Court also had to decide the important issue of whether a manufacturer can be held responsible for the use of its devices by consumers for copyright infringement. Here, the Court said, a manufacturer could not be held contributorily liable for copyright infringement by users of a device if the device has substantial other legitimate, non-infringing uses, which the VCR has. This principle is an important one, and resurfaces in discussing the Digital Millenium Copyright Act.

The Napster music file swapping service, which appeared on the Web in 1999, created a situation in which many of the questions in the Sony case were revisited in an entirely new technological context. Were the people using Napster to share music files engaging in a fair use? If not, was Napster liable for its users' infringement? In February, 2001 in A&M Records v. Napster, a federal District Court had little trouble in finding that the swapping of copyrighted music was not fair use. Users were not engaging in "personal" use of the music files, they were trading them with thousands of strangers. Music is creative, strongly protected expression and the users were copying complete songs. Perhaps most importantly, the Court accepted the plaintiff's argument that online music swapping was dramatically affecting the market for music compact disks.

Regarding Napster's liability for copyright infringement by its users, Napster argued that, like Sony's VCR, Napster's music distribution service had significant non-infringing uses, such as promoting new bands and artists who were unable to obtain contracts with traditional record companies. The Court saw a difference between Napster and Sony. The latter's relationship with its customers – and therefore Sony's control over the uses of an individual's VCR – ended when the customer purchased the device. Napster, on the other hand, had a continuing relationship with its users that gave Napster the opportunity to control their ongoing activities. Napster was therefore liable for its users' copyright infringement.

3A.4.3 THE DIGITAL MILLENNIUM COPYRIGHT ACT

In response to pressure from the motion picture industry, record labels, software publishers, and other entities with a major stake in profits to be made from copyrighted material, Congress passed the Digital Millennium Copyright Act (DMCA) in 1998, incorporating it as part of the Copyright Act. The DMCA criminalizes making, distributing, or using tools (e.g. software) to circumvent technological protection measures used by copyright owners to prevent access to copyrighted material. Criminal penalties can be severe – up to five years in prison and $500,000 in fines for a first offense.

The DMCA has created enormous controversy, as well as confusion among the courts, resulting in a flurry of seemingly conflicting decisions. Many of these decisions are still under review, and many more cases involving the DMCA are sure to follow. Opponents argue that the DMCA effectively negates the fair use provisions of the Copyright Act and that enforcement of the DMCA is often an unconstitutional suppression of free speech. Two recent high-profile cases illustrate the issues.

The motion picture industry developed a protection scheme for movies distributed on DVDs. This encryption code, called the Content Scrambling System (CSS), restricted play of the DVDs to approved devices – and also protected the movie from digital copying. A 15-year-old Norwegian teenager named Jon Johansen cracked the CSS encryption scheme, allegedly so that he could play his DVDs on a computer running the Linux operating system as opposed to Windows. Johansen made the "DeCSS" code available on the Web and it rapidly spread around the world. Motion picture studios filed suits against Web site owners who posted the DeCSS code or even posted links to other sites that made the code available.

Universal City Studios v. Corley was one such case. Corley operated a Web site called "2600.com," which was primarily frequented by hackers and those interested in such activities. He had posted the DeCSS code on his site, as well as links to other sites that made it available. Several movie studios sued

Corley under the DMCA provision prohibiting the distribution of tools to circumvent technological protection measures to force him to remove the DeCSS code and the links to other sites. Corley argued both that computer code is speech and that DeCSS could be used to obtain access to copyrighted material that would be used in a manner consistent with the fair use provisions of the Copyright Act. He claimed that the DMCA was therefore unconstitutional on two grounds: that it suppressed free speech and that it violated the Constitution's Copyright Clause by unduly obstructing the fair use of copyrighted materials.

However, the Court of Appeals for the Second Circuit agreed in May 2001 with the District Court for the Southern District of New York's reasoning that although computer code is "speech" within the meaning of the First Amendment, the DMCA targeted only the "functional" aspects of the speech in question here, was therefore "content-neutral" regulation of speech, and survived constitutional scrutiny. The Court dismissed Corley's arguments regarding the Copyright Clause and fair use by saying that, "to whatever extent the argument might have merit at some future time in a case with a properly developed record, the argument is entirely premature and speculative at this time on this record. There is not even a claim, much less evidence, that any Plaintiff has sought to prevent copying of public domain works, or that the injunction prevents the Defendants from copying such works. As [District Court] Judge Kaplan noted, the possibility that encryption would preclude access to public domain works 'does not yet appear to be a problem, although it may emerge as one in the future.' " The injunction against Corley's posting of the code and links to sites providing it was affirmed. Nonetheless, the DeCSS code was soon being distributed around the Internet in forms more traditionally protected as "speech," such as poetry and songs.

At about the same time as the Corley decision, the California Court of Appeals reviewed a lower court decision in the case of DVDCCA v. Bunner. The DeCSS was available on Bunner's Web site and the DVD Copy Control Association, representing owners of copyrighted material distributed on DVDs, sued to prevent disclosure of the CSS code, which was disclosed every time the DeCSS code was published, and which the DVDCCA claimed was a protected trade secret. The lower court granted the requested injunction.

The Court of Appeals reversed, ruling that the code was constitutionally protected speech and that its suppression would violate the First Amendment. The court specifically rejected the DVDCCA's argument that because of the code's "functional" nature it fell outside the protection of the First Amendment. Revealing a trade secret, of course, can have civil legal consequences as a form of trade secret misappropriation. The Court emphasized that the First Amendment did not shield Bunner, or anyone else who revealed a trade secret while under a contractual obligation not to disclose it, or who disclosed it after acquiring it knowing that it had been obtained by improper means, from the possibility of a misappropriation suit by the trade secret's owner. Thus, there is a "right" to speak, but not always without consequences.

From the invention of the printing press, which first created the ability to make multiple, inexpensive copies of a written work, and thus an enhanced need to protect an author's right to profit from that work, to inexpensive audio and video tape recorders, to computers permitting easy and digitally perfect copying of software, music, and video, technological innovation has been a constant source of concern to the creators of intellectual property. The continued popularity of new music file-swapping services designed for pure "peer-to-peer" file exchange, and which thus avoid the contributory liability trap into which Napster fell, underlines what seems to be a fascinating fundamental reality of human behavior, namely that the more easily and anonymously people can engage in a prohibited activity, the more likely they are to do it, particularly if it is a "victimless" prohibited activity, such as copyright infringement.

The Internet and related computing and networking technologies have created a crisis in copyright law. However, as in the past, it is likely that over time, copyright law and its interpretation will be reshaped to adjust to each technological advance, and that copyright owners will develop new business models that will protect their interests and maximize their profits to the greatest extent possible.

PART B COMPUTER MISUSE IN AMERICA

Eoghan Casey

The Computer Fraud and Abuse Act (CFAA) was enacted in 1984 and was amended by the Computer Fraud and Abuse Act of 1986 (this act has been amended several times since). Unfortunately, the CFAA has not been very useful – it has only been used a few times since its enactment. Richard Morris is one of the few individuals to be prosecuted under the CFAA for releasing his infamous Internet worm.

CASE EXAMPLE (UNITED STATES v. MORRIS 1991):
In 1988, Robert Morris, a graduate student at Cornell University and the son of the National Computer Security Center's chief scientist, made history by creating and letting loose a computer program that replicated itself repeatedly on thousands of machines on the Internet. This program, called a worm, exploited vulnerabilities in

a widely used operating system called BSD UNIX. Although this worm automatically broke into computers and made efforts to hide itself, it made no explicit attempt to steal from or damage the computers it infected. In essence, its only purpose was to break into as many computers as it could. Morris later claimed that he was simply experimenting, trying to add to his already formidable knowledge of computers. Unfortunately, the experiment went terribly wrong. The worm was so successful at replicating itself that it overloaded the Internet bringing more than 6,000 installations to a grinding halt (Spafford 1989).

After a few days, the worm was eradicated, but the aftermath was even more dramatic than the event itself. The worm had demonstrated, more than any other single event, that the Internet was not secure and that trust alone was not sufficient protection against attack. Anger and fear overshadowed the trust that had made the Internet possible. People were out for blood and Morris made history once again by being the first person to be convicted by jury under the Computer Fraud and Abuse Act (CFAA) of 1986 (two others had been convicted under the CFAA but not by a jury). He was required to pay the maximum allowable under the CFAA ($10,000), serve three years probation, and contribute 400 hours of community service.

The CFAA was primarily designed to protect national security, financial, and commercial information, medical treatment, and interstate communication systems. The CFAA protects these systems against a wide range of malicious acts, including unauthorized access. In this statute, access to a computer is considered to be unauthorized if it is without permission, or it exceeds the permission originally granted. Therefore, authorized users can be liable under this statute if they do something that they were not permitted to do. In addition to addressing intrusion and damage, this statute prohibits denial of service attacks that cause a loss of $1,000 or more. Additionally, the CFAA allows any person who suffers a loss as a result of one of the actions covered by the Act to bring a civil action against the violator to obtain compensation.

An overview of this statute is provided in Table 3.1 with a summary of the most interesting portions.

It is worth noting that the CFAA is not designed to exclude other laws. Therefore, the CFAA can be used to bring additional charges against an individual for a single crime as two members of a group called the Legion of Doom discovered.

CASE EXAMPLE (UNITED STATES v. RIGGS 1990):

In 1988, Robert Riggs gained unauthorized access to the computer system of a telephone company named Bell South and downloaded information describing an enhanced 911 system for handling emergency services in municipalities (e.g. police, fire, and ambulance calls). Riggs then gave the materials to Craig Neidorf who published them in an online newsletter called PHRACK. Riggs and Neidorf were charged under three separate laws: the CFAA; a federal wire fraud statute; and a statute prohibiting interstate transportation of stolen property. The court specifically noted that the CFAA could be used in conjunction with other laws.

Riggs was convicted for breaking into the Bell South computer system. The charges against Neidorf were dropped after it transpired that the materials he published were not as private as Bell South had claimed – they were selling copies to anyone who requested them.

Table 3.1

Summary of the Computer Fraud and Abuse Act of 1986.

SECTION	SUMMARY	PENALTIES
Section (a)(1)	Obtaining unauthorized access to information regarding national defense, foreign relations, and atomic energy.	A fine and/or up to 10 years imprisonment for a first offense and up to 20 years for subsequent offenses
Section (a)(2)	Obtaining unauthorized access to records from a financial institution, credit card issuer, or consumer-reporting agency.	A fine and/or up to 1 year imprisonment for a first offense and up to 10 years for subsequent offenses
Section (a)(3)	Interfering with government operations by obtaining unauthorized access to their computers or computers that they use.	A fine and/or up to 1 year imprisonment for a first offense and up to 10 years for subsequent offenses
Section (a)(4)	Obtaining unauthorized access to a Federal interest computer to commit fraud or theft unless the object of the fraud and the thing obtained consists only of the use of the computer.	A fine and/or up to 5 years imprisonment for a first offense and up to 10 years for subsequent offenses
Section (a)(5)(A)	"Whoever ... through means of a computer used in interstate commerce or communications, knowingly causes the transmission of a program, information, code, or command to a computer or computer system if the person causing the transmission intends that such transmission will damage, or cause damage to, a computer, computer system, network, information, data, or program; or withhold or deny, or cause the withholding or denial, of the use of a computer, computer services, system or network, information, data, or program" provided the access is unauthorized and causes loss or damage of $1,000 or more over a one year period or "modifies or impairs, or potentially modifies or impairs, the medical examination, medical diagnosis, medical treatment, or medical care of one or more individuals."	A fine and/or up to 5 years imprisonment for a first offense and up to 10 years for subsequent offenses
Section (a)(5)(B)	"Whoever ... through means of a computer used in interstate commerce or communications, knowingly causes the transmission of a program, information, code, or command to a computer or computer system with reckless disregard of a substantial and unjustifiable risk that the transmission will damage, or cause damage to, a computer, computer system, network, information, data, or program; or withhold or deny, or cause the withholding or denial, of the use of a computer, computer services, system or network, information, data, or program" provided the access is unauthorized and causes loss or damage of $1,000 or more over a one year period or "modifies or impairs, or potentially modifies or impairs, the medical examination, medical diagnosis, medical treatment, or medical care of one or more individuals."	A fine and/or up to 1 year imprisonment
Section (a)(6)	Trafficing in passwords that affect interstate commerce or involve the password to a computer that is used by or for the US government	A fine and/or up to 1 year imprisonment for a first offense and up to 10 years for subsequent offenses

A *Federal interest computer* is a computer used exclusively by a financial institution or the US Government, used on a nonexclusive basis but where the conduct affects use by the financial institution or government or which is one of two or more computers used in committing the offense, not all of which are located in the same state.

Another noteworthy ruling involving the CFAA occurred when a recently dismissed bank employee named Bernadette Sablan was charged with damaging her employer's records [United States v. Sablan, 92 F.3d 865 (9th Cir. 1995)]. Sablan claimed to be drunk at the time and argued that she did not intend to do any damage. However, Sablan was convicted after the court determined that the CFAA only requires intent to gain unauthorized access to the computer and does not require intent to do damage.

All states except Vermont have additional computer crime statutes that extend the CFAA. These state statutes apply to all computers, not just government, financial, or communication systems. Also, many of these state statutes make it illegal to break into a computer (even if no damage is done), alter or destroy data (even if the damage is recoverable), steal services, deny another person access, or use the computer with intent to commit a variety of crimes. However, as with the CFAA, these state computer crime statutes are used infrequently. Because these laws are new and are often vaguely worded, it can be difficult to find attorneys who understand the issues and procedures. Also, few organizations (including law enforcement agencies) are willing to spend the time and resources necessary to investigate a computer crime when they are uncertain of the results.

PART C TECHNOLOGY AND CRIMINAL LAW – A EUROPEAN PERSPECTIVE

Tessa Robinson

This part of the chapter presents a European perspective of computer misuse, computer-generated evidence, jurisdiction and procedure, in the main setting down the relevant legislation and using case examples from England and Ireland to illustrate key points. The first section of this part considers how the criminal law in Europe deals with computer related crime. The elements to be discussed are the types of offenses involved, rules with respect to search and seizure warrants, and sentencing policy in respect of child pornography offences.

The European Union (EU) is a political union between 15 European countries, 12 of which make up the Euro zone. Its common objective is to offer a single market. The union is currently under expansion (Turkey and some former Eastern bloc countries are in negotiation over accession) and at the time of writing is undergoing constitutional change. The union can be compared to the Federal and State legal systems in the United States, although EU member states would hold that they enjoy a much greater degree of sovereignty. While EU legislation (directives, decisions, recommendations and opinions) emanates from the European Institutions (the Parliament, Council and Commission), it is incorporated by member state governments into domestic law and applied by national courts. In summary, the European Courts located in Strasbourg (the Court of First Instance (CFI) and the European Court of Justice (ECJ)) derive jurisdiction under the European Treaties in the following circumstances:

- Preliminary rulings – Disputes involving Community law are heard by domestic courts which apply the relevant EU law principles. The Treaty provisions provide for a mechanism whereby a national court may request the ECJ to rule on questions of Community law that arise in the course of domestic litigation. Note that the CFI does not have jurisdiction to give preliminary rulings.
- Direct actions against member states – Where member states are alleged to have infringed their Community obligations, actions may be brought by the Commission or by another member state.
- Judicial review of Community acts – Where the legality of acts of the European Institutions intended to produce legal effects are in dispute, actions may be brought by the Council, Commission, member states and natural and legal persons (i.e. individuals or companies) – where they can establish the requisite interest in the impugned act to achieve standing.
- Plenary jurisdiction – Natural and legal persons may bring actions against the Commission to review penalties imposed. In limited circumstances expressed in the Treaties, there lie actions in contract and tort.

While EU member states have created a single economic area without borders, they currently maintain separate national criminal jurisdictions and policing systems. The move towards greater Europe-wide harmonization of these areas has however been facilitated by provisions in the Treaty of Amsterdam regarding police and judicial cooperation (Articles 29–42), and is on the agenda of the working group involved in putting together a new constitution for the EU.

3C.1 OVERVIEW OF CRIMINAL OFFENSES

Although technology creates new challenges that require new legislation, in some instances existing laws may apply. For instance, in *AG's Reference No. 5 of 1980* ([1980] 3 All E.R. 88), the court was asked to decide whether a person who provides screen images derived from a videotape "publishes an obscene article" contrary to section 2 of the Obscene Publications Act, 1959. The defense counsel submitted that these words should not be applied to a piece of electronic equipment that Parliament could not have conceived of when the law was enacted. However, the court ensured that the new technology came within the meaning of the Act, holding at p. 92 that:

> if the clear words of the statute are sufficiently wide to cover the kind of electronic device with which we are concerned in this case the fact that that particular form of electronic device was not in the contemplation of Parliament in 1959 is an immaterial consideration.

This same approach was endorsed by the Court of Appeal in R. v. Fellows, Arnold ([1997] 2 All E.R. 548), a case concerning the distribution of indecent photographs of children over the Internet where defense counsel argued that an image consisting of computer data was not a photograph.

In 2001, realizing that certain computer-related offenses required special consideration, 26 member countries convened in Budapest and signed the Council of Europe Convention on Cybercrime to create "a common criminal policy aimed at the protection of society against cybercrime, *inter alia*, by adopting appropriate legislation and fostering international co-operation" (paragraph 4 of the preamble to the convention). Several other countries subsequently signed the Convention. Although the COE Convention on Cybercrime represents an aspirational policy document, a country that ratifies the Convention commits to putting in place a legislative framework that deals with cybercrime according to Convention requirements. Within this commitment, each country is given discretion in relation to the full scope, say, of a criminal offense, by defining its particular elements of dishonest intent or requiring that serious harm be done before an offense is deemed to have been committed.

Despite a clear need for consistent legislation in Europe to facilitate cross-border investigations, there are major differences between the legal systems and cultures in European countries, making legislative consistency difficult. The COE Convention on Cybercrime has already been faulted by some for not taking due account of privacy rights. Also, there are discrepancies between the Convention and existing laws in some European countries that will take time to resolve. To appreciate these differences, it is instructive to compare categories of offenses set out by the Convention with related offenses in English law.

3C.1.1 FRAUD AND FORGERY

Fraud and forgery are traditional offenses that may be facilitated by the use of technology. The Convention describes computer-related fraud and forgery offenses as follows.

- computer-related forgery, that is, the intentional input, alteration, deletion or suppression of computer data resulting in inauthentic data with the intent that it be considered or acted upon for legal purposes as if it were authentic, regardless of whether or not the data are directly readable and intelligible (Article 7); and
- computer-related fraud, the intentional causing of a loss of property to another by any input, alteration, deletion or suppression of computer data or any interference with the functioning of a computer system with fraudulent or dishonest intent of procuring, without right, an economic benefit for oneself or for another (Article 8).

Existing legislation is, in most cases, fit to deal with their commission. An example of the need for new legislation to combat computer-related crime is the situation in England where, for a fraud to be committed, it must be shown that a *person* was deceived (Section 15 of the Theft Act 1968). Where the process is automated, the element of deception of a person may be missing, and thus, no offense proved. It may be necessary to widen the meaning of deception to include deception of machines, or to introduce new legislation directed at this computer-related mischief.

Section 9 of the Irish Criminal Justice (Theft and Fraud Offences) Act, 2001, tackles computer-related fraud and forgery by creating the offense of unlawful use of a computer in the following terms:

> A person who dishonestly, whether within or outside the State, operates or causes to be operated a computer within the State with the intention of making a gain for himself or herself or another, or of causing loss to another, is guilty of an offence.

Another area of growing concern is identity fraud – effectively stealing an individual's virtual identity for financial gain. In an effort to address this and other computer-related fraud and forgery, Section 25 of the Irish Electronic Commerce Act, 2000 (an Act that provides for the legal recognition of electronic contracts, electronic writing, electronic signatures and original information in electronic form in relation to commercial and non-commercial transactions, the admissibility of evidence in relation to such matters, the accreditation, supervision and liability of certification service providers and the registration of domain names) prohibits fraud and misuse of electronic signatures and signature creation devices by creating offenses in the following terms:

> 25.—A person or public body who or which—
>
> (a) knowingly accesses, copies or otherwise obtains possession of, or recreates, the signature creation device of another person or a public body, without the authorisation of that other person or public body, for the purpose of creating or allowing, or causing another person or public body to create, an unauthorised electronic signature using the signature creation device,
>
> (b) knowingly alters, discloses or uses the signature creation device of another person or a public body, without the authorisation of that other person or public body or in excess of lawful authorisation, for the purpose of creating or allowing, or causing another person or public body to create, an unauthorised electronic signature using the signature creation device.
>
> (c) Knowingly creates, publishes, alters or otherwise uses a certificate or an electronic signature for a fraudulent or other unlawful purpose,
>
> (d) Knowingly misrepresents the person's or public body's identity or authorisation in requesting or accepting a certificate or in requesting suspension or revocation of a certificate,
>
> (e) Knowingly accesses, alters, discloses or uses the signature creation device of a certification service provider used to issue certificates, without the authorisation of the certification service provider or in excess of lawful authorisation, for the

purpose of creating, or allowing or causing another person or a public body to create, an unauthorized electronic signature using the signature creation device, or

(f) Knowingly publishes a certificate, or otherwise knowingly makes it available to anyone likely to rely on the certificate or on an electronic signature that is verifiable with reference to data such as codes, passwords, algorithms, public cryptographic keys or other data which are used for the purposes of verifying an electronic signature, listed in the certificate, if the person or public body knows that –

(i) the certification service provider listed in the certificate has not issued it,

(ii) the subscriber listed in the certificate has not accepted it, or

(iii) the certificate has been revoked or suspended, unless its publication is for the purpose of verifying an electronic signature created before such revocation or suspension, or giving notice of revocation or suspension, is guilty of an offence.

These kinds of offenses are likely to arise more frequently with the increased use of digital certificates and other digital identification mechanisms.

3C.1.2 CHILD PORNOGRAPHY

Offenses relating to the possession and distribution of child pornography are probably the most litigated and certainly the most notorious of cyber offenses. The Convention addresses this complex area in the following suggestions.

1 Each party shall adopt such legislative and other measures as may be necessary to establish as criminal offences under its domestic law, when committed intentionally and without right, the following conduct:

a. producing child pornography for the purpose of its distribution through a computer system;

b. offering or making available child pornography through a computer system;

c. distributing or transmitting child pornography through a computer system;

d. procuring child pornography through a computer system for oneself or for another;

e. possessing child pornography in a computer system or on a computer-data storage medium.

2 For the purpose of paragraph 1 above "child pornography" shall include pornographic material that visually depicts:

a. a minor engaged in sexually explicit conduct;

b. a person appearing to be a minor engaged in sexually explicit conduct;

c. realistic images representing a minor engaged in sexually explicit conduct.

3 For the purpose of paragraph 2 above, the term "minor" shall include all persons under 18 years of age. A party may, however, require a lower age-limit, which shall be not less than 16 years.

4 Each party may reserve the right not to apply, in whole or in part, paragraph 1(d) and 1(e), and 2(b) and 2(c).

The associated law in England predates the Convention and did not specifically mention computers. Section 1(1) of the Protection of Children Act, 1978 as amended by the Criminal Justice and Public Order Act, 1994 makes it an offense:

> (a) to take, or permit to be taken, an indecent photograph of a child (a person under the age of 16); or
>
> (b) to distribute or show such indecent photographs or pseudo-photographs; or
>
> (c) to have in his possession such indecent photographs or pseudo-photographs with a view to their being distributed or shown by himself or others ...

By virtue of the amendment made by the 1994 Act, the term *photograph* includes data stored on a computer disk or by other electronic means which are capable of conversion into a photograph, including graphic images (Section 7.4(b)). The test, therefore, is that if data can be converted into an indecent image it will be deemed a photograph for the purposes of the section. In addition, Section 160 of the English Criminal Justice Act, 1988 provides inter alia that:

> 1 It is an offence for a person to have any indecent photograph or pseudo-photograph of a child in his possession.
>
> 2 Where a person is charged with an offence under subsection (1) it shall be a defence for him to prove –
>
> (a) that he had a legitimate reason for having the photograph or pseudo-photograph in his possession; or
>
> (b) that he had not himself seen the photograph or pseudo-photograph and did not know nor had any cause to suspect, it to be indecent; or
>
> (c) that the photograph or pseudo-photograph was sent to him without any prior request made by him or on his behalf and that he did not keep it for any unreasonable time.

The Court of Appeal case of R. v. Fellows, Arnold ([1997] 2 All E.R. 548) is a leading English case on the interpretation of Section 1 of the Protection of Children Act, 1978, and specifically on the question of what might constitute the "distributing" or "showing" of offending material.

CASE EXAMPLE (R. v. FELLOWS 1997):
Alban Fellows and Stephen Arnold were arrested after a large amount of child pornography was found on an external hard drive attached to a computer belonging to Fellows's employer, Birmingham University. Fellows and Arnold were convicted of distributing the child pornography in this archive to others on the Internet. In appeal, defense counsel submitted to the court, *inter alia*, that the data were not "distributed or shown" merely by reason of its being made available for downloading by other computer users, since the recipient did not view the material held in the archive file, but rather a reproduction of that data which were then

held in the recipient's computer after transmission had taken place. The Court of Appeal rejected this argument, holding at p. 558 that:

> the fact that the recipient obtains an exact reproduction of the photograph contained in the archive in digital form does not mean, in our judgment, that the (copy) photographs in the archive are not held in the first appellant's possession with a view to those same photographs being shown to others. The same data are transmitted to the recipient so that he shall see the same visual reproduction as is available to the sender whenever he has access to the archive himself.

Fellows was sentenced to three years in prison and Arnold to six months.

In another English case, R. v. Bowden ([2000] 1 Crim.App.R. 438), the Court of Appeal considered the question of whether the downloading and/or printing out of computer data of indecent images of children from the Internet was capable of amounting to the offense of making child pornography.

CASE EXAMPLE (R. v. BOWDEN 2000):
The facts of the case as set out in the judgment of Otton L.J. are that the defendant took his computer hard drive in for repair. While examining the computer, the repairer found indecent material on the hard drive. As a result of a subsequent investigation, police seized a computer and equipment including hard disk and floppy disks from the defendant. They examined the disks, which contained indecent images of young boys. The defendant had downloaded the photographs from the Internet, and either printed them out himself, or stored them on his computer disks. It was not contested that all the photographs were indecent and involved children under sixteen years. When arrested and interviewed, the defendant accepted that he had obtained the indecent material from the Internet and downloaded it onto his hard disk in his computer for his own personal use. He did not know it was illegal to do this. He admitted that he had printed out photographs from the images he had downloaded.

At first instance, defense counsel submitted that the defendant was not guilty of "making" photographs contrary to the section. He submitted that the defendant was in possession of them but nothing more. The Court of Appeal held that despite the fact that he made the photographs and the pseudo-photographs for his "own use", the defendant's conduct was clearly caught by the Act, stating at p. 444:

> Section 1 is clear and unambiguous in its true construction. Quite simply, it renders unlawful the making of a photograph or a pseudo-photograph … the words "to make" must be given their natural and ordinary meaning … As a matter of construction such a meaning applies not only to original photographs but, by virtue of section 7, also to negatives, copies of photographs and data stored on computer disk". The court adopted the prosecution's submissions, reported at pp. 444 to 445 of the judgment that: "a person who either downloads images onto a disk or who prints them off is making them. The Act is not only concerned with the original creation of images, but also their proliferation. Photographs or pseudo-photographs found on the Internet may have originated from outside the United Kingdom; to download or print within the jurisdiction is to create new material which hitherto may not have existed therein.

By equating downloading a file from the Internet with making it, the court concluded that Bowden had violated Section 1(1)(a) of the Protection of Children Act 1978.

To avoid any ambiguity, the Convention independently addresses producing and procuring child pornography using a computer. Also, in its definition of child pornography, the Convention includes images rendered using a computer that appear to contain minors but do not depict actual children. It should also be noted that in addition to these direct offenses, the convention recommends offenses concerning ancillary liability, that is, attempting, aiding and abetting.

3C.1.3 COMPUTER MISUSE

The Convention introduces the following five offenses against the confidentiality, integrity and availability of computer data and systems.

1. illegal access, that is, intentional access to the whole or any part of a computer system without right (Article 2);
2. illegal interception, being the intentional interception without right made by technical means of non-public transmissions of computer data to, from or within a computer system (Article 3);
3. data interference, that is, the intentional damaging, deletion, deterioration, alteration or suppression of computer data without right (Article 4);
4. system interference, being intentionally seriously hindering without right the functioning of a computer system by inputting, transmitting, damaging, deleting, deteriorating, altering or suppressing computer data (Article 5); and
5. misuse of devices, that is, the production, sale, procurement for use, import, distribution or otherwise making available of a device or password or access code with the intent that it be used for the purpose of committing any of the offenses established in articles 2–5 (Article 6).

In 1990, England became the first European country to enact a law to address computer crime specifically. The Computer Misuse Act introduced three new offenses: unauthorized access to a computer; unauthorized access with intent to commit or facilitate the commission of further offenses; and unauthorized modification of computer material (ss. 1, 2, and 3).

3C.1.3.1 UNAUTHORIZED ACCESS

The first offense under the Computer Misuse Act is your basic computer intrusion offense, which one commentator compares with breaking and entering (Gringas 2002, p. 285). Section 1(1) provides that:

A person is guilty of an offense if –

(a) he causes a computer to perform any function with intent to secure access to any program or data held in any computer;

(b) the access he intends to secure is unauthorised; and

(c) he knows at the time when he causes the computer to perform the function that that is the case.

So, the elements to be proved are that the perpetrator intended to break into the computer in the knowledge that he/she did not have authority so to do. The *actus reus* (the act or omissions that comprise the physical elements of a crime as required by law) is the action of breaking in (causing a computer to perform any function); the *mens rea* (literally: guilty mind) is the dishonest intent with knowledge of no authority. The definition of *unauthorized access* in the Act is quite literal and, as a result, is limiting.

CASE EXAMPLE (D.P.P. v. BIGNELL 1998):
In this case, the court was concerned with a situation where police officers secured access to the police national computer for a non-police but rather personal use. The question was whether this amounted to commission of an offense contrary to section 1 of the 1990 Act. The court held that the defendants had authority to access the police computer even though they did not do so for an authorized purpose. Therefore, they did not commit an offense contrary to section 1 of the Act. The court noted in its judgment that the 1990 Act was enacted to criminalize the act of breaking into computer systems. Thus, once the access was authorized, the Act did not look at the purpose for which the computer was accessed.

In this case, the defendant used the police computer in relation to two motor vehicles, the property of the defendant's former wife and her new partner. While this may have been a reprehensible infringement on their privacy, it did not constitute the crime of unauthorized access. Furthermore, the defendant narrowly avoided another crime when the innocent parties denied that he stalked them. The case, nonetheless, gave rise to the question of whether the offense of unauthorized access might be extended to a situation of improper or illegal use by an authorized user. This question was considered by the House of Lords in the case of R. v. Bow Street Magistrate *(ex parte* US Government, Allison [1999] 3 W.L.R. 620) where they refined interpretation of the notion of authorized or unauthorized access.

CASE EXAMPLE (R. v. BOW STREET MAGISTRATE – ALLISON 1997):
The defendant was accused of conspiring with legitimate employees of American Express to secure access to the American Express computer system with intent to commit theft and fraud, and to cause a modification of the contents of the American Express computer system. The Court of Appeal held that access was unauthorized under the Computer Misuse Act if
(a) the access to the particular data in question was intentional; (b) the access in question was unauthorized by a person entitled to authorize access to that particular data; (c) knowing the access to that particular data was unauthorized.
The court explained the decision as follows:

> the evidence concerning [the American Express employee]'s authority to access the material data showed that she did not have authority to access the data she used for this purpose. At no time did she have any blanket authorisation to access any account or file not specifically assigned to her to work on. Any access by her to an account which she was not authorised to be working on would be considered a breach of company policy and ethics and would be considered an unauthorised access by the company. The computer records showed that she accessed 189 accounts that did not fall within the scope of her duties. Her accessing of these accounts was unauthorised. ... The proposed charges against Mr. Allison therefore involved his alleged conspiracy with [the employee] for her to secure unauthorised access to data on the American Express computer with the intent to commit the further offences of forging cards and stealing from that company. It is [the employee]'s alleged lack of authority which is an essential element in the offences charged.

The House of Lords noted that the court at first instance had felt constrained by the strict definition of unauthorized access in the Act and the interpretation put upon them by the court in D.P.P. v. Bignell. The House of Lords went on to assert that the definition of unauthorized access in section 17 of the Act was open to interpretation, clarifying the offense as follows.

> Section 17 is an interpretation section. Subsection (2) defines what is meant by access and securing access to any programme or data. It lists four ways in which this may occur or be achieved. Its purpose is clearly to give a specific meaning to the phrase "to secure access". Subsection (5) is to be read with subsection (2). It deals with the relationship between the widened definition of securing access and the scope of the authority which the relevant person may hold. That is why the subsection refers to "access of any kind" and "access of the kind in question". Authority to view data may not extend to authority to copy or alter that data. The refinement of the concept of access requires a refinement of the concept of authorisation. The authorisation must be authority to secure access of the kind in question. As part of this refinement, the subsection lays down two cumulative requirements of lack of authority. The first is the requirement that the relevant person be not the person entitled to control the relevant kind of access. The word "control" in this context clearly means authorise and forbid. If the relevant person is so entitled, then it would be unrealistic to treat his access as being unauthorised. The second is that the relevant person does not have the consent to secure the relevant kind of access from a person entitled to control, i.e., authorise, that access.
>
> Subsection (5) therefore has a plain meaning subsidiary to the other provisions of the Act. It simply identifies the two ways in which authority may be acquired – by being oneself the person entitled to authorise and by being a person who has been authorised by a person entitled to authorise. It also makes clear that the authority must relate not simply to the data or programme but also to the actual kind of access secured. Similarly, it is plain that it is not using the word "control" in a physical sense of the ability to operate or manipulate the computer and that it is not derogating from the requirement that for access to be authorised it must be authorised to the relevant data or relevant programme or part of a programme. It does not introduce any concept that authority to access one piece of data should be treated as authority to access other pieces of data "of the same kind" notwithstanding that the relevant person did not in fact have authority to access that piece of data. Section 1 refers to the intent to secure

> unauthorised access to any programme or data. These plain words leave no room for any suggestion that the relevant person may say: "yes, I know that I was not authorised to access that data but I was authorised to access other data of the same kind."
> (pp. 626–627)

It is not clear how the COE Convention of Cybercrime defines "without right" and the same issue may arise. This situation is explicitly addressed by the US Computer Fraud and Abuse Act using the language "accessed a computer without authorization or exceeding authorized access".

3C.1.3.2 FACILITATING THE COMMISSION OF OTHER OFFENSES

The second of the Computer Misuse offenses has the additional element of an intent to commit or facilitate the commission of further offenses, such as the theft of or damage to data or the system in the previous case example (R. v. Bow Street Magistrate – Allison). It should be noted that a perpetrator may be guilty of this offense even where he/she has not in fact committed a further offense or indeed where the intended further offense would have been impossible to commit. It is the intention that offends. Section 2(3) of the Act states that, "It is immaterial for the purposes of this section whether the further offence is to be committed on the same occasion as the unauthorised access or on any future occasion."

For instance, the case of R. v. Governor of Brixton Prison (*ex parte* Levin) ([1997] 3 All E.R. 289), would come under section 2(3), if committed in England. In that case, Levin used a computer terminal in Russia to gain unauthorized access to the computerized fund transfer service of Citibank in the United States and made fraudulent transfers of funds from the bank to accounts that he or his associates controlled.

The COE Convention does not clearly address this offense.

3C.1.3.3 UNAUTHORIZED MODIFICATION OF COMPUTER MATERIAL

The third Computer Misuse offense involves unauthorized modification of the contents of any computer. The offender must intend to cause the modification and the knowledge that such modification is unauthorized as stated in Section 3(2):

> ... the requisite intent is an intent to cause modification of the contents of any computer and by so doing:
>
> (a) to impair the operation of any computer;
>
> (b) to prevent or hinder access to any program or data held in any computer; or
>
> (c) to impair the operation of any such program or the reliability of any such data.

The kinds of activities envisaged in this offense include denial of service attack and the spreading of malicious code such as viruses or worms. The

COE Convention describes this type of offense in two sections – data interference and system interference – using more general terms like "deterioration, alteration or suppression of computer data". These terms may be too general for legislative purposes. Despite being a decade older than the Convention, the Computer Misuse Act addresses the same offenses in a more concise and clear manner.

It should be noted that the possession or production of code that could be used to cause modification is not an offense. Nevertheless, its discovery on a computer may be useful in investigating cases of unauthorized modification. To prove this offense, it is necessary to show that modification occurred as a result of the acts of the defendant.

CASE EXAMPLE (R. v. WHITELEY 1991):
This case occurred prior to the Computer Misuse Act and was prosecuted under the Criminal Damage Act, 1971. The defendant had broken into the Joint Academic Network system, a network of connected ICL mainframe computers at universities, polytechnics and science and engineering research institutions. The defendant deleted and added files, put on messages, made sets of his own users and operated them for his own purposes, changed the passwords of authorized users and deleted files that would have recorded his activity. He successfully attained the status of systems manager of particular computers, enabling him to act at will without identification or authority.

Under the Criminal Damage Act, the defendant was charged with causing criminal damage to the computers by bringing about temporary impairment of usefulness of them by causing them to be shut down for periods of time or preventing them from operating properly and, distinctly, with causing criminal damage to the disks by way of alteration to the state of the magnetic particles on them so as to delete and add files – the disks and the magnetic particles on them containing the information being one entity and capable of being damaged. The jury acquitted the defendant of the first charge and convicted on the second. The defense appealed the conviction to the Court of Appeal on the basis that a distinction had to be made between the disk itself and the intangible information held upon it which, it was contended, was not capable of damage as defined in law (at that time).

The Court of Appeal held that what the Criminal Damage Act required to be proved was that tangible property had been damaged, not necessarily that the damage itself should be tangible. There could be no doubt that the magnetic particles on the metal disks were a part of the disks and if the defendant was proved to have intentionally and without lawful excuse altered the particles in such a way as to impair the value or usefulness of the disk, it would be damage within the meaning of the Act. The fact that the damage could only be detected by operating the computer did not make the damage any less within the ambit of the Act.

3C.2 SEARCH AND SEIZURE

In England and Ireland, law enforcement must obtain legal authorization to search a location and seize evidence. Part II of the English Police and Criminal Evidence Act, 1984, sets out a statutory framework governing powers of entry, search and seizure. Specifically, Section 8(1) lists the following requirements for obtaining a search warrant.

> If on an application made by a constable, a justice of the peace is satisfied that there are reasonable grounds for believing –
>
> (a) that a serious arrestable offence has been committed; and
>
> (b) that there is material on premises specified in the application which is likely to be of substantial value (whether by itself or together with other material) to the investigation of the offence; and
>
> (c) that the material is likely to be relevant evidence; and
>
> (d) that it does not consist of or include items subject to legal privilege, excluded material or special procedure material; and
>
> (e) that any of the conditions specified in subsection (3) below applies,
>
> he may issue a warrant authorising a constable to enter and search the premises.

In making a request for a search warrant, law enforcement officers in England are required to state the grounds for their application including the law that has been broken. Also, as in the United States, the application must specifically describe the premises that will be searched and, as much as possible, the items or individuals that are being sought. Additionally, computer information is specifically mentioned in Section 19(4) of the Police and Criminal Evidence Act.

> The constable may require any information which is contained in a computer and is accessible from the premises to be produced in a form in which it can be taken away and in which it is visible and legible if he has reasonable grounds for believing –
>
> (a) that –
>
> (i) it is evidence in relation to an offence which he is investigating or any other offence; or
>
> (ii) it has been obtained in consequence of the commission of an offence; and
>
> (b) that it is necessary to do so in order to prevent it being concealed, lost, tampered with or destroyed.

Additionally, Section 20(1) authorizes law enforcement to collect digital evidence.

> Every power of seizure which is conferred by an enactment to which this section applies on a constable who has entered premises in the exercise of a power conferred by

> an enactment shall be construed as including a power to require any information contained in a computer and accessible from the premises to be produced in a form in which it can be taken away and in which it is visible and legible.

3C.3 JURISDICTION AND EXTRADITION

3C.3.1 VENUE

In the UK Computer Misuse Act, 1990, the test for jurisdiction in cases where borders are transversed in the commission of an offense under the Act is whether there is a "significant link" with the domestic jurisdiction. Thus, the UK courts will claim jurisdiction where the perpetrator was in the United Kingdom when he/she caused the computer to perform the offending function or when the computer used was in the United Kingdom or when the victim computer was in the United Kingdom or if the defendant accessed a computer and his/her intention was to commit a further offense in the United Kingdom (see s.4)

3C.3.2 EXTRADITION

Extradition between European states is governed by the European Convention on extradition and related legislation.

It should be noted that, while one of the basic tenets of the EU is the free movement of persons across the internal borders, extradition in the European context is thwarted somewhat by traditional principles. The principle of territoriality holds that a state's criminal law only applies to criminal activity occurring within its jurisdiction. Pursuant to what is known as the nationality principle, states do not extradite their own citizens, but rather retain the power to institute domestic proceedings against them. Under Article 7 of the Convention on Extradition a requested party may refuse to extradite a person sought for an offense which is regarded by its law as having been committed in whole or in part in its territory. The refusal may be made mandatory in domestic legislation. Thus, for example, Section 15 of the Irish Extradition Act 1965 provides:

> Extradition shall not be granted where the offence for which it is requested is regarded under the law of the State as having been committed in the State.

There is an extradition treaty between the United Kingdom and the United States which has been given effect by Order in Council, the United States of America (Extradition) Order, 1976, made under Section 2 of the Extradition Act 1870. The procedure for extradition to the United States is, therefore, governed by the provisions of that Act which have been consolidated in schedule 1 to the Extradition Act 1989. (See Levin [1997] 3 All E.R. 289.)

Article III of the treaty provides:

> (1) Extradition shall be granted for an act or omission the facts of which disclose an offence within any of the descriptions listed in the Schedule annexed to this Treaty, which is an integral part of the Treaty, or any other offence, if:
>
> (a) the offence is punishable under the laws of both parties by imprisonment or other form of detention for more than one year or by the death penalty;
>
> (b) the offence is extraditable under the relevant law, being the law of the UK or other territory to which this Treaty applies by virtue of sub-para (1)(a) of article II; and
>
> (c) the offence constitutes a felony under the law of the United States of America.
>
> (2) Extradition shall also be granted for any attempt or conspiracy to commit an offence within paragraph (1) of this article if such attempt or conspiracy is one for which extradition may be granted under the laws of both parties and is punishable under the laws of both parties by imprisonment or other form of detention for more than one year or by the death penalty.

The schedule annexed to the Treaty does not include any reference to computer crime. Therefore, if an offence under the Computer Misuse Act, 1990 is to be pleaded within the terms of the Treaty it will come in as some "other offense."

CASE EXAMPLE (R. v. GOVERNOR OF BRIXTON PRISON, *ex parte* LEVIN 1997): Levin was charged before the Federal District Court for the Southern District of New York with the Federal offenses of wire fraud and bank fraud and certain offenses relating to the misuse of computers. The allegation was that he used a computer terminal in St. Petersburg to gain unauthorized access to the computerized fund transfer service of Citibank NA (Citibank) in Parsipanny, New Jersey, and fraudulently made 40 transfers of funds from the accounts of clients of Citibank to accounts which he or his associates controlled.

Levin was arrested in the transfer lounge at Stansted Airport in London on a provisional warrant issued at the request of the US government. The Secretary of State signified to the metropolitan magistrate that a requisition for Levin's surrender had been made by the government of the United States, stating that he was accused of various extradition crimes within the jurisdiction of the United States.

As stated by the court at p. 292 of the report:

it thereupon become the duty of the metropolitan magistrate, pursuant to paragraph 6(1) of schedule 1, to hear the case in the same manner as if the defendant were charged with an indictable offence committed in this country. Paragraph 7(1) provides that if—

> "such evidence is produced as ... would, according to the law of England and Wales, justify the committal for trial of the prisoner if the crime of which he is accused had been committed in England or Wales, the metropolitan magistrate shall commit him to prison, but otherwise shall order him to be discharged."

The magistrate found that the evidence justified the defendant's committal for trial for 66 offences. These included four counts of theft and numerous counts of forgery, false accounting and computer misuse. Accordingly, he ordered the defendant's committal to prison to await the decision of the Secretary of State as to whether he should be surrendered.

CASE EXAMPLE

R. v. Bow Street Magistrates, *ex parte* U.S. Government, Allison [1999] 3 W.L.R. 620 is authority for the proposition that offenses under the Computer Misuse Act, 1990, are extradition crimes. The facts of the case are summarized above. Following committal by the court of first instance, the accused brought *habeas corpus* proceedings challenging the view than any of the offenses alleged (unauthorised access with intent to commit theft and forgery, and intent to cause an unauthorized modification of the contents of the victim computer system) were "extradition crimes" under the Extradition Act, 1989 and the US Extradition Order in Council, 1976. This submission was rejected by the court, noting that section 15 of the 1990 Act provides that: "The offences to which an Order in Council under Section 2 of the Extradition Act, 1870, can apply shall include – (a) offences under Section 2 or 3; (b) any conspiracy to commit such an offense; and (c) any attempt to commit an offense under Section 3".

3C.4 PENALTIES

In recognition of the growing problem, penalties for computer-related crimes are being made more severe. For instance, the English Criminal Justice and Court Services Act, 2000 increased the maximum penalty for offenses contrary to Section 1(1) of the Protection of Children Act, 1978 from 3 to 10 years imprisonment. Anyone convicted of or pleading guilty to an offense involving child pornography might be subject to a range of other legal consequences including registration under the Sex Offenders Act, 1997, disqualification from working with children under the Criminal Justice and Court Services Act, 2000 and being barred or restricted from employment as a teacher or worker with persons under the age of 19.

The English Sentencing Advisory Panel (SAP) is a body established to advise the Court of Appeal. In August 2002, it published its advice on offenses involving child pornography. (See Gillespie, Alisdair A. "Sentences for Offences Involving Child Pornography," [2003] Crim.L.R. 81.)

The SAP's advice was discussed in the case of R. v. Oliver, Hartrey and Baldwin [2003] Crim.L.R. 127 where the English Court of Appeal dealt with three appeals together for the purpose of giving sentencing guidelines for offenses involving indecent photographs and pseudo-photographs of children. The court agreed with the panel that the two primary factors which determined the seriousness of a particular offense were the nature of the indecent material and the extent of the offender's involvement with it. The seriousness of an individual offense increased with the offender's proximity to and responsibility for the original abuse. Any element of commercial gain would place an offence at a high level of seriousness. Swapping of images could properly be regarded as a commercial activity, albeit without financial gain, because it fuelled demand for such material. Widespread distribution was intrinsically more harmful than a transaction limited to two or three individuals. Merely locating an image on the Internet would generally be less serious than downloading it. Downloading would generally be less serious than taking an original photograph. Possession, including downloading, of artificially created pseudo-photographs and the making of such images should generally be treated as being at a lower level of seriousness than the making and possessing of images of real children. The court noted, however, that although pseudo-photographs lacked the historical element of likely corruption of real children depicted in photographs, pseudo-photographs might be as likely as real photographs to fall into the hands of or to be shown to the vulnerable, and therefore to have an equally corrupting effect.

The SAP categorized the increasing seriousness of material into five levels, characterized by the court, in making certain amendments, as follows:

1 images depicting erotic posing with no sexual activity;
2 sexual activity between children or solo masturbation by a child;
3 non-penetrative sexual activity between adults and children;
4 penetrative sexual activity between adults and children;
5 sadism or bestiality.

The court held that a fine would normally be appropriate in a case where (i) the offender was merely in possession of material solely for his own use, including cases where material was downloaded from the Internet but was not further distributed, (ii) the material consisted entirely of pseudo-photographs, the making of which had involved no abuse or exploitation of children, or (iii) there was no more than a small quantity of material at level 1.

The court agreed with the SAP's recommendation that in any case which was close to the custody threshold, the offender's suitability for treatment should be assessed with a view to imposing a community rehabilitation order with a requirement to attend a sex offender treatment program. With regard to custodial sentences, in summary, the court found as follows:

- a sentence of up to six months would be appropriate in a case where the offender was in possession of a large amount of material at level 2 or a small amount at level 3 or the offender had shown, distributed or exchanged indecent material at level 1 or 2 on a limited scale and without financial gain;

- a sentence of between six and twelve months would be appropriate for showing or distributing a large number of images at level 2 or 3 or possessing a small number of images at level 4 or 5;
- a sentence between twelve months and three years would be appropriate for possessing a large quantity of material at level 4 or 5, showing or distributing a large number of images at level 3 or producing or trading in material at level 1, 2 or 3;
- sentences longer than three years should be reserved for cases where images at level 4 or 5 had been shown or distributed, the offender was actively involved in the production of images at level 4 or 5, especially where that involvement included breach of trust and whether or not there was an element of commercial gain, or the offender had commissioned or encouraged the production of such images;
- sentences approaching the ten year maximum would be appropriate in very serious cases where the defendant had a previous conviction either for dealing in child pornography or for abusing children sexually or with violence.

The court set out specific factors which were capable of aggravating the seriousness of a particular offense:

1 the images had been shown or distributed to a child;

2 there were a large number of images;

3 the way in which a collection of images was organized on a computer might indicate a more or less sophisticated approach on the part of the offender to, say, trading;

4 images posted on a public area of the Internet;

5 if the offender was responsible for the original production of the images, especially if the child or children were family members or located through abuse of the offender's position of trust, for example, as a teacher;

6 the age of the children involved.

So far as mitigation was concerned, the court agreed with the SAP that some weight might be attached to good character, but not much. A plea of guilty was a statutory mitigating factor; the extent of the sentencing discount to be allowed for a plea of guilty would vary according to the timing and circumstances of the plea.

Applying these principles to the instant cases, the court imposed a sentence of 8 months imprisonment with an extension of 28 months in the case of a man of previous good character who had pleaded guilty to six offenses of making indecent photographs or pseudo-photographs of a child, his computer and some floppy disks having been found to contain some 20,000 images at levels 3 and 4. The court imposed a sentence of three years on a guilty plea in the case of a man who had distributed and made photographs of children at level 4, his computer systems having been found to contain a total of 20,000 indecent images and 500 movie files of child abuse. In the third case, the court imposed a sentence of 2.5 years for the offenses of making indecent photographs. A concurrent sentence of 3 years was imposed for indecent assault on a girl aged 8 or 9 years, a video recording depicting the defendant committing the assault having been found in the home of another person.

It should be noted that a new criminal justice bill is at the time of writing before the English Parliament which may propose reform of sentencing guidelines.

3C.5 PRIVACY

> There is no clear definition of what constitutes privacy or the legal right to privacy. What constitutes data protection can be more readily answered by the circular device of saying that it is the application of privacy principles to the collection, retention, use and disclosure of information about individual human beings, especially in a computerized environment. (Regan, p. 134)

One commentator proposes the equation of the notion of "privacy" with that of "control of personal information," arguing that

> placing control of information at the heart of our deliberations about privacy achieves what the orthodox analysis has conspicuously failed to do: it postulates a presumptive entitlement accorded to all individuals that their personal data may be collected only lawfully or fairly and that once obtained, may not be used, in the absence of the individual's consent, for a purpose other than that for which it was originally given." (Wacks, p.4)

European Directive 95/46/EC on the protection of individuals with regard to the processing of personal data and on the free movement of such data of October 1995 ("the data protection directive") notes in its preamble the dual purpose of ensuring that personal data should be able to flow freely from one EU member state to another, but also that the fundamental rights of individuals, notably the right to privacy, must be safeguarded. These fundamental rights are recognized in the constitution and laws of member states and in the European Convention for the Protection of Human Rights and Fundamental Freedoms (ECHR). Article 8(1) of the ECHR provides that "everyone has the right to respect for his private and family life, his home and his correspondence".

Article 5 of the data protection directive sets out strict principles relating to data quality, requiring that data must be:

(a) processed fairly and lawfully;

(b) collected for specified, explicit and legitimate purposes and not further processed in a way incompatible with those purposes. Further processing of data for historical, statistical or scientific purposes shall not be considered as incompatible provided that member states provide appropriate safeguards;

(c) adequate, relevant and not excessive in relation to the purposes for which they are collected and/or further processed;

(d) accurate and, where necessary, kept up to date; every reasonable step must be taken to ensure that data which are inaccurate or incomplete, having regard to the purposes for which they were collected or for which they are further processed, are erased or rectified;

(e) kept in a form which permits identification of data subjects for no longer than is necessary for the purposes for which the data were collected or for which they are further processed. Member states shall lay down appropriate safeguards for personal data stored for longer periods for historical, statistical or scientific use.

Article 5 further sets out binding criteria for making data processing legitimate, requiring EU member states to provide that personal data may be processed only if:

(a) the data subject [i.e. the identifiable person to whom the information relates] has unambiguously given his consent; or

(b) processing is necessary for the performance of a contract to which the data subject is party or in order to take steps at the request of the data subject prior to entering into a contract; or

(c) processing is necessary for compliance with a legal obligation to which the controller [i.e. the legal person who determines the purposes and means of processing the data] is subject; or

(d) processing is necessary in order to protect the vital interests of the data subject; or

(e) processing is necessary for the performance of a task carried out in the public interest or in the exercise of official authority vested in the controller or in a third party to whom the data are disclosed; or

(f) processing is necessary for the purposes of the legitimate interests pursued by the controller or by the third party or parties to whom the data are disclosed, except where such interests are overridden by the interests for fundamental rights and freedoms of the data subject which require protection under Article 1(1).

Article 25 of the data protection directive governs principles concerning the transfer of data to third countries, that is, to non-EU member states. Such transfers of data may only take place if the third country "*ensures an adequate level of protection*". Where third countries do not ensure an adequate level of protection, member states are required to take all measures necessary to prevent the transfer of data to that country. Article 25(2) provides that:

> the adequacy of the level of protection afforded by a third country shall be assessed in the light of all the circumstances surrounding a data transfer operation or set of data transfer operations; particular consideration shall be given to the nature of the data, the purpose and duration of the proposed processing operation or operations, the country of origin and country of final destination, the rules of law, both general and sectoral, in

force in the third country in question and the professional rules and security measures which are complied with in that country.

Clearly, these provisions may create a barrier to, say, the sharing of information by law enforcement agencies within and outside of the EU.

Notably, in response to the September 11 terrorist attacks on the United States in 2001, there have been significant efforts in Europe to give state agencies greater access to data relating to personal communications, including telephone records and Internet usage. These efforts have encountered strong opposition because of the concerns over invasion of privacy. Although legislation such as the UK Regulation of Investigatory Powers Act 2000 (RIPA) and the Data Protection Act 1992 in Ireland permit authorities to access personal data under certain circumstances, proposed anti-terrorist legislation could require communication service providers to retain usage records for longer periods of time and give more agencies access to this data.

3C.6 SUMMARY

The law in relation to computer related crime analyzed above is a discreet field within the criminal law. While its application and interpretation are becoming more sophisticated as technology develops, it is clear that it is digital evidence that has the wider scope of application to traditional criminal offenses. Indeed, digital evidence may be adduced in any type of criminal proceeding. Where before handwritten lists or diary entries may have been the vital link between a perpetrator and a crime (be it armed robbery, murder, etc.), now such vital evidence is more likely to be discovered on a perpetrator's personal computer, mobile telephone, or electronic organizer. Where before eye-witness evidence may have been required to secure a conviction, now security camera footage or till-roll evidence will act as proof. It may not be an exaggeration to suggest that the survival of the criminal justice system rests with the requisite expertise of individuals in anticipating, collecting and presenting computer-generated evidence.

UNITED STATES PERSPECTIVE BIBLIOGRAPHY

CASES

A&M Records v. Napster (2001) Appeals Court, 9th Circuit, Case number 00-16401

Ashcroft v. American Civil Liberties Union (2002) US Supreme Court (Available online at http://www.aclu.org/Cyber-Liberties/Cyber-Liberties.cfm?ID=12039&c=59)

Bensusan Restaurant Corporation v. King (1997) Appeals Court, 2nd Circuit, Case Number 96-9344 (Available online at http://laws.findlaw.com/2nd/969344.html)

California v. Greenwood (1987) US Supreme Court, Case number 86-684 (Available online at http://laws.findlaw.com/us/486/35.html)

DVDCCA v. Bunner (2001) Appeals Court, Califorina, Case number CV786804 (Available online at http://www.eff.org/sc/20011101_bunner_appellate_decision.html)

FCC v. Pacifica (1978) US Supreme Court, Case number 77-528 (Available online at http://www.eff.org/Legal/Cases/FCC_v_Pacifica/)

Free Speech Coalition v. Ashcroft (2002) Appeals Court, 9th Circuit, Case number 00-795 (Available http://laws.findlaw.com/us/000/00-795.html)

Inset Systems, Inc. v. Instruction Set, Inc. (1996) District Court, Connecticut, Case Number CV-3 : 95CV-01314

Katz v. United States (1967) US Supreme Court, Case number 35 (Available online at http://laws.findlaw.com/us/389/347.html)

Kyllo v. United States (2001) US Supreme Court, Case number 99-8508 (Available online at http://laws.findlaw.com/us/000/99-8508.html)

Miller v. California (1973) Supreme Court, California, Case number 70-73 (Available online at http://laws.findlaw.com/us/413/15.html)

Minnesota v. Granite Gate Resorts, Inc. (1996) Appeals Court, Minnesota, Case Number C6-95-7227

New York v. Ferber (1982) Supreme Court, New York, Case number 81-55 (Available online at http://laws.findlaw.com/us/458/747.html)

Playboy Enterprise Inc. v. Frena (1993) District Court, Florida, Case number 93-489-Civ-J-20

Reno v. American Civil Liberties Union (1997) US Supreme Court, Case number 96-511 (Available online at http://www.aclu.org/Privacy/Privacy.cfm?ID=13904&c=252)

Sony Corporation of America v. Universal City Studios, Inc. (1984) US Supreme Court, Case number 81-1687 (Available online at http://laws.findlaw.com/us/464/417.html)

United States v. Hilton (1999) Appeals Court, 1st Circuit, Case number 98-1513 (Available online at http://laws.findlaw.com/1st/981513.html)

United States v. Morris (1991) Appeals Court, 2nd Circuit (928 F.2d 504), Case number 90-1336

United States v. Thomas (1996) Appeals Court, 6th Circuit, Case Number 94-6648/6649 (Available online at http://www.eff.org/Legal/Cases/AABBS_Thomases_Memphis/)

Universal City Studios v. Corley (2001) Appeals Court, 2nd Circuit, Case number 00-9185 (Available online at http://laws.findlaw.com/2nd/009185.html)

LEGISLATION

Child Online Protection Act of 1998 (COPA)

Child Pornography Prevention Act of 1996 (CPPA)

Child Protection Act, 1997

Child Protection and Obscenity Enforcement Act, 1988

Child Protection Restoration and Penalties Enforcement Act, 1990

Child Sexual Abuse and Pornography Act, 1986

Computer Fraud and Abuse Act of 1984, 1986 (CFAA)

Digital Millenium Copyright Act, 1998

Electronic Communications Privacy Act, 1986 (ECPA)

Fair Credit Reporting Act, 2002

Family Educational Rights & Privacy Act, 1974 (FERPA)

Federal Right to Financial Privacy Act, 1978

Federal Cable Communications Policy Act, 1984

No Electronic Theft Act, 1997

Privacy Act 1974

Protection of Children Against Sexual Exploitation Act of 1977

Protection of Children Act of 1978

Telecommunications Act of 1996 (Title V: Communications Decency Act)

USA PATRIOT Act, 2001

Video Privacy Protection Act, 1988

REFERENCES

Egan T. (2000) "Wall Street Meets Pornography", New York Times, October 23

Rosenblatt K. S. (1995) High-Technology Crime: Investigating Cases Involving Computes, San Jose, CA: KSK Publications.

Spafford G. H. (1989) "The Internet Worm: Crisis and After math" Communications of the ACM, 32(6), 678–87.

USDOJ (2002) "Searching and Seizing Computers and Obtaining Electronic Evidence in Criminal Investigations" (Available online at http://www.usdoj.gov/criminal/cybercrime/s&smanual2002.htm)

EUROPEAN PERSPECTIVE BIBLIOGRAPHY/ SUGGESTED FURTHER READING

CASES

AG's Reference (No. 5 of 1980) (1980) 3 All E.R. 88.

Atkins, Goodland v. Director of Public Prosecutions (2000) 2 Cr.App.R. 248; [2000] 2 All E.R. 425.

Director of Public Prosecutions v. Bignell (1998) 1 Cr.App.R.1.

Director of Public Prosecutions v. McKeown, Jones (1997) 1 All E.R. 737.

R. v. Bow Street Magistrates, ex parte US Government, Allison (1999) 3 W.L.R. 620.

R. v. Bowden (2000) 1 Cr.App.R 438.

R. v. Chesterfield Justices ex parte Bramley (2000) 2 W.L.R. 409.

R. v. Cochrane (1993) Crim.L.R. 48.

R. v. Fellows, Arnold (1997) 2 All E.R. 548.

R. v. Governor of Brixton Prison, ex parte Levin (1997) 3 All E.R. 289.

R. v. Oliver, Hartrey and Baldwin (2003) Crim.L.R. 127.

R. v. Whiteley (1991) 93 Cr.App.R 25.

LEGISLATION

Child Trafficking and Pornography Act, 1998 (IRE).

Computer Misuse Act, 1990 (ENG).

Council of Europe, Convention on Cyber-Crime, Budapest, 23.XI.2001.

Criminal Damage Act, 1971 (ENG).

Criminal Evidence Act, 1992 (IRE).

Criminal Justice Act, 1988 (ENG).

Criminal Justice and Court Services Act, 2000 (ENG).

Criminal Justice and Public Order Act, 1994 (ENG).

Criminal Justice (Theft and Fraud Offences) Act, 2001 (IRE).

Electronic Commerce Act, 2001 (IRE).

Extradition Act, 1870 (ENG).

Extradition Act, 1989 (ENG).

Obscene Publications Act, 1959 (ENG).

Order in Council, the United States of America (Extradition) Order, 1976 (ENG).

Police and Criminal Evidence Act, 1984 (ENG).

Protection of Children Act, 1978 (ENG).

Sex Offenders Act, 1997 (ENG).

Theft Act, 1968 (ENG).

Directive 95/46/EC on the protection of individuals with regard to the processing of personal data and on the free movement of such data, October 1995 (O.J. – L. 281 p. 31). "First Orientations on Transfers of Personal Data to Third Countries" Discussion Document adopted by the Working Party on 26 June 1997 (see: www.datenschutz-berlin.de/doc/eu/gruppe29/wp04_en.htm).

European Convention on Human Rights and Fundamental Freedoms (1950).

REFERENCES

ARTICLES

Gillespie, A. A. (2003) "Sentences for Offences Involving Child Pornography", *Criminal Law Review* 81.

Gindin, S. E. (1998) "As the Cyber-World Turns", [www.info-law.com/eupriv.html].

Gindin, S. E. (1997) "Lost and Found in Cyberspace: Informational Privacy in the Age of the Internet", [www.info-law.com/lost.html].

Harding, C. (2000) "Exploring the Intersection of European Law and National Criminal Law," *European Law Review* 25, 374.

Kelleher, D. (2000) "The Council of Europe's Draft Convention on Cyber-Crime", *Technology and Law Journal* 12.

Sommer, P. (1998) "Digital Footprints: Assessing Computer Evidence", *Criminal Law Review* Special Edition, pp. 61–78.

Taylor, M., Quayle, E. and Holland, G. "*Child Pornography, the Internet and Offending*", [http://www.isuma.net/v02n02/taylor/taylor_e.shtml].

Wacks, R. (1999) "Towards a New Legal and Conceptual Framework for the Protection of Internet Privacy", *Irish Intellectual Property Review* 1.

TEXTS

Brown, L. N. and Kennedy, T. (2000) *The Court of Justice of the European Communities,* London: Sweet & Maxwell.

Gringas, C. (2002) *The Laws of the Internet,* London: Butterworths.

Hollinger, R. C. and Lanza-Kaduce, L. (1988) "The Process of Criminalization: The Case of Computer Crime Law," *Criminology* 26(1), 104.

Kelleher, D. and Murray, K. (1999) *IT Law in the European Union,* London: Sweet & Maxwell.

Reed, C. and Angel, J. (eds) (2000) *Computer Law,* London: Blackstone Press.

Regan, E. (editor) (2000) *The New Third Pillar: Cooperation Against Crime in the European Union,* Dublin: Institute of European Affairs.

Smith, G. J. H. (1999) *Internet Law and Regulation,* London: Sweet & Maxwell.

Wyngaert Van den (1993) *Criminal Procedure Systems in the European Community,* London: Butterworths.

CHAPTER 4

THE INVESTIGATIVE PROCESS

Eoghan Casey and Gary Palmer

...the law and the scientific knowledge to which it refers often serve different purposes. Concerned with ordering men's conduct in accordance with certain standards, values, and societal goals, the legal system is a prescriptive and normative one dealing with the "ought to be". Much scientific knowledge, on the other hand, is purely descriptive; its "laws" seek not to control or judge the phenomenon of the real world, but to describe and explain them in neutral terms.

(Korn 1966)

The goal of any investigation is to uncover and present the truth. Although this chapter will deal primarily with truth in the form of digital evidence, this goal is the same for all forms of investigation whether it be in pursuit of a murderer in the physical world or trying to track a computer intruder online. As noted in the Introduction, when evidence is presented as truth of an allegation it can impact on whether people are deprived of their liberties, and potentially whether they live or die. This is reason enough to use trusted methodology and technology to ensure that the processing, analysis, and reporting of evidence are reliable and objective. This chapter describes such a methodology, based on the scientific method, to help investigators uncover truths to serve justice. This methodology is designed to assist in the development of case management tools, Standard Operating Procedures (SOPs), and final investigative reports. This methodology has grown out of experiences and discussions in the field, and is believed to be complete and sufficient in scope. However, every investigation is unique and can bring unforeseeable challenges, so this methodology should not be viewed as an end-point but rather as a framework or foundation upon which to build.

The investigative process is part of a larger methodology most often associated with courts of law shown in Figure 4.1. The process of determining if wrongdoing has occurred and if punitive measures are warranted is complex and goes beyond investigative steps normally referred to as "forensic."

Digital Evidence and Computer Crime Second Edition
ISBN: 0-12-163104-4

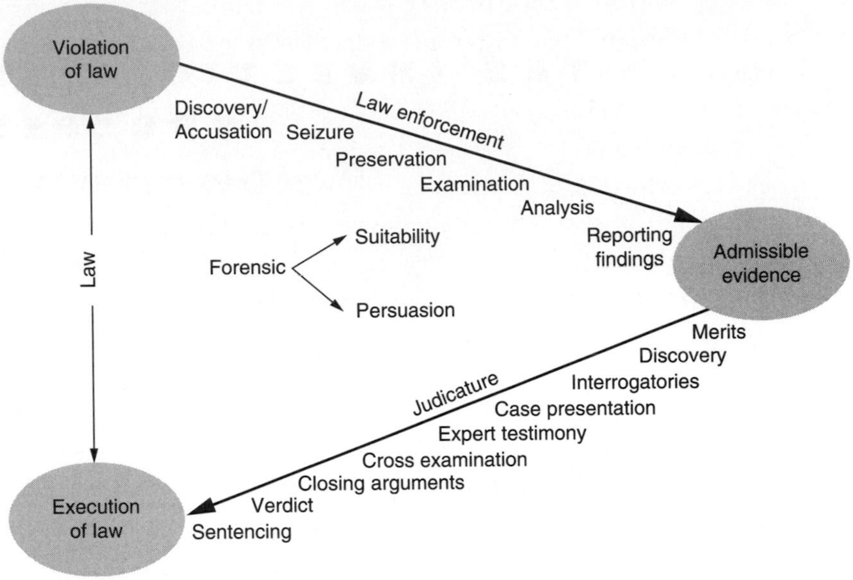

Figure 4.1
Overview of case/incident resolution process.

By *forensic* we mean a characteristic of evidence that satisfies its suitability for admission as fact and its ability to persuade based upon proof (or high statistical confidence).

The simplified methodology depicted in Figure 4.1 is provided to help investigators see the placement of their activities relative to other necessary events. The investigative process begins with an accusation and progresses through evidence handling to a clear and precise explanation of facts and techniques in expert testimony. This linear representation is useful for structuring procedures and a final report that describes each step of an investigation to decision makers. In practice, investigations can be non-linear, such as performing some basic analysis in the collection stage, or returning to the collection step when analysis leads to additional evidence. Before delving into this investigative methodology in detail, there are some fundamental concepts that must be understood.

Trained, experienced investigators will begin by asking themselves a series of questions aimed at deciding if a crime or infraction has actually occurred. The answer to these questions will help determine whether or not a full investigation will proceed or if valuable and limited investigative resources are better applied to other matters. For instance, when log files indicate that an employee misused a machine but he adamantly denies it, a digital investigator should carefully examine the logs for signs of error. Similarly, when a large amount of data are missing on a computer and an intruder is suspected, digital investigators should determine if the damage is more consistent with disk corruption than an intrusion. In one case, a suicide note on a computer raised concern because it had a creation date after the victim's death. It transpired that the computer clock was incorrect and the note was actually written before the suicide.

When these questions are answered affirmatively, the focus shifts toward determining what happened, where, when, how, who was involved, and why. The process by which digital evidence is uncovered and applied to these issues is composed of several steps each employing strict protocols, proven methods, and, in some cases, trusted tools. More importantly, the success of this process depends heavily on the experience and skill of the investigators, evidence examiners and crime scene technicians who must collaborate to piece the evidence together and develop a convincing account of the offense.

The effectiveness of the investigative process depends upon high levels of objectivity applied at all stages. Some cases and the nature of the evidence uncovered (digital or otherwise) will take investigators and forensic examiners to emotional limits, testing their resolve. Computer security professionals in the private sector often have to investigate long-time coworkers and cases in all sectors can involve brutal abuse of innocent victims, inciting distraught individuals and communities to strike out at the first available suspect. A good investigator can remain objective in the most trying situations.

The very traits that make a good investigator or forensic examiner may lead us to depend on experience in place of individual case-related facts, resulting in unfounded conclusions. Individuals with inquiring minds and an enthusiasm for apprehending offenders begin to form theories about what may have occurred the moment they learn about an alleged crime, even before examining available evidence. Even experienced investigators are prone to forming such preconceived theories because they are inclined to approach a case in the same way as they have approached past cases, knowing that their previous work was upheld.

Hans Gross, one of this century's preeminent criminologists, put it best in the following quotation:

> Nothing can be known if nothing has happened; and yet, while still awaiting the discovery of the criminal, while yet only on the way to the locality of the crime, one comes unconsciously to formulate a theory doubtless not quite void of foundation but having only a superficial connection with the reality; you have already heard a similar story, perhaps you have formerly seen an analogous case; you have had an idea for a long time that things would turn out in such and such a way. This is enough; the details of the case are no longer studied with entire freedom of mind. Or a chance suggestion thrown out by another, a countenance which strikes one, a thousand other fortuitous incidents, above all losing sight of the association of ideas end in a preconceived theory, which neither rests on juridical reasoning nor is justified by actual facts. (Gross 1924, pp. 10–12)

As experience increases and methods employed are verified, the accuracy of these "predictions" may improve. Conjecture based upon experience has its place in effective triage but should not be relied upon to the exclusion of rigorous investigative measures. The investigative process demands that each case be viewed as unique with its own set of circumstances and exhibits. Letting the evidence speak for itself is particularly important when offenders take steps to misdirect investigators by staging a crime scene or concealing evidence.

The main risk of developing full hypotheses before closely examining available evidence is that investigators will impose their preconceptions during evidence collection and analysis, potentially missing or misinterpreting a critical clue simply because it does not match their notion of what occurred. For instance, when recovering a deleted file named "σorn1yr5.gif" depicting a naked baby, an investigator might impose a first letter of the file that indicates "porn1yr5.gif" rather than "born1yr5.gif". Instead, if the original file name is not recoverable, a neutral character such as "_" should be used to indicate that the first letter is unknown.

This caveat also applies to the scientific method from which the investigative process borrows heavily. At the foundation of both is the tenet that no observation or analysis is free from the possibility of error. Simply trying to validate an assertion increases the chance of error – the tendency is for the analysis to be skewed in favor of the hypothesis. Conversely, by developing many theories, an investigator is owned by none and by seeking evidence to disprove each hypothesis, the likelihood of objective analysis increases (Popper 1959). Therefore, the most effective way to counteract preconceived theories is to employ a methodology that compels us to find flaws in our theories, a practice known as falsification.

Generally, in the prosecutorial environment, scientific truth is subordinate to legal truth and investigators must accept the ruling of the court. Similarly, investigators must generally accept an attorney's decision not to take a case. However, in some instances, investigators will face an ethical dilemma if they feel that a miscarriage of justice has occurred. An investigator may be motivated to disclose information to the media or assist in a follow-up investigation but such choices must be made with great care because a repeated tendency to disagree with the outcome of an investigation will ruin an investigator's credibility and even expose him/her to legal action.

As an example, as an investigation progresses a prime suspect may emerge. Although it is an investigator's duty to champion the truth, investigators must resist the urge to formally assert that an individual is guilty. A common misdeed is to use a verification methodology, focusing on a likely suspect and trying to fit the evidence around that individual. When a prime suspect has been identified and a theory of the offense has been formed, experienced investigators will try to prove themselves wrong. Implicating an individual is not the job of investigators – this is for the courts to decide and unlike scientific truth, legal truth is negotiable.

For instance, in common law countries, the standard of proof for criminal prosecutions is *beyond a reasonable doubt* and for civil disputes it is the *balance of probabilities*. Legal truth is influenced by ideas like fairness and justice, and the outcome may not conform to the scientific truth. A court may convict an individual even if the case is weak or some evidence suggests innocence.

> Most forensic scientists accept the reality that while truthful evidence derived from scientific testing is useful for establishing justice, justice may nevertheless be negotiated. In these negotiations, and in the just resolution of conflict under the law, truthful evidence may be subordinated to issues of fairness, and truthful evidence may be manipulated by forces beyond the ability of the forensic scientist to control or perhaps even to appreciate fully. (Thornton 1997)

Galileo Galilei's experiences provide us with an illustrative example of the power of the scientific method in discovering the truth and the cost of ignoring the reality that scientific truth may be subordinated to other truths. By observing the motion of stellar objects, Galileo gathered evidence to support Copernicus's theory that the Earth revolved around the Sun. Although Galileo was correct and was widely respected as a scientist and mathematician, he was unable to dislodge the heliocentric conception of the Solar system that had persisted since Aristotle proposed it in the fourth century B.C. It seemed absurd to claim that the Earth was in motion when anyone could look at the ground and see that it was still. Also, the most vehement opponents of the idea felt that it contradicted certain passages in the Holy Scripture and thus threatened the already wavering authority of the Catholic Church (Sobel 1999).

The issue came to a head in 1616 when Pope Paul V appointed a panel of theologians to decide the matter. Despite its widespread acceptance and Galileo's efforts to present supporting evidence, the panel concluded that certain aspects of Copernican astronomy were heretical. In essence, scientific truth was subordinated to a religious truth. Although Galileo was instructed not to present his opinions about the Solar system as fact, he was not specifically named as a heretic, one of the most grave crimes of the time. Almost twenty years later, by claiming that he had abandoned his belief in the Copernican model as instructed but wanted to demonstrate to the world that he and the Church fully understood all of the scientific arguments, Galileo obtained permission to publish his observations and theories in *Dialogue of Galileo Galilei.* However, the *Dialogue* quickly generated outrage and, in 1633, the book was banned and the 70-year-old Galileo was imprisoned for heresy and compelled to formally renounce his belief that the Earth rotated around the Sun.

There are a few valuable lessons here. The employment of a rigorous investigative process may uncover unpopular or even unbelievable truths subject to rejection unless properly and clearly conveyed to the intended audience. Investigators may be faced with a difficult choice – renounce the truth or face the consequences of holding an unpopular belief. It is the duty of investigators to unwaveringly assert the truth even in the face of opposition.

This account of Galileo is not intended to suggest that science is infallible. The fact is that science is still advancing and previous theories are being replaced by better ones. For instance, DNA analysis has largely replaced blood typing in forensic serology, and although the technique of blood typing was valid, it was not conclusive enough to support some of the convictions based upon evidence derived from that analysis alone. This weakness can be shown in dramatic fashion by the existence and success of the Innocence Project,[1] which is using results of DNA analysis to overturn wrongful convictions based on less than conclusive ABO Blood Typing and enzyme testing.

[1] *http://www.innocenceproject.org*

While preparing for the final step of the investigative process (the decision or verdict) it is important to keep in mind that discrepancies between scientific and legal truth may arise out of lack of understanding on the part of the decision makers. This is different from scientific peer review, where reviewers are qualified to understand and comment on relevant facts and methods with credibility. When technical evidence supporting a scientific truth is presented to a set of reviewers who are not familiar with the methods used, misunderstandings and misconceptions may result. To minimize the risk of such misunderstandings, the investigative process and the evidence uncovered to support prosecution must be presented clearly to the court. A clear presentation of findings is also necessary when the investigative process is applied to support decision makers who are in charge of civilian and military network operations. However, investigators may find this situation easier since decision makers in these domains often have some familiarity with methods and tools employed in forensic investigations for computer and network defense.

4.1 THE ROLE OF DIGITAL EVIDENCE

One of the main goals in an investigation is to attribute the crime to its perpetrator by uncovering compelling links between the offender, victim, and crime scene. Witnesses may identify a suspect but evidence of an individual's involvement is usually more compelling and reliable. According to Locard's Exchange Principle, anyone, or anything, entering a crime scene takes something of the scene with them, and leaves something of themselves behind when they leave. In the physical world, an offender might inadvertently leave fingerprints or hair at the scene and take a fiber from the scene. For instance, in a homicide case the offender may attempt to misdirect investigators by creating a suicide note on the victim's computer, and in the process leave fingerprints on the keyboard. With one such piece of evidence, investigators can demonstrate the strong possibility that the offender was at the crime

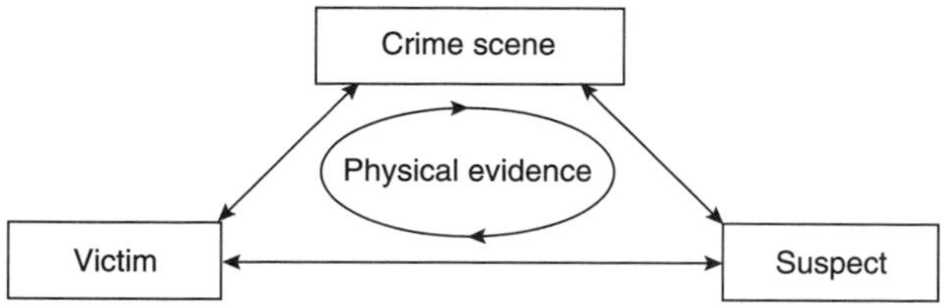

Figure 4.2
Locard's Exchange Principle.

scene. With two pieces of evidence the link between the offender and crime scene becomes stronger and easier to demonstrate (Figure 4.2).

This type of exchange produces evidence belonging in one of two general categories: (i) evidence with attributes that fit in the group called *class characteristics,* and (ii) exhibits with attributes that fall in the category called *individual characteristics.* As detailed in Chapter 9, class characteristics are common traits in similar items whereas individual characteristics are more unique and can be linked to a specific person or activity with greater certainty. Consider the physical world example from Chapter 1 of a shoe print left under a window at a crime scene. Forensic analysis of those impressions might only reveal the make and model of the shoe, placing it in the class of all shoes with the same make and model. Therefore, if a suspect were found to be in possession of a pair with the same manufacturer and model, a tenuous circumstantial link can be made between the suspect and the wrongdoing. If forensic analysis uncovers detailed wear patterns in the shoe prints and finds identical wear of the suspect's soles, a much stronger link is possible. The margin of error has just been significantly reduced by the discovery of an individual characteristic making the link much less circumstantial and harder to refute.

In the digital realm, we move into a more virtual and less tangible space. The very notion of individual identity is almost at odds with the philosophy of openness and anonymity associated with many communities using the Internet. However, similar exchanges of evidence occur in the digital realm, such as data from an offender's computer recorded by a server or data from servers stored on the offender's computer. Such links have been used to demonstrate that a specific individual was involved. When all of this evidentiary material does not conclusively link a suspect with the computer, the evidence is still individual relative to the computer.

Browsing the Web provides another example of Locard's Exchange Principle in the digital realm. If an individual sends a threatening message via a Web-based e-mail service such as Hotmail, his/her browser stores files, links, and other information on the hard disk along with date–time related information. Investigators can find an abundance of information relating to the sent message on the offender's hard drive including the original message. Additionally, investigators can find related information on the

Web server used to send the message including access logs, e-mail logs, IP addresses, browser version, and possibly the entire message in the Sent mail folder of the offender's e-mail account.

Akin to categories of evidence in the "traditional" forensic sense, digital equipment and their attributes can be categorized into class and individual groups. Scanners, printers, and all-in-one office devices may exhibit or leave discernible artifacts that lead to common class characteristics allowing the identification of an Epson, Canon, or Lexmark device. The more conclusive individual characteristics are more rare but not impossible to identify through detailed analysis. Unique marks on a digitized photograph might be used to demonstrate that the suspect's scanner or digital camera was involved. Similarly, a specific floppy drive may make unique magnetic impressions on a floppy disk, helping establish a link between a given floppy disk and the suspect's computer.

These are examples of the more desirable category of evidence because of their strong association with an individual source. Generally, however, the amount of work required to ascertain this level of information is significant and may be for naught, especially if a proven method for its recovery has not been researched and accepted in the community and used to establish precedent in the courts. This risk coupled with the fact that the objects of analysis change in design and complexity at such a rapid pace, makes it difficult to remain current.

Preview (Chapter 9): Interestingly, the MD5 computation is an example of a derived attribute that can be useful as a class or individual characteristic depending on its application. For instance, the MD5 value of a common component of the Windows 2000 operating system (e.g. kernel32.dll) places a file in a group of all other similar components on all Windows 2000 installations but does not indicate that the file came from a specific machine. On the other hand, when the MD5 computation is computed for data that are or seem to be unique, such as an image containing child pornography or suspect steganographic data, the hash value becomes an individual characteristic due to the very low probability that any other data (other than an exact copy) will compute to the same hash value. Therefore, MD5 values are more trustworthy than filenames or file sizes in the comparison of data.

Class characteristics can enable investigators to determine that an Apache Web server was used, a particular e-mail encapsulation scheme (e.g. MIME) was employed, or that a certain manufacturer's network interface card was the source. Categorization of characteristics from various types of digital components has yet to be approached in any formal way but the value of this type of information cannot be underestimated. Class characteristics can be used collectively to determine a probability of involvement and the preponderance of this type of evidence can be a factor in reaching conclusions about guilt or innocence.

> The value of class physical evidence lies in its ability to provide corroboration of events with data that are, as nearly as possible, free of human error and bias. It is the thread that binds together other investigative findings that are more dependent on human judgements and, therefore, more prone to human failings. (Saferstein 1998)

To better appreciate the utility of Locard's Exchange Principle, class characteristics, and individual characteristics in the digital realm, consider a computer intrusion. When an intruder gains unauthorized access to a UNIX system from his/her personal computer using a stolen Internet dial-up account, and uploads various tools to the UNIX machine via FTP (file transfer protocol), the tools are now located on both the Windows and UNIX systems. Certain characteristics of these tools will be the same on both systems, including some of the date–time stamps and MD5 hash values (described in Chapter 9).

The Windows application used to connect to the UNIX system (e.g. Telnet, SecureCRT, SSH) may have a record of the target IP address/hostname. Directory listings from the UNIX system may be found on the intruder's hard drive if they were swapped to the disk while being displayed on screen by Telnet, SecureCRT, SSH, or another program as shown in Figure 4.3. The stolen account and password is probably stored somewhere on the intruder's system, possibly in a sniffer log or in a list of stolen accounts from various systems. The FTP client used (e.g. WS_FTP) may create a log of the transfer of tools to the server.

The UNIX system may have login records and FTP transfer logs showing the connection and file transfers. Additionally, some of the transferred files may carry characteristics from the source computer (e.g. TAR files contain user and group information from UNIX systems). These types of digital evidence transfer can be used to establish the continuity of offense in a connect-the-dots manner. In the threatening e-mail example above, the information on the sender's hard disk along with the date and time it was created can be compared with data on the server and the message received by the target to demonstrate the continuity of the offense. To establish continuity of offense investigators should seek the sources, conduits, and

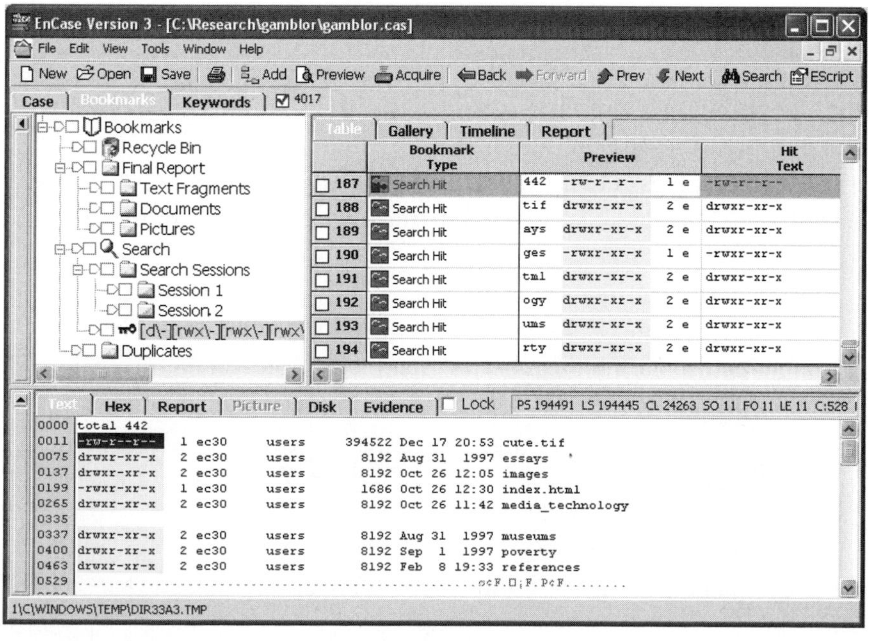

Figure 4.3

Remnants of a directory listing from a UNIX system found on a Windows computer using the grep feature in EnCase to search for the pattern "[d\-][rwx\-][rwx\-][rwx\-][rwx\-][rwx\-][rwx\-][rwx\-][rwx\-][rwx\-](space)."

targets of an offense. Each of these three areas can have multiple sources of digital evidence and can be used to establish the continuity of offense. Additional systems may be peripherally involved in an offense (e.g. for storage, communication, or information retrieval) and may contain related evidence. For instance, in a computer intrusion investigation, there may be related digital evidence on intrusion detection system, NetFlow logs, and other intermediate systems discussed in later chapters.

The more corroborating evidence that investigators can obtain, the greater weight the evidence will be given in court and the more certainty they can have in their conclusions. In this way, investigators can develop a reconstruction of the crime and determine who was involved. The addition of a mechanism or taxonomy to categorize digital evidence as described would benefit the investigator by allowing them to present the relative merits of the evidence and help them maintain the objectivity called for by the investigative process.

As another example, take a case of downloading child pornography from an FTP server on the Internet via a dial-up connection as depicted in Figure 4.4. The date–time stamps of the offending files on the suspect's personal computer show when the files were downloaded. Additionally, logs created by the FTP client may show when each file was downloaded and from where. The following log entry created by WS_FTP shows an image being downloaded from an FTP server with IP address 192.168.1.45 on November 12, 1998, at 1953 hours from a remote directory on the FTP server named "/home/johnh".

```
98.11.12 19:53 A C:\download\image12.jpg ,<-- 192.168.1.45 /home/johnh image12.jpg
```

Modem logs on the computer may show that the computer was connected to the Internet at the time in question.

Dial-up server logs at the suspect's Internet Service Provider (ISP) may show that a specific IP address was assigned to the suspect's user account at the time.

Figure 4.4

Potential sources of evidence useful for establishing continuity of offense.

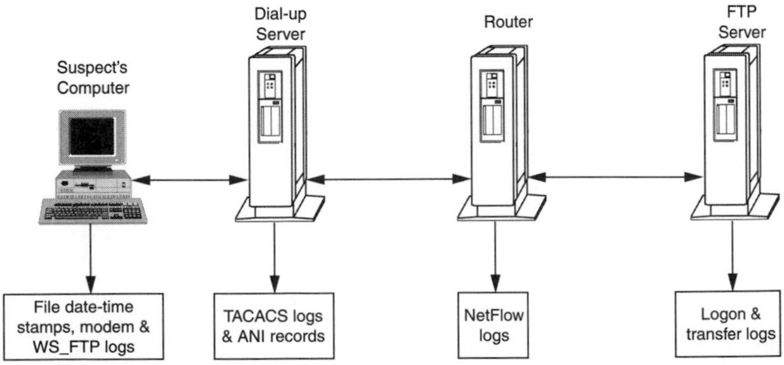

The ISP may also have Automatic Number Identification (ANI) logs – effectively Caller-ID – connecting the suspect's home telephone number to the dial-up activity. Routers connecting the suspect's computer to the Internet may have associated NetFlow logs containing additional information about the suspect's connection to the FTP server.

Logs on the FTP server may confirm that files were downloaded to the suspect's IP address at the time in question. For instance, the following FTP server transfer log entry shows a file with the same name and size as that found on the suspect's computer being downloaded to the IP address that was assigned to the suspect's account at the time in question.

```
Nov 12 19:53:23 1998 15 216.58.30.131 780800 /home/johnh/image12.jpg a _ o r user
```

CASE EXAMPLE (UNITED STATES v. HILTON 1997):
In United States v. Hilton, the forensic examiner was asked to justify transport charges by explaining his conclusion that pornographic images on the suspect's computer had been downloaded from the Internet. The examiner explained that the files were located in a directory named MIRC (the name of an Internet chat client) and that the date–time stamps of the files coincided with the time periods when the defendant was connected to the Internet. The court was satisfied with this explanation and accepted that the files were downloaded from the Internet.

These examples describe suspected offenses and allude to types and locations of potential evidentiary material. This section also introduced the established forensic concepts of class and individual characteristics and how to apply them to digital evidence, helping investigators and prosecutors assess the suitability and persuasive strength of the evidence. These are essential elements of any investigation but only represent the highlights of the structured process detailed in the following sections.

4.2 INVESTIGATIVE METHODOLOGY

The investigative process, depicted as a sequence of ascending stairs in Figure 4.5, is structured to encourage a complete, rigorous investigation, ensure proper evidence handling, and reduce the chance of mistakes created by preconceived theories and other potential pitfalls. This process applies to criminal investigations as well as military and corporate inquiries dealing with policy violations or system compromise.

The categories in Figure 4.5 are intended to be as generic as possible. The unique methods and tools employed in each category tie the investigative process to a particular forensic domain. The terms located on the riser of each step are those more closely associated with the law enforcement

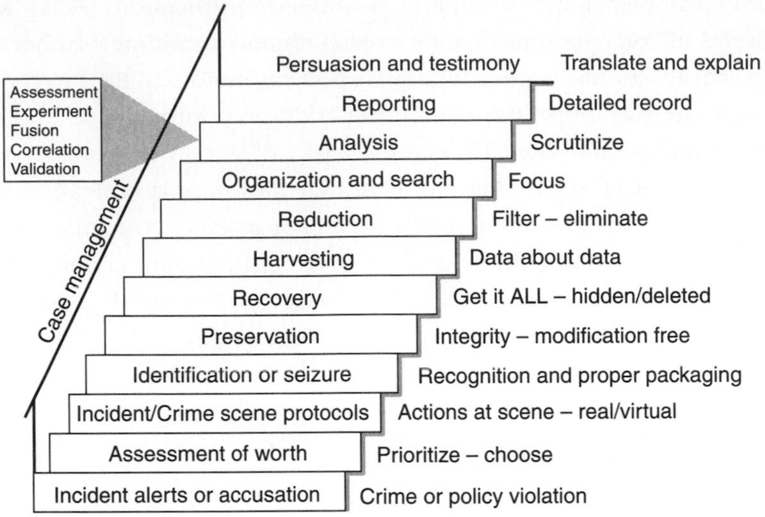

Figure 4.5
Categories of the Investigative Process Model (depicted as a flight of stairs).

perspective. To the right of each term is a more general descriptor that may help to express the essence of each step of the process.

Investigators and examiners work together to scale these steps from bottom to top in a systematic, determined manner in an effort to present a compelling story after reaching the landing (persuasion/testimony). There they will pass their hard work on to prosecutors or other decision makers who scrutinize the findings and decide whether to continue or refocus resources to solving other matters. In the case of the courts, investigators will present their findings to the trier-of-fact who will decide if the merits of the evidence make a strong enough case to proceed to trial. In civilian and military operational communities, facts are presented to resource managers who will rely on the confidence and accuracy of the information before taking corrective action. Often, in this operational environment the mission or business objectives are of primary concern with possible prosecution left as a secondary consideration.

Two items of particular note and special importance stand out in our depiction. First, Case Management plays a vital role and spans across all the steps in the process model. It provides stability and enables investigators effectively to tie all relevant information together, allowing the story to be told clearly. In many cases the mechanisms used to structure, organize, and record pertinent details about all events and physical exhibits associated with a particular investigation is just as important as the information presented. Second, the term *analysis* is used rather loosely in many implementations of the investigative process. Our intent is to attach a more precise definition to this term so that it can be properly placed within the steps of our model. The analysis phase of the investigative process borrows heavily on the long-standing scientific method, beginning with fact gathering and validation, proceeding to hypothesis formation and testing, actively seeking evidence that disproves the hypothesis, and revising conclusions as new evidence emerges.

In general, this model affords investigators and examiners a logical flow of events that, taken together, seek to provide:

1. Acceptance – the steps and methods have earned professional consensus.
2. Reliability – the methods employed can be proven (trusted) to support findings.
3. Repeatability – the process can be applied by all, independent of time and place.
4. Integrity – the state of evidence is proven (trusted) to be unaltered.
5. Cause and effect – logical connection between suspected individuals, events, and exhibits.
6. Documentation – recordings essential for testimonial evidence (expert testimony).

All six tenets have a common purpose – to form the most persuasive argument possible based upon facts, not supposition, and to do so considering the legal criteria for admissibility.

As noted at the beginning of this chapter, although depicted as a linear progression of events in Figure 4.5, the stages in this process are often intertwined and those professionals who participate may find the need to revisit steps after it was thought to be complete. This "feedback" cannot be avoided nor should it be. It is often essential to make improvements and enhancements to methods and tools used in each step. Also, most steps are not only "digital forensic" in nature – many parts of the process function by applying and integrating methods and techniques in police science and criminalistics as aids. Finally, as with most processes, there is a relationship between successive steps. That relationship can often be described by the input and output expected at each stage, with products of one step feeding into the steps that follow.

With that said, let us take a closer look at each step along with details of the processing required in each and the associated inputs and outputs.

4.2.1 ACCUSATION OR INCIDENT ALERT

Every process has a starting point – a place, event, or for lack of a better term, a "shot from a starting gun" that signals the race has begun. This step can be signaled by an alarm from an intrusion detection system, a system administrator reviewing firewall logs, curious log entries on a server, or some combination of indicators from multiple security sensors installed on networks and hosts. This initial step can also be triggered by events in more traditional law enforcement settings. Citizens reporting possible criminal activity will lead to investigative personnel being dispatched to a physical scene. That scene will likely contain exhibits of which some may be electronic, requiring part of the investigation to take a digital path. The prevalence of computers makes it increasingly likely that even traditional crimes will have related information derived from digital sources that require close scrutiny.

When presented with an accusation or automated incident alert, it is necessary to consider the source and reliability of the information. An individual making a harassment complaint because of repeated offensive messages appearing on her screen might actually be dealing with a computer worm/virus. An intrusion detection system alert may only indicate an attempted, unsuccessful intrusion or might be a false alarm. Therefore, it is necessary to weigh the strengths, weakness, and other known nuances related to the sources and include human factors as well as digital.

In addition, thoroughly to assessing an accusation or alert, some initial fact gathering is usually necessary before launching a full-blown investigation. Even technically proficient individuals sometimes misidentify normal system activity as a computer intrusion. Initial interviews and fact checking can correct such misunderstandings, clarify what happened, and help develop an appropriate response. To perform this fact gathering and initial assessment, it is usually necessary to enter a crime scene and scan or very carefully sift through a variety of data sources looking for items that may contain relevant information.

This is a very delicate stage in an investigation because every action in the crime scene may alter evidence. Additionally, delving into an investigation prematurely, without proper authorization or protocols, can undermine the entire process. Therefore, an effort should be made to perform only the minimum actions necessary to determine if further investigation is warranted. Although an individual investigator's experience or expertise may assist in forming internal conclusions that may have associated confidence levels, at this stage few firm, evidence-based conclusions are being drawn about whether a crime or an offence was actually committed.

4.2.2 ASSESSMENT OF WORTH

Those involved in investigative activities are usually busy with multiple cases or have competing duties that require their attention. Given that investigative resources are limited, they must be applied where they are needed most. How this step in the process is handled varies with the associated investigative environment. Applied in law enforcement environments, all suspected criminal activity must be investigated. In civil, business, and military operations, suspicious activity will be investigated but policy and continuity of operations often replaces legalities as the primary concern. Regardless of environment, a form of triage is performed at this step in the process. Questions are asked that try to focus vital resources on the most severe problems or where they are most effective.

Factors that contribute to the severity of a problem include threats of physical injury, potential for significant losses, and risk of wider system compromise or disruption. If a problem can be contained quickly, if there is little or no damage, and if there are no exacerbating factors, a full investigation may not be warranted. The output of this step in the investigative process is a decision that will fit into two basic categories.

- No further action is required – suspicion proved unwarranted. Available data and information are sufficient to indicate no wrongdoing. Document decision with detailed justification, report, and reassign resources.
- Continue to apply investigative resources based upon the merits of evidence examined to this point with priority based on initial available information. All incidents or accusations deserve detailed initial investigation. This category aims to inform about discernment based on practical as well as legal precedent coupled with the informed experience of the investigative team.

Expertise from a combination of on-the-job and certified training plays a tremendous role in effective triage.

4.2.3 INCIDENT/CRIME SCENE PROTOCOLS

When a full investigation is warranted the first challenge is to retain and document the state and integrity of items (digital or otherwise) at the crime scene. Protocols, practices, and procedures are employed at this critical juncture to minimize the chance of errors, oversights, or injuries. Whoever is responsible for securing a crime scene, whether first responders or digital evidence examiners, should be trained to follow accepted protocols. These protocols should address issues such as health and safety (limiting exposure to hazardous materials such as chemicals in drug labs or potentially infectious body fluids), what other authorities are informed, and what must be done to secure the scene.

Preventing people from disturbing a single computer or room is relatively straightforward but, when networks are involved, a crime scene may include sources of evidence in several physically distant locations. Assuming investigators can determine where these locations are, they may not be able to reach them to isolate and preserve associated evidence. This raises the issues of evidence collection on a network, which are discussed in Part 3 of this book.

The product or output of this stage is a secure scene where all the contents are mapped and recorded, with accompanying photographs and basic diagrams to document important areas and items. The evidence is, in essence, frozen in place. This pristine environment is the foundation for all successive steps and provides the "ground truth" for all activities to follow. Items discovered in this initial phase remain an ever present and unchanging part of the case ahead. Steps that follow will serve to add items as well as the attributes of detail, connection, and validation so vital in building event reconstruction, timelines, and motive.

Importantly, the information gathered during this step regarding the state of a crime scene is at the highest level. This means that potential elements of a crime or incident are usually being scrutinized at the macro level. For the most part, investigators are observing "surface details" of potential evidence that may be indicative but are rarely conclusive.

4.2.4 IDENTIFICATION OR SEIZURE

Once the scene is secured, potential evidence of an alleged crime or incident must be seized. Clear procedures and understanding of necessary legal criteria are essential before activity can proceed successfully. The goal here for trained and experienced investigators is not to seize everything at a scene (physical or virtual) but to make informed, reasoned decisions about just what to seize and be prepared to document and justify the action.

Documentation permeates all steps of the investigative process but is particularly important in the digital evidence seizure step. It is necessary to record details about each piece of seized evidence to help establish its authenticity and initiate chain of custody. For instance, numbering items, photographing them from various angles, recording serial numbers, and documenting who handled the evidence helps keep track of where each piece of evidence came from and where it went after collection. Standard forms and procedures help in maintaining this documentation, and experienced investigators and examiners keep detailed notes to help them recall important details. Any notebook that is used for this purpose should be solidly bound and have page numbers that will indicate if a page has been removed.

In a traditional investigative context, seizure implies "to confiscate" or "to take possession of" material, physical items for detailed scrutiny of the items' state and character at some later time in a controlled facility by proven, prescribed means. In the digital realm, unlike most of the traditional forensic disciplines, the seizure of material items occur but all or part of the state and character of some material evidence may be lost almost immediately upon seizure by virtue of the volatility of electronic devices and their design.

Many modern computers have large amounts of Random Access Memory (RAM) where process context information, network state information, and much more are maintained. Once a system is powered down the immediate contents of that memory is lost and can never be completely recovered. So, when dealing with a crime or incidents involving digital evidence, it may be necessary to perform operations on a system that contains evidence, especially in network connected environments.

The output of this phase follows clearly from the triage stage. Inventories, not only of physical electronic components but also attributes of those components that indicate possible networking between local and remote devices and other locations should be cataloged. This recognition is vital because it will allow investigators the opportunity to capture important state and character information before power down and seizure are accomplished. Therefore, even if the investigation warrants the seizure of electronic components, methods and techniques that allow "confiscation" of certain volatile system and network information, even in part, should be considered.

At this step, properly trained first responders might be instructed to find and physically seize evidence for later processing by a digital evidence examiner. Two useful documents outlining effective practices for seizing digital evidence are mentioned here briefly and details of this process are presented in later chapters. This information can be adapted to conform to an organization's policies and should be used to create memory aids for investigators and examiners such as procedures, checklists, and forms.

The Good Practices Guide for Computer Based Electronic Evidence, published by the Association of Chief Police Officers in the United Kingdom (NHCTU 2003), provides a starting point for the discussion of the initial step of digital evidence handling. This guide is designed to cover the most common types of computers: electronic organizers and IBM compatible laptops or desktops with a modem. In addition to practical advice, this guide provides the following four overarching principles that are useful for anyone handling digital evidence.

> *Principle 1*: No action taken by the police or their agents should change data held on a computer or other media that may subsequently be relied upon in court.
>
> *Principle 2*: In exceptional circumstances where a person finds it necessary to access original data held on a target computer that person must be competent to do so and to give evidence explaining the relevance and the implications of their actions.
>
> *Principle 3*: An audit trail or other record of all processes applied to computer-based evidence should be created and preserved. An independent third party should be able to examine those processes and achieve the same result.
>
> *Principle 4*: The officer in charge of the case is responsible for ensuring that the law and these principles are adhered to. This applies to the possession of and access to

> information contained in a computer. They must be satisfied that anyone accessing the computer, or any use of a copying device, complies with these laws and principles.

The US Department of Justice created a useful guide called *Electronic Crime Scene Investigation: A Guide for First Responders* (USDOJ 2001). This guide discusses various sources of digital evidence, providing photographs to help first responders recognize them, and describes how they should be handled. These documents are useful for developing a standard operating procedure (SOP) that covers simple investigations involving a few computers. An SOP is necessary to avoid mistakes, ensure that the best available methods are used, and increase the probability that two forensic examiners will reach the same conclusions when they examine the evidence.

Keep in mind that digital evidence comes in many forms including audit trails, application logs, badge reader logs, biometrics data, application metadata, Internet service provider logs, intrusion detection system reports, firewall logs, network traffic, and database contents and transaction records (i.e. Oracle NET8 or 9 logs). Given this variety, identifying and seizing all of the available digital evidence are challenging tasks. More technically involved procedures are required to deal with large servers or evidence spread over a network. Also, situations will arise that are not covered by any procedure. This is why it is important to develop a solid understanding of forensic science and to learn to apply general principles creatively. Initial interviews should be performed to determine who is involved, what people know, what is not known, and what other information needs to be gathered.

4.2.5 PRESERVATION

Working from the known inventory of confiscated or seized components investigators must act to make sure that potentially volatile items remain unchanged. Another way to put it is that proper actions must be taken to ensure the integrity of potential evidence, physical and digital. The methods and tools employed to ensure integrity are key here. Their accuracy and reliability as well as professional acceptance may be subject to question by opposing council if the case is prosecuted. These same criteria will give decision makers outside of court the necessary confidence to proceed on recommendations from their investigators.

To many practitioners in our field this is where digital forensics begins. It is generally the first stage in the process that employees commonly used tools of a particular type. The output of this stage is usually a set of duplicate copies of all sources of digital data. This output provides investigators with two categories of exhibits. First, the original material is cataloged and stored in a proper environmentally controlled location, in an unmodified state.

Second, an exact copy of the original material that will be scrutinized as the investigation continues.

4.2.6 RECOVERY

Prior to performing a full analysis of preserved sources of digital evidence, it is necessary to extract data that have been deleted, hidden, camouflaged, or that are otherwise unavailable for viewing using the native operating system and resident file system. In some instances, it may also be necessary to reconstitute data fragments to recover an item. Whenever feasible, this process is performed on copies of original digital evidence from the preservation step – this may not be possible in the case of embedded systems.

At this step in the process the focus is on the recovery of all unavailable data whether or not they may be germane to the case or incident. The objective is to identify, and if possible make visible, all data that can be recognized as belonging to a particular data type. The output provides the maximum available content for the investigators and enables them to move to the next phase of the process. It provides the most complete data timeline and may provide insight into the motives of an offender if concrete proof of purposeful obfuscation is found and recorded.

4.2.7 HARVESTING

By the start of this phase all the potential digital evidence associated with a case or incident is available for investigation. Activities designed to gather data and metadata (data about data) about all objects of interest may now proceed. This stage in the process is where the actual reasoned scrutiny begins, where concrete facts begin to take shape that support or falsify hypotheses built by the investigative team. Working from the preserved, recovered source material the investigation proceeds to gather descriptive material about the contents. This gathering will typically proceed with little or no discretion related to the data content, its context, or interpretation. Rather, the investigator will look for categories of data that can be harvested for later analysis – groupings of data with certain class characteristics that, from experience or training, seem or are known to be related to the major facts of the case or incident known to this point in the investigation.

For example, an accusation related to child pornography requires visual digital evidence most likely rendered in a standard computer graphics format like GIF or JPEG. Therefore, the investigators would likely be looking for the existence of files exhibiting characteristics from these graphic formats. That would include surface observables like the objects file type (expressed as a three-character alphanumeric designator in MS Windows based file systems) or more accurately a header and trailer unique to a specific graphical format. In the case of incidents related to hacking investigators might focus some attention on the collection of files or objects associated with particular rootkits or sets of executables, scripts, and interpreted code that are known to aid crackers in successfully compromising systems as discussed in Chapter 19.

A familiarity with the technologies and tools used, coupled with an understanding of the underlying mechanisms and technical principles involved are of more importance in this step. The general output expected here are large organized sets of digital data that have the potential for evidence. It is the first layer organizational structure that the investigators and examiners will start to decompose in steps that follow.

4.2.8 REDUCTION

This step involves activities that help eliminate or target specific items in the collected data as potentially germane to an investigation. This process is analogous to separating the wheat from the chaff. The decision to eliminate or retain is made based on external data attributes such as hashing or checksums, type of data (after type is verified), etc. In addition, material facts associated with the case or incidents are also brought to bear to help eliminate data as potential evidence. This phase remains focused primarily on the overall structure of the object and very likely does not consider content or context apart from examination of fixed formatted internal data related to standards (like headers and trailers). The result (output) of the work in this stage of the investigative process is the smallest set of digital information that has the highest potential for containing data of probative value. This is the answer to the question: "Where's the beef?" The criteria used to eliminate certain data are very important and might possibly be questioned by judge, jury, or any other authorized decision maker.

4.2.9 ORGANIZATION AND SEARCH

To facilitate a thorough analysis, it is advisable to organize the reduced set of material from the previous step, grouping, tagging, or otherwise placing them into meaningful units. At this stage it may be advantageous to actually group certain files physically to accelerate the analysis stage. They may be placed in groups using folders or separate media storage or in some instances a database system may be employed to simply point to the cataloged file system objects for easy, accurate reference without having to use rudimentary search capability offered by most host operating systems.

The primary purpose of this activity is to make it easier for the investigator to find and identify data during the analysis step and allow them to reference these data in a meaningful way in final reports and testimony. This activity may incorporate different levels of search technology to assist investigators in locating potential evidence. A searchable index of the data can be created to enable efficient review of the materials to help identify relevant, irrelevant, and privileged material. Any tools or technology used in this regard should be understood fully and the operation should follow as many accepted standards as exist. The results of this stage are data organization attributes that enable repeatability and accuracy of analysis activities to follow.

4.2.10 ANALYSIS

This step involves the detailed scrutiny of data identified by the preceding activities. The techniques employed here will tend to involve review and study of specific, internal attributes of the data such as text and narrative meaning of readable data, or the specific format of binary audio and video data items. Additionally, class and individual characteristics found in this step are used to establish links, determine the source of items, and ultimately locate the offender. Generally, analysis includes these subcategories (including but not limited to):

- *Assessment (content and context)* – Human readable (or viewable) digital data objects have content or substance that can be perceived. That substance will be scrutinized to try to determine factors such as means, motivation, opportunity.
- *Experimentation* – A very general term but applied here to mean that unorthodox or previously untried methods and techniques might be called for during investigations. All proven methodologies began as experiments so this should come as no surprise especially when applying the scientific method. What remains crucial is that all experimentation be documented rigorously so that the community, as well as the courts, have the opportunity to test it. Eventually, experimentation leads to falsification or general acceptance.
- *Fusion and correlation* – These terms are subtly distinct. During the course of the investigation, data (information) have been collected from many sources (digital and non-digital). The likelihood is that digital evidence alone will not tell the full tale. The converse is also true. The data must be fused or brought together to populate structures needed to tell the full story. An example of Fusion would be the event timeline associated with a particular case or incident. Each crime or incident has a chronological component where event or actions fill time slices. This typically answers the questions where, when, and sometimes how? Time slices representing all activities will likely be fused from a variety of sources such as digital data, telephone company records, e-mail transcripts, suspect and witness statements. Correlation is related but has more to do with reasoned cause and effect. Do the data relate? Not only does event B follow event A chronologically, but the substance (e.g. narrative, persons, or background in a digital image) of the events shows with high probability (sometimes intuition) that they are related contextually.
- *Validation* – This is the output or result of the Analysis stage. It is the reasoned findings that investigators propose to submit to jurists or other decision makers as "proof positive" for prosecution or acquittal.

A failure objectively to assess digital evidence and to utilize experimentation, fusion, and correlation to validate it can lead to false conclusions and personal liability as demonstrated in the following examples.

CASE EXAMPLE (LISER v. SMITH 2003):
Investigators thought they have found the killer of a 54-year-old hotel waitress Vidalina Semino Door when they obtained a photograph of Jason Liser from an ATM where the victim's bank card had been used. Despite the bank manager's warning that there could be a discrepancy between the time indicated on the tape and the actual time, Liser's photograph was publicized and he was subsequently arrested but denied any involvement in the murder. A bank statement confirmed that Liser had been at the ATM earlier that night but that he had used his girlfriend's card, not the murder victim's. Investigators made an experimental withdrawal from the ATM and found that the time was significantly inaccurate and that Liser had used the ATM before the murder took place. Eventually, information relating to the use of the victim's credit card several days after her death implicated two other men who were convicted for the murder. Liser sued the District of Columbia and Jeffrey Smith, the detective responsible for the mistaken arrest, for false arrest and imprisonment, libel and slander, negligence, and providing false information to support the arrest. The court dismissed all counts except the negligence charge. The court felt that Smith should have made a greater effort to determine how the bank surveillance cameras operated or consulted with someone experienced with this type of evidence noting, "The fact that the police finally sought to verify the information – and quickly and readily learned that it was inaccurate – *after* Liser's arrest certainly does not help their cause". Liser's lawsuit against Bank of America for negligence and infliction of emotional distress due to the inaccuracy in the timing mechanism was dismissed.

4.2.11 REPORTING

To provide a transparent view of the investigative process, final reports should contain important details from each step, including reference to protocols followed and methods used to seize, document, collect, preserve, recover, reconstruct, organize, and search key evidence. The majority of the report generally deals with the analysis leading to each conclusion and descriptions of the supporting evidence. No conclusion should be written without a thorough description of the supporting evidence and analysis. Also, a report can exhibit the investigator or examiner's objectivity by describing any alternative theories that were eliminated because they were contradicted or unsupported by evidence.

4.2.12 PERSUASION AND TESTIMONY

In some cases, it is necessary to present the findings outlined in a report and address related questions before decision makers can reach a conclusion. A significant amount of effort is required to prepare for questioning and to convey technical issues in a clear manner. Therefore, this step in the process includes techniques and methods used to help the analyst and/or domain expert translate technological and engineering detail into understandable narrative for discussion with decision makers.

4.3 SUMMARY

This chapter provided a formalized process to help investigators reach conclusions that are reliable, repeatable, well documented, as free as possible from error, and supported by evidence. Heavy reliance on the scientific method helps overcome preconceived theories, encouraging investigators to validate their findings by trying to prove themselves wrong, leading to well-founded conclusions that support expert testimony. Fundamental concepts such as Locard's Exchange Principle, class and individuating characteristics, and establishing continuity of offense were discussed. The important concepts of case management and analysis were discussed along with each discrete step in the investigative process. The ultimate aim of this investigative model is to help investigators and examiners ascend a sequence of steps that are generally accepted, reliable, and repeatable, and lead to logical, well documented conclusions of high integrity. All six tenats have a common purpose – to form the most persuasive argument possible based upon facts, not supposition, and to do so considering the legal criteria for admissibility.

The success of each step of the investigative process is dependent on preparation in the form of policies, protocols, procedures, training, and experience. Anyone responding to an accusation or incident should already have policies and protocols to follow and should have the requisite knowledge and training to follow them. Similarly, anyone processing and analyzing digital evidence should have standard operating procedures, necessary tools, and the requisite training to implement them.

REFERENCES

Carrier B. and Spafford E. H. (2003) "Getting Physical with the Digital Investigation Process", International Journal of Digital Evidence, Volume 2, Issue 2 (Available online at http://www.ijde.org/docs/03_fall_carrier_Spa.pdf)

Gross H. (1924) Criminal Investigation, London: Sweet & Maxwell

Korn H. (1966) "Law, Fact, and Science in the Courts", 66 Columbia Law Review 1080, 1093–94

Popper K. R. (1959) Logic of Scientific Discovery, London: Hutchinson

Saferstein R. (1998) Criminalistics: An Introduction to Forensic Science, Sixth Edition. Upper Saddle River, NJ: Prentice Hall

Sobel D. (1999) "Galileo's Daughter: A Drama of Science, Faith, and Love", London: Fourth Estate

Thornton J. I. (1997) "The General Assumptions and Rationale of Forensic Identification", for David L. Faigman, David H. Kaye, Michael J. Saks, & Joseph Sanders, Editors, Modern Scientific Evidence: The Law and Science of Expert Testimony, Volume 2, St. Paul, MN: West Publishing Company

United Kingdom Association of Chief Police Officers (2003) "The Good Practices Guide for Computer Based Electronic Evidence", National High-tech Crime Unit (Available online at http://www.nhtcu.org/ACPO Guide v3.0.pdf)

United States Department of Justice (2001) "Electronic Crime Scene Investigation: A Guide for First Responders", National Institute of Justice, NCJ 187736 (Available online at http://www.ncjrs.org/pdffiles1/nij/187736.pdf)

CASES

Liser v. Smith (2003) District Court, District of Colombia, Case Number 00-2325 (Available online at http://www.dcd.uscourts.gov/Opinions/2003/Huvelle/00-2325.pdf)

United States v. Hilton (1997) District Court, Maine, Case Number 97-78-P-C (Available online at http://www.med.uscourts.gov/opinions/carter/2000/gc_06302000_2-97cr078_us_v_hilton.pdf)

CHAPTER 5

INVESTIGATIVE RECONSTRUCTION WITH DIGITAL EVIDENCE

Eoghan Casey and Brent Turvey

Reconstructing human behavior from physical evidence is a multidimensional jigsaw puzzle. Pieces of the puzzle are missing, damaged, and some are even camouflaged. The puzzle pieces come in seemingly incompatible data types – some are visual, some are in such microscopic form that it takes days of specialized analysis to show their existence and in some cases the evidence is intangible, such as oral testimony. But practitioners of these two disciplines, each for totally different reasons, sit at their desks and doggedly persist in completing these puzzles – archaeologists and forensic scientists

(Scott and Conner, 1997)

Crime is not always committed in a straightforward or easily decipherable manner. Nor is it always possible for the investigator to prove what they suspect occurred with the evidence left behind. A crime can involve multiple victims, multiple crime scene locations, and offenders engaging in various degrees of planning, aggression, fantasy, concealment, victim response, and a multitude of other behavioral interactions. Only the offender knows the full story of their involvement in a crime, and it can be difficult to establish their associated motives, movements, interactions, sequences, and timing using the fragmentary clues.

Reconstruction refers to the systematic process of piecing together evidence and information gathered during an investigation to gain a better understanding of what transpired between the victim and the offender during a crime. A core tenet of this process is that, when they commit a crime, criminals leave an imprint of themselves at the scene. This is provided by Locard's Exchange Principle, which states that when any two objects come in contact, there is a cross-transfer. Footwear impressions, fingerprints, and DNA from bloodstain patterns are clear examples of imprints left by an offender at a crime scene. Reconstruction involves taking physical imprints a step further, using them to infer offense related behavior, or *behavioral*

Digital Evidence and Computer Crime Second Edition
ISBN: 0-12-163104-4

imprints. For example, footwear impressions show who walked on a particular surface (and perhaps even when), fingerprints show who touched a particular object, and DNA from bloodstain patterns can demonstrate who bled where, when, and in what sequence.

Taken together, the behavioral imprints established at a particular crime scene can be used to provide who did what, when, where, and how. Taken together, a connected series of behavioral imprints can also be used to establish an offender's *modus operandi*, their knowledge of the crime scene, their knowledge of the victim, and even their motivation. This is as true in digital crime scenes as it is in corporeal world; digital crime scene evidence contains behavioral imprints. For example, the words that an offender uses on the Internet may disclose precious details, the tools that an offender uses online can be significant, and how an offender conceals their identity and criminal activity can be telling.

Take the issue of toolkits as an example. Some computer intruders use toolkits that automate certain aspects of their *modus operandi.* Any customization of a toolkit may say something about the offender and the absence of a toolkit is also worth pondering. Did the offender erase all signs of the tool kit? Is the tool kit so effective that it is undetectable? Was the offender skilled enough not to need a toolkit? Perhaps the offender had legitimate access to the system and would ordinarily be overlooked as a suspect, making a toolkit unnecessary. On this one issue alone we find enough of a behavioral imprint from the digital evidence to build a healthy list of questions that require investigation.

Therefore, creating as complete a reconstruction of the crime as possible using available evidence is a crucial stage in an investigation. The basic elements of an investigative reconstruction include equivocal forensic analysis, victimology, and crime scene characteristics. Although investigative reconstruction is presented as a stage that follows the initial investigation, in practice, a basic reconstruction should be developed concurrently. When investigators are collecting evidence at a crime scene, they should be performing some of the reconstructive tasks detailed in this chapter to develop leads and determine where additional sources of evidence can be found. Once investigators are confident that they have enough evidence to start building a solid case, a more complete reconstruction should be developed.

In addition to helping develop leads and locating additional evidence, investigative reconstruction has a number of other uses. It can be used to:

- Develop an understanding of case facts and how they relate. Getting the big picture can help solve a case and can be useful for explaining events to decision makers.
- Focus the investigation by exposing important features and fruitful avenues of inquiry.

- Locate concealed evidence.
- Develop suspects with motive, means, and opportunity.
- Prioritize investigation of suspects.
- Establish evidence of insider or intruder knowledge.
- Anticipate intruder actions and assess potential for escalation. This can prompt investigators to implement safeguards to protect victims and install monitoring to gather more evidence.
- Link related crimes with the same behavioral imprints. This is a contentious area and care is required to rely on evidence rather than speculation to establish connections between crimes.
- Give insight into offender fantasy, motives, intents, and state of mind.
- Guide suspect interview or offender contact.
- Case presentation in court.

Because investigative reconstruction is used to learn more about a particular offender in a particular case, the arrows may begin to point in a specific direction. Subsequently, the temptation to point a finger at a specific individual may become unbearable. However, great care must be taken not to implicate a specific individual until enough evidence exists to support an arrest. Even then it is not advisable to make public declarations of guilt or innocence. Recall the discussion in the previous chapter regarding legal truth versus scientific truth. An investigator's job is to present the facts of a case objectively and it is up to the courts to decide if the defendant is guilty. If investigators make any statements naming or implicating a specific individual, their objectivity is immediately compromised, casting a fog of doubt over their work.

Investigators can avoid this pitfall by concentrating on the evidence rather than the suspect. For instance, in an intrusion investigation, one might assert, "the files found on the suspect's computer are consistent with those found on the compromised server." However, this does not imply that the suspect broke into the server to obtain the files. Someone else may have gained unauthorized access to the files and given them to the suspect. In a child pornography case, one might assert, "the files found on the suspect's computer were last accessed on November 18, 2001" but this does not imply that the images were viewed at this time, only that the files were accessed in some way. For instance, the files may simply have been moved or copied from another disk, changing file creation and access times.

Making objective statements becomes more challenging when a suspect appears to be implicated by evidence such as a photograph. For instance, in an online child pornography investigation one might state, "the images found on the suspect's computer were also found on the Internet." However, a claim that "the images found on the suspect's computer depicting the suspect and victim engaged in sexual acts were also found on the Internet" could be inaccurate if the suspect's face was morphed into the images. Alternatively, a claim that "the images found on the suspect's computer were distributed by the suspect on the Internet" could be inaccurate if someone else distributed the images and the suspect obtained them from the Internet.

CASE EXAMPLE (CALIFORNIA v. WESTERFIELD 2002):
There was much confusion in the murder trial of David A. Westerfield regarding whether he or his son (David N. Westerfield) viewed specific pornographic images on a given computer. Efforts to attribute specific computer activities to one or the other caused both the prosecution and defense to overstate or incorrectly interpret the digital evidence. For instance, one forensic examiner did not initially realize that the date–time stamps in an important e-mail were in GMT rather than local time. The opposing expert did not realize that an important CD-ROM attributed to the son was assigned the name "Spectrum" when it was created. The name of the defendant's company was Spectrum, suggesting that he created the CD-ROM.

The challenge for investigators is to stay within the confines of the evidence when forming conclusions about the established case facts and making subsequent comments. This requires no small amount of investigative objectivity, and a certain amount of immunity from the zeal and personal motives that often accompany those who desire justice to be more swift than accurate.

Note that some Web browsers retain a history of the pages visited, when they were first viewed, and how many times they were accessed. Although it is tempting to attribute such activities to an individual, several people may share systems and even passwords. Therefore, great care must be taken to avoid jumping to the incorrect conclusions. Since seemingly minor variations in language can make a major difference in an investigator's notebook or final report, it is important to become adept at stating only what is known and questioning all underlying assumptions.

The mark of the truly objective forensic investigator is objectivity. In report writing and testimony alike, casual use of inflammatory, editorial or partial language signals either a lack of training, a lack of experience, or a personal agenda. This should be kept in mind not only when forming opinions, but when reviewing the work of others.

5.1 EQUIVOCAL FORENSIC ANALYSIS

The *corpus delicti*, or body of the crime, refers to those essential facts that show a crime has taken place. If these basic facts do not exist, it cannot be reliably established that there was indeed a crime. For example, to establish that a computer intrusion has taken place, investigators should look for evidence such as a point of entry, programs left behind by the criminal, destroyed or altered files, and any other indication of unauthorized access to a computer. Even if investigators can establish that a crime has been committed, it may become clear that there is not enough evidence to identify suspects, link suspects to the victim, link suspects to the crime scene, link similar cases to the same perpetrator, and/or disprove or support witness testimony. In extreme situations, there may not even be enough evidence to generate leads sufficient to engine the investigation forward. Such cases are rare, and present the investigator with the prospect of a case growing cold. When this occurs, investigators must bear down and re-investigate each piece of evidence collected until it is an exhausted possibility.

Equivocal refers to anything that can be interpreted in more than one way or where the interpretation is open to question. An equivocal forensic analysis is one in which the conclusions regarding the physical and digital evidence are still open to interpretation.

Before relying on evidence gathered by others, it is imperative to assess its reliability and significance. Witness statements may be inaccurate or contradictory, evidence may have been overlooked or processed incorrectly, or there may be other complexities that only become apparent upon closer inspection. Equivocal forensic analysis is the process of objectively evaluating available evidence, independent of the interpretations of others, to determine its true meaning. The goal is to identify any errors or oversights that may have already been made.

Difficult as it may be, it is critical to examine incoming evidence as objectively as possible, questioning everything and assuming nothing. In many situations, evidence will be presented to an investigator along with an interpretation (e.g. this is the evidence of a computer intrusion or death threat). Investigators should not accept another person's interpretation without question but should instead verify the origins and meanings of the available evidence themselves to develop their own hypotheses and opinions.

From one perspective, an equivocal forensic analysis is necessary for self-preservation. When investigators render opinions in a case, they are staking their reputations on the veracity of these opinions. An investigator who does not base his/her conclusions on sound evidence will have a short career.

From a less selfish perspective, investigators should want to be sure that everything they assert is accurate because it will be used to determine an individual's innocence or guilt and deprive them of their liberty or, in extreme cases, their life.

In essence, an equivocal forensic analysis is somewhat of a repetition of the investigative process detailed in Chapter 4. The reason for this repetition is that several people with varying degrees of expertise may have investigated different aspects of the crime at different times (e.g. first responders, system administrators) and a full analysis of the evidence is required to ensure that prior investigations were complete and sound. If digital evidence was overlooked, altered, processed inadequately, or misunderstood, this may become apparent when viewed by a critical mind in the context of other evidence. A side benefit of an equivocal forensic analysis is that the investigator becomes familiar with the entire body of evidence in a case.

In addition to physical and digital evidence, an equivocal forensic analysis should include information sources such as suspect, victim and witness statements, other investigators' reports, and crime scene documentation. A sample of the information sources that are used at this stage to establish a solid basis of fact is provided here:

- known facts and their sources;
- suspect, victim and witness statements, including information technology staff with knowledge of the crime or systems involved;
- first responder and investigator reports, and interviews with everyone who handled evidence;
- crime scene documentation, including photos or video of the crime scene;
- original media for re-examination;
- network map, network logs, backup tapes;
- usage and ownership history of computer systems;
- results of Internet searches for related information;
- badge/biometric sensors, cameras;
- traditional physical evidence;
- fingerprints, DNA, fibers, etc.

Basic goals of an equivocal forensic analysis involving a computer should include addressing fundamental issues such as, where the computer came from, who used it in the past, how was it used and what data it contained, and whether a password was required. If a computer was handed down from father to son, transferred from one employee to another, or used by multiple individuals, this can make a difference when attempting to attribute activities. Failure to establish any of these circumstances should seriously reduce the confidence of any theories regarding the *corpus delicti* and subsequent offender identity. Similarly, in an apparent intrusion investigation, interviews with system administrators may reveal that one of their co-workers was fired recently and threatened to damage the system. Close examination of a network map or statements made by network administrators may reveal another potential source of digital evidence that was previously overlooked.

5.1.1 RECONSTRUCTION

As the following quotation explains, evidence that is used to reconstruct crimes falls into three categories: relational, functional, and temporal.

> Most evidence is collected with the thought that it will be used for identification purposes, or its ownership property. Fingerprints, DNA, bullets, casing, drugs, fibers, and safe insulation are examples of evidence used for establishing source or ownership. These are the types of evidence that are brought to the laboratory for analysis to establish the identification of the object and/or its source. The same evidence at the crime scene may be the evidence used for reconstruction. We use the evidence to sequence events, determine locations and paths, establish direction or establish time and/or duration of the action. Some of the clues that are utilized in these determinations are relational, that is, where an object is in relation to the other objects or to the crime; functional, the way something works or how it was used or temporal, things based on the passage of time. (Chisum, in Turvey 2002)

Even within the limitations already discussed, digital evidence is a rich and often unexplored source of information. It can establish action, position, origin, associations, function, sequence, and more enabling an investigator to create an incredibly detailed picture of events surrounding a crime. Log files are a particularly rich source of behavioral evidence because they record so many actions. Piecing together the information from various log files, it is often possible to determine what an individual did or was trying to achieve with a high degree of detail.

Temporal aspects of evidence, or when events occurred, are obviously important. Since computers often note the time of specific events, such as the time a file was created or the time a person logged on using a private password, digital evidence can be very useful for reconstructing the sequence of events. Less obviously, the position of digital evidence in relation to other objects can be very informative. For instance, the geographic location of computers in relation to suspects and victims, or the locations of files or programs on a computer can be important. Determining where a computer intruder hides files can help reconstruct a crime and can help investigators of similar crimes discover similar hiding places.

Missing items are also important but the presence must be inferred from other events. For example, if there is evidence that a certain program was used but the program cannot be found, it can be inferred that the program was removed after use. This could have significant implications in the context of a crime, since covering behavior is very revealing about criminals, as is *what* they want to hide. The functionality of a piece of digital evidence can shed light on what happened. Of course, knowing what a program does is crucial for reconstruction, but if a computer program has options that determine what it does, then the options that are selected to commit a crime are also very telling, potentially revealing skill level, intent, and concealment behavior.

Individual pieces of digital data might not be useful on their own, but patterns may emerge when they are combined. If a victim checks e-mail at

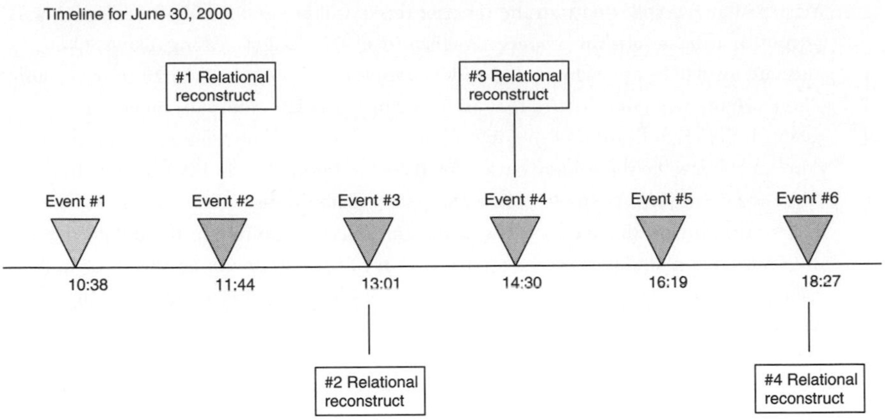

Figure 5.1

Conceptual view of timeline and relational reconstructions.

a specific time or frequents a particular area on the Internet, a disruption in this pattern could be an indication of an unusual event. An offender might only strike on weekends, at a certain location, or in a unique way. With this in mind, there are three forms of reconstruction that should be performed when analyzing evidence to develop a clearer picture of the crime and see gaps or discrepancies (Figure 5.1):

- Temporal (when): helps identify sequences and patterns in time of events;
- Relational (who, what, where): components of crime, their positions and interactions;
- Functional (how): What was possible and impossible.

5.1.2 TEMPORAL ANALYSIS

Creating a chronological list of events can help an investigator gain insight into what happened and the people involved in a crime. Such a timeline of events can help an investigator identify patterns and anomalies, shedding light on a crime and leading to other sources of evidence. For instance, a computer log file with a large gap or entries that are out of sequence may be an indication that the log was tampered with.

There are other approaches to analyzing temporal information and identifying patterns. Creating a histogram of times can reveal a period of high activity that deserves closer inspection. Arranging times in a grid with days on the horizontal axis and hours on the vertical axis can highlight repeated patterns and deviations from those regular events. Examples of these and other temporal analysis techniques are provided in Chapter 9 and subsequent chapters.

5.1.3 RELATIONAL ANALYSIS

Determining where an object or person was in relation to other objects or people is very useful when investigating crimes involving networked computers. In large computer fraud cases, thousands of people and computers can be involved, making it difficult to keep track of the many relationships between objects. Creating a diagram depicting the associations between the people and computers can clarify what had occurred. Similarly, when dealing with large telephone call records or network traffic logs, creating a diagram of connections can reveal patterns as discussed in Chapter 15.

Take a simple computer intrusion scenario for example. Suppose a computer intruder obtained unauthorized access to a computer behind an organization's firewall and then broke into their accounting system. However, to obtain access to the accounting system, the intruder had to know a password that is only available to a few employees. A simple relational reconstruction of the computers and individuals involved is provided in Figure 5.2. This diagram can also be useful for locating potential sources of digital evidence such as firewall, intrusion detection, and router error logs. Firewall and intrusion detection system logs show that the intruder initially scanned the network for vulnerabilities. Although the firewall and intrusion detection system do not contain any other relevant data, network traffic logs show the intruder targeting one system on the network. Deleted log files recovered from that system confirm that the intruder gained unauthorized access using a method designed to bypass the intrusion detection system. Network traffic logs also show connections between the compromised machine and the accounting server.

In a cyberstalking case, a link analysis may reveal how the offender is obtaining information about the victim (e.g. by accessing the victim's computer or through a friend). Investigators might use this knowledge to prevent the offender from obtaining additional information to protect the victim, feed the offender false information in an effort to identify him, or simply monitor the connection to gather evidence.

Be warned that, with enough information, anything can appear to be connected. It is possible that the suspect went to school with the victim's

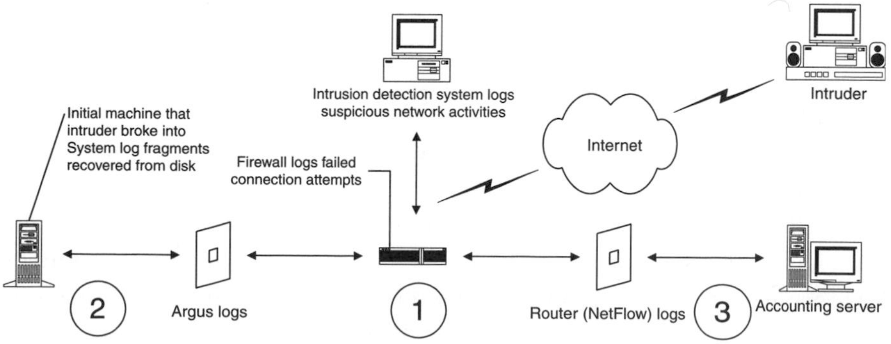

Figure 5.2

Diagram depicting intruder gaining access to accounting server.

brother-in-law but this may be coincidence. Investigators must decide how much weight to give to any relationships that they find. Creating a relational reconstruction works best for a small number of entities. As the number of entities and links increase, it becomes increasingly harder to identify important connections. To address this issue, some software tools have a facility to assign weights to each connection in a relational reconstruction diagram. Additionally, techniques are being developed to perform relational analyses on large amounts of digital evidence using sophisticated computer algorithms.

5.1.4 FUNCTIONAL ANALYSIS

When reconstructing a crime, it is often useful to consider what conditions were necessary for certain aspects of the crime to be possible. For instance, it is sometimes useful to perform some functional testing of the original hardware to ensure that the system was capable of performing basic actions, such as a floppy drive's ability to write and to read from a given evidentiary diskette.

> It is critical to answer any questions on the stand from the defense regarding the capabilities of the system available to the suspect. The defense attorney could inquire how you know the suspected file or picture on the disk or CD you found could even be read or created on the computer. If you have not verified drive operation, especially for external drives, you could leave a hole in your testimony large enough to create that "reasonable doubt" that could lead to a weakening of the case. (Flusche 2001)

Similarly, it is useful to perform functional testing to determine if the suspect's computer was capable of downloading and displaying the graphics files that are presented as incriminating evidence.

Keep in mind that the purpose of functional reconstruction is to consider all possible explanations for a given set of circumstances, not simply to answer the question as asked. For instance, when asked if a defendant's computer could download a group of incriminating files in one minute as indicated by their date–time stamps, an examiner might determine that the modem was too slow to download the files so quickly. However, the examiner should not be satisfied with this answer and should determine how the files were placed on the computer. Further testing and analysis may reveal that the files were copied from a compact disk, which begs the questions, where did that compact disk come from and where can it be found.

If a firewall was configured to block direct access to a server from the Internet, such as the accounting server in Figure 5.2, it was functionally impossible to connect directly from the Internet and, therefore, investigators must determine how the intruder actually gained access to the server. This realization may lead investigators to other sources of evidence such as the internal system that the intruder initially compromised and used to launch an attack against the accounting server.

It may also be necessary to determine how a program or computer was configured to gain a better understanding of a crime or a piece of digital evidence. For instance, if a password was required to access a certain computer or program, this functional detail should be noted. Knowing that an e-mail client was configured to automatically check for new messages every 15 minutes can help investigators differentiate human acts from automated acts. If a program was purposefully created to destroy evidence, this can be used to prove willfulness on the part of the offender to conceal his activities. This is especially the case when dealing with computer intrusions – the tools used to break into a computer deserve close study.

Even in comparatively non-technical cases, determining how a given computer or application functions can shed light on available digital evidence and can help investigators assess the reliability and meaning of the digital evidence. For instance, if an examination of a computer shows that the system time drifts significantly, losing 2 minutes every hour, this should be taken into account when developing the temporal reconstruction in a case.

If the computer has been reconfigured since the crime or a software configuration file is not available, a direct examination might not be possible. However, it might still be possible to make an educated guess based on associated evidence. For instance, if a log file shows that the e-mail client checked for new messages precisely every 15 minutes for an entire day, an educated guess is that it was automated as opposed to manual.

During an equivocal forensic analysis, potential patterns of behavior may begin to emerge and gaps in the evidence may appear. The hope is that evidence will begin to fit together into a coherent whole, like pieces of a jigsaw puzzle combining to form a more complete picture and holes in this picture will become more evident. Realistically, investigators can never get the entire picture of what occurred at a crime. Forensic analysis and reconstruction only includes evidence that was left at a crime scene and is intrinsically limited.

5.2 VICTIMOLOGY

Victimology is the investigation and study of victim characteristics. Conducting a thorough victimology leads to understanding why an offender chose a specific victim and what risks the offender took to gain access to that victim. This information can be used to identify possible links between the victim and offender. If investigators can understand how and why an offender has selected a particular victim or target, they may also be able to establish a link of some kind between them. These links may be geographical, work-related, schedule-oriented, school-related, hobby-related, or even more substantial.

Keep in mind that victims can include individuals, organizations, or an industry as a whole. For instance, pharmaceutical companies that test their products on animals are targeted by animal rights groups in various ways, including denial of service attacks against their Web and e-mail servers to disrupt their daily operations. One of the most important things to establish when a computer is directly involved in the commission of a crime is who or what was the intended target or victim. Although many computer intrusions are not intended to impact a specific individual, always consider the possibility that the intruder intended a particular individual or organization to suffer as a result of the crime. Also recall from Chapter 2 that the particular victim or target may not be as significant to the offender as what they symbolize. In this type of situation, there may not be any connection between the offender and victim prior to the offense.

When looking for offender–victim links, it is helpful to create a timeline of the period leading up to the crime. In crimes against individuals, the 24 hours leading up to the crime often contains the most important clues regarding the relationship between the offender and the victim. Cyberstalking and computer intrusions can extend over several weeks or months, in which case investigators can look for pivotal moments and focus on those during the reconstruction process. Crimes against an organization often involve significant planning, in which case it may be necessary to consider events in the months preceding the crime.

Such a timeline can organize the many details of a day, week, or month and thus clarify how a victim came into contact with an offender. When reconstructing the period before the crime, include any Internet activity. For instance, when investigating a computer intrusion, log files on the victimized organization's network may reveal reconnaissance or other related activity. When investigating crimes against individuals, their Internet activities may reveal whom they communicated with and where they were planning on going. In such situations, it can be fruitful to question individuals with whom a victim or suspect interacted on the Internet.

In addition to reconstructing the recent past of a victim or target, try to imagine how the crime might have been committed. For instance, in a child exploitation case, consider that the offender may have spent time grooming the victim as discussed in Chapter 20. Investigators should also ask themselves whether or not the offender performed surveillance on the victim or target. For example, computer intruders may have probed the target system for the information necessary to gain access. Signs of probing and failed attempts to access the computer suggest that the offender is not very familiar with the computer. A lack of probing indicates that the intruder either knows the system intimately or is very good at removing signs of probing and intrusion.

5.2.1 RISK ASSESSMENT

Among the most informative aspects of offender–victim relationships are victim risk and the effort that an offender was willing to make to access a specific victim. Offenders who go to great lengths to target a specific, well-protected individual have specific reasons for doing so – these reasons are key to understanding an offender's intent, motives, and even identity. Conversely, if a victim did not employ any self-protective measures, an offender may have selected them for convenience and may not have a prior relationship with the victim. Also, the circumstances surrounding a crime can contribute to victim risk. If an attack against an organization occurred during a labor dispute, investigators should consider the possibility that a disgruntled employee is responsible. If the risks present during the crime are not understood, the relationship between the offender and victim cannot be well understood.

Keep in mind that the operative question when assessing risk is: Risk of what? A woman who is new to the Internet, uses her real name online, puts personal information in her AOL profile, and tries to meet people in non-sexual online chat rooms may be at high risk of cyberstalking but not necessarily of sexual assault. However, if the same woman participates in sexually oriented discussions and meets men to have sex is at higher risk of sexual assault.

Because the Internet can significantly increase a victim's risk, victimology should include a thorough search for cybertrails, even in traditional criminal investigations. It might not be obvious that a victim used the Internet but if a thorough search of the victim's computer and Internet activities is not performed, information that could drastically change victimology might be missed. Consider Sharon Lopatka, the woman who traveled from Maryland to North Carolina to meet her killer. Friends described Lopatka as a normal woman who loved children and animals. However, Lopatka's activities on the Internet give a very different impression. Lopatka was evidently interested in sex involving pain and torture. Victimology that did not include her Internet activities would have been incomplete, lacking the aspects of her character most relevant to the crime being investigated and would probably describe her as a low-risk victim when in fact she was quite a high-risk victim.

When a computer is the target of an attack, it is also useful to determine if the system was at high or low risk of being targeted. For instance, a machine with an old operating system, no patches, many services running (some with well-known vulnerabilities), located on an unprotected network, containing valuable information, and with a history of intrusions or intrusion attempts is at high risk of being broken into. Since computer security professionals often make risk assessments of computer networks, they may already have information that is useful for developing a risk assessment in an investigation involving one of their systems.

While assessing the risk of a target computer, investigators should ask themselves: Did the offender need a high level of skill and, if so, who possesses such talents? Similarly, if an offender required a significant amount of knowledge about the target system to commit the crime, investigators should try to determine how this knowledge was obtained. Was it only available to employees of an organization? Could the offender have obtained the information through surveillance and if so, what skill level and equipment was required to perform the surveillance?

5.3 CRIME SCENE CHARACTERISTICS

As investigators systematically analyze crime scenes, certain aspects and patterns of the criminal's behavior should begin to emerge. Specifically, the behaviors that were necessary to commit the crime (*modus operandi* oriented behavior) and behaviors that were not necessary to commit the crime (motive or signature oriented behavior) may become evident if enough evidence is available. These characteristics can be used investigatively to link crimes that may have been committed by a single offender, thus changing investigators' understanding of the crime and offender. They can also lead to additional evidence and insights. For instance, realizing that an intruder broke into multiple computers on a network can result in more evidence, and the type of information on these systems can reveal an offender's true motive.

Most investigators are familiar with the concept of MO but may not realize that it is derived from a careful reconstruction of crime scene characteristics.

> Crime scene characteristics are the distinguishing features of a crime scene as evidenced by an offender's behavioral decisions regarding the victim and the offense location, and their subsequent meaning to the offender. (Turvey 2002)

Such characteristics are derived from the totality of choices an offender makes during the commission of a crime. In addition to choosing a specific victim and/or target, an offender chooses (consciously or unconsciously) a location and time to commit the crime and a method of approaching the victim/target, a method of controlling the victim/target, whether or not tools will be brought or left behind, whether or not items will be taken from the scene, a method of leaving the location, and whether or how to conceal their actions. Each of these kinds of choices, and the skill with which they are carried out, evidence characteristics that establish an offender's *modus operandi.*

When offenders plan their crimes, they can have in mind a specific victim (someone who has wronged them), a type of victim (someone who represents a group that has wronged them), or depend on acquiring a victim of a convenient victim (someone who they can easily find and control with limited fear of detection and subsequent consequences). The amount of planning related to victim selection, approach, and control varies depending on victim type; specific victims tend to involve the most planning and victims of opportunity tend to involve the least. The victim type becomes evident after a careful study of the location that was selected to commit the crime, as well as a careful study of the victim themselves. For example:

- With a specific victim in mind, an offender needs to plan around a specific set of pre-established variables. To complete a successful attack, the offender must know where the victim will be at a certain time, whether or not they are prepared for an attack, and how to exploit their particular set of vulnerabilities. For example, a woman who walks the same route after work, a bank that opens its vaults at a set time, or an organization that makes certain bulk transactions every evening can all be easily targeted by someone who has observed their schedule.
- With a general type of victim in mind, an offender may regularly troll specific types of locations. Some sexual predators frequent playgrounds and online chat rooms to acquire children and others hang out at singles bars to acquire women. Still other sexual predators will troll a location of convenience, perhaps constrained by an inability to travel, and victimize family members, a neighbor, or neighbor's child.
- When any victim will fulfill an offender's needs, an offender might trawl a convenient or comfortable location hosting a variety of victim types until a victim happens to come along. This includes shopping malls, parking lots, public parks, and individuals simply walking on the roadside. Alternatively, the offender might, on an impulse, attack the nearest available person. In such cases, the location of choice would be a reflection of the offender's regular habits and patterns.

In all of the above scenarios, the crime scene has certain characteristics that appeal to the offender. When performing an investigative reconstruction, it is important to examine carefully these characteristics and determine why they appealed to the offender. Neglecting to analyze the characteristics of a crime scene, or failing to identify correctly the significance of a crime scene can result in overlooked evidence and grossly incorrect conclusions.

Networks add complexity to crime scene analysis by allowing offenders to be in a different physical location than their victims or targets and furthermore allow them to be in multiple places in cyberspace. In essence, criminals use computer networks as virtual locations thus adding new characteristics and dimensions to the crime scene. For example, chat rooms and newsgroups are the equivalent of town squares on the Internet providing a venue for meetings, discussions, and exchanges of materials in digital form. Criminals use these areas to acquire victims, convene with other criminals, and coordinate with accomplices while committing a crime.

CASE EXAMPLE

Some groups of computer intruders meet on IRC to help each other gain unauthorized access to hosts on the Internet. If the owner of a system that has been broken into does not notice the intrusion, word gets around and other computer intruders take advantage of the compromised system. Thus, a group of computer intruders become squatters, using the host as a base of operations to experiment and launch attacks against other hosts. IRC functions as a staging area for this type of criminal activity and investigators sometimes can find relevant information by searching IRC using individualizing characteristics of the digital evidence that the intruders left at the primary crime scene; the compromised host.

Criminals choose specific virtual spaces that suit their needs and these choices and needs provide investigators with information about offenders. An offender might prefer a particular area of the Internet because it attracts potential victims or because it does not generate much digital evidence. Another offender might choose a virtual space that is associated with their local area to make it easier to meet victims in person. Conversely, an offender might select a virtual space that is far from their local area to make it more difficult to find and prosecute them (Figure 5.3).

When a crime scene has multiple locations on the Internet, it is necessary to consider the unique characteristics of each location to determine their significance, such as where they are geographically, what they were used for, and how they were used. An area on the Internet can be the point of contact between the offender and victim and can be the primary scene where the crime was committed, or secondary scene used to facilitate a crime or avoid apprehension. The type of crime scene will dictate how much evidence it contains and how it will be searched. For example, a primary scene on a local area network will contain a high concentration of evidence (many bits per

Figure 5.3

Offender in Europe, victim in the United States, crime scenes spread around the world on personal computers and servers (AOL in Virginia).

square inch) and can be searched thoroughly and methodically. Conversely, when secondary scenes are on the Internet, evidence might be scattered around the globe making a methodical search impractical and making any investigative direction towards a competent reconstruction all the more valuable.

5.3.1 METHOD OF APPROACH AND CONTROL

How the offender approaches and obtains control of a victim or target is significant, exposing the offender's confidences, concerns, intents, motives, etc. For example, an offender might use deception rather than threats to approach and obtain control because he/she does not want to cause alarm. Another offender might be less delicate and simply use threats to gain complete control over a victim quickly.

An offender's choice of weapon is also significant. For practical or personal reasons an offender might choose a lead pipe, a gun, or a computer connected to a network to get close to and gain control over a victim or target. Criminals use computer networks like a weapon to terrorize victims and break into target computer systems. Although a criminal could visit the physical location of their victims or targets, using a network is easier and safer, allowing a criminal to commit a crime from home (for comfort) or from an innocuous Internet cafe (for anonymity).

When an offender uses a network to approach and control a victim, the methods of approach and control are predominantly verbal since networks do not afford physical access/threats. These statements can be very revealing about the offender so investigators should make an effort to ascertain exactly what the offender said or typed. The way a computer intruder approaches, attacks, and controls a target can give investigators a clear sense of the offender's skill level, knowledge of the computer, intents, and motives. Crime scene characteristics of computer intrusions are described more fully in Chapter 19.

Different offenders can use the same method of approach or control for very different reasons. Subsequently, it is not possible to make reliable generalizations based on individual crime scene characteristics. For example, one offender might use threats to discourage a victim from reporting the crime whereas another offender might simply want control over the victim regardless of the surrounding circumstances. Therefore, it is necessary to examine crime scene characteristics in unison, determining how they influence and relate to each other.

5.3.2 OFFENDER ACTION, INACTION AND REACTION

Seemingly minor details regarding the offender can be important. Therefore, investigators should get in the habit of contemplating what the offender brought to, took from, changed or left at the crime scene. For instance, investigators might determine that an offender took valuables from a crime scene, indicating a profit motive. Alternatively, investigators might determine that an offender took a trophy or souvenir to satisfy a psychological need. In both cases, investigators would have to be perceptive enough to recognize that something was taken from the crime scene.

Although it can be difficult to determine if someone took a copy of a digital file (e.g. a picture of a victim or valuable data from a computer), it is possible to do so. Investigators can use log files to glean that the offender took something from a computer and might even be able to ascertain what was taken. Of course, if the offender did not delete the log files investigators should attempt to determine why the offender left such a valuable source of digital evidence. Was the offender unaware of the logs? Was the offender unable to delete the logs? Did the offender believe that there was nothing of concern in the logs? Small questions like these are key to analyzing an offender's behavior.

5.4 EVIDENCE DYNAMICS AND THE INTRODUCTION OF ERROR

Investigators and digital evidence examiners will rarely have an opportunity to examine a digital crime scene in its original state and should therefore expect some evidence dynamics. Evidence dynamics are any influence that changes, relocates, obscures, or obliterates evidence, regardless of intent between the time evidence is transferred and the time the case is resolved. Offenders, victims, first responders, digital evidence examiners, and anyone else who had access to digital evidence prior to its preservation can cause evidence dynamics.

For instance, responding to a computer intrusion, a system administrator deleted an account that the intruder had created and attempted to preserve digital evidence using the standard backup facility on the system. This backup facility was outdated and had a flaw that caused it to change the times of the files on the disk before copying them. Thus, the date–time stamps of all files on the disk were changed to the current time making it nearly impossible to reconstruct the crime. As another example, during an investigation involving several machines, a first responder did not follow standard operating procedures and failed to collect important evidence. Additionally, evidence collected from several identical computer systems was not thoroughly documented making it very difficult to determine which evidence came from which system.

Media containing digital evidence can deteriorate over time or when exposed to fire, water, jet fuel, and toxic chemicals. Errors can also be introduced during the examination and interpretation of digital evidence. Digital evidence examination tools can contain bugs that cause them incorrectly to represent data, and digital evidence examiners can misinterpret data. For instance, while a digital evidence examiner was examining several log files, transcribing relevant entries for later reference, he transcribed several dates and IP addresses incorrectly. For instance, he misread 03:13 as 3:13 P.M., resulted in the wrong dial-up records being retrieved, implicating the wrong individual. Similarly, he transcribed 192.168.1.54 as 192.168.1.45 in a search warrant and implicated the wrong individual.

These examples are only a small sampling – there are many other ways that evidence dynamics can occur.

CASE EXAMPLE (UNITED STATES v. BENEDICT):
Lawrence Benedict was accused of possessing child pornography found on a tape that he exchanged with another individual named Mikel Bolander who had been previously convicted of sexual assault of a minor and possession of child pornography. Benedict claims that he was exchanging games with many individuals and did not realize that the tape contained child pornography. Although Benedict initially pleaded guilty purportedly based on advice from his attorney, he changed his plea when problems were found in digital evidence relating to his case. A computer and disks that the defense claimed could prove Benedict's innocence were stored in a post office basement that experienced several floods. The water damage caused the computers to rust and left a filmy white substance encrusted on the disks (McCullagh 2001). Furthermore, after Bolander's computer was seized, police apparently copied child pornography from

the tape allegedly exchanged by Bolander and Benedict onto Bolander's computer for examination. Police also apparently installed software on Bolander's computer to examine its contents and files on the computer appeared to have been added, altered, and deleted while it was in police custody. According to the defense:

> On February 2, 1995, Robert Davis of the San Diego Police Department, while examining the computer evidence, placed computer programs and evidentiary files onto the Bolander C-drive. The programs, which Davis supplied himself, were used to download the evidentiary files from tape onto the computer for examination. As I discussed in my previous affidavits, this is an unacceptable practice since it destroys the integrity of the original evidence. Davis's excuse was that he had no other computer available to perform a forensic analysis. However, it can be shown that files were also deleted from the Bolander C-drive while said evidence was in custody in San Diego. Not only were the files that Davis downloaded onto Bolander's drives deleted, but also a large number of files that he did not download were deleted while said drives were in the custody of the San Diego Police. In addition, attempts were made to completely "wipe" (obliterate all evidence of previous existence) these files from the computer. Among these files were "MB" letters, including MB626, MB57, MB51, and M425. (Littlefield 2002)

Bolander's computer was destroyed before Benedict's sentencing. Additionally, a floppy disk containing evidence was mostly overwritten, presumably by accident. The evidence dynamics in this case created a significant amount of controversy.

Evidence dynamics creates investigative and legal challenges, making it more difficult to determine what occurred and making it more difficult to prove that the evidence is authentic and reliable. Additionally, any conclusions that a forensic examiner reaches without the knowledge of how evidence was changed will be open to criticism in court, may misdirect an investigation, and may even be completely incorrect.

5.5 REPORTING

Writing a report is one of the most important stages of the investigative reconstruction process because, unless findings are communicated clearly in writing, others are unlikely to understand or make use of them.

The two types of reports most commonly associated with an investigative reconstruction are Threshold Assessments and Full Investigative Reports. A Threshold Assessment is an investigative report that reviews the initial physical evidence of crime related behavior, victimology, and crime scene characteristics for a particular unsolved crime, or a series of potentially related unsolved crimes, to provide immediate investigative direction. This type of report is more common because it requires less time and is often sufficient to bring an investigation to a close. Although a Threshold Assessment is a preliminary report, it still involves the employment of scientific principles and knowledge, including Locard's Exchange Principle, critical thinking, analytical logic, and evidence dynamics.

A Full Investigative Report follows the same structure as a Threshold Assessment but includes more details and has firmer conclusions based on all available evidence. A full report is useful in particularly complex cases and can be useful when preparing for trial because it highlights many of the weaknesses that are likely to be questioned in court. Additionally, a Full Investigative Report provides the foundation for further analysis such as criminal profiling.

A common format for these reports are provided here:

1. Abstract: summary of conclusions;
2. Summary of examinations performed:
   - examination of computers, log files, etc.
   - victim statements, employee interviews, etc.
3. Detailed Case Background;
4. Victimology/Target Assessment;
5. Equivocal Analysis of others' work:
   - missed information or incorrect conclusions;
6. Crime Scene Characteristics:
   - may include offender characteristics;
7. Investigative Suggestions.

Two fictitious Threshold Assessments are provided here to demonstrate their structure and purpose. The first involves a homicide involving computers, very loosely based on *The Name of the Rose* by Umberto Eco. The second involves a computer intrusion.

5.5.1 THRESHOLD ASSESSMENT: QUESTIONED DEATHS OF ADELMO OTRANTO, VENANTIUS SALVEMEC, AND BERENGAR ARUNDEL

Complaint received: November 25, 1323

Investigating Agencies: Papal Inquisition, Avignon, Case No. 583

Report by: William Baskerville, Independent Examiner, appointed by Emperor Louis of Germany

For: Abbot of the Abbey

After reviewing case materials detailed below, this examiner has determined that insufficient investigation and forensic analysis have been performed in this case. That is to say, many of the suggested events and circumstances in this case require verification through additional investigation before reliable inferences about potentially crime related activity and behavior can be made. To assist the successful investigation and forensic analysis of the material and evidence in this case, this examiner prepared a Threshold Assessment.

EXAMINATIONS PERFORMED

The examiner made this Threshold Assessment of the above case based upon a careful examination of the following case materials:

- IBM laptop and associated removable media formerly the property of Adelmo Otranto;
- Solaris workstation belonging to the Abbey, formerly used by Venantius Salvemec;
- personal digital assistant formerly the property of Adelmo Otranto.
- mobile telephone formerly the property of Venantius Salvemec;
- various log files relating to activities on the Abbey network;
- interviews with the abbot and other members of the Abbey;
- postmortem examination reports by Severinus Sankt Wendel.

CASE BACKGROUND

All deaths in this case occurred in an Abbey inhabited by monks who cannot speak, having sworn an oath of silence before cutting off their own tongues. On November 21, Adelmo Otranto went missing and his body was found on November 23 by a goatherd at the bottom of a cliff near the Abbey and postmortem examination revealed anal tearing but no semen. Biological evidence may have been destroyed by a heavy snowfall on the night of his disappearance. On November 26, Venantius Salvemec's body was found partially immersed in a barrel of pig's blood that swineherds had preserved the previous day for food preparation. However, the cellarer later admitted to finding Salvemec's corpse in the kitchen, but moved the body to avoid questions about his nocturnal visits to the kitchen. A postmortem examination indicated that Salvemec had died by poison but the type of poison was not known. On November 27, Berengar Arundel's body was found immersed in a bath of water but the cause of death appeared to be poison versus drowning.

VICTIMOLOGY

All victims were Caucasian male monks residing at the Abbey in cells, working in the library translating, transcribing, and illuminating manuscripts.

Details relating to each victim obtained during the investigation are summarized here.

Adelmo Otranto
Age: 15
Height: 5′ 2″
Weight: 150 lbs.
Relationship Status: According to written statements made by Berengar Arundel, he pressured Adelmo into having sexual intercourse the night before his body was found at the bottom of the cliff.
Social history: According to the abbot, Adelmo had problems socializing with children his own age.
Family history: Unknown
Medical and medical health history: Adelmo was known to chew herbs that induced visions.
Lifestyle risk: This term refers to ... Based on even the limited information available to this examiner, Adelmo was at a high overall lifestyle risk of being the victim of sexual exploitation. In addition to taking drugs and being sexually active in the Abbey, Adelmo participated in relationship-oriented online chat and communicated with adult males who were interested in him sexually. During these sexually explicit exchanges, he revealed personal, identifying information including pictures of himself. At least one adult on the Internet sent Adelmo child pornography in an effort to break down his sexual inhibitions.
Incident risk: High risk of sexual assault because fellow monks and adults via the Internet were grooming him. Unknown risk of exposure to poison without understanding of how poison got into his system.

Venantius Salvemec
Age: 16
Height: 5′ 5″
Weight: 145 lbs.
Relationship Status: According to interviews, Venentius accepted presents from older monks and received packages from individuals outside the Abbey. Additionally, he received frequent messages and photographs on his mobile phone, some of a sexual nature.
Social history: Well liked by all and close friends with Adelmo and Berengar.
Family history: Unknown
Medical and medical health history: None available
Lifestyle risk: Insufficient information available to determine lifestyle risk

Incident risk: Medium to high risk of sexual assault and poisoning given his close friendship with the other victims, older monks, and individuals outside the Abbey.

Berengar Arundel
Age: 15
Height: 5′ 4″
Weight: 130 lbs.
Relationship Status: Sexually active with other young monks in the Abbey
Social history: According to the abbot, problems socializing with children his own age.
Family history: According to interviews with other monks, Berengar lived alone with his mother prior to coming to the Abbey. Berengar expressed disdain for his parents and was sent to the Abbey after setting fire to a local landlord's barn. His father moved away from the area after being accused of physically and sexually abusing Berengar.
Medical and medical health history: According to Severinus Sankt Wendel, Berengar made regular visits to the Abbey infirmary for various ailments. Severinus believes that Berengar had Attention Deficit Disorder (ADD).
Lifestyle risk: Based on the likelihood of sexual abuse by his father, sexual activities with other monks, and behavioral and medical problems, Berengar was at a high overall lifestyle risk of being the victim of sexual exploitation.
Incident risk: Medium to high risk of sexual assault and poisoning given his close friendship with the other victims, older monks, and individuals outside the Abbey.

EQUIVOCAL ANALYSIS

Given the exigent circumstances surrounding this investigation, this examiner has only made a preliminary examination of digital evidence relating to this case. A summary of findings is provided here and details of this preliminary examination are provided in a separate report "Digital Evidence Examination for Case No. 583".

- Each victim communicated with many individuals on the Abbey network and Internet, resulting in a significant amount of digital evidence. Some of these communications were of a sexual nature. Additional analysis is required to determine if any of these communications are relevant to this case.
- Adelmo's laptop contained child pornography that was sent to him by an individual on the Internet using the nickname dirtymonky69@yahoo.com. The originating IP address in e-mail messages from this address corresponds

to the Abbey's Web proxy. An examination of the Web proxy access logs revealed that several computers in the Abbey were used to access Yahoo.com around the times the messages were sent. Additionally, log files from the Abbey e-mail server show that all of the victims received messages from this address.

- Adelmo's personal digital assistant contained contact and schedule information, in addition to what appears to be a personal diary. Unfortunately, entries in this diary appear to be encoded and have not been deciphered.
- Venentius's mobile phone contained images of other monks in the nude. It is not clear whether these photographs were taken with the monks' knowledge and additional analysis of the telephone and associated records are required to determine if these photographs were taken using the digital camera, on the telephone, or downloaded from somewhere else.
- Exhume Adelmo's body to determine if he died by poison.

CRIME SCENE CHARACTERISTICS

Location and type: The specific locations of the primary scenes where the victims were exposed to poison are unknown. The victim's bodies were found in locations that were frequented by others in the Abbey.

Point of contact: Unknown

Use of weapons: Poison

Victim resistance: None apparent

Method of approach, attack, and control: How the victims were exposed to poison is unknown, and the existence of an offender in this case had not been firmly established.

Sexual acts: Unknown

Verbal behavior: Requires further analysis of online communications

Destructive acts: None

Evidence of planning and precautionary acts: Insufficient evidence to make a determination

Motivational aspects: Insufficient evidence to make a determination

OFFENDER CHARACTERISTICS

Sex: Investigative assumptions in this case to date have included the preconceived theory (treated as fact) that there was only one offender involved in these crimes and that this offender must be male. The first part of this assumption may not be correct. Berengar's lack of knowledge of and access to poisons weakens the hypothesis that he murdered Adelmo and Venentius, and that he committed suicide. The second part of this assumption cannot be supported or falsified using available evidence. The anal tearing could have occurred during sexual intercourse that might not be associated with the crimes. Even if the anal tearing were associated with the crimes, this would not be definitive proof of a male attacker since no semen was found.

Knowledge of/familiarity with location: It is still unclear if all of these deaths were caused by exposure to poison, and whether this exposure was accidental or malicious. If the exposure were malicious, the perpetrator would not necessarily require knowledge of the Abbey. A valuable item coated with or containing poison could have been delivered to one of the victims in any number of ways and may have subsequently found its way into the hands of the other victims.

Skill level: The fact that no apparent effort was made to conceal the bodies could be interpreted as low homicide-related skill because it increases the chances that the crime would be discovered. However, the offender has some skill administering poison.

Knowledge of/familiarity with victims: There is insufficient evidence to make a determination on this matter. Based on the available evidence, the targeting of victims in this case could be either targeted or random.

INVESTIGATIVE SUGGESTIONS

The following is a list of suggestions for further investigation and establishing the facts of this case:

1 Examine Macintosh desktop belonging to the Abbey, formerly used by Berengar Arundel.

2 After obtaining necessary authorization, examine all computers in the Abbey that were used to access Yahoo.com around the times that messages from dirtymonky69@yahoo.com were sent.

3 After obtaining necessary authorization, perform keyword searches of all computers in the Abbey to determine whether the victims used computers other than those already seized.

4 Using MD5 hash values of the image files, search all computers in the Abbey for copies of the child pornography found on Adelmo's laptop and for copies of the naked monks found on Venantius's mobile phone in an effort to determine their origin.

5 Obtain Venantius Salvemec's mobile telephone records to determine who sent him text messages and photographs.

6 Attempt to decipher Adelmo's diary.

7 Look for hiding places in the victim's cells, library desks, and other locations they had access to in an effort to further develop victimology.

8 Attempt to determine how Venantius gained access to the kitchen on the night of his death. The kitchen and adjoining buildings are locked in the evening and only the abbot, cellarer and head librarian have keys.

9 Perform full investigative reconstruction using digital evidence and information from interviews to determine where the victims were and whom they communicated with between November 15 and November 27.

The same type of analysis and report structure can be used in computer intrusion investigation. For instance, the following report pertains to an intrusion into an important system (project-db.corpX.com) containing proprietary information.

5.5.2 THRESHOLD ASSESSMENT: UNAUTHORIZED ACCESS TO project-db.corpX.com

Complaint received: February 28, 2003
Investigating Agencies: Knowledge Solutions, Case No. 2003022801
Report by: Eoghan Casey
For: CIO, Corporation X

CASE BACKGROUND AND SUMMARY OF FINDINGS

On February 28, an intruder gained unauthorized access to project-db.corpX.com and Corporation X is concerned that the intruder stole valuable proprietary information. Based on an analysis, the available digital evidence in this case, this examiner has determined that the attack against project-db.corpX.com was highly targeted. The amount and type of information accessed by the intruder suggests that intellectual property theft is likely. The perpetrator had a significant amount of knowledge of the computer systems involved and information they contained, suggesting insider involvement. The intruder used an internal system to perpetrate this attack – this system should be examined.

EXAMINATIONS PERFORMED

The examiner made this Threshold Assessment of the above case based upon a careful examination of the following case materials.

- target computer system (project-db.corpX.com);
- various log files relating to activities on target network;
- configuration files of firewalls and routers on the target network;
- memos and media reports describing organizational history and situation;
- interviews with system administrators familiar with the target network and system.

VICTIMOLOGY OF TARGET ORGANIZATION

Organization name: Corporation X

Real space location: 1542 Charles Street, Suite B, Baltimore, MD, 21102

Purpose/role: Software development and sales

Type of product/service: Banking software

Operational risk: High risk because Corporation X has the largest market share in a highly competitive area. As a result, the value of Corporation X's products is high. Additionally, knowledge of the internal workings of this software might enable a malicious individual to manipulate banking systems for financial gain.

Incident risk: High risk because Corporation X recently went public and has received extensive media attention.

VICTIMOLOGY OF TARGET COMPUTER

Computer name: project-db.corpX.com

IP address: 192.168.1.45

Hardware: Sun Enterprise server

Operating system: Solaris 9

Real space location: Machine room, Corporation X

Purpose/role: Programming, file sharing, and project management

Contents (type of data on system): Design documents and source code for Corporation X's main products, along with project schedules and other project related information.

Physical assessment: Locked cabinet in machine room. Only two individuals have a key to the cabinet (the machine room operator and CIO).

Network assessment: Highly secure. All network services are disabled except for Secure Shell (SSH). Logon access only permitted using SSH keys. Protected by firewall that only permits network connection to server on port 22 (SSH) from computers on the Corporation X network.

Operational risk: Low–medium risk because project-db.corpX.com is physically secure, has a good patch and configuration history, no prior intrusions, and is well configured services. However, over one hundred (100) employees have authorized access to the system and database.

Incident risk: Low–medium risk because, although project-db.corpX.com contains valuable data, it is well patched and protected by configuration and hardware firewall.

EQUIVOCAL ANALYSIS OF NETWORK RELATED DATA

An examination of the digital evidence in this case provided additional details of the intruder's activities and revealed several discrepancies that had been overlooked. The main findings are summarized here and a detailed description of the digital evidence examination is provided in a separate report "Digital Evidence Examination for Case No. 2003022801".

- An examination of the system indicates that most activity occurred on February 28 with many files accessed.
- Although server logs indicate that the intruder connected from an IP address in Italy, an examination of the Internet firewall configuration revealed that only internal connections are permitted. A connection from Italy would have been blocked indicating that the server logs have been altered.
- NetFlow logs confirm that the unauthorized access occurred on February 28 between 18:57 and 19:03 hours and that this was a focused attack on the target system. However, the source of the attack was from another machine on the Corporation X network (workstation13.corpX.com), indicating that the intruder altered logs files on the server to misdirect investigators.

CRIME SCENE CHARACTERISTICS

Location and type: The primary scene is project-db.corpX.com. Secondary scenes in this crime include the Corporation X network and the other computer that the intruder used to perpetrate this attack. This other computer (workstation13.corpX.com) will contain digital evidence relating to the intrusion such as SSH keys, tools used to commit or conceal the crime, and data remnants from the primary scene (project-db.corpX.com) transferred during the commission of the crime. If workstation13.corpX.com was compromised, there will be another secondary crime scene – the computer that the intruder used to launch the attack. Once the original source of the attack is found, the computer and surrounding workspace should be searched thoroughly because this crime scene will contain the most digital evidence of the intruder's activities.

Point of contact: SSH daemon on project-db.corpX.com

Use of weapons/exploits: Legitimate user account and SSH key

Method of approach: Through workstation13.corpX.com

Method of attack: Gained target's trust using legitimate user account and SSH key

Method of control: Altering log files to misdirect investigators

Destructive/precautionary acts: Altered log files to misdirect investigators

OFFENDER CHARACTERISTICS

Knowledge of/familiarity with target system: The intruder had knowledge of, and authentication tokens for, an authorized account on the system. However, the intruder did appear to know that the firewall was configured to block external connections (e.g. from Italy). Additionally, the intruder did not appear to know that Corporation X maintained NetFlow logs that could be used to determine the actual source of the intrusion.

Knowledge of/familiarity with target information: There is no indication that the intruder scanned the network or probed any other machines prior to breaking into the target system. Once the intruder gained access to the target, very little time was spent exploring the system. The direct, focused nature of this attack indicates that the intruder knew what information he/she was looking for and where to find it.

Skill level: Any regular user of the target computer would have the necessary skills to access the system as the intruder did. However, the intruder was also capable of altering log files to misdirect investigators, indicating a higher degree of technical skill than an average user.

INVESTIGATIVE SUGGESTIONS

It is likely that the intruder is within the organization or had assistance from someone in the organization. The following is a list of suggestions for further investigation and establishing the facts of this case:

- After obtaining necessary authorization, seize and examine the internal system that the intruder used to perpetrate this attack.
- Interview the owner of the user account that the intruder used to gain access to project-db.corpX.com. Do not assume that this individual is directly responsible. Examine this individual's workstation for signs of compromise and try to determine if the intruder could have obtained this individual's SSH key and associated passphrase.
- Find the original source of the attack and search the associated computer and workspace thoroughly. This secondary crime scene will contain the most digital evidence of the intruder's activities.
- Determine how the intruder was capable of altering log files on the target system. This usually requires root access unless there is a system vulnerability or misconfiguration.
- After obtaining necessary authorization, examine all computers on the Corporation X network for the stolen information.

It is worth reiterating that all conclusions should be based on fact and supporting evidence should be referenced in and attached to the report.

5.6 SUMMARY

Investigative reconstruction provides a methodology for gaining a better understanding of a crime and focusing an investigation. Great clarity can emerge from objectively reviewing available evidence, performing temporal, relational, and functional analyses, and studying the victims and crime scenes. Although investigative reconstruction is an involved process, it can save time and effort in the long run by focusing an investigation from the outset. Furthermore, in many cases, a Threshold Assessment is sufficient, requiring less time than a full investigative reconstruction. However, in complex cases or when preparing a case for trial, a Full Investigative Report can be more useful.

REFERENCES

Flusche K. J. (2001) Computer Forensic Case Study: Espionage, Part 1 Just Finding the File is Not Enough!, *Information Systems Security*, March–April 2001, Auerbach.

Geberth V. (1996) *Practical Homicide Investigation*, 3rd edn. New York, NY: CRC Press.

Horvath F. and Meesig R. (1996) "The Criminal Investigation Process and the Role of Forensic Evidence: A Review of Empirical Findings", *Journal of Forensic Sciences* 41 (6), 963–969.

Holmes R. (1996) *Profiling Violent Crimes: An Investigative Tool*, 2nd edn, Sage Publications.

Scott D. and Conner M. (1997) (Haglund and Sorg eds), *Forensic Taphonomy: The Postmortem Fate of Human Remains*, and, Chapter 2, Boca Raton, FL: CRC Press.

Turvey B. (2002) *Criminal Profiling: An Introduction to Behavioral Evidence Analysis,* London: Academic Press.

CASES

California v. Westerfield (2002) Case No. CD165805, Superior Court of California, County of San Diego Central Division.

CHAPTER 6

MODUS OPERANDI, MOTIVE, AND TECHNOLOGY

Brent E. Turvey

"All our lauded technological progress – our very civilization – is like the axe in the hand of the pathological criminal."

(Albert Einstein)

The purpose of this chapter is to discuss the development of computer and Internet technologies as they relate to both offender *modus operandi* and offender motive. That is to say, their impact on how and why criminals commit crimes. The context of this effort is informed by a historical perspective, and by examples of how computer and Internet technologies may and have influenced criminal behavior. It is hoped through this brief rendering that readers may come to appreciate that while technology and tools change, as does their language, the underlying psychological needs, or motives, for criminal behavior remain historically unchanged.

6.1 AXES TO PATHOLOGICAL CRIMINALS, AND OTHER UNINTENDED CONSEQUENCES

What the Internet is today was never intended or imagined by those who broke its first ground.

In 1969 the US Department of Defense's research arm, ARPA (the Advanced Research Projects Agency) began funding what would eventually evolve to become the technological basis for the Internet.[1]

[1] *The development of the Internet is discussed in more detail in Chapter 15.*

Their intent was to create a mechanism for ensured communication between military installations. It was not their intent to provide for synchronous and asynchronous international person-to-person communication between private individuals, and the beginnings of a pervasive form of social-global connectedness. It was not their intent to create venues for trade and commerce in a digital-international marketplace. Nor was it their intent to place axes in the hands of pathological criminals in the form of robust and efficient tools for stealing information, monitoring individual activity, covert

Digital Evidence and Computer Crime Second Edition
ISBN: 0-12-163104-4

communication, and dispersing illicit material. Regardless, that technology, and every related technology subsequent to its evolution, provides for these things and much more.

The Internet began as an endeavor to help one group within the US Government share information and communicate within its own ranks on a national level. It has evolved into a system that provides virtually any individual with some basic skills and materials the ability to share information and contact anyone else connected to that system on an international level. Without exaggeration, the Internet and its related technologies represent nothing short of historically unparalleled global, trans-social, and trans-economic connectedness. In every sense it is a technological success.

However, history is replete with similar examples of sweet technological success followed by deep but unintended social consequences:

- The American businessman, Eli Whitney, invented the cotton gin in 1793, which effectively cleaned the seeds from green-seeded inland cotton, bringing economic prosperity to the South and revitalizing the dying slave trade. This added much fuel to the engines which were already driving the United States towards civil war.
- The American physician, Dr Richard J. Gatling, invented the hand crank operated rapid fire multi-barreled Gatling gun in 1862, which he believed would decrease the number of lives lost in battle through its efficiency. This led the way for numerous generations of multi-barreled guns with increased range and extremely high rates of fire. Such weapons have been employed with efficient yet devastating results against military personnel and civilians in almost every major conflict since. The efficiency of such weapons to discharge projectiles has not been the life saving element that Dr Gatling had hoped, but rather has significantly compounded the lethality of warfare.
- The American theoretical physicist, Robert J. Oppenheimer, director of the research laboratory in Los Alamos, New Mexico, headed the US Government's Manhattan Project in the mid-1940s with the aim of unlocking the power of the atom, which resulted in the development of the atomic bomb. The atomic bomb may have been intended to end World War II and prevent the loss of more soldiers in combat on both sides. However its use against the citizens of Japan in 1945 arguably signaled the official beginning of both the Cold War and the arms race between the United States and the Soviet Union. Not to mention the devastation it caused directly, the impact of which is still felt today.

These simple examples do us the service of demonstrating that, historically, no matter what objective a technology is designed to achieve, and no matter what intentions or beliefs impel its initial development, technology is still subordinate to the motives and morality of those who employ it. Technology helps to create more efficient tools. Any tool, no matter how much technology goes into it, is still only an extension of individual motive and intent. Invariably, some individuals will be driven to satisfy criminal motives and intents.

Either through fear or misunderstanding, there are those who believe and argue that technology is to blame for its misuse. This is a misguided endeavor, and one that shifts the responsibility for human action away from human hands:

> "It's something I call 'technophobia,' " says Paul McMasters, First Amendment ombudsman at the Freedom Forum in Arlington, Virginia. "Cyberpanic is all about the demonization of a new form of technology, where that technology is automatically perceived as a crime or a criminal instrument." (Shamburg 1999)

In the process of demonizing technology, it may be suggested that there are new types of crimes and criminals emerging. This is not necessarily the case. It is more often that computer and Internet technologies merely add a new dimension to existing crime. As Meloy (1998) points out, "The rather mundane reality is that every new technology can serve as a vehicle for criminal behavior." McPherson (2003) discusses the issue as it relates to computer fraud and forensic accounting:

> Technology simply enables people to commit fraud on a larger scale.
>
> ...
>
> "The computer has just given fraud another dimension."
>
> In relation to computers, forensic accountants look for electronic footprints of people's actions. Previously, people created hard copies – it was easier to shred them and to interrupt an investigator's trail or auditing procedures. Now people try to delete files or keep them on other disks or hard drives.

Computers and the Internet are no different from other technologies adapted by the criminal. With this simple observation in mind we can proceed towards understanding how it is that criminals employ technology in the commission of their crimes.

6.2 *MODUS OPERANDI*

Modus operandi (MO) is a Latin term that means "a method of operating." It refers to the behaviors that are committed by a criminal for the purpose of successfully completing an offense. A criminal's MO reflects *how* they committed their crimes. It is separate from their motives, which have to do with *why* they commit their crimes (Burgess 1997; Turvey 2002).

A criminal's MO has traditionally been investigatively relevant for the case linkage efforts of law enforcement. However, it is also investigatively relevant because it can involve procedures or techniques that are characteristic of a particular discipline or field of knowledge. This can include behaviors that are reflective of both criminal and non-criminal expertise (Turvey 2002).

A criminal's MO consists of learned behaviors that can evolve and develop over time. It can be refined, as an offender becomes more experienced, sophisticated, and confident. It can also become less competent and less skilful over time, decompensating by virtue of a deteriorating mental state, or increased used of mind-altering substances (Turvey 2002).

In either case, an offender's MO behavior is functional by its nature. It most often serves (or fails to serve) one or more of three purposes (Turvey 2002):

- protects the offender's identity;
- ensures the successful completion of the crime;
- facilitates the offender's escape.

Examples of MO behaviors related to computer and Internet crimes include, but are most certainly not limited to (Turvey 2002):

- Amount of planning before a crime, evidenced by behavior and materials (i.e. notes taken in the planning stage regarding location selection and potential victim information, found in e-mails or personal journals on a personal computer).
- Materials used by the offender in the commission of the specific offense (i.e. system type, connection type, software involved, etc.).
- Presurveillance of a crime scene or victim (i.e. monitoring a potential victim's posting habits on a discussion list, learning about a potential victim's lifestyle or occupation on their personal website, contacting a potential victim directly using a friendly alias or a pretense, etc.).
- Offense location selection (i.e. a threatening message sent to a Usenet newsgroup, a conversation had in an Internet Relay Chat room to groom a potential victim, a server hosting illicit materials for covert distribution, etc.).
- Use of a weapon during a crime (i.e. a harmful virus sent to a victim's PC as an e-mail attachment, etc.).
- Offender precautionary acts (i.e. the use of aliases, stealing time on a private system for use as a base of operations, IP spoofing, etc.).

6.3 TECHNOLOGY AND *MODUS OPERANDI*

As already alluded to at that beginning of this chapter, technology has long shared a relationship with criminal behavior. For example, without notable exception each successive advance in communications technology (including most recently the proliferation of portable personal computers and Internet related technologies) has been adopted for use in criminal activity, or has acted as a vehicle for criminal behavior. Some prominent examples include, but are not limited to:

- *Spoken language* has been used to make threats of violence and engage in perjury.
- *Paper and pencil* have been used to write notes to tellers during bank robberies, to write ransom notes in kidnappings, and to falsify financial documents and records.

- *The postal system* has been used for selling non-existent property to the elderly, distributing stolen or confidential information, distributing illicit materials such as drugs and illegal pornographic images, the networking of criminal subcultures, and the delivery of lethal explosive devices to unsuspecting victims.
- *Telephones* have been used for anonymous harassment of organizations and individuals, the networking of criminal subcultures, and for credit card fraud involving phony goods or services.
- *Fax machines* have been used for the networking of criminal subcultures, distributing stolen or confidential information, and the harassment of organizations and individuals.
- *E-mail* has been used for anonymous harassment of organizations and individuals, the networking of criminal subcultures, for credit card fraud involving phony goods or services, distributing stolen or confidential information, and distributing illicit materials such as illegal pornographic images.
- *Web sites* have also been used for anonymous harassment of organizations and individuals, the networking of criminal subcultures, and for credit card fraud involving phony goods or services, distributing stolen or confidential information, and distributing illicit materials such as illegal pornographic images.

The proactive aspect of this relationship has been that criminals can borrow from existing technologies to enhance their current modus operandi to achieve their desired ends, or to defeat technologies, and circumstances that might make the completion of their crime more difficult. If dissatisfied with available or existing tools, and sufficiently skilled or motivated, criminals can also endeavor to develop new technologies.

The result is a new technological spin on an existing form of criminal behavior.

In a variety of forms, computer, and Internet technologies may be used on their own to facilitate of accomplish the following types of criminal activities:

- victim selection;
- victim surveillance;
- victim contact/grooming;
- stalking/harassment;
- theft of assets such as money from bank accounts, intellectual property, identity, and server time;
- destruction of assets such as money from bank accounts, intellectual property, identity, and network functions;
- locating confidential and/or illicit materials;
- gathering and storing confidential and/or illicit materials;
- narrow dissemination of confidential and/or illicit materials;
- broad dissemination of confidential and/or illicit materials.

The following examples are provided to illustrate some of these situations:

CASE EXAMPLE 1 (REUTERS 1997):
In August of 1997, a Swiss couple, John (52 years old) and Buntham (26 years old) Grabenstetter, were arrested at the Hilton in Buffalo, New York and accused of smuggling thousands of computerized pictures of children having sex into the United States.

The couple were alleged by authorities to have sold wholesale amounts of child pornography through the Internet, and carried with them thousands of electronic files of child pornography to the United States from their Swiss home. They were alleged to have agreed over the Internet to sell child pornography to US Customs agents posing as local US porn shop owners. They were alleged to have agreed to sell 250 CD-ROMs to US investigators for $10,000. According to reports, one CD-ROM had over 7,000 images.

It is further alleged that their two-year-old daughter, who was traveling with them at the time of their arrest, is also a victim. Authorities claim that photographs of their daughter are on the CD-ROMs her parents were distributing.

In Case Example 1, digital imaging technology and the Internet allegedly enhanced an existing MO, which consisted of manufacturing and marketing child pornography to other distributors. Alleged contact with international buyers was first made using Internet technologies, through which communications resulted in an agreement for sale of illicit materials. The illicit images were then alleged to have been digitized for transport, ease of storage, and ease of duplication once in the United States.

CASE EXAMPLE 2 (WIRED NEWS 1998):
From an article in *Wired* magazine from February 1998:

Police in four states say they're the victims of what amounts to a cybersex sting in reverse, the latest in a string of Internet pornography cases getting headlines around the United States.

The *News & Observer* of Raleigh, North Carolina, reports that the officers encountered a 17-year-old Illinois girl in chat rooms – and that their e-mail relationships quickly became sexually explicit. The girl then told her mother about the contacts with deputies in Virginia, North Carolina, Georgia, and Texas, and her mother informed authorities in those states. Discipline followed.

The chain of events – which included one North Carolina deputy sending the girl a photograph of his genitals – led an attorney for one of the officers to decry what he suggests was a setup.

"This young woman has gone around the country, as best we can determine, and made contact with a very vulnerable element of our society – police officers – and then drawn them in and alleged some type of sexual misconduct," said Troy Spencer, the attorney for one suspended Virginia officer. "She's a cyberspider."

The same teenager from the above instances, who acted under the alias "Rollerbabe," was connected to other similar incidents which were published in

The News Observer of North Carolina in November of 1998 (Jarvis 1998):

"… Earlier this year, Wake County sheriff's deputies were accused of taking advantage of a Midwestern teenager in an Internet sex scandal that eventually snared law enforcement officers in several states.

Now another officer has been caught in the Web, raising questions about who is snaring whom. A rural county sheriff in Illinois said this week that he had been enticed into a romantic e-mail correspondence with "Rollerbabe" – who claimed to be an athletic, 18-year-old blonde from suburban Chicago named Brenda Thoma. The summer relationship surfaced this month when her mother complained to county officials about it.

That pattern also emerged in Wake County and in three other states – prompting one officer's attorney to call the young woman a "*cyberspider*" – where e-mail friendships between law enforcement officers and Rollerbabe escalated into sexually explicit electronic conversations. Scandals broke out when her mother, Cathy Thoma, 44, complained to the officers' superiors. One officer whose career was ruined by the encounter, former Chesapeake, Va., police detective Bob Lunsford, said Friday that he is convinced the young woman's mother is involved with the e-mail. No one has brought criminal charges against the pair, nor has any one claimed that the women did anything illegal.

In March, Mrs Thoma insisted her daughter was courted by the police officers whom she trusted after meeting them online. She said she wasn't troubled by her daughter's computer habits. The Thoma family – a husband and wife and several children – was living in Manhattan, Ill., until several weeks ago when they moved to Lansing, Mich. An e-mail request for comment about the incident with the sheriff brought a brief response Friday, signed by someone identifying herself as Brenda Thoma.

… Earlier this week, (Paul) Spaur, 56, a Clinton County, Ill., sheriff, acknowledged carrying on an Internet romance with Rollerbabe from his county computer this summer. When Mrs Thoma complained to county officials, Spaur said he had done nothing wrong but offered to pay $1,222 for 679 hours worth of phone bills spent on the computer.

… In January, Wake County Sheriff John H. Baker Jr. suspended seven deputies and demoted one of them because some of the officers had e-mail conversations with Rollerbabe while on duty; their supervisors were punished because it happened on their watch. Mrs Thoma said the deputy who was demoted had initiated the relationship and sent nude photos of himself over the Internet, but Baker said there was no way to prove who was depicted in the photos.

… Shortly afterward, it was discovered that officers in Virginia, Texas, and Georgia had had similar encounters with Rollerbabe. An officer in Richland, Texas, resigned after Mrs Thoma complained about the relationship.

Lunsford, the Virginia detective, was publicly humiliated when he was suspended and a local TV station referred to the investigation as a child pornography case, because the girl was then 17. Before that he had won several commendations, including for saving another police officer's life. In May, the Chesapeake Police Department formally cleared Lunsford, who had been on leave because of a stress-related illness; he eventually resigned. His marriage also broke apart.

In Case Example 2, we have the MO of what might be referred to as a female law enforcement "groupie." Arguably, she is responding to what is referred to by some in the law enforcement community as the *Blue Magnet.* This term is derived from the reality that some individuals are deeply attracted to those in uniform, and who, by extension, have positions of perceived authority. In the past, their have been cases where law enforcement groupies have obsessively made contact with those in blue through seductive letter writing, random precinct house telephone calling, the frequenting of "cop bars," and participation in law enforcement conferences or fund raisers. Now, law enforcement e-mail addresses and personal profiles can be gathered quickly and easily over the Internet on personal and department websites, and in online chat rooms, making them more easily accessible to those attracted to the blue magnet. And the truth is that some officers provide this information, and seek out these online chat areas, with the overt intention of attracting just these types of individuals (i.e. registered IRC chat rooms such as #COPS, dedicated to "Cops Who Flirt"; AOL chat rooms such as "Cops who flirt," etc.).

It is important to keep in mind, however, that law enforcement groupies are not necessarily individuals engaged in criminal activity. That is, unless they attempt to blackmail an officer in some fashion after they get them to engage in some kind of compromising circumstance, or engage in harassment and/or stalking behavior, all of which can and does happen. The criminal activity in these instances (if there is any at all), as in the example above, can actually come from the law enforcement officers involved. This can take the form of misusing and abusing department resources and violating the public trust, including but not limited to things like inappropriate telephone charges, vehicle use, and desertion of one's assigned duties. And we are not talking about small misallocations, but rather large ones such as in the example, which are symptomatic of ongoing patterns of departmental resource misuse and abuse.

As in Case Example 2, criminal activity in these instances can also take on the form of the distribution of pornographic materials (an officer allegedly e-mailed a digital photograph of his genitals to the 17-year-old girl), which, depending on the circumstances, can have serious legal consequences.

In both examples, technology facilitated criminal behavior in terms of providing both the mechanisms for initial contact between the involved parties, and a means for communication and illicit materials sharing between the parties over great distances. But as we have shown, less complex and "immediate" technologies do exist which have facilitated the same type of behavior in the past.

A more reactive aspect of the relationship between MO and technology, from the criminal's point of view, involves the relationship between the advancement of crime detection technologies in the forensic sciences, and a criminal's knowledge of them.

Successful criminals are arguably those who avoid detection and identification, or at the very least capture. The problem for criminals is that as they incorporate new and existing technologies into their MO to make their criminal behavior or identity more difficult to detect, the forensic sciences can make advances to become more competent at crime detection. Subsequently, criminals that are looking to make a career, or even a hobby, for themselves in the realm of illegal activity must rise to the meet that challenge. That is to say, as criminals learn about new forensic technologies and techniques being applied to their particular area of criminal behavior, they must be willing to modify their MO, if possible, to circumvent those efforts.

But even an extremely skilful, motivated, and flexible offender may only learn of a new forensic technology when it has been applied to one of their crimes and resulted in their identification and/or capture. While this encounter can teach them something that they may never forget in the commission of future crimes, in such cases the damage will already have been done.

MAURY ROY TRAVIS

A glaring example of this type of inadvertent slip-up occurred in a recent case out of St Louis, Missouri, resulting in the apprehension of alleged serial killer Maury Roy Travis, a 36-year-old hotel waiter. In May of 2002, angered by a news story sympathetic to one of his victims, an unidentified serial killer wrote the publication in question to let his dissatisfaction be known. So that he would be believed, he provided details regarding location of an undiscovered victim. According to Bryan (2002):

> In the letter that arrived Friday at the *Post-Dispatch*, the writer said human remains would be found within "a 50-yard radius from the X" that had been inscribed on an accompanying map of the West Alton area. Police followed up on Saturday and found a human skull and bones at that location, just off of Highway 67. The remains were unidentified on Monday.
>
> The letter writer said the remains belonged to another victim, and the author indicated that the locations of even more bodies might be divulged to the newspaper at a later time. St Louis police, who are spearheading a multi-jurisdictional investigation, have refused to talk about the letter.
>
> "The letter writer believes he is brilliant," Turvey added. "And the letter writer has a proficient knowledge of evidence," illustrated by the fact that the letter was typed.
>
> "There's only been a couple of serial killers like this person," Turvey said. "One was the Zodiac killer in the San Francisco area in the '70s who was never caught."
>
> ... The remains found Saturday were within 300 yards of where the bodies of Teresa Wilson, 36, and Verona "Ronnie" Thompson, also 36, were found just a few yards apart in May and June of last year.

> In October, detectives from several jurisdictions in the St Louis area began comparing notes after they realized that the deaths of six prostitutes whose bodies were found mostly alongside roadways might be the work of a serial killer or killers. The prostitutes were drug users, and most had ties to a trucking area in the Baden neighborhood.
>
> This year, the skeletal remains of three unidentified women were found alongside roadways in the Metro East area. Those cases added to the list of the existing six cases.
>
> Turvey... said it was fortunate that a police task force had already been looking into the killings here and warned not to make the letter writer angry.

The offender's map turned out to be a crucial form of previously untapped digital evidence. The online service that Mr Travis used to render his map had logged his IP address. A description of the technology involved in associating Mr Travis with the map he generated online, and his subsequent identification and apprehension, is provided in (Robinson 2002):

> "Basically, whenever you go online, you're leaving a track," said Peter Shenkin, professor of Computer Information Systems in Criminal Justice and Public Administration at John Jay College in New York. "For instance, when I log on, I have unique number, an IP address, assigned to me by the Internet service provider, and I have that address as I go from one site to another. If I access a site, that site makes a record of my IP address. They know when I was online, how long I was on the site, what pages I looked at."
>
> Accused serial killer Maury Troy Travis had no idea that he would leave police a virtual trail when he allegedly sent a letter to a St Louis *Post-Dispatch* reporter. The letter was sent in response to an article about a slain prostitute believed to be one of the victims of a serial killer in Missouri and Illinois. The note to the reporter read, "Nice sob story. I'll tell you where many others are. To prove im real here's directions to number seventeen. [sic]"
>
> The second part of the letter contained a downloaded map of West Alton, Ill., marked with an X. Police went to the spot marked by the X and found a woman's skeleton. But that was not the only information the map provided. By surfing on different travel sites, Illinois State police found out the map had been downloaded from Expedia.com. After receiving a federal subpoena from investigators, Expedia.com pulled up the IP address of every user that had looked at the map in recent days. There was only one person.
>
> The FBI subpoenaed the Internet service provider to find out who had been assigned the IP address. That user, ISP records indicated, turned out to be Travis, who resided in St Louis County. FBI agents searched Travis' home and found blood spatters and smears throughout his home and on belts and other things used to tie people up.
>
> Travis was arrested and charged with two counts of kidnapping. Officials suspected him in the killings of six prostitutes and four unidentified women found in the St Louis area between April 2001 and May 2002 and were reportedly planning additional charges for murder.

However, before Mr Travis could be brought to trial, let alone be charged with murder, he committed suicide in custody. According to Clubb (2002):

> The suicide Monday night of Maury Troy Travis, 36, of Ferguson, sent shock waves Tuesday through the law enforcement community and the St Louis area media. Officials from the Clayton Police Department held a news conference late Tuesday to answer questions about how Travis managed to hang himself in his cell, despite being under a suicide watch.
>
> … Travis had not yet been charged with murder, which is usually prosecuted as a state crime. The federal case kept him in custody while prosecutors in at least three jurisdictions considered additional charges.
>
> However, one law enforcement source close to the investigation told *The Telegraph* that police already had discovered evidence that would have incriminated Travis in multiple torture-killings of women.
>
> The source said the FBI found the evidence when it searched Travis' house in Ferguson last Friday. Investigators found videotapes concealed inside walls at the home, the source said. Police viewed the videotapes this week and found they showed a number of torture-killings of women known to be victims, including some who identified themselves on the tapes by name.

By comparison with other serial murderers, Mr Travis was not foolish, impulsive, or unskilled. In fact, the evidence shows just the opposite: a patient and meticulous offender, conscious of the need for a disposable victim population and nurturing a specific set of sexual control oriented fantasies that required a specific methods of control and "props." According to reports (Home Movies 2003), Mr Travis was among other things sadistic in nature:

> Police believe Travis picked up prostitutes along a strip of Broadway just north of St Louis that is riddled with crack houses and prostitution, then took them to his ranch-style home in Ferguson, a nearby suburb.
>
> They found numerous videotapes in Travis' home showing him giving the prostitutes crack cocaine to smoke, then having consensual sex with them. He apparently let some of the women leave at that point.
>
> The "wedding" tape included similar scenes – including a shot of a woman sitting on Travis' bed after an introductory caption "ANOTHER CRACKHEAD HO." But it showed that in some cases – police are not sure how he chose his victims – Travis would start asking the women to engage in bizarre rituals, such as having them dance in white clothes or wear sunglasses with the lenses blackened so they could not see.
>
> Then he would take them captive, binding them with ropes and handcuffs and covering their eyes with duct tape. He would then begin to torment them, either in the bedroom, or after dragging them downstairs to the basement and shackling them to a wooden post.
>
> The excerpts the police released to Primetime show Travis tormenting the women verbally, taunting them about their fate and haranguing some of them over how

> they had abandoned their children for crack. One exchange, with an unidentified victim, went as follows:
>
> > Travis: You want to say something to your kids?
> >
> > Victim: I'm sorry.
> >
> > Travis: Who's raising your kids?
> >
> > Victim: Me, my mom and dad.
> >
> > Travis: You ain't raising s---, b---. You over here on your back smoking crack. You ain't going home tomorrow. I'm keeping you about a week. Is that all right?
>
> He forced one victim to say to him, "You are the master. It pleases me to serve you." When he didn't like the way she said it, he yelled at her, "Say it clearer!"
>
> When another victim tried to remove the duct tape covering her eyes and knocked his camera out of focus, he told her: "You don't need to see s--- ... Lay down on your back. Shut your eyes."
>
> At one point, a woman can be heard gasping in agony as he orders her, "Sit still!"

There is no question regarding the skill and care taken by Mr Travis in the commission of his crimes. There is further no question that police had failed to link him with all of his crimes prior to his capture, let alone link all of his crimes together. In fact, police had few tangible leads, and the case was apparently growing cold. The only question that remains is whether police would have linked him to his crimes without his inadvertent cybertrail and the work of diligent local investigators examining his correspondence for clues. The most reasonable answer is no.

6.4 MOTIVE AND TECHNOLOGY

The term *motive* refers to the emotional, psychological, or material need that impels, and is satisfied by, a behavior (Turvey 2002). Criminal motive is generally technology independent. That is to say, the psychological or material needs that are nurtured and satisfied by a criminal's pattern of behavior tend to be separate from the technology of the day. The same motives that exist today have arguably existed throughout recorded history, in one form or another. However, it may also be argued that existing motives (i.e. sexual fetishes) can evolve with the employment of, or association of, offense activities with specific technologies. Towards understanding these issues, this section demonstrates how an existing behavioral motivational typology may be applied within the context of computer and Internet related criminal behavior.

In 1979, A. Nicholas Groth, an American clinical psychologist working with both victims and offender populations, published a study of over 500 rapists. In his study, he found that rape, like other crimes involving behaviors that satisfy emotional needs, is complex and multi-determined. That is to say, that the act of rape itself serves a number of psychological needs and purposes (motives) for the offender. The purpose of his work was clinical, to understand the motivations of rapists for the purpose of the development of effective treatment plans (Groth 1979).

Eventually, the Groth rapist motivational typology was taken and modified by the FBI's National Center for the Analysis of Violent Crime (NCAVC) and its affiliates (Hazelwood, *et al.* 1991; Burgess and Hazelwood 1995).

This author has found through casework, that this behaviorally based motivational classification system, with some modifications, is useful for understanding the psychological basis for most criminal behavior. The basic psychological needs, or motives, that impel human criminal behaviors remain essentially the same across different types of criminals, despite their behavioral expression, which may involve computer crimes, stalking, harassment, kidnapping, child molestation, terrorism, sexual assault, homicide, and/or arson. This is not to say that the motivational typology presented here should be considered the final word in terms of all *specific* offender motivations. But in terms of general types of psychological needs that are being satisfied by offender behavior, they are fairly inclusive, and fairly useful.

Below, the author gives a proposed behavioral motivational typology (Turvey 2002), and examples, adapted from Burgess (1995). This author takes credit largely for the shift in emphasis from classifying *offenders* – to classifying *offense behaviors* (turning it from an inductive labeling system to a deductive tool). They include the following types of behaviors: *Power Reassurance, Power Assertive, Anger Retaliatory, Sadistic, Opportunistic, and Profit oriented.*[2]

[2] *Sections of text in this typology are taken directly from Turvey (2002).*

6.4.1 POWER REASSURANCE (COMPENSATORY)

These include criminal behaviors that are intended to restore the criminal's self-confidence or self-worth through the use of low aggression means. These behaviors suggest an underlying lack of confidence and a sense of personal inadequacy. This may manifest itself in a misguided belief that the victim desires the offense behavior, and is somehow a willing or culpable participant. In may also manifest itself in the form of self-deprecating or self-loathing behavior which is intended to garner a response of pity for sympathy from the victim.

The belief motivating this behavior is often that the victim will enjoy and eroticize the offense behavior, and may subsequently fall in love with the offender. This stems from the criminal's own fears of personal inadequacy. The offense behavior is restorative of the offender's self doubt, and therefore emotionally reassuring. It will occur as his need for that kind of reassurance arises.

CASE EXAMPLE (DURFEE 1996):
The following is a media account of the circumstances surrounding Andrew Archambeau, a man who pled no contest to harassing a woman via e-mail and the telephone:

> ... Archambeau, 32, was charged with a misdemeanor almost two years ago for stalking the Farmington Hills woman ... Archambeau met the woman through a computer dating service. He messaged her by computer and (they) talked on the phone.
>
> The couple met in person twice. After the second meeting, the woman dumped Archambeau by e-mail. He continued to leave phone messages and e-mail the woman (urging her to continue dating him), even after police warned him to stop. Archambeau was charged in May 1994 under the state's stalking law, a misdemeanor.
>
> "Times have changed. People no longer have to leave the confines and comfort of their homes to harass somebody," (Oakland County Assistant Prosecutor Neal) Rockind said.

In this example, the offender was unwilling to let go of the relationship, perceiving a connection to the victim that he was unwilling to relinquish. The content of the messages that he left was not described as violent, or threatening, merely persistent. While it is possible that this could have eventually escalated to more *retaliatory* behaviors, the behaviors did not appear to be coming from that emotion.

6.4.2 POWER ASSERTIVE (ENTITLEMENT)

These include criminal behaviors that are intended to restore the offender's self-confidence or self-worth through the use of moderate to high aggression means. These behaviors suggest an underlying lack of confidence and a sense of personal inadequacy, that are expressed through control, mastery, or humiliation of the victim, while demonstrating the offender's perceived sense of authority.

Offenders evidencing this type of behavior exhibit little doubt about their own adequacy and masculinity. In fact, they may be using their attacks as an expression of their own virility. In their perception, they are entitled to the fruits of their attack by virtue of being a male and being physically stronger.

Offenders evidencing this type of behavior may grow more confident over time, as their egocentricity may be very high. They may begin to do things that can lead to their identification. Law enforcement may interpret this as a sign that the offender desires to be caught. What is actually true is that the offender has no respect for law enforcement, has learned that they can commit their offenses without the need to fear identification or capture, and subsequently they may not take precautions that they have learned are generally unnecessary.

This type of behavior does not evidence a desire to harm the victim, necessarily, but rather to posses them. Demonstrating power over their victims is their means of expressing mastery, strength, control, authority, and identity to themselves. The attacks are therefore intended to reinforce the offender's inflated sense of self-confidence or self-worth.

CASE EXAMPLE (ASSOCIATED PRESS 1997b):
The following is taken from a media account of the circumstances surrounding the Dwayne and Debbie Tamai family of Emeryville, Ontario. This case of electronic harassment involved their 15-year-old son, Billy, who took control of all of the electronic devices in the family's home, including the phone, and manipulated them to distress of other family members for his own amusement. The incidents began in December of 1996, when friends of the family complained that phone calls to the Tamai home were repeatedly being waylaid and cut off:

... Police confirmed that the sabotage was an inside job, but refused to name the culprit and said nothing would be gained by filing charges against him. Dwayne and Debbie Tamai issued a statement saying that their son, Billy, had admitted to making the mysterious calls.

The interruptions included burps and babbling and claims of control over the inner workings of the Tamais' custom-built home, including what appeared to be the power to turn individual appliances on and off by remote control.

"It started off as a joke with his friends and just got so out of hand that he didn't know how to stop it and was afraid to come forward and tell us in fear of us disowning him," the Tamais said in their statement, which was sent to local news media.

On Saturday, the Tamais said they were planning to take their son to the police to defend him against persistent rumors that he was responsible. Instead, he confessed to being the intruder who called himself Sommy.

"All the crying I heard from him at night I thought was because of the pain he was suffering caused by Sommy," the letter said. "We now realize it was him crying out for help because he wanted to end all this but was afraid because of how many people were now involved."

... "We eliminated all external sources and interior sources," Babbitt said.

A two-day sweep by a team of intelligence and security experts loaded with high-tech equipment failed to locate "Sommy" on Friday. The team was brought in by two television networks.

... missed messages and strange clickings seemed minor when a disembodied voice, eerily distorted by computer, first interrupted a call to make himself known.

After burping repeatedly, the caller told a startled Mrs Tamai, "I know who you are. I stole your voice mail."

Mocking, sometimes menacing, the high-tech stalker became a constant presence, eavesdropping on family conversations, switching TV channels and shutting off the electricity.

> "He would threaten me," Mrs Tamai said last week. "It was very frightening: 'I'm going to get you. I know where you live.'
>
> "I befriended him, because the police asked me to, and he calmed down and said he wasn't going to hurt me. The more I felt I was kissing his butt, the safer I felt."

In this case, the son repeatedly made contact with the victims (his parents), and made verbal threats in combination with the electronic harassment, all in an effort to demonstrate his power and authority over them. The victims were not physically harmed, though they were in fear and greatly inconvenienced by the fact that an unknown force appeared to have control over a great many aspects of their lives.

6.4.3 ANGER RETALIATORY (ANGER OR DISPLACED)

These include criminal behaviors that suggest a great deal of rage, either towards a specific person, group, institution, or a symbol of either. These types of behaviors are commonly evidenced in stranger-to-stranger sexual assaults, domestic homicides, work-related homicide, harassment, and cases involving terrorist activity.

Anger retaliation behavior is just what the name suggests. The offender is acting on the basis of cumulative real or imagined wrongs from those that are in their world. The victim of the attack may be one of these people such as a relative, a girlfriend, or a coworker. Or the victim may symbolize that person to the offender in dress, occupation, and/or physical characteristics.

The main goal of this offender behavior is to service their cumulative aggression. They are retaliating against the victim for wrongs or perceived wrongs, and their aggression can manifest itself spanning a wide range, from verbally abusive epithets to hyper-aggressed homicide with multiple collateral victims. In such cases, even sexual acts can be put into the service anger and aggression (this is the opposite of the sadistic offender, who employs aggression in the service of sexual gratification).

It is important not to confuse retaliatory behavior with sadistic behavior. Although they can share some characteristics at first blush, the motivations are wholly separate. Just because a crime is terrible or brutal does not confirm that the offender responsible was a sadist, and tortured the victim. Reliance upon a competent reconstruction by the appropriate forensic scientists is requisite.

> CASE EXAMPLE (ASSOCIATED PRESS 1997a):
> The following is a media account of the circumstances surrounding the homicide of Marlene Stumpf. Her husband, Raymond Stumpf, who was host and producer of a home shopping show that aired in Pottstown, Pennsylvania, allegedly stabbed her to death. He was known as "Mr Telemart," and also worked full-time as a manager at a fast-food restaurant.

> A woman who received flowers from a man she corresponded with on the Internet has been slain, and her husband has been charged with murder.
>
> The dozen roses were sent several days ago to "Brandis," the online name used by Marlene Stumpf, 47, police said. Her son found her body Monday night on the kitchen floor with three blood-covered knives nearby.
>
> Raymond Stumpf, 54, her husband of 13 years and host of a local cable television show, was found in the dining room, bleeding from arm and stomach wounds that police consider self-inflicted.
>
> "It was a particularly gruesome scene with a lot of blood that showed evidence of extreme violence," prosecutor Bruce Castor Jr. said Wednesday. "(Stumpf) tried to kill himself, presumably because he felt bad he had killed his wife."
>
> Stumpf told police his wife started slapping him during an argument Monday night and he "just went wild." Police said he couldn't remember what happened.
>
> Detectives hope Mrs Stumpf's computer and computer files will provide information about her online relationships and people who could help prosecutors with a motive, Castor said.

In this example, it is alleged that the husband killed his wife after an argument over her Internet romance, and then tried to kill himself. The fact that there is digital evidence related to this crime, and that the Internet is somehow involved, is incidental to the husband's motive for killing her. Instances of similar domestic murder-suicides involving real or perceived infidelity are nothing new in the history of human relationships, and are always tragic.

The retaliatory aspect of this case comes from the description of the nature and extent of the injuries to the victim (i.e. that Mr Stumpf "just went wild," and that there was "extreme violence").

The retaliatory aspect of this case is further evidenced by circumstances that support the context of that retaliatory behavior, including:

- the argument;
- the use of available materials;
- the use of multiple weapons;
- the relatively short duration of the attack.

6.4.4 ANGER EXCITATION (SADISTIC)

These include criminal behaviors that evidence offender sexual gratification from victim pain and suffering. The primary motivation for the behavior is sexual, however the sexual expression for the offender is manifested in physical aggression, or torture behavior, toward the victim.

This offense behavior is perhaps the most individually complex. This type of behavior is motivated by intense, individually varying fantasies that involve inflicting brutal levels of pain on the victim solely for offender sexual pleasure. The goal of this behavior is total victim fear and submission for the purposes of feeding the offender's sexual desires. Aggression services sexual gratification. The result is that the victim must be physically or psychologically abused and humiliated for this offender to become sexually excited and subsequently gratified.

Examples of sadistic behavior must evidence sexual gratification that an offender achieves by witnessing the suffering of their victim, who must requisitely be both living and conscious. Dead or unconscious victims are incapable of suffering in the manner that gives the necessary sexual stimulation to the sadist. For an example of such a case involving the use of the Internet and a subsequent cybertrail, see the previous discussion regarding serial murderer Maury Roy Travis in this chapter.

6.4.5 PROFIT ORIENTED

These include criminal behaviors that evidence an offender motivation oriented towards material or personal gain. These can be found in all types of homicides, robberies, burglaries, muggings, arsons, bombings, kidnappings, and fraud, just to name a few.

This type of behavior is the most straightforward, as the successful completion of the offense satisfies the offender's needs. Psychological and emotional needs are not necessarily satisfied by purely profit motivated behavior (if one wants to argue that a profit motivation is also motivated by a need for reassurance that one is a good provider, that would have to be followed by a host of other reassurance behaviors). Any behavior that is not purely profit motivated, which satisfies an emotional or psychological need should be examined with the lens of the other behavior motivational types.

CASE EXAMPLE (PIPER 1998):
The following is excepted from a media account regarding the circumstances surrounding the activities of Valdimir Levin in St Petersburg, Russia:

> Vladimir Levin, a computer expert from Russia's second city of St Petersburg, used his skills for ill-gotten gains. He was caught stealing from Citibank in a fraud scheme and said he used bank customer passwords and codes to transfer funds from their accounts to accounts he controlled in Finland, the Netherlands, Germany, Israel and the United States.

In this example, regardless of any other motivation that may be evident in this offender's behavioral patterns, the desire for profit is clearly primary.

6.5 CURRENT TECHNOLOGIES

Perhaps the best way to finalize our exploration of how criminals engage and adapt computer and Internet technology is by discussing a couple of examples. The technologies discussed are only a very small sample of what is available to the cyber criminal. Of these technologies, only a few of the many criminal adaptations are illustrated.

6.5.1 A COMPUTER VIRUS

A computer virus is a foreign program that is designed to enter a computer system with the purpose of executing one or more particular functions without the knowledge or consent of the system administrator. The function of a virus is specified by its creator. The criminal applications of viruses in the cyberverse are almost without limits. They are typically used to steal, broadcast, and/or destroy information (examples include computer files containing personal contact information, credit card numbers, and passwords).

- A thief can program and disseminate a virus on a given network that is designed to locate and gather victim password information used in online banking.
- A stalker can program and disseminate a virus to a particular victim's PC via anonymous personal e-mail designed to locate and gather sensitive personal information including address books, financial files, and digital images.
- A terrorist can program and disseminate a virus on a particular network that is designed to delete or alter specific files essential to that network's function. In doing so, they can alter or disrupt that function.

6.5.2 A PUBLIC E-MAIL DISCUSSION LIST

Individuals may develop and maintain or join one of the many public e-mail discussion lists available via the Internet to share the details and experiences of their lives with others. They are also a way to meet and learn from people with similar experiences and interests. The content of an e-mail discussion list is dependent on the list topic, and the types of posts that are sent by subscribers. However, any e-mail discussion list represents a captive audience susceptible to individual and multiple broadcasts of information over that list.

- A thief may use information (personal details elicited from text and photographs) gathered from a victim's posts on an e-mail discussion list to plan a burglary, targeting specific valuables in specific rooms.
- An ex-intimate may join a discussion list to which their former intimate subscribes. Once subscribed, they may publicly harass and defame their former intimate with a mixture of true and false information. This can be accomplished by the distribution of explicit and/or invasive personal images, as well as the dissemination of false accusations of child abuse, sex crimes, or other criminal conduct.

6.6 SUMMARY

As this chapter has illustrated, technology is generally developed for one purpose, but is often harnessed or adapted for another by those with criminal motive and intent. It can also have unintended consequences within the criminal and forensic communities. So long as technology evolves, criminal enterprise will evolve to incorporate and build upon it.

REFERENCES

Associated Press (1997a) "Wife's Internet friendship may have led to her death", January 23.

Associated Press (1997b) "High-tech 'stalking' of Canadian family linked to teen-aged son", April 20.

Bryan, B. (2002) "Letter writer is serial killer, concludes criminal profiler", *St. Louis Post-Dispatch*, May 28.

Burgess A., Burgess A., Douglas J., and Ressler R. (1997) *Crime Classification Manual*, San Francisco, CA: Jossey-Bass, Inc.

Burgess A. and Hazelwood R. (eds) (1995) *Practical Aspects of Rape Investigation: A Multidisciplinary Approach*, 2nd edn, New York, NY: CRC Press.

Clubb S. (2002) "Police explain suspect's suicide," *The Illinois River Bend Telegraph*, June 12. (Available online at http://www.zwire.com/site/news.cfm?newsid=4412382&BRD=1719&PAG=461&dept_id=25271&rfi=8).

Durfee D. (1996) "Man pleads no contest in stalking case", *The Detroit News*, January 25.

Groth A.N. (1979) *Men Who Rape: The Psychology of the Offender*, New York, NY: Plenum.

Hazelwood R., Reboussin R., Warren J.I., and Wright J.A., (1991) "Prediction of Rapist Type and Violence from Verbal, Physical, and Sexual Scales", *Journal of Interpersonal Violence*, 6(1), 55–67.

Jarvis C. (1998) "Teen again linked to e-mail affair", *The News Observer*, North Carolina, November 28.

McPherson T. (2003) "Sherlock Holmes' modern followers", *The Advertiser*, May 31.

Meloy J.R. (ed) (1998) *The Psychology of Stalking: Clinical and Forensic Perspectives*, San Diego, CA: Academic Press.

Piper E. (1998) "Russian cybercrime flourishes: deteriorating economic conditions have brought pirating and cracking mainstream", *Reuters*, December 30.

Reuters Information Service (1997) "Swiss couple charged in U.S. child pornography sting", August 22.

Robinson B. (2002) "Taking a byte out of cybercrime", *ABC News*, July 15.

Shamburg R. (1999) "A tortured case", *Net Life*, April 7.

Turvey B. (2002) *Criminal Profiling: An Introduction to Behavioral Evidence Analysis*, 2nd edn, London: Academic Press.

Wired News (1998) "Cops 'lured' into net sex", February 16.

CHAPTER 7

DIGITAL EVIDENCE IN THE COURTROOM

In this age of science, science should expect to find a warm welcome, perhaps a permanent home, in our courtrooms. The reason is a simple one. The legal disputes before us increasingly involve the principles and tools of science. Proper resolution of those disputes matters not just to the litigants, but also to the general public – those who live in our technologically complex society and whom the law must serve. Our decisions should reflect a proper scientific and technical understanding so that the law can respond to the needs of the public.

(Breyer 2000)

Individuals processing evidence must realize that, in addition to being pertinent, evidence must meet certain standards to be admitted. It is easy enough to claim that a bloody glove was found in a suspect's home, but it is another matter to prove it. When guilt or innocence hangs in the balance, the proof that evidence is authentic and has not been tampered with becomes essential. The US Federal Rules of Evidence, the UK Police and Criminal Evidence Act (PACE) and Civil Evidence Act, and similar rules of evidence in other countries were established to help evaluate evidence. For instance, before admitting evidence, a court will generally ensure that it is relevant and evaluate it to determine if it is what its proponent claims, if the evidence is hearsay, and if the original is required or a copy is sufficient. There are many other issues that a court must consider to determine if evidence is admissible and a failure to consider these issues from the outset may cause evidence to be excluded, potentially losing the case.

One of the most important aspects of authentication is maintaining and documenting the chain of custody (a.k.a. continuity of possession) of evidence. Each person who handled evidence may be required to testify that the evidence presented in court is the same as when it was processed during the investigation. Although it may not be necessary to produce at trial every individual who handled the evidence, it is best to keep the number to a minimum and

Digital Evidence and Computer Crime Second Edition
ISBN: 0-12-163104-4

maintain documentation to demonstrate that digital evidence has not been altered since it was collected. Without a solid chain of custody, it could be argued that the evidence was handled improperly and may have been altered, replaced with incriminating evidence, or contaminated in some other fashion.

Having someone on the search team who is trained to handle digital evidence can reduce the number of people who handle the evidence, thus streamlining the presentation of the case, and minimizing the defense opportunities to impugn the integrity of the evidence. Additionally, having standard operating procedures, continuing education, and clear policies help to maintain consistency and prevent contamination of evidence. Given the ease with which digital evidence can be altered, the importance of procedures and the use of only trained personnel to handle and examine cannot be overstated.

This chapter provides an overview of the major issues that arise when digital evidence is presented in court, including admissibility, uncertainty, and presentation of digital evidence. The process of preparing a case for trial is time consuming, expensive, and may not result in a satisfactory outcome, particularly if there is insufficient evidence or evidence was handled improperly. Also, before deciding to take legal action, organizations should consider the impact if they are required to disclose information about their systems that may be sensitive (e.g. network topology, system configuration information, source code of custom monitoring tools) and other details about their operations that they may not want to make public.

7.1 ADMISSIBILITY – WARRANTS

The most common mistake that prevents digital evidence from being admitted by courts is that it was obtained without authorization. Generally, a warrant is required to search and seize evidence. The main exceptions are plain view, consent, and exigency. If investigators see evidence in plain view, they can seize it provided they obtained access to thc area validly. By obtaining consent to search, investigators can perform a search without a warrant but some care must be employed when obtaining consent to reduce the chance of the search being successfully challenged in court.

CASE EXAMPLE (UNITED STATES v. TURNER 1999):
Law enforcement officers obtained permission from the defendant to search his home for evidence relating to a sexual assault of one of his neighbors. During the search, an investigator looked at Turner's computer and identified child pornography. Turner was indicted for possessing child pornography but filed a suppression hearing to exclude the computer files on the ground that he had not consented to the search of his computer and it was not objectively reasonable for the detective to have concluded that evidence of the sexual

assault – the stated object of the consent search – would be found in files with such labels as "young" or "young with breasts."

Regarding exigency, a warrantless search can be made for any emergency threatening life and limb. It is difficult to imagine a case in which a computer could be collected under exigent circumstances. Even in a homicide, a warrant is required for an in-depth search of the suspect's possessions.

There are four questions that investigators must ask themselves when searching and seizing digital evidence:

1 Does the Fourth Amendment and/or ECPA apply to the situation?
2 Have the Fourth Amendment and/or ECPA requirements been met?
3 How long can investigators remain at the scene?
4 What do investigators need to re-enter?

When asking answering these questions, remember that the ECPA prohibits anyone, not just the government, from unlawfully accessing or intercepting electronic communications, whereas the Fourth Amendment only applies to the government. Recall that the Fourth Amendment requires that a search warrant be secured before law enforcement officers can search a person's house, person, papers, and effects. To obtain a warrant, investigators must demonstrate probable cause and detail the place to be searched and the persons or things to be seized. More specifically, investigators have to convince a judge or magistrate that:

1 a crime has been committed;
2 evidence of crime is in existence;
3 the evidence is likely to exist at the place to be searched.

Even when investigators are authorized to search a computer, they must maintain focus on the crime under investigation. For instance, in United States v. Carey (case ref), the investigator found child pornography on a machine while searching for evidence of drug related activity but the images were inadmissible because they were outside of the scope of the warrant. The proper action when evidence of another crime is discovered is to obtain another search warrant for that crime.

CASE EXAMPLE (UNITED STATES v. GRAY 1999):
During an investigation into Montgomery Gray's alleged unauthorized access to National Library of Medicine computer systems, the FBI obtained a warrant to seize four computers from Gray's home and look for information downloaded from the library. While examining Gray's computers, a digital evidence examiner found pornographic images in directories named "teen" and "tiny teen," halted the search and obtained a second warrant to search for pornography.

CASE EXAMPLE (WISCONSIN v. SCHROEDER):
While investigating an online harassment complaint made against Keith Schroeder, a digital evidence examiner found evidence relating to the harassment complaint on his computer and noticed some pornographic pictures of children. A second warrant was obtained, giving the digital evidence examiner authority to look for child pornography on Schroeder's computer. Schroeder was charged with 19 counts of possession of child pornography and convicted on 18 counts after a jury trial. For the harassment, Schroeder was tried in a separate proceeding for unlawful use of a computer and disorderly conduct.

The other common mistake that prevents digital evidence from being admitted by courts is improper handling. Although courts were somewhat lenient in the past, as more judges and attorneys become familiar with digital evidence, more challenges are being raised relating to evidence handling procedures.

7.2 AUTHENTICITY AND RELIABILITY

The process of determining whether evidence is worthy is called authentication.

> Authentication means satisfying the court that (a) the contents of the record have remained unchanged, (b) that the information in the record does in fact originate from its purported source, whether human or machine, and (c) that extraneous information such as the apparent date of the record is accurate. As with paper records, the necessary degree of authentication may be proved through oral and circumstantial evidence, if available, or via technological features in the system or the record. (Reed 1990–91)

Authentication is actually a two-step process, with an initial examination of the evidence to determine that it is what its proponent claims and, later, a closer analysis to determine its probative value. In the initial stage, it may be sufficient for an individual who is familiar with the digital evidence to testify to its authenticity. For instance, the individual who collected the evidence can confirm that the evidence presented in court is the same as when it was collected. Alternately, a system administrator can testify that log files presented in court originated from her/his system.

In some cases, the defense will cast doubt on more malleable forms of digital evidence, such as logs of online chat sessions.

CASE EXAMPLE (MICHIGAN v. MILLER 2002):
In 2000, e-mail and AOL Instant Messages provided the compelling evidence to convict Sharee Miller of conspiring to kill her husband and abetting the suicide of the admitted killer (Jerry Cassaday) she had seduced with the assistance of the Internet. Miller carefully controlled the killer's perception of her husband, going so far as to masquerade as her husband to send the killer

offensive messages. In this case, the authenticity of the AOL Instant Messages was questioned in light of the possibility that such an online conversation could be staged (Bean 2003).

CASE EXAMPLE (UNITED STATES v. TANK):
In United States v. Tank, a case related to the Orchid/Wonderland Club investigation, the defendant argued that the authenticity and relevance of Internet chat logs was not adequately established. One of the points the defense argued was that the chat logs could be easily modified. The prosecution used a number of witnesses to establish that the logs were authentic. The court held that "printouts of computer-generated logs of 'chat room' discussions may be established by evidence showing how they were prepared, their accuracy in representing the conversations, and their connection to the defendant."

This case is significant because it is one of the first to deal with the authentication of chat logs. However, some feel that there are still questions about the authenticity and reliability of Internet chat logs that have not been addressed. On IRC, for example, in addition to the chat channel window, there may be important information in other areas of an IRC client such as the status window and in private chat or fserve windows. Since it is not possible for one investigator simultaneously to view every window, we must rely heavily on the logs for an account of what occurred. In some instances, investigators have been able to compensate for a lack of documentation by testifying that the evidence being presented is authentic and reliable. Of course, it is best to have solid documentation.

To authenticate digital evidence, it may also be necessary to demonstrate that a computer system or process that generated digital evidence was working properly during the relevant time period. For instance, the section in the Federal Rules of Evidence 901(b)(9) titled "Requirement of Authentication or Identification" includes "evidence describing a process or system used to produce a result and showing that the process or system produces an accurate result." In the United Kingdom, under Section 69 of the PACE, there is a formal requirement for a positive assertion that the computer systems involved were working properly.

CASE EXAMPLE (R. v. COCHRANE 1993, UNITED KINGDOM):
The accused was convicted of theft by fraudulent use of his cash card, withdrawing sums that his building society inadvertently credited to his account. The issue before the court was whether the trial judge should have admitted evidence in the form of computer printouts or till rolls. The evidence before the court was that two computers were involved in the relevant process. The person using the cash-point machine provided certain information which was relayed to the branch computer, which retained a back-up in its memory before transmitting it to the central mainframe computer. The court found that none of the prosecution witnesses had

> any knowledge of the actual working of the mainframe computer in that part of its operation, and none of them was able to supply affirmative information that the mainframe computer was operating correctly at the relevant time. As such the prosecution had failed to adduce adequate evidence to enable the court to properly rule that the till rolls were admissible evidence; in the absence of the till rolls the prosecution's case could not be proved.

The increasing variety and complexity of computer systems makes this type of evaluation increasingly difficult leading the UK Law Commission to recommend the repeal of Section 69 of PACE (Law Commission 1997). Requiring programmers and system designers to establish that computer systems are reliable at the lowest level is untenable, "overburdening already crowded courts with hordes of technical witnesses" (People v. Lugashi 1998). Therefore, US and UK courts have accepted the testimony of individuals who are familiar with the operation of computer systems. For instance, in R. v. Shephard (1993), The House of Lords held that Section 69(1) can be satisfied by the oral evidence of a person familiar with the operation of the computer who can give evidence of its reliability and the person need not be a computer expert. In United States v. Miller, telephone company records were admitted after a telephone-billing supervisor authenticated them. In a sexual assault case, the manager of the Southwestern Bell's security office testified that their telephone billing records were reliable as noted in the following quote.

> Figlio's testimony was sufficient to confirm the reliability of the telephone records. She explained that entries in the record were made instantaneously with the making of the calls and that AT&T would send Southwestern Bell the billing tapes, which established when the call took place, the originating number and the terminating number. She explained that the source of the information was a computer, which monitored Southwestern Bell's switching operations. The circuit court was correct in concluding that these records were uniquely reliable in that they were computer-generated rather than the result of human entries. (Missouri v Dunn 1999)

Once digital evidence is admitted, its reliability is assessed to determine its probative value. For instance, if there is concern that the evidence was tampered with prior to collection, these doubts may reduce the weight assigned to the evidence. In several cases, attorneys have argued that digital evidence was untrustworthy simply because there was a theoretical possibility that it could have been altered or fabricated. However, as judges become more familiar with digital evidence, they are requiring evidence to support claims of untrustworthiness. As noted in the US Department of Justice Searching and Seizing Computers and Obtaining Electronic Evidence in Criminal Investigations:

> Absent specific evidence that tampering occurred, the mere possibility of tampering does not affect the authenticity of a computer record. See Whitaker, 127 F.3d at 602

(declining to disturb trial judge's ruling that computer records were admissible because allegation of tampering was "almost wild-eyed speculation... [without] evidence to support such a scenario"); United States v. Bonallo, 858 F.2d 1427, 1436 (9th Cir. 1988) ("The fact that it is possible to alter data contained in a computer is plainly insufficient to establish untrustworthiness."); United States v. Glasser, 773 F.2d 1553, 1559 (11th Cir. 1985) ("The existence of an air-tight security system [to prevent tampering] is not, however, a prerequisite to the admissibility of computer printouts. If such a prerequisite did exist, it would become virtually impossible to admit computer-generated records; the party opposing admission would have to show only that a better security system was feasible.") ... the government may need to disclose "what operations the computer had been instructed to perform [as well as] the precise instruction that had been given" if the opposing party requests. United States v. Dioguardi, 428 F.2d 1033, 1038 (C.A.N.Y. 1970). Notably, once a minimum standard of trustworthiness has been established, questions as to the accuracy of computer records "resulting from... the operation of the computer program" affect only the weight of the evidence, not its admissibility. United States v. Catabran, 836 F.2d 453, 458 (9th Cir. 1988). (USDOJ 2002)

Even when there is a reasonable doubt regarding the reliability of digital evidence, this does not necessarily make it inadmissible, but will reduce the amount of weight it is given by the court.

7.3 CASEY'S CERTAINTY SCALE

Computers can introduce errors and uncertainty in various ways, making it difficult to assess the trustworthiness of digital evidence meaningfully. Although courts are warned to consider the computer systems involved carefully, little guidance is provided.

> Business records that are generated by computers present structural questions of reliability that transcend the reliability of the underlying information that is entered into the computer. Computer machinery may make errors because of malfunctioning of hardware, the computer's mechanical apparatus. Computers may also make errors that arise out of defects in the software, the input procedures, the database, and the processing program. In view of the complex nature of the operation of computers, courts have been cautioned to take special care to be certain that the foundation is sufficient to warrant a finding of trustworthiness and that the opposing party has full opportunity to inquire into the process by which information is fed into the computer. (American Oil Co. v. Valenti 1979).

Computer networks complicate reliability considerations because multiple systems and mechanisms are involved. Possibly because of the complexity and multiplicity of computer systems, there is a lack of consistency in the way that the reliability of digital evidence is assessed. To improve our ability to assess the reliability of digital evidence, we need a consistent method of referring to the relative certainty of different types of digital evidence.

The scale in Table 7.1 is proposed when attempting to assess the probative value of digital evidence (Casey 2002).

The certainly values (C-values) in Table 7.1 provide a method for a digital evidence examiner to denote the level of certainty he/she has in a given piece of evidence in a given context. This scale is not intended to be used rigidly to categorize types of evidence in general – it is not valid to claim that all NT Event logs have C3 certainly level because in some cases there may be signs of tampering such as deleted log entries, reducing the certainly level of the log to C1. The primary purpose of this Certainty Scale is to help others understand how much weight an examiner has given pieces of digital evidence when making a conclusion based on that evidence. Without these C-values, one might wonder how a digital evidence examiner reached his/her conclusion, particularly if there is disagreement over the certainty assigned to a given piece of evidence. For instance, two digital evidence

Table 7.1

A proposed scale for categorizing levels of certainty in digital evidence.

CERTAINTY LEVEL	DESCRIPTION/INDICATORS	COMMENSURATE QUALIFICATION	EXAMPLES
C0	Evidence contradicts known facts	Erroneous/incorrect	Examiners found a vulnerability in Internet Explorer (IE) that allowed scripts on a particular Web site to create questionable files, desktop shortcuts, and IE favorites. The suspect did not purposefully create these items on the system
C1	Evidence is highly questionable	Highly uncertain	Missing entries from log files or signs of tampering
C2	Only one source of evidence that is not protected against tampering	Somewhat uncertain	E-mail headers, sulog entries, and syslog with no other supporting evidence
C3	The source(s) of evidence are more difficult to tamper with but there is not enough evidence to support a firm conclusion or there are unexplained inconsistencies in the available evidence	Possible	An intrusion came from Poland suggesting that the intruder might be from that area However, a later connection came from South Korea suggesting that the intruder might be elsewhere or that there is more than one intruder
C4	(a) Evidence is protected against tampering or (b) evidence is not protected against tampering but multiple, independent sources of evidence agree	Probable	Web server defacement probably originated from a given apartment since tcpwrapper logs show FTP connections from the apartment at the time of the defacement and Web server access logs show the page being accessed from the apartment shortly after the defacement
C5	Agreement of evidence from multiple, independent sources that are protected against tampering. However, small uncertainties exist (e.g. temporal error, data loss)	Almost certain	IP address, user account, and ANI information lead to suspect's home. Monitoring Internet traffic indicates that criminal activity is coming from the house
C6	The evidence is tamper proof and unquestionable	Certain	Although this is inconceivable at the moment, such sources of digital evidence may exist in the future

examiners might make the following conclusions about the same case:

1 Log entries from System 2 indicate that Suspect B was logged in at the time of the crime and is almost certainly the offender.

2 The wtmp log on trusted System 1 (C4) indicates that the offender logged in from System 2. The wtmp log on untrusted System 2 (C2) indicates that two potential suspects were logged in at the time of the crime. However, RADIUS logs (C4) relating to Suspect A's PPP connection show that she disconnected from the Internet long before the crime, indicating that the associated wtmp entry on untrusted System B was not terminated properly, probably due to an abrupt disconnection on her part. Therefore, only Suspect B was logged onto System 2 at the time of the crime. The pacct logs on System 2 (C4) show that Suspect B was using Secure Shell (SSH) at the time of the crime. Although the pacct entry does not indicate which system Suspect B was connecting to using SSH, an examination of his command history (C2) shows that he was connecting to System 1. Based on this evidence, it is *probable* that Suspect B is the offender.[1]

[1]*Observe that the use of the word "probable" here corresponds to the C4 level in the certainty scale*

It is difficult to assess the validity of the first conclusion because the examiner does not explicate his thought process. Conversely, the though process leading to the second conclusion is clear and easier to access. For instance, another digital evidence examiner might argue that the wtmp log on System 2 is highly questionable (C1) given the erroneous entry associated with Suspect B's logon and the fact that several individuals, including both suspects, had root access to the machine and could have modified the logs. Similarly, it can be argued that anyone with root access to System 2 could have altered the pacct logs, reducing their C-value to C2. Based on these revised certainty values, it is *possible* (not probable) that Suspect B is the offender but a more reliable source of digital evidence is required to be more certain because any of the (preferably few) people with root access to System 2 could have altered the wtmp, pacct, and command history logs after the crime to implicate Suspect B.

Notably, these certainty values are not simply additive – the circumstances of a case, the questions at issue, and the types of digital evidence involved will determine how much weight each C-value is given and how they are combined. Digital evidence examiners must use their judgment when weighing and combining certainty values.

One major advantage of this Certainty Scale is that it is flexible enough to assess the evidential weight of both the process that generated a piece of digital evidence and its contents, which may be documents or statements. For instance, an e-mail header may be assigned a C-value of C2 in a specific case but the contents may only be assigned a C-value of C1 because there are signs of tampering. In another case, the C-value of an e-mail header may drop to C1 if any inconsistencies or signs of forgery are detected.

Another major advantage of this Certainty Scale is that it is non-technical and therefore easily understood by non-technical people such as those found in most juries. Although it may be necessary at some stage to ask the court to consider the complexities of the systems involved, it is invaluable to give them a general sense of the level of certainty they are dealing with and to help them decide what evidential weight to give the evidence. Only focusing on the complexities, without providing a non-technical overview, can lead to confusion and poor decisions.

Ultimately, it is hoped that this Certainty Scale will point to areas that require additional attention in digital evidence research. Debate over C-values in specific cases may reveal that certain types of evidence are less reliable than was initially assumed. For some types of digital evidence, it may be possible to identify the main sources of error or uncertainty and develop analysis techniques for evaluating or reducing these influences. For other types of digital evidence, it may be possible to identify all potential sources of error or uncertainty and develop a more formal model for calculating the level of certainty for this type of evidence.

7.4 BEST EVIDENCE

When dealing with the contents of a writing, recording, or photograph courts sometimes require the original evidence. This was originally intended to prevent a witness from misrepresenting such materials by simply accepting their testimony regarding the contents. With the advent of photocopiers, scanners, computers, and other technology that can create effectively identical duplicates, copies became acceptable in place of the original, unless "a genuine question is raised as to the authenticity of the original or the accuracy of the copy or under the circumstances it would be unfair to admit the copy in lieu of the original" (Best Evidence Rule).

Because an exact duplicate of most forms of digital evidence can be made, a copy is generally acceptable. In fact, presenting a copy of digital evidence is usually more desirable because it eliminates the risk that the original will bc accidentally altered. Even a paper printout of a digital document may be considered equivalent to the original unless important portions of the original are not visible in printed form. For example, a printed Microsoft Word document does not show all of the data embedded within the original file such as edits and notes.

7.5 DIRECT VERSUS CIRCUMSTANTIAL EVIDENCE

Direct evidence establishes a fact. Circumstantial evidence may suggest one. It is a common misconception that digital evidence cannot be direct evidence because of its separation from the events it represents as discussed in Chapter 1. However, digital evidence can be used to prove facts. For example, if the reliability of a computer system is at issue, showing the proper functioning of that specific system is direct evidence of its reliability, whereas showing the proper functioning of an identical system is circumstantial.

Although digital evidence is generally only suggestive of human activities, circumstantial evidence may be as weighty as direct evidence and digital evidence can be used to firmly establish facts. For example, a computer logon record is direct evidence that a given account was used to log into a system at a given time but is circumstantial evidence that the individual who owns the account was responsible. Someone else may have used the individual's account and other evidence would be required to prove that he actually logged into the system. It may be sufficient to demonstrate that nobody else had access to the individual's computer or password. Alternately, other sources of digital evidence such as building security logs may indicate that the account owner was the only person in the vicinity of the computer at the time of the logon.

Consider intellectual property theft as another example. Even if nobody saw the defendant take the proprietary data, it may be sufficient to show that the data in his possession are the same as the proprietary data and that he had the opportunity for access. So, there is nothing inherently wrong with circumstantial evidence. Given enough circumstantial evidence, the court may not require direct evidence to convict an individual of a crime.

7.6 HEARSAY

Digital evidence might not be admitted if it contains hearsay because the speaker or author of the evidence is not present in court to verify its truthfulness.

> Evidence is hearsay where a statement in court repeats a statement made out of court in order to prove the truth of the content of the out of court statement. Similarly, evidence contained in a document is hearsay if the document is produced to prove that statements made in court are true. The evidence is excluded because the crucial aspect of the evidence, the truth of the out of court statement (oral or documentary), cannot be tested by cross-examination. (Hoey 1996)

For instance, an e-mail message may be used to prove that an individual made certain statements but cannot be used to prove the truth of the statements its contains. Therefore, although Larry Froistad sent a message to an e-mail list indicating that he killed his daughter, investigators needed a confession and other evidence to prove this fact (see Chapter 18 for case details). The Canadian case against Pecciarich provides an interesting example of what may be considered hearsay in the context of online activities.

CASE EXAMPLE (REGINA v PECCIARICH):
Pecciarich was initially charged with one count of distributing obscene pictures and one count of distributing child pornography by using his personal computer to upload files to computer bulletin boards where others could download the files. The bulletin board was examined remotely, only allowing investigators to testify that they had seen many files on the bulletin board that contained the suspect's code name "Recent Zephyr" and had downloaded a few of them.

> Mr Blumberg testified that the graphic, or pictorial files Moppet 1.GIF through Moppet 4.GIF were downloaded by him on September 20, 1993, all exhibiting on screen a printed statement that they were uploaded by Recent Zephyr on dates in August and September, 1993. A sample description of MOPPET 01 was "A Gateway original GIF! Two with girls fully nude and a younger one without panties, and just pulling off the top!" He testified that all remaining files specified in count 2 of the information were seen on either the Gateway or another bulletin board such as "Scruples," and all were identified as having been uploaded by Recent Zephyr on August 3, 1993. Only certain ones were downloaded and stored, due to time and space limitations… Other files purportedly uploaded by Recent Zephyr were seen on many bulletin boards, and sometimes identified as associated with the company names "Yes Software" and "UCP Software."

On appeal the judge overturned the distribution charges stating that, "the statements from the bulletin 'uploaded by Recent Zephyr' accompanied by a date in August or September 1993, are pure hearsay and therefore not evidence of uploading or of the date specified." This decision appears to have been influenced by the description of the bulletin board, leading the court to believe that the data could not be relied upon. In cross-examination, Blumberg acknowledged that even if a subscriber to the bulletin board uploaded the images, the systems operator could alter any data on the system, including removing clothing, "drawing in" body parts including genitalia, and inserting the words "uploaded by Recent Zephyr." Blumberg even acknowledged that an imposter could upload materials onto the bulletin board in the name of another subscriber, using his telephone number without his knowledge; however, in testimony, which was less than crystal clear, Blumberg explained that a system of call back verification may or may not pick up on the false identity of the uploader.

The court upheld the charge of possession despite the defense argument that the evidence used to attribute the documents to Pecciarich was also hearsay.

> Defense counsel argues that proof of authorship is not possible unless the documents are used in violation of the hearsay rule – namely to prove the truth of their message that the creator is "Recent Zephyr." However, rather than for truth, I have used the documents as pieces of original circumstantial evidence that the accused and the name

> "Recent Zephyr" are so frequently linked in a meaningful way as to create the logical inference that they are the same person.

Proving that someone distributed materials online is challenging and generally requires multiple data points that enable the court to connect the dots back to the defendant beyond a reasonable doubt. In Regina v. Pecciarich, although there was only a theoretical possibility of evidence tampering, the judge had little confidence in the digital evidence and believed that the date–time stamps on the bulletin board were hearsay even though the computer probably generated them (technically, hearsay only applies to human statements). The judge may have been skeptical of these date–time stamps because they were observed remotely through the bulletin board interface rather than collected directly from the system's hard drive. More corroborating evidence such as creation and modification times of the relevant files on the bulletin board system's hard drive and telephone records showing when the suspect had accessed the bulletin board may have helped prove distribution to the satisfaction of the court. A list of bulletin board user names with associated addresses and telephone numbers was presented to show that the defendant's telephone number was associated with the Recent Zephyr user name. However, the court determined that could not be used "to show that the accused and Recent Zephyr have the same telephone number and city of residence. Such use would clearly be for the truth of the contents, and thus would violate the hearsay rule." Furthermore, lists of users cannot demonstrate that the defendant had connected to the bulletin board at the times the images in question were uploaded.

7.6.1 HEARSAY EXCEPTIONS

There are several exceptions to the hearsay rule to accommodate evidence that portrays events quite accurately and that is easier to verify than other forms of hearsay. For instance, the US Federal Rules of Evidence specify that records of regularly conducted activity are not excluded by the hearsay rule:

> A memorandum, report, record, or data compilation, in any form, or acts, events, conditions, opinions or diagnoses, made at or near the time by, or from information transmitted by a person with knowledge, if kept in the course of a regularly conducted business activity, and if it was the regular practice of that business activity to make the memorandum, report, record, or data compilation, all as shown by the testimony of the custodian or other qualified witness, unless the source of the information or the method or circumstances of preparation indicate lack of trustworthiness. The term "business" as used in this paragraph includes business, institution, association, profession, occupation, and calling of every kind, whether or not conducted for profit.

The Irish Criminal Evidence Act, 1992, has a similar exception in Section 5(1):

> ... information contained in a document shall be admissible in any criminal proceedings as evidence of any fact therein of which direct oral evidence would be admissible if the information
>
> (a) was compiled in the ordinary course of a business,
>
> (b) was supplied by a person (whether or not he so compiled it and is identifiable) who had, or may reasonably be supposed to have had, personal knowledge of the matters dealt with, and
>
> (c) in the case of information in non-legible form that has been reproduced in permanent legible form, as reproduced in the course of the normal operation of the reproduction system concerned.

Although some courts evaluate all computer-generated data as business records under the hearsay rule, this approach may be inappropriate when a person was not involved. In fact, computer-generated data may not considered hearsay at all because they do not contain human statements or they do not assert a fact but simply document an act. The USDOJ manual (USDOJ 2002) clearly described the difference between digital evidence that is computer-generated versus computer-stored:

> The difference hinges upon whether a person or a machine created the records' contents. Computer-stored records refer to documents that contain the writings of some person or persons and happen to be in electronic form. E-mail messages, word processing files, and Internet chat room messages provide common examples. As with any other testimony or documentary evidence containing human statements, computer-stored records must comply with the hearsay rule ... In contrast, computer-generated records contain the output of computer programs, untouched by human hands. Log-in records from Internet service providers, telephone records, and ATM receipts tend to be computer-generated records. Unlike computer-stored records, computer-generated records do not contain human "statements," but only the output of a computer program designed to process input following a defined algorithm ... The evidentiary issue is no longer whether a human's out-of-court statement was truthful and accurate (a question of hearsay), but instead whether the computer program that generated the record was functioning properly (a question of authenticity).

As an example, in the English case of R. v. Governor of Brixton Prison, *ex parte* Levin ([1997] 3 All E.R. 289) the House of Lords considered whether computer printouts were inadmissible because they were hearsay. In this case Levin was charged with unauthorized access to the computerized fund transfer service of Citibank in New Jersey, USA, and making fraudulent transfers of funds from the bank to accounts that he or his associates controlled.

Lord Hoffman concluded that the printouts were not hearsay:

> The hearsay rule, as formulated in Cross and Tapper on Evidence (8th Ed., 1995) p. 46, states that "an assertion other than one made by a person while giving oral evidence in the proceedings is inadmissible as evidence of any fact asserted." The print-outs are tendered to prove the transfers of funds which they record. They do not assert that such transfers took place. They record the transfers themselves, created by the interaction between whoever purported to request the transfers and the computer programme in [New Jersey]. The evidential status of the print-outs is no different from that of a photocopy of a forged cheque. (p. 239)

However, data that depend on humans for their accuracy, such as entries in a database that are derived from information provided by an individual, are covered under the business record exception if they meet the above description.

More courts are likely to acknowledge the distinction between computer-generated and computer-stored records as they become familiar with digital evidence and as more refined methods for evaluating the reliability of computer-generated data become available, such as the Certainty Scale.

7.7 SCIENTIFIC EVIDENCE

In addition to challenging the admissibility of digital evidence directly, tools and techniques used to process digital evidence have been challenged by evaluating them as scientific evidence. Because of the power of science to persuade, courts are careful to assess the validity of a scientific process before accepting its results. If scientific process is found to be questionable, this may influence the admissibility or weight of the evidence, depending on the situation.

In the United States, scientific evidence is evaluated using four criteria developed in Daubert v. Merrell Dow Pharmaceuticals, Inc., 1993. These criteria are:

1. whether the theory or technique can be (and has been) tested;
2. whether there is a high known or potential rate of error, and the existence and maintenance of standards controlling the technique's operation;
3. whether the theory or technique has been subjected to peer review and publication;
4. Whether the theory or technique enjoys "general acceptance" within the relevant scientific community.

Thus far, digital evidence processing tools and techniques have withstood scrutiny when evaluated as scientific evidence. However, testing techniques or tools and determining error rates is challenging, not just in the digital realm. Although many types of forensic examinations have been evaluated using the criteria set out in Daubert, the testing methods have been weak.

"The issue is not whether a particular approach has been tested, but whether the sort of testing that has taken place could pass muster in a court of science." (Thornton 1997). Also, error rates have not been established for most types of forensic examinations, largely because there are no good mechanisms in place for determining error rates. Fingerprinting, for example, has undergone recent controversy (Specter 2002). Although the underlying concepts are quite reliable, in practice, there is much room for error. Therefore, errors are not simply caused by flaws in underlying theory but also in its application. This problem applies to the digital realm and can be addressed with increased standards and training.

One approach to validating tools is to examine the source code. However, as noted earlier, many commercial developers are unwilling to disclose this information. When the source code is not available, another form of validation is performed – verifying the results by examining evidence using another tool to ensure that the same results are obtained. Formal testing is being performed by the National Institute of Standards and Technology (NIST) and some organizations and individuals perform informal tests. However, given the rate at which computer technology is changing, it is difficult for testers to keep pace and establish error rates for the various tools and systems. Additionally, tool testing does not account for errors introduced by digital investigators through misapplication or misinterpretation. Therefore, the most effective approach to validating results and establishing error rates is through peer review – that is to have another digital investigator double check findings using multiple tools to ensure that the results are reliable and repeatable.

7.8 PRESENTING DIGITAL EVIDENCE

Preparation is one of the most important aspects of testifying in court (National Center for Forensic Science 2003). Scripting direct examination and rehearsing it with the attorney ahead of time provides an opportunity to identify areas that need further explanation and to anticipate questions that the opposition might raise during cross-examination. Conclusions should be stated early in testimony rather than as a punch line at the end because there is a risk that the opportunity will not arise later. During cross-examination, attorneys often attempt to point out flaws and details that were overlooked by the digital investigator. The most effective response to this type of questioning is to be prepared with clear explanations and supporting evidence.

It is advisable to pause before answering questions to give your attorney time to express objections. When objections are raised, carefully consider why the attorney is objecting before answering the question. If prompted to answer a complex question with simply "Yes" or "No," inform the court that you do not feel that you can adequately address the question with such a simplistic answer but follow the direction of the court. Above all, be honest.

In addition to presenting findings, it is necessary to explain how the evidence was handled and analyzed to demonstrate chain of custody and thoroughness of methods. Also, expect to be asked about underlying technical aspects in a relatively non-technical way, such as how files are deleted and recovered and how tools acquire and preserve digital evidence. Simple diagrams depicting these processes are strongly recommended.

It can be difficult to present digital evidence in even the simplest of cases. In direct examination, the attorney usually needs to refer to digital evidence and display it for the trier of fact (e.g. judge, jury). This presentation can become confusing and counterproductive, particularly if materials are voluminous and not well arranged. For instance, referring to printed pages in a binder is difficult for each person in a jury to follow, particularly when it is necessary to flip forwards and backwards to find exhibits and compare items. Such disorder can be reduced by arranging exhibits in a way that facilitates understanding and by projecting data onto a screen to make it visible to everyone in the court.

Displaying digital evidence with the tools used to examine and analyze it can help clarify details and provide context, taking some of the weight of explaining off the examiner. Some examiners place links to exhibits in their final reports, enabling them to display the reports onscreen and efficiently display relevant evidence when required. However, it is important to become familiar with the computer that will be used during the presentation to ensure a smooth testimony. Visual representations of timelines, locations of computers, and other fundamental features of a case also help provide context and clarity. Also, when presenting technical aspects of digital evidence such as how files are recovered or how logon records are generated, first give a simplified, generalized example and then demonstrate how this applies to the evidence in the case.

The risk of confusion increases when multiple computers are involved and it is not completely clear where each piece of evidence originated. Therefore, make every effort to maintain the context of each exhibit, noting which computer or floppy disk it came from and the associated evidence number. Also, when presenting reconstructions of events based on large amounts of data such as server logs or telephone records, provide simplified visual depictions of the main entities and events rather than just presenting the complex data. It should not be necessary to fumble through pages of notes to determine the associated computer or evidence number. Also, refer to exhibit numbers during testimony rather than saying, "this e-mail" or "that print screen."

Digital investigators are often required to provide all notes related to their work and possibly different versions of an edited/corrected report. Therefore, organize any screenshots or printouts (initialed, dated, and numbered) of important items found during examination. For instance, create a neatly written index of all screenshots and printouts.

7.9 SUMMARY

The foundation of any case involving digital evidence is proper evidence handling. Therefore, the practice of seizing, storing, and accessing evidence must be routine to the point of perfection. Standard operating procedures with forms are a key component of consistent evidence handling, acting as both memory aids for digital investigators and documentation of chain of custody. Also, training and policies should provide digital investigators with a clear understanding of acceptable evidence handling practices and associated laws.

Verifying that evidence was handled properly is only the first stage of assessing its reliability. Courts may also consider whether digital evidence was altered before, during, or after collection, and whether the process that generated the evidence is reliable. Claims of tampering generally require some substantiation before they are seriously considered. Someone familiar with the system in question, who can testify that the computer was operating normally at the time, can generally address questions regarding the process that generated a given piece of digital evidence. Digital evidence examiners are encouraged to state clearly their certainty in each piece of digital evidence that they use to reach their conclusions. A proposed Certainty Scale is provided in Table 7.1 for this purpose. If there are significant doubts about the reliability of relevant computer systems and processes, the court may decide to give the associated digital evidence less weight in the final decision.

On the stand, digital investigators may be asked to testify to the reliability of the original evidence, the collection and analysis systems and processes, assert that they personally collected and verified the data, and established the chain of custody. An unexplained break in the chain of custody could be used to exclude evidence. An understanding of direct versus circumstantial evidence, hearsay, and scientific evidence is necessary to develop solid conclusions and to defend those conclusions and the associated evidence on the stand. A failure to understand these concepts can weaken an examiner's conclusions and testimony. For instance, interpreting circumstantial evidence as though it were direct evidence, or basing conclusions on hearsay, could undermine an examiner's findings and credibility.

Ultimately, digital evidence examiners must present their findings in court to a non-technical audience. As with any presentation, the key to success is preparation, preparation, and more preparation. Be familiar with all aspects of the case, anticipate questions, rehearse answers, and prepare visual presentations to address important issues. Although this requires a significant amount of effort, keep in mind that someone's liberty might be at stake.

REFERENCES

Breyer S. (2000) "Reference Manual on Scientific Evidence", 2nd Ed., Federal Judicial Center (Available online at http://www.fjc.gov/public/pdf.nsf/lookup/sciman00.pdf)

Casey E. (2002) "Error, Uncertainty and Loss in Digital Evidence", International Journal of Digital Evidence, Volume 1, Issue 2, 2002 (Available online at http://www.ijde.org/archives/docs/02_summer_art1.pdf)

Gahtan A. (1999) "Electronic Evidence", Ontario: Carswell Legal Publications

Guidance Software (2001–2002) "EnCase Legal Journal" 2nd Ed. (Available online at http://www.guidancesoftware.com/support/downloads/LegalJournal.pdf)

Hoey A. (1996) "Analysis of The Police and Criminal Evidence Act, s.69 – Computer Generated Evidence", Web Journal of Current Legal Issues, in association with Blackstone Press Ltd.

Law Commission (1997) Evidence in Criminal Proceedings: Hearsay and Related Topics, Law Commission Report 245 (Available online at http://www.lawcom.gov.uk/231.htm#lcr245)

Mattei M., Blawie J. F. and Russell A. (2000) "Connecticut Law Enforcement Guidelines for Computer Systems and Data Search and Seizure", State of Connecticut Department of Public Safety and Division of Criminal Justice

National Center for Forensic Science (2003) "Digital Evidence in the Courtroom: A Guide for Preparing Digital Evidence for Courtroom Presentation", Mater Draft Document, U.S. Department of Justice, National Institute of Justice, Washington (Available online at http://www.ncfs.org/DE_courtroomdraft.pdf)

Reed C. (1990–91) 2 CLSR 13–16 as quoted in Sommer, P. "Downloads, Logs and Captures: Evidence from Cyberspace Journal of Financial Crime", October, 1997, 5JFC2 138–152

Specter M. (2002) "Do Fingerprints Lie?: The gold standard of forensic evidence is now being challenged", The New Yorker Issue of 2002-05-27 (Available online at http://www.newyorker.com/printable/?fact/020527fa_FACT)

Thornton J. I. (1997) "The General Assumptions and Rationale of Forensic Identification," for David L. Faigman, David H. Kaye, Michael J. Saks, & Joseph Sanders, Editors, Modern Scientific Evidence: The Law and Science of Expert Testimony, Volume 2, St. Paul, MN: West Publishing Company

United States Department of Justice (2002) "Searching and Seizing Computers and Obtaining Electronic Evidence in Criminal Investigations" (Available online at http://www.usdoj.gov/criminal/cybercrime/s&smanual2002.htm)

CASES

American Oil Co. v. Valenti (1979) 179 Connecticut 349, 358, 426 A.2d 305

Bean M. (2003) "Mich. v. Miller: Sex, lies and murder", Court TV (Available online at http://www.courttv.com/trials/taped/miller/background.html)

Daubert v. Merrell Dow Pharmaceuticals, Inc. (1993) 509 U.S. 579, 113 S.Ct. 2786, 125 L.Ed.2d 469.

Michigan v. Miller (2001) 7th Circuit Court, Michigan

Missouri v. Dunn (1999) Appeals Court, Western District of Missouri, Case Number 56028 (Available online at http://www.missourilawyersweekly.com/mocoa/56028.htm)

People v. Lugashi, (1988) Appeals Court, California (205 Cal.App.3d 632) Case Number B025012

R. v. Cochrane (1993) Crim. L. R. 48

R. v. Governor of Brixton Prison, *ex parte* Levin (1997) 3 All E. R. 289

R. v. Shephard (1993) 1 All E. R. 225

Regina v. Pecciarich (1995) 22 O.R. (3d) 748, Ontario Court, Canada (Available online at http://www.efc.ca/pages/law/court/R. v.Pecciarich.html)

United States v. Gray (1999) District Court, Eastern District of Virginia, Alexandria Division, Case Number 99-326-A

United States v. Miller (1985) 771 F.2d 1219, 1237 (9th Cir.)

United States v. Tank (1998) Appeals Court, 9th Circuit, Case Number 98-10001 (Available online at http://laws.findlaw.com/9th/9810001.html)

United States v. Turner (1999) Appeals Court, 1st Circuit, Case Number 98-1258 (Available online at http://laws.lp.findlaw.com/1st/981258.html)

Wisconsin v Schroeder (1999) Appeals Court, Wisconsin, Case Number 99-1292-CR (Available online at http://www.courts.state.wi.us/html/ca/99/99-2264.HTM)

PART 2

COMPUTERS

## CHAPTER 8

COMPUTER BASICS FOR DIGITAL INVESTIGATORS

Although digital investigators can use sophisticated software to recover deleted files and perform advanced analysis of computer hard disks, it is important to understand what is happening behind the scenes. A lack of understanding of how computers function and the processes that sophisticated tools have automated make it more difficult for digital investigators to explain their findings in court and can lead to incorrect interpretations of digital evidence. For instance, when recovering deleted directories, there is a chance that two deleted directories occupied the same space at different times. Additionally, every tool has its limitations that a competent digital investigator should recognize and address. For instance, an automated tool may only be able partially to recover a deleted file – a digital evidence examiner may be able to locate the remainder of the file.

This chapter provides an overview of how computers developed, how they operate, and how they store data. This basic information is necessary to understand how digital evidence is collected from computers and how deleted data can be recovered and examined.

8.1 A BRIEF HISTORY OF COMPUTERS

The development of the modern computer is not an easy one to trace because of the many concepts that it combines. In the early 1800s, Jacquard developed ideas of Falcon and Vaucanson (who may have been influenced by second century Chinese looms) to create an automated loom that used sequences of wooden/cardboard cards punched with holes to create specific patterns in the woven fabric, resembling punch cards used to program computers in the twentieth century. Less than a decade later, Babbage conceived of a steam powered "difference engine" that could perform arithmetic operations and some consider him to be the father of the computer. Later in the 1800s Augusta Ada suggested a binary system rather than decimal and George Boole developed Boolean logic.

Digital Evidence and Computer Crime Second Edition
ISBN: 0-12-163104-4

Even the more recent developments of the computer are contested. From 1940 onwards, George Stiblitz of the Bell Atlantic Laboratories developed several computing machines including The Model 5 and demonstrated one simple relay computing machine (not completely electronic) using a remote terminal in Dartmouth connected via modified telephone lines to the main computer in New York City. Then, in 1941, a German engineer named Konrad Zuse apparently created an electronic binary computer called the Z3 that used old movie film to store his programs and data.

At around the same time the electronic digital Atanasoff–Berry Computer (ABC), named after its inventors, was built with vacuum tubes, capacitors, and punch cards (Figure 8.1). Shortly after, the Electronic Numerical Integrator and Computer (ENIAC) was created by Eckert and Mauchly but the patent was later voided as a derivative of the ABC (Honeywell v. Rand 1973).

> ENIAC was comprised of thousands of electric vacuum tubes, filled a 30 by 50 foot room, generated vast quantities of heat, weighed 30 tons, and possessed less computing power than today's basic hand-held calculator. It was a second technological breakthrough, however, that insured the future viability of the electronic computer; namely, the invention of the solid-state transistor one year later in 1947. (Hollinger 1997)

Many others played a role in the development of the modern computer and there have been revolutionary developments in computer technology since the 700-pound ABC and 30-ton ENIAC that have made the most significant impact on crime and digital evidence. In particular, personal computers enable individuals to own and command a powerful machine that only a nation could afford 50 years ago. The mass availability of

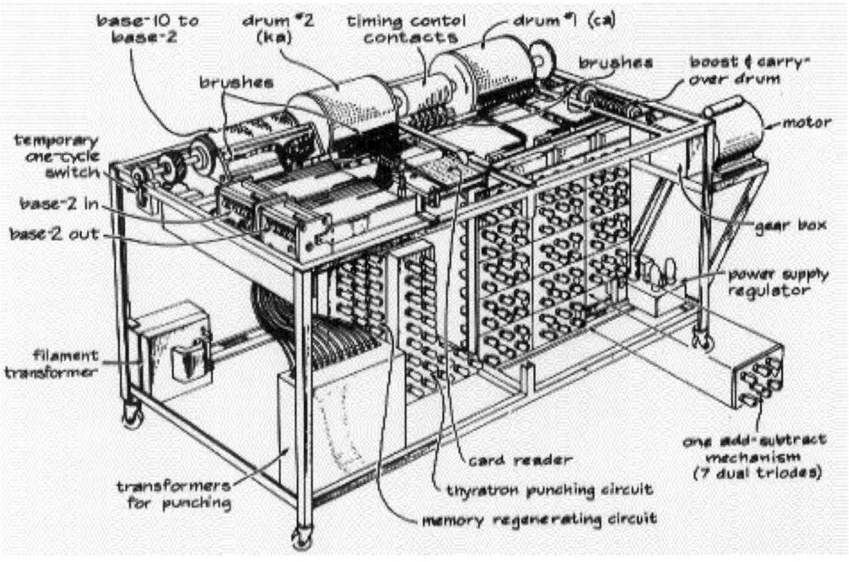

Figure 8.1

Diagram of the Atanasoff–Berry Computer (ABC). Image from http://www.scl.ameslab.gov/ABC/Progress.html (reproduced with permission).

computers has caused significant changes in the way that criminals operate and evidence is conceived of – and the courts are still grappling with these changes.

The personal computer became possible in 1974 when a small company named Intel started selling inexpensive computer chips called 8080 microprocessors. A single 8080 microprocessor contained all of the electronic circuits necessary to create a programmable computer. Almost immediately, a few primitive computers were developed using this microprocessor. By the early 1980s, Steve Jobs and Steve Wozniak were mass marketing Apple computers and Bill Gates was working with IBM to mass market IBM personal computers. In England, the Acorn and the Sinclair computers were being sold. The Sinclair, a small keyboard that plugged into a standard television and audio cassette player for memory storage, was revolutionary in 1985. By supplanting expensive, centralized mainframes, these small, inexpensive computers made Bill Gates's dream of putting a computer in every home a distinct possibility. Additionally, the spread of these computers around the world made a global network of computers the next logical step.

8.2 BASIC OPERATION OF COMPUTERS

Each time a computer is turned on, it must familiarize itself with its internal components and the peripheral world. This start-up process is called the *boot process*, because it is as if a computer has to pull itself up by its bootstraps. The boot process has three basic stages: the Central Processing Unit (CPU) reset, the Power-On Self Test (POST), and the disk boot.

8.2.1 CENTRAL PROCESSING UNIT

The CPU is the core of any computer. Everything depends on the CPU's ability to process instructions that it receives. So, the first stage in the boot process is to get the CPU started – *reset* – with an electrical pulse. This pulse is usually generated when the power switch or button is activated but can also be initiated over a network on some systems. Once the CPU is reset it starts the computer's basic input and output system (BIOS) (Figure 8.2).

8.2.2 BASIC INPUT AND OUTPUT SYSTEM

The BIOS deals with the basic movement of data around the computer. Every program run on a computer uses the BIOS to communicate with the

Figure 8.2

An electrical pulse resets the CPU, which, in turn, activates the BIOS.

CPU. Some BIOS programs allow an individual to set a password and then until the password is typed in the BIOS will not run and the computer will not function.

8.2.3 POWER-ON SELF TEST AND CMOS CONFIGURATION TOOL

The BIOS contains a program called the Power-On Self Test (POST) that tests the fundamental components of the computer. When the CPU first activates the BIOS, the POST program is initiated. To be safe, the first test verifies the integrity of the CPU and POST program itself. The rest of the POST verifies that all of the computer's components are functioning properly, including the disk drives, monitor, RAM, and keyboard. Notably, after the BIOS is activated and before the POST is complete, there is an opportunity to interrupt the boot process and have it perform specific actions. For instance, Intel-based computers allow the user to open the Complementary Metal Oxide Silicon (CMOS) configuration tool at this stage. Computers use CMOS RAM chips to retain the date, time, hard drive parameters, and other configuration details while the computer's main power is off. A small battery powers the CMOS chip – older computers may not boot even when the main power is turned on because this CMOS battery is depleted, causing the computer to "forget" its hardware settings.

Preview (Chapter 9): BIOS passwords can present a significant barrier when digital investigators need to boot a computer from a floppy disk to collect evidence from a computer. In many cases, it is possible to circumvent the password by resetting the CMOS or having a data recovery expert manually control the read/write heads to overwrite the password. However, these processes can alter the system settings significantly and cause more problems than they solve and should only be used as a last resort. Therefore, when prompted for a BIOS password, try to obtain the password from the user along with all other passwords for the system and its contents. Alternatively, remove the hard drive from the computer and copy it using an evidence collection system as described in later chapters. Some systems, such as IBM ThinkPads, associate the hard drive, motherboard, and BIOS in a way that makes it very difficult to get around the BIOS password. Again, the easiest way to deal with this type of situation is to obtain the password from the user but there are some organizations such as Nortek (www.nortek.on.ca/nortek) that can physically manipulate the drive to overwrite the BIOS passwords.

Using the CMOS configuration tool, it is possible to determine the system time, ascertain if the computer will try to find an operating system on the primary hard drive or another disk first, and change basic computer settings as needed. When collecting digital evidence from a computer, it is often necessary to interrupt the boot process and examine CMOS setting such as the system date and time, the configuration of hard drives, and the boot sequence. In some instances it may be necessary to change the CMOS settings to ensure that the computer will boot from a floppy diskette rather than the evidentiary hard drive (see Section 8.2.4).

CASE EXAMPLE (UNITED STATES v. ZACARIAS MOUSSAOUI 2003):
During the trial of convicted terrorist Zacarias Moussaoui, a question arose regarding the original CMOS settings of his laptop. The laptop had lost all power by the time the government examined its contents, making it more difficult to authenticate the associated digital evidence.

> The loss of all power means that the original date and time settings cannot be retrieved, and that other settings, such as how the computer performed its boot sequence, the types of ports and peripherals enabled, and the settings regarding the hard disk and the controller, are all lost as well. All of this is essential information on how the laptop was set up. (United States v. Moussaoui 2003)

Fortunately, the CMOS settings were recorded when the laptop was originally processed by a Secret Service Agent on September 11, 2001 before the power was lost.

In many computers, the results of the POST are checked against a permanent record stored in the CMOS microchip. If there is a problem at any stage in the POST, the computer will emit a series of beeps and possibly an error message on the screen. The computer manual should explain the beep combinations for various errors. When all of the hardware tests are complete, the BIOS instructs the CPU to look for a disk containing an operating system.

Sun and Macintosh computers follow slightly different boot sequences and terminology. For instance, newer Macintosh computers call the CMOS chip Parameter RAM (PRAM). After the POST, a program called Open Firmware (similar to the PC-BIOS) initializes and attempts to locate attached hardware. Open Firmware then performs a sequence of operations to load the Macintosh operating system (Mac OS). Sun systems have an initial low-level POST that tests the most basic functions of the hardware. After Sun machines perform this initial POST, they send control to the OpenBoot PROM (OBP) firmware (similar to the PC-BIOS) and perform additional system tests and initialization tasks.

8.2.4 DISK BOOT

An operating system extends the functions of the BIOS, and acts as an interface between a computer and the outside world. Without an operating system it would be very difficult to interact with the computer – basic commands would be unavailable, data would not be arranged in files and folders, and software would not run on the machine.

Most computers expect an operating system to be provided on a floppy diskette, hard disk, or compact disk. So, when the computer is ready to load an operating system, it looks on these disks in the order specified by the boot sequence setting mentioned in the previous section. The computer loads the first operating system it finds. This fact allows anyone to preempt a computer's primary operating system by providing an alternate operating system on another disk. For instance, a floppy diskette containing an operating system can be inserted into an Intel-based computer to prevent the operating system on the hard disk from loading. The Macintosh Open Firmware can be instructed to boot from a CD-ROM by holding down the "c" key. The Sun OBP can be interrupted by depressing the "Stop" and "A" keys simultaneously and the boot device can be specified at the `ok` prompt (e.g. `boot cdrom`).

This ability to prevent a computer from using the operating system on the hard disk is important when the disk contains evidence. For instance, in one case a technician was asked to note system time of a Macintosh iBook before removing its hard drive. He booted the system and tried to interrupt the boot process to access the CMOS, not realizing that this feature does not exist on Macintosh. As a result, the system booted from the evidentiary hard drive, altering date–time stamps of files and other potentially useful data on the disk.

8.3 REPRESENTATION OF DATA

All digital data are basically combinations of ones and zeros, commonly called *bits*. It is often necessary for digital investigators to deal with data at the bit level, requiring an understanding of how different systems represent data. For instance, the number 511 is represented as 00000001 11111111 on *big-endian* systems (e.g. computers with Motorola processors such as Macintosh; RISC-based computers such as Sun). The same number is represented as 11111111 00000001 on *little-endian* systems such as Intel-based computers. In other words, big-endian architectures place the most significant bytes on the left (putting the big end first) whereas little-endian architectures place the most significant bytes on the right (putting the little end first).[1]

[1]The terms big-endian and little-endian are based on the story in Gulliver's Travels, in which the Lilliputians' main political conflict was whether soft-boiled eggs should be opened on the big end or the little end.

Whether little- or big-endian, this binary representation of data (ones and zeros) is cumbersome. Instead, digital investigators often view the hexadecimal representation of data. Another commonly used representation of data is ASCII. The ASCII standard specifies that certain combinations of ones and zeros represent certain letters and numbers. Table 8.1 shows the ASCII and hexadecimal values of capital letters.

Table 8.1

ASCII and hexadecimal values of some capital case letters.

Letter	Hexadecimal	ASCII
A	41	65
B	42	66
C	43	67
D	44	68
E	45	69
F	46	70
G	47	71
H	48	72
I	49	73
J	4A	74
K	4B	75
L	4C	76
M	4D	77
N	4E	78
O	4F	79
P	50	80
Q	51	81
Y	59	89
Z	5A	90

Conceptually, programs that display each byte of data in hexadecimal and ASCII format are like microscopes, allowing digital investigators to view features that are normally invisible. For instance, Word documents contain data that are not generally visible but can be displayed using a program like WinHex[2] as shown in Table 8.2 with hexadecimal on the left and ASCII on the right.

Table 8.2

Segment of a Word document shown in hexadecimal and ASCII format.

Hexadecimal	ASCII
1e 10 00 00 01 00 00 00 0a 00 00 00 43 68 61 70	............Chap
74 65 72 20 38 00 0c 10 00 00 02 00 00 00 1e 00	ter 8...........
00 00 06 00 00 00 54 69 74 6c 65 00 03 00 00 00	......Title.....
01 00 00 00 98 00 00 00 03 00 00 00 00 00 00 00	....
20 00 00 00 01 00 00 00 36 00 00 00 02 00 00 00	.......6.......
3e 00 00 00 01 00 00 00 02 00 00 00 0a 00 00 00	>...............
5f 50 49 44 5f 47 55 49 44 00 02 00 00 00 10 27	_PID_GUID......'
00 00 41 00 00 00 4e 00 00 00 7b 00 30 00 43 00	..A...N...{.0.C.
33 00 37 00 34 00 46 00 30 00 30 00 2d 00 42 00	3.7.4.F.0.0.-.B.
37 00 30 00 30 00 2d 00 31 00 31 00 44 00 32 00	7.0.0.-.1.1.D.2.
2d 00 38 00 46 00 43 00 46 00 2d 00 39 00 35 00	-.8.F.C.F.-.9.5.
46 00 39 00 43 00 38 00 34 00 37 00 41 00 31 00	F.9.C.8.4.7.A.1.
33 00 30 00 7d 00 00 00 00 00 00 00 00 00 00 00	3.0.}...........

The difference between little- and big-endian representations is most apparent when converting data from their computer representation into a more readable form. For instance, Table 8.3 shows the first two lines of a tcpdump file created on an Intel-based computer (left) compared with a tcpdump file created at the same time on a Sun computer (right). As discussed in Chapter 11, UNIX represents the date "Sat, 10 May 2003 08:37:01 GMT" using the sequence of bytes shown in Table 8.3 in bold – the different byte order on both systems is clearly visible.

Table 8.3

Viewing two tcpdump files created on Intel-based and Sun systems shows the difference between little- and big-endian representations of the same UNIX date (in bold).

Linux on Intel (little-endian)	Solaris on Sun (big-endian)
D4C3B2A1 02000400 00000000 00000000	A1B2C3D4 00020004 00000000 00000000
60000000 01000000 **2DBABC3E** 46C30500	00000044 00000001 **3EBCBA2D** 0004BFF0

[2]http://www.winhex.com

An awareness of byte order is also required when searching through digital evidence for specific combinations of bytes.

8.4 STORAGE MEDIA AND DATA HIDING

> [On binary systems] each data element is implemented using some physical device that can be in one of two stable states: in a memory chip, for example, a transistor switch may be on or off; in a communications line, a pulse may be present or absent at a particular place and at a particular time; on a magnetic disk, a magnetic domain may be magnetized to one polarity or to the other; and, on a compact disk, a pit may be present or not at a particular place. (Sammes and Jenkinson 2000)

Although storage media come in many forms, hard disks are the richest sources of digital evidence on computers. Even modern photocopy machines have hard drives and can be augmented by connecting external controllers with a CPU, RAM, and high capacity hard drives to accommodate more complex printing more quickly. Understanding how hard drives function, how data are stored on them, and where data can be hidden can help digital investigators deal with hard drives as a source of evidence.

There are several common hard drive technologies. Integrated Disk Electronics (IDE) drives – also called Advanced Technology Attachment (ATA) drives – are simpler, less expensive, and therefore more common than higher performance SCSI drives. Firewire is an adaptation of the SCSI standard that provides high-speed access to a chain of devices without many of the disadvantages of SCSI such as instability and expense. Regardless of which technology is used, all hard drives contain spinning platters made of a light, rigid material such as aluminum, ceramic, or glass. These platters have a magnetic coating on both sides and spin between a pair of read/write heads – one head on each side of a platter. These heads, moving over a platter like the needle of a record player but floating above the surface of a spinning platter on a cushion of air created by the rotation of the disk, can align particles in the magnetic media (called writing) and conversely, can detect how the particles on the platter are aligned (called reading). Particles aligned one way signify a binary one (1) and particles aligned the other way signify a binary zero (0) as shown in Figure 8.3.

Data are recorded on a platter in concentric circles (like the annual rings of a tree trunk) called *tracks*. The term *cylinder* is effectively synonymous with track, collectively referring to tracks with the same radius on all platters in a hard drive. Each track is further broken down into *sectors*, usually big enough to contain 512 bytes of information (512 × 8 ones and zeros).[3] Many file systems use two or more sectors, called a *cluster*, as their basic storage unit of a disk. For instance, Figure 8.4 shows a disk with 64 sectors per cluster, resulting in 32 kbytes per cluster (64 sectors × 512 bytes/sector ÷ 1024 bytes).

[3]Sectors are actually 557 bytes but only 512 bytes are used to store data. The additional space is used for low-level encoding data. A discussion of the low-level encoding schemes on magnetic media such as Frequency Modulation (FM), Modified Frequency Modulation (MFM), Run Length Limited (RLL), and Advanced Run Length Limited (ARLL) encoding methods is available in (Sammes and Jenkinson 2000).

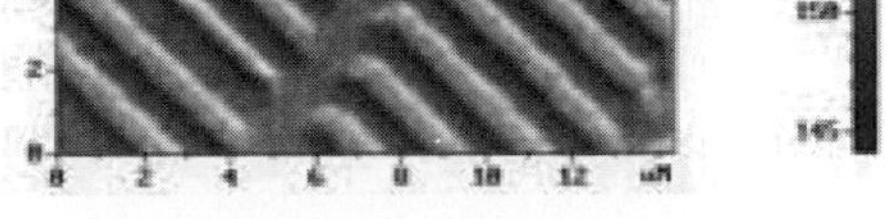

Figure 8.3

Magnetic patterns on a hard disk as seen through a magnetic force microscope. Peaks indicate a one (1) and troughs signify a zero (0). Image from http://www.ntmdt.ru/applicationnotes/MFM/ (reproduced with permission).

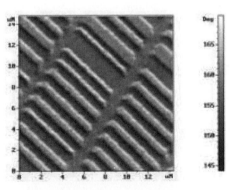

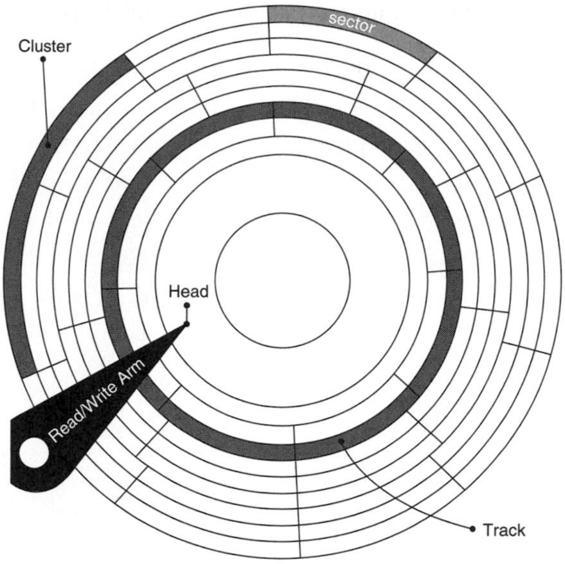

Figure 8.4
A depiction of platters, tracks, sectors, clusters, and heads on a computer disk.

As shown in Figure 8.4, the location data on a disk can be determined by which cylinder they are on, which head can access them, and which sector contains them; this is called CHS addressing. Therefore, the capacity of a hard disk can be calculated by multiplying the number of cylinders, heads, and sectors by 512 bytes. The numbers of cylinders, heads, and sectors per track are often printed on the outside of the hard drive and the calculated capacity ($C \times H \times S \times 512$ bytes) can be compared with the amount of data extracted from a hard drive to ensure that all evidence has been obtained. For instance, a hard drive with 1024 cylinders, 256 heads, and 63 sectors contains 8455716864 bytes ($1024 \times 256 \times 63 \times 512$ bytes). This equates to 8.4 Gbytes (8455716864 bytes ÷ 1024 bytes ÷ 1024 bytes) where 1 Gbyte can contain about one billion characters.

There are a few nuances to hard drives that enable a wily individual to conceal the presence of large amounts of data on them. The first cylinder on a disk (a.k.a. the maintenance track) is used to store information about the drive such as its geometry and the location of bad sectors. By intentionally marking portions of the disk as bad, an individual can conceal data in these areas from the operating system. The evidence collection tools described in this text are not fooled by this technique and some utilities such as Anadisk[4] can copy the maintenance track of a floppy disk. Another potential area for data hiding is the Protected Area on post 1998-ATA disks. As the name suggests, most programs cannot access this area but tools such as BXDR[5] have been developed to detect and copy this area.

[4] http://www.forensics-intl.com/anadisk.html

[5] http://www.sandersonforensics.co.uk/BXDR.htm

8.5 FILE SYSTEMS AND LOCATION OF DATA

File systems such as FAT16, FAT32, NTFS, HFS (Macintosh Hierarchical Filesystem), HFS+, Ext2 (Linux), and UFS (Solaris) keep track of where data are located on a disk, providing the familiar file and folder structure. Before a file system can be created, a partition must be created to specify how much of the hard drive it will occupy. The first sector of a hard disk contains the Master Boot Record (MBR) containing a partition table to tell the operating system how the disk is divided. Figure 8.5 shows the general structure of a disk with two partitions.

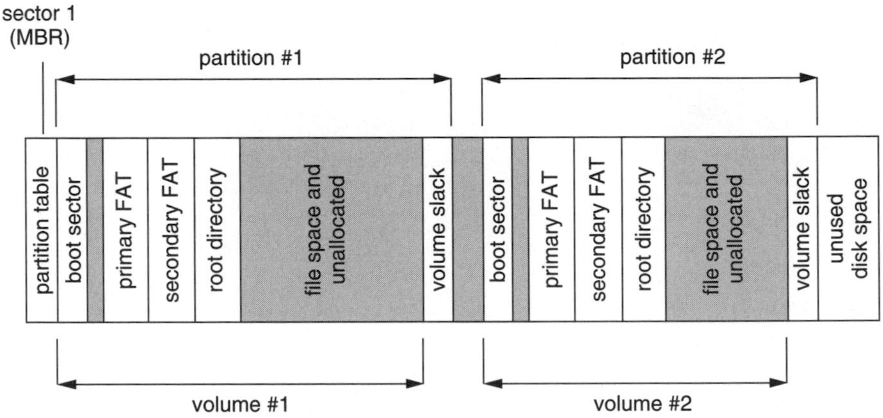

Figure 8.5

Simplified depiction of disk structure with two partitions, each containing a FAT formatted volume.

The partition table specifies the first and last sectors in each partition, as well as additional information about the partition. The simplest example of creating or viewing a partition is using the fdisk command. The following example shows output from the Linux fdisk command run on a Dell computer with two hard drives – one hard drive has a small partition for recovery purposes and a larger partition containing an NTFS file system (Windows NT/2000/XP), and the other hard drive has several partitions containing an ext2 file system (Linux).

```
# /sbin/fdisk -1
Disk /dev/hdc: 255 heads, 63 sectors, 9726 cylinders
Units = cylinders of 16065 * 512 bytes
  Device    Boot    Start      End      Blocks      Id       System
/dev/hdc1               1        4      32098+      de       Dell Utility
/dev/hdc2     *         5     9725   78083932+       7        HPFS/NTFS
Disk /dev/hdd: 255 heads, 63 sectors, 7476 cylinders
Units = cylinders of 16065 * 512 bytes
  Device    Boot    Start      End      Blocks      Id       System
/dev/hdd1     *         1        6      48163+      83       Linux
/dev/hdd2               7     7346   58958550       83       Linux
/dev/hdd3            7347     7476    1044225       82       Linux swap
```

A failure to realize that this system has two hard drives could result in lost digital evidence.

As another example, the following output from the Windows **fdisk** command shows a hard drive with one *primary* partition and an *extended* partition that is subdivided into four smaller partitions. The use of extended partitions is necessary because the partition table only has room for four primary partitions – an extended partition can be subdivided into additional partitions without entries in the partition table.

```
                    Display Partition Information
Current fixed disk drive: 2
Partition    Status   Type      Volume Label    Mbytes  System   Usage
  D: 1         A      PRI DOS   MELPOMENE        4910   FAT32     25%
     2                EXT DOS                   14614             75%
Total disk space is 19532 Mbytes (1 Mbyte = 1048576 bytes)
The Extended DOS Partition contains Logical DOS Drives.
Do you want to display the logical drive information (Y/N)......?[Y]
    Display Logical DOS Drive Information
Drv Volume Label    Mbytes   System     Usage
E:  CLIO             4871    FAT32       33%
F:  ERATO            4903    FAT32       34%
G:  TERPSICHORE      4840    FAT32       33%
  Total Extended DOS Partition size is 14614 Mbytes (1 MByte = 1048576 bytes)
```

Once a partition has been created it can be formatted with any file system. For instance, a FAT file system can be created using the format command on Windows. The area occupied by the file system is called a *volume*, which is assigned a letter such as C: by the operating system. Contrary to popular belief, the format command does not erase data from the volume – it is possible to recover data from a hard drive after it has been formatted.[6] Comparing volumes to bookcases in a library, file systems are analogous to library catalogs, providing an efficient way to locate a particular item. Formatting a volume is like destroying the card catalog in a library but leaving the books on the shelves. It is still possible to find a particular book but it takes more time.

[6]This does not apply to low-level formatting. The format command can perform a low-level format on floppy diskettes prior to creating a file system, thus destroying all information on the floppy. To low-level format a hard drive it is necessary to obtain a special program from the vendor. For example, IBM provides the Drive Fitness Test utility (www.storage.ibm.com) to help individual maintain disks in IBM systems.

The first sector on each volume, called the *boot sector* (a.k.a. boot record or boot block), contains important file system information. For instance, Figure 8.6 shows the boot sector of a Windows 95 machine. It shows that two (2) copies of the file allocation table (FAT) are available – this table is the equivalent of the library card catalog and a backup copy is maintained in case the primary one is damaged or destroyed. This figure also shows that each cluster on the disk is quite large (64 sectors/cluster $\times$ 512 bytes/sector $=$ 32 kbytes).

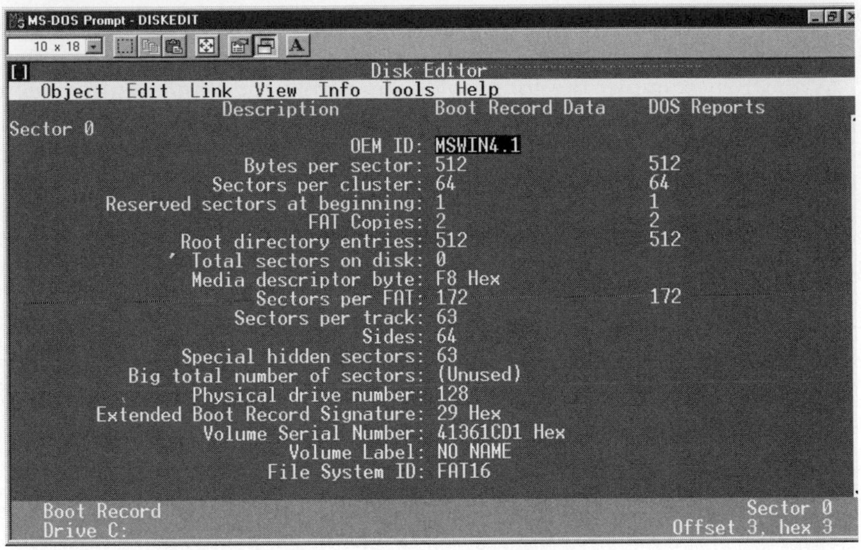

Figure 8.6

Windows 95 boot sector viewed using Norton Diskedit.

Be aware that a file system may not use an entire partition, leaving space between the end of the volume and the end of the partition, an area called *volume slack* that can be used to hide data. Figure 8.7 shows remnants of the Form virus stored in volume slack.

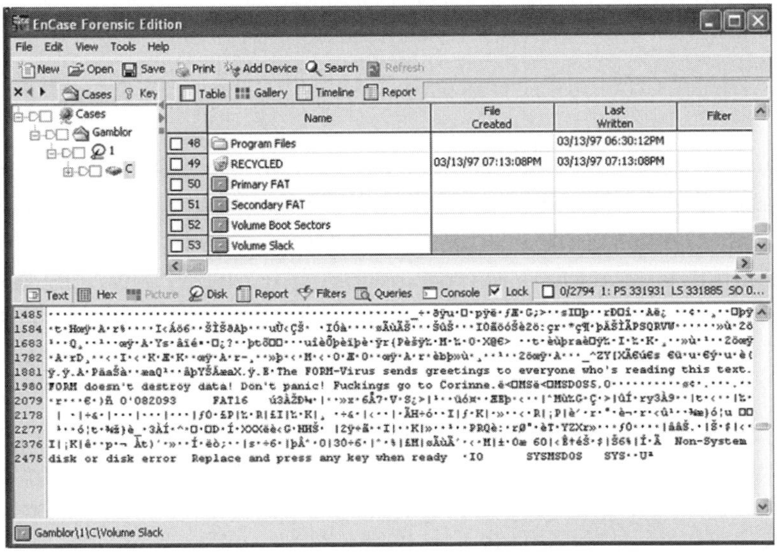

Figure 8.7

Volume slack containing remnants of Form virus viewed using EnCase.

Also be aware that partitions typically start at the beginning of a cylinder resulting in unused space between the end of one partition and the beginning of the next.

There are several features of file systems that are useful from a data recovery standpoint. When a file takes up less than one cluster, other files will not use the additional space in that cluster. In short, once a cluster contains data, the entire cluster is reserved. This is similar to the situation in most restaurants. If three people are sitting at a table that seats four, the additional seat remains empty until the three people have finished using the table. The idea is that a fourth stranger might interfere with these three people's meal. Similarly, if a computer tried to squeeze extra data into the unused part of a cluster, the new data might interfere with the old. The extra sectors in a cluster are called *file slack* space. When a file does not end on a sector boundary, operating systems prior to Windows 95a fill the rest of the sector with data from RAM, giving it the name *RAM slack*. Later versions of Windows fill this space with zeros.

When a file is deleted, its entry in the file system is updated to indicate its deleted status and the clusters that were previously allocated to storing are *unallocated* and can be reused to store a new file. However, the data are left on the disk and it is often possible to retrieve a file immediately after it has been deleted. The data will remain on the disk until a new file overwrites them (Figure 8.8). However, if the new file does not take up the entire cluster, a portion of the old file might remain in the slack space. In this case, a portion of a file can be retrieved long after it has been deleted and partially overwritten. The process of recovering deleted or partially overwritten data from a disk is described in later chapters.

Having large clusters such as those in Figure 8.6 results in large amounts of slack space. More modern file systems are designed to limit slack space because it is wasted from a file system viewpoint.

Notably, not all storage devices have file systems. For instance, data can be written to backup tapes in a simple way that does not require a file system. This approach maximizes the amount of space used for data storage and minimizes the amount used for data organization. Also, on UNIX machines, swap partitions do not have file systems. A swap partition or file acts as virtual memory, enabling a computer to run more processes than can fit within a computer's physical memory (RAM). This illusion of extra memory is achieved by either swapping or paging data into and out of RAM as required. Swapping replaces a complete process with another in memory whereas paging removes a "page" (usually 2–4 kbytes) of a process and replaces it with a page from another process.

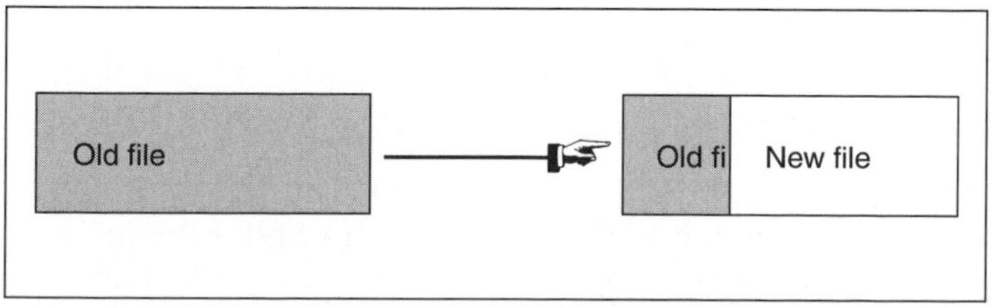

Figure 8.8

When old data are overwritten with new data, some of the old data can remain.

8.6 OVERVIEW OF ENCRYPTION

Encryption is a process by which a readable digital object (plaintext) is converted into an unreadable digital object (ciphertext) using a mathematical function. Strong encryption schemes use the equivalent of a password, called a *key*. However, there are simple, keyless encoding systems. For instance, ROT13 is a simple code that substitutes each letter in the plaintext message with the letter that is 13 letters further along in the alphabet (A is followed by Z). So, a becomes n, b becomes o, and so on.

ROT13 is sometimes used in public discussion forums on the Internet to obfuscate potentially objectionable messages, allowing the reader to decide whether to decrypt the message. The following Usenet message demonstrates this application of ROT13.

> From: AndrewB (andrewbee@my_deja.com)
>
> Subject: Sexual differences [view thread]
>
> Newsgroups: soc.religion.christian
>
> Date: 2000-10-02 20:58:37 PST
>
> [This posting asks advice on a sexually explicit topic. My first reaction is that it's a troll, but perhaps I'm just narrow-minded. To avoid offending people, the body of the posting has been translated using rot13.]
>
> Uv,
>
> Sbetvir zr sbe orvat irel senax urer. Zl jvsr naq V unir n ceboyrz. V nz cerggl "bhg gurer" jura vg pbzrf gb zl frkhny cersreraprf. Zl jvsr, ubjrire, vf n irel pbafreingvir fznyy-gbja tvey jura vg pbzrf gb gung. Fcrpvsvpnyyl, V nz irel ghearq ba zl fcnaxvat, jurernf zl jvsr frrf ab cynpr sbe vg ng nyy va gur orqebbz.
>
> V xabj gung gurer ner thlf nebhaq jub tvir cevingr fcnaxvatf. Ab frkhny pbagnpg; whfg gur tengvsvpngvba vaurerag gurerva. Vs V pbhyq qb guvf, vg jbhyq zrrg zl arrq, naq rnfr zl sehfgengvba. Bayl ceboyrzf ner: zl jvsr rdhngrf vg gb purngvat, ba gur tebhaqf gung vg vaibyirf obqvyl pbagnpg sbe frkhny tengvsvpngvba, naq V unir qbhogf nobhg jurgure vg'f ernyyl BX sbe n Tbq-srnevat Puevfgvna gb qb gung. V jnag gb yvir va chevgl orsber Tbq, ohg V nyfb unir guvf fgebat hetr naq qrfver. Nalbar unq nal fvzvyne rkcrevraprf be pna bssre nal uryc?
>
> Gunaxf
>
> N.O.
>
> Frag ivn Qrwn.pbz uggc://jjj.qrwn.pbz/
>
> Orsber lbh ohl.

Many newsreaders have the ability to decrypt ROT13, saving individuals from the tedious, manual process. Windows 2000/XP use similar encoding schemes to obfuscate values in some Registry keys as discussed in Chapter 10.

8.6.1 PRIVATE KEY ENCRYPTION

Private key encryption (a.k.a. symmetric key encryption) is conceptually straightforward – the key that is used to encrypt a message is also used to decrypt it. This encryption method is analogous to a lock in the physical world but, in actuality, the lock is a mathematical function. Since it is not safe to rely on the secrecy of the mathematical function used to encrypt the data, most popular encryption schemes utilize mathematical functions that are difficult to reverse. In this way, even if the mathematical function is known, it is difficult to decrypt data without knowing the key. Some commonly used symmetric key encryption algorithms are DES, IDEA, and Blowfish. For example, taking the text "This is a secret message" and encrypting it using the key "eoghan" and the DES algorithm gives the following ciphertext.


```
---ENCRYPTED---
BFADQGxwAwYABz2FQEz0E3C3QF3zB11BAz43VGBFE4GxI8GADBoub8EWE0YF
+Wk9OpfbGxVgix+Hr6mXKzSHRX54jDvtvQWNQ6VBv9JD/dMZqsYAHnHPa4XJ
pC4jnAF8VWgfSIPJnyGBlUVpuVWiUlmjO1Qfu3O+FE753JZxXFhXd5ivslVY
RsxEJFY/Nx1FRu/2r1+dYFrknA0m8ihJJHs+ARss+GjzjDtagw9emTyed0Kb
mMwo1BQyKKsiiqzvoD4rNs2bSZslQ6mJMxonlJnST9ruH/25XmK1uXpr2rK8
hJ1DT8UEKW1z4ylKkAWS3sSf5/v96t6sSOhDP+2mkAxdELL7PNb46g16Aeth
f3j/3GkYCz5jT793t3sO+aa+MQhIEPRA2/2QYpfO7boVViXJp3pRS6w1bdwL
o3sbeUvIQcEZnx5bgCK7CTI+aAS4x62jMIiMQ6CXEfAAwjzE5XaibgK/NcP4
3cdsst/kvSzmVjsah671.
------END------
```


Since the key is known, it is possible to decrypt this ciphertext using a program that implements the DES algorithm.

8.6.2 PUBLIC KEY ENCRYPTION

One of the main difficulties with symmetric key encryption arises when people want to encrypt their communications. Both people must have the key that encodes and decodes the data. For instance, if two people want to exchange encrypted e-mail, how do they exchange the key to decrypt the message? Should they send the key in one message and then the encrypted data separately? If the concern is that the e-mail will be intercepted, then the key could just as easily be intercepted. Should they send the key on a disk by regular mail? This is slow and not very secure since a determined adversary could intercept the disk.

The answer to this apparent riddle is public key encryption. Continuing the lock analogy, imagine that an individual could make thousands of identical padlocks and distribute them to anyone who wanted to send him/her a private message. In the 1970s, clever mathematicians finally developed a mechanism to implement this idea, allowing an individual to disseminate a piece of information called a *public key* that anyone could use to encrypt a message and only the intended recipient who possessed the corresponding *private key* could decrypt the message. Two commonly used public key algorithms are RSA and DSA. For an excellent account of the history of cryptography and simplified descriptions of these algorithms see Singh (2000). More technical coverage of cryptography can be found in Schneier (1996).

8.6.3 PRETTY GOOD PRIVACY

One program that uses both private and public key cryptography is Pretty Good Privacy (PGP).[7] Although it is possible to just use a public key algorithm like RSA to encrypt messages, this would be slow when dealing with large messages. Private key encryption is significantly more efficient. Therefore, PGP took the best of both methods and combined them. PGP encrypts a message using a private key algorithm like DES using a randomly generated private key and encrypts that private key using a public key algorithm like RSA (this last step requires the intended recipient's public key). PGP then sends both the encrypted text and the encrypted private key to the recipient. Thus, when the recipient receives the encrypted message, he/she uses his/her personal private key to decrypt the randomly generated private key and uses the randomly generated private key to decrypt the message.

[7] http://www.pgpi.com

Criminals have not overlooked the power of encryption and are using it to protect data stored on their computer and conceal their activities on the Internet. For instance, in 2001, when the Earth Liberation Front (ELF[8]) was placed at the top of the FBI's list of North American terrorist threats, their Web site instructed ELF members to maintain a high level of secrecy and security using PGP.

[8] http://www.earthliberationfront.com

8.7 SUMMARY

Digital investigators require a basic understanding of how computers operate and how data are stored on media. A failure to understand and control the boot process can result in changes being made to an evidentiary hard drive. To recover data, digital investigators must know how data are arranged on a disk. To analyze data, digital investigators must know how to view them and interpret them. Details of the collection, recovery, and analysis of digital evidence are elaborated on in the next chapter.

Observing the life of a file is an illustrative way to summarize some of the important concepts presented in this chapter. When a program instructs the operating system to create a file, the first step is to find an available space on the disk where the data can be stored. The file system serves this purpose, reserving the necessary clusters. Then the read/write heads of the hard drive are moved to the proper track and, when the disk spins to the correct sector, a binary representation of the data is created by altering the surface of the disk. When the file is deleted, the space in unallocated – the file system is updated to indicate that the clusters are available for new data. However, until these clusters are reused, the original data remain. Even when one of the clusters is reused, some of the original data will remain in file slack space.

REFERENCES

Hollinger R. C. (1997) *Crime, Deviance and the Computer*. Brookfield, VT: Dartmouth Publishing Company.

Sammes T. and Jenkinson B. (2000) *Forensic Computing: A Practitioner's Guide*, London: Springer.

Schneier B. (1996) *Applied Cryptography: Protocols, Algorithms, and Source Code in C*. New York: John Wiley & Sons.

Singh S. (2000) *The Code Book: The Science of Secrecy from Ancient Egypt to Quantum Cryptography*. New York: Anchor Books.

CASES

United States v Moussaoui (2003) "Government's opposition to standby counsel's reply to the government's response to court's order on computer and e-mail evidence" (Available online at http://cryptome.org/usa-v-zm-email.htm).

## CHAPTER 9

APPLYING FORENSIC SCIENCE TO COMPUTERS

Like a detective, the archaeologist searches for clues in order to discover and reconstruct something that happened. Like the detective, the archaeologist finds no clues too small or insignificant. And like the detective, the archaeologist must usually work with fragmentary and often confusing information. Finally, the detective and the archaeologist have as their goal the completion of a report, based on a study of their clues, that not only tells what happened but proves it.

(Meighan 1966)

Digital evidence examination is analogous to diamond cutting. By removing the unnecessary rough material, the clear crystal beneath is revealed. The diamond is then carved and polished to enable others to appreciate its facets. Similarly, digital evidence examiners extract valuable bits from large masses of data and present them in ways that decision makers can comprehend. Flaws in the underlying material or the way it is processed reduce the value of the final product.[1]

[1] *Digital evidence examination is also analogous to an autopsy in that some skill is required to operate on the system and determine what occurred.*

Stretching the analogy, digging rough diamonds from the earth requires one set of skills, whereas a diamond cutter requires another set of skills entirely. A jeweler who examines gems closely to assess their worth and combines them to create a larger piece requires yet another set of skills. Digital investigators often perform all of the requisite tasks from collecting, documenting, and preserving digital evidence to extracting useful data and combining them to create an increasingly clearer picture of the crime as a whole. Digital investigators need a methodology to help them perform all of these tasks properly, find the scientific truth, and ultimately have the evidence admitted in court.

This is where forensic science is useful, offering carefully tested methods for processing and analyzing evidence and reaching conclusions that are reproducible and free from distortion or bias. Concepts from forensic science

Digital Evidence and Computer Crime Second Edition
ISBN: 0-12-163104-4

can also help digital investigators take advantage of digital evidence in ways that would otherwise not be possible. For example, scientific techniques such as comparing features of digital evidence with exemplars can be used to discern minor details that would escape the naked eye.

This chapter applies the methodologies covered in Chapter 4 (Investigative Process) and Chapter 5 (Investigative Reconstruction) to single, non-networked computers. These methodologies incorporate principles and techniques from forensic science, including comparison, classification, individualization, and evaluation of source. Each stage of the process is detailed in the following sections.

- Authorization and Preparation.
- Identification.
- Documentation, Collection (Seizure), and Preservation.
- Examination and Analysis.
- Reconstruction.
- Reporting Results.

These stages service the ultimate goals of discovering the truth (based upon proof or high statistical confidence) and presenting evidence in a way that helps decision makers reach a verdict.

9.1 AUTHORIZATION AND PREPARATION

Before approaching digital evidence there are several things to consider. One should be certain that the search is not going to violate any laws or give rise to liability. As noted in Chapter 3, there are strict privacy laws protecting certain forms of digital evidence like stored e-mail. Unlike the Fourth Amendment, which only applies to the government, privacy laws such as the ECPA also apply to non-government individuals and organizations. If these laws are violated, the evidence can be severely weakened or even suppressed.

Computer security professionals should obtain instructions and written authorization from their attorneys before gathering digital evidence relating to an investigation within their organization. An organization's policy largely determines whether the employer can search its employees' computers, e-mail, and other data. However, a search warrant is usually required to access areas that an employee would consider personal or private unless the employee consents. There are some circumstances that permit warrantless searches in a workplace but corporate security professionals are best advised to leave this determination to their attorneys. If a search warrant is required to search an employee's computer and related data, it may be permissible to seize the computer and secure it from alteration until the police arrive.

As a rule, law enforcement should obtain a search warrant if there is a possibility that the evidence to be seized requires a search warrant. Although obtaining a search warrant can be time consuming, the effort is well spent if it avoids the consequences of not having a warrant when one is required. Sample language for search warrants and affidavits relating to computers is provided in the United States Department of Justice's (USDOJ) search and seizure manual to assist in this process. However, competent legal advice should be sought to address specifics of a case and to ensure that nuances of the law are considered.

For a search warrant to be valid, it must both particularly describe the property to be seized and establish probable cause for seizing the property. Although some attempt should be made to describe each source of digital evidence that might be encountered, it is generally recommended to use language that is defined in the relevant statutes of the jurisdiction. For example, sample language to describe a search in Connecticut for digital evidence related to a financial crime is provided here. This example is only provided to demonstrate the use of terms defined in Connecticut General Statutes (C.G.S.) and is not intended as legal advice.

> A "computer system" (as defined by C.G.S. §53a-250(7)) that may have been used to "access" (as defined by C.G.S. §53a-250(1)) "data" (as defined by C.G.S. §53-250(8)) relating to the production of financial documents; computer related documentation, whether in written or data form; other items related to the storage of financial documents; records and data for the creation of financial documents; any passwords used to restrict access to the computer system or data and any other items related to the production of fraudulent documents; to seize said items and transport the computer system, computer system documentation and data to the State Police Computer Crimes and Electronic Evidence Unit for forensic examination and review. The forensic examination will include making true copies of the data and examining the contents of files. (Mattei *et al.* 2000)

Digital investigators are generally authorized to collect and examine only what is directly pertinent to the investigation, as established by the probable cause in an affidavit. Even in the simple case of a personal computer, digital investigators have been faulted for searches of a hard drive that exceeded the scope of a warrant.

CASE EXAMPLE (UNITED STATES v. CAREY 1998):
Although investigators may seize additional material under the "plain view" exception to search warrant requirements, it is not always clear what "plain view" means when dealing with computers. This is demonstrated in the precedent setting case of United States v. Carey that has made digital investigators more cautious in their search methods.

> Mr Carey had been under investigation for some time for possible sale and possession of cocaine. Controlled buys had been made from him at his residence, and six weeks after the last purchase, police obtained a warrant to arrest him. During the course of the arrest, officers observed in plain view a "bong," a device for smoking marijuana, and what appeared to be marijuana in defendant's apartment.
>
> Alerted by these items, a police officer asked Mr Carey to consent to a search of his apartment. The officer said he would get a search warrant if Mr Carey refused permission. After considerable discussion with the officer, Mr Carey verbally consented to the search and later signed a formal written consent at the police station…
>
> Armed with this consent, the officers returned to the apartment that night and discovered quantities of cocaine, marijuana, and hallucinogenic mushrooms. They also discovered and took two computers, which they believed would either be subject to forfeiture or evidence of drug dealing. (United States v. Carey 1998)

Investigators obtained a warrant that authorized them to search the files on the computers for "names, telephone numbers, ledger receipts, addresses, and other documentary evidence pertaining to the sale and distribution of controlled substances." However, during the examination of the computer investigators found files with sexually suggestive titles and the label ".jpg" that contained child pornography. At this stage, the detective temporarily abandoned his search for evidence pertaining to the sale and distribution of controlled substances to look for more child pornography, and only "went back" to searching for drug-related documents after conducting a five-hour search of the child pornography files. Mr Carey was eventually charged with one count of child pornography.

In appeal, Carey challenged that the child pornography was inadmissible because it was taken as the result of a general, warrantless search. The government argued the warrant authorized the detective to search any file on the computer because any file might have contained information relating to drug crimes and claimed that the child pornography came into plain view during this search. The court concluded that the investigators exceeded the scope of the warrant and reversed Carey's conviction, noting that the Supreme Court has instructed, "the plain view doctrine may not be used to extend a general exploratory search from one object to another until something incriminating at last emerges."

The main issue in this case was that the investigator acknowledged abandoning his authorized search and did not obtain a new warrant before conducting a new search for additional child pornography.

The issue of broad versus narrow searches becomes even more problematic when dealing with multi-user systems that many organizations have come to rely on. These systems may contain information belonging and relating to individuals who are not involved with the crime that is under investigation. To address these concerns, courts are becoming more restrictive and are putting time constraints on the examination, acknowledging that the bulk of information on a hard disk may have no bearing on a case and that businesses rely on these systems.

When creating an affidavit for a search warrant, it is recommended to describe how the search will be conducted. For instance, if hardware is going to be seized, this should be noted and explained why it is necessary to perform an offsite examination to protect against later criticisms that taking the hardware was unauthorized. Also, when possible, the affidavit should detail how the digital evidence examination will be performed. As stated in the USDOJ Manual, "[w]hen the agents have a factual basis for believing that they can locate the evidence using a specific set of techniques, the affidavit should explain the techniques that the agents plan to use to distinguish incriminating documents from commingled documents."

Planning is especially important in cases that involve computers. Whenever possible, while generating a search warrant, the search site should be researched to determine what computer equipment to expect, what the systems are used for, and if a network is involved. If the computers are used for business purposes or to produce publications, this will influence the authorization and seizure process. Also, without this information, it is difficult to know what expertise and evidence collection tools are required for the search. If a computer is to be examined on-site, it will be necessary to know which operating system the computer is running (e.g. Mac OS, UNIX, Windows). It will also be necessary to know if there is a network involved and if the cooperation of someone who is intimately familiar with the computers will be required to perform the search.

> Before the search begins, the search leader should prepare a detailed plan for documenting and preserving electronic evidence, and should take time to brief carefully the entire search team to protect both the identity and integrity of all the data. At the scene, agents must remember to collect traditional types of evidence (e.g. latent fingerprints off the keyboard) before touching anything. (USDOJ 1994)

If the assistance of system administrators or other individuals who are familiar with the system to be searched is required, they should be included in a pre-search briefing. They might be able to point out oversights or potential pitfalls. One person should be designated to take charge of all evidence to simplify the chain of custody. Such coordination is especially valuable when dealing with large volumes of data in various locations, ensuring that important items are not missed. In situations where there is only one chance to collect digital evidence, the process should be practised beforehand under similar conditions to become comfortable with it.

A final preparatory consideration is proper equipment. Most plans and procedures will fail if adequate acquisition systems and storage capacity are not provided.

9.2 IDENTIFICATION

Identification of digital evidence is a two-fold process. First, digital investigators have to recognize the hardware (e.g. computers, floppy disks, network cables) that contains digital information. Second, digital investigators must be able to distinguish between irrelevant information and the digital data that can establish that a crime has been committed or can provide a link between a crime and its victim or a crime and its perpetrator. During a search, manuals and boxes related to hardware and software can give hints of what hardware, software, and Internet services might be installed/used.

9.2.1 IDENTIFYING HARDWARE

There are many computerized products that can hold digital evidence such as telephones, hand held devices, laptops, desktops, larger servers, mainframes, routers, firewalls, and other network devices. There are also many forms of storage media including compact disks, floppy disks, magnetic tapes, high capacity flip, zip and jazz disks, memory sticks, and USB storage devices (Figure 9.1).

In addition, wires, cables, and the air can carry digital evidence that, with the proper tools, can be picked out of the ether and stored for future examination.

Examples of various computer systems with photographs are available in USDOJ (2001). This guide also provides useful checklists of digital evidence to look for in certain types of investigations, including online auction fraud, child exploitation/abuse, computer intrusion, death investigation, domestic violence, economic fraud, e-mail threats/harassment/stalking, extortion, gambling, identity theft, narcotics, prostitution, software piracy, and telecommunications fraud.

Exposure to different kinds of computing environments is essential to develop expertise in dealing with digital evidence. Local organizations (especially local Computer Science departments and Internet Service Providers) may provide a tour of their facilities. Visits can be made to local computer stores, university computer labs, and Internet cafes. Whenever possible, ask people about their systems. Most system administrators are delighted to talk about their networks if asked. Also, many computer manufacturers and suppliers have Web sites with detailed pictures and functional specifications of their products. Digital investigators can use this information to become more familiar with a variety of hardware.

Before approaching a crime scene, try to determine which types of hardware might be encountered since different equipment and expertise is required for terabytes of storage versus miniature systems.

9.2.2 IDENTIFYING DIGITAL EVIDENCE

Different crimes result in different types of digital evidence. For example, cyberstalkers often use e-mail to harass their victims, computer crackers sometimes inadvertently leave evidence of their activities in log files, and child pornographers sometimes have digitized images stored on their computers. Additionally, operating systems and computer programs store digital evidence in a variety of

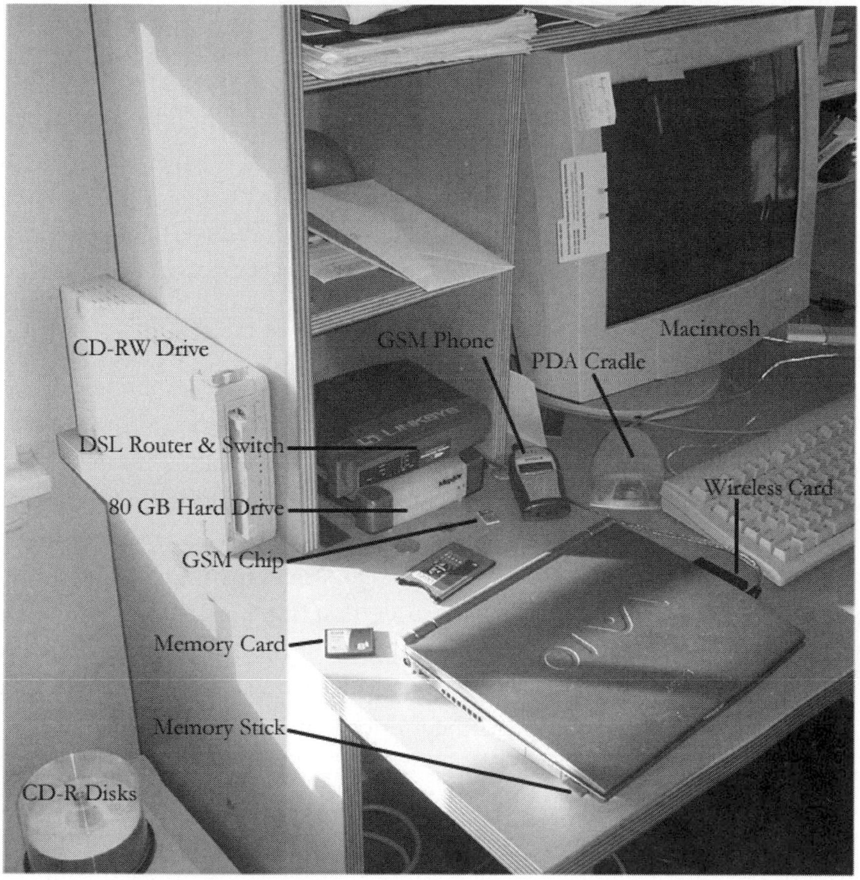

Figure 9.1

A selection of storage media and computerized devices.

places. Therefore, the ability to identify evidence depends on a digital investigator's familiarity with the type of crime that was committed and the operating system(s) and computer program(s) that are involved.

9.3 DOCUMENTATION

Documentation is essential at all stages of handling and processing digital evidence. Documenting who collected and handled evidence at a given time is required to maintain the chain of custody. It is not unusual for every individual who handled an important piece of evidence to be examined on the witness stand.

> Continuity of possession, or the chain of custody, must be established whenever evidence is presented in court as an exhibit... Frequently, all of the individuals involved in the collection and transportation of evidence may be requested to testify in court. Thus, to avoid confusion and to retain complete control of the evidence at all times, the chain of custody should be kept to a minimum. (Saferstein 1998)

A videotape or similar visual representation of dynamic onscreen activities is often easier for non-technical decision makers (e.g. attorney, jury, judge, manager, military commander) to understand than a text log file. Although it may not be feasible to videotape all sessions, important sessions may warrant the effort and expense. Also, software such as Camtasia, Lotus ScreenCam, and QuickTime can capture events as they are displayed on the computer screen, effectively creating a digital video of events. One disadvantage of this form of documentation is that it captures more details that can be criticized. Therefore, digital investigators must be particularly careful to follow procedures strictly when using this approach.

So, careful note should be made of when the evidence was collected, from where, and by whom. For example, if digital evidence is copied onto a floppy diskette, the label should include the current date and time, the initials of the person who made the copy, how the copy was made, and the information believed to be contained on the diskette. Additionally, MD5 values of the original files should be noted before copying. If evidence is poorly documented, an attorney can more easily shed doubt on the abilities of those involved and convince the court not to accept the evidence.

Documentation showing evidence in its original state is regularly used to demonstrate that it is authentic and unaltered. For instance, a video of a live chat can be used to verify that a digital log of the conversation has not been modified – the text in the digital log should match the text on the screen. Also, the individuals who collected evidence are often called upon to testify that a specific exhibit is the same piece of evidence that they originally collected. Since two copies of a digital file are identical, documentation may be the only thing that a digital investigator can use to tell them apart. If a digital investigator cannot clearly demonstrate that one item is the original and the other is a copy, this inability can reflect badly on the digital investigator. Similarly, in situations where there are several identical computers with identical components, documenting serial numbers and other details is necessary to specifically identify each item.

Documenting the original location of evidence can also be useful when trying to reconstruct a crime. When multiple rooms and computers are involved, assigning letters to each location and numbers to each source of digital evidence will help keep track of items. Furthermore, digital investigators may be required to testify years later or, in the case of death or illness, a digital investigator may be incapable of testifying. So, documentation should provide everything that someone else will need in several years time to understand the evidence. Finally, when examining evidence, detailed notes are required to enable another competent investigator to evaluate or replicate what was done and interpret the data.

It is prudent to document the same evidence in several ways. If one form of documentation is lost or unclear, other backup documentation can be invaluable. So, the computer and surrounding area, including the contents of nearby drawers and shelves, should be photographed and/or videotaped to document evidence *in situ*. Detailed sketches and copious notes should be made that will facilitate an exact description of the crime scene and evidence as it was found.

9.3.1 MESSAGE DIGESTS AND DIGITAL SIGNATURES

For the purposes of this text, a message digest algorithm can be thought of as a black box that accepts a digital object (e.g. a file, program, or disk) and

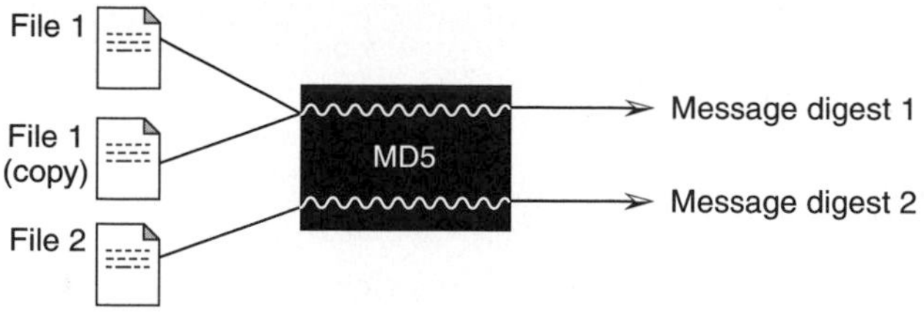

Figure 9.2

Black box concept of the message digest.

produces a number (Figure 9.2). A message digest algorithm always produces the same number for a given input. Also, a good message digest algorithm will produce a different number for different inputs. Therefore, an exact copy will have the same message digest as the original but if a file is changed even slightly it will have a different message digest from the original.

Currently, the most commonly used algorithm for calculating message digests is MD5. There are other message digest algorithms such as SHA, HAVAL, and SNEFRU. SHA is very similar to MD5 and is currently the US government's message digest algorithm of choice.

> The [MD5] algorithm takes as input a message of arbitrary length and produces as output a 128-bit "fingerprint" or "message digest" of the input. It is conjectured that it is computationally unfeasible to produce two messages having the same message digest, or to produce any message having a given prespecified target message digest. (RFC1321 1992)

Note the use of the word "fingerprint" in the above paragraph. The purpose of this analogy is to emphasize the near uniqueness of a message digest calculated using the MD5 algorithm. Basically, the MD5 algorithm uses the data in a digital object to calculate a combination of 32 numbers and letters. This is actually a 16 character hexadecimal value, with each byte represented by a pair of letters and numbers. Like human fingerprints and DNA, it is highly unlikely that two items will have the same message digest unless they are duplicates.

> It is conjectured that the difficulty of coming up with two messages having the same message digest is on the order of 2^{64} operations, and that the difficulty of coming up with any message having a given message digest is on the order of 2^{128} operations. (RFC1321 1992)

This near uniqueness makes message digest algorithms like MD5 an important tool for documenting digital evidence. For instance, by computing the MD5 value of a disk prior to collection, and then again after collection, it can be demonstrated that the collection process did not change the data. Similarly, the MD5 value of a file can be used to show that it has not changed since it was collected. Table 9.1 shows that changing one letter in a sentence changes the message digest of that sentence.

Table 9.1

Two files on a Windows machine that differ by only one letter have significantly different MD5 values.

DIGITAL INPUT	MD5 OUTPUT
The suspect's name is John	c52f34e4a6ef3dce4a7a4c573122a039
The suspect's name is Joan	c1d99b2b4f67d5836120ba8a16bbd3c9

In addition to making minor changes clearly visible, message digests can be used to search a disk for a specific file – a matching MD5 value indicates that the files are identical even if the names are different. Notably, an MD5 value alone does not indicate that the associated evidence is reliable, since someone could have modified the evidence before the MD5 value was calculated. Ultimately, the trustworthiness of digital evidence comes down to the trustworthiness of the individual who collected it.

Digital signatures provide another means of documenting digital evidence by combining a message digest of a digital object with additional information such as the current time. This bundle of information is then encrypted using a signing key that is associated with an individual or a small group. The resulting encrypted block is the signature – showing that the digital data is intact (e.g. an MD5 value), when the object was signed, and who performed the operation, that is, the owner(s) of the signing key.

9.4 COLLECTION AND PRESERVATION

Once identified, digital evidence must be preserved in such a way that it can later be authenticated as discussed in Chapter 7. A major aspect of preserving digital evidence is collecting it in a way that does not alter it. Imagine for a moment a questioned death crime scene with a suicide note on the computer screen. Before considering what the computer contains, the external surfaces of the computer should be checked for fingerprints and the contents of the screen should be photographed. It would then be advisable to check the date and time of the system for accuracy and save a copy of the suicide note to a sanitized, labeled floppy diskette.

CASE EXAMPLE

In one homicide case, law enforcement seized the victim's computer but instead of treating it as they would any other piece of evidence, they placed the computer in an office, turned it on and operated it to see what they could find thus altering the system and potentially destroying useful date–time stamp information and other data. Additionally, they connected to the victim's Internet account, thus altering data on the e-mail server and creating log entries that alarmed other investigators because they did not know who had accessed the victim's account after her death.

In a child pornography investigation, papers, photographs, videotapes, digital cameras, and all external media should be collected. At the very least, hardware should be collected that may help determine how child pornography was obtained, created, viewed, and or distributed. In one case, investigators found a scrapbook of newspaper articles concerning sexual assault trials and pending child pornography legislation as well as a hand-drafted directory of names, addresses and telephone numbers of children in the local area (R. v. Pecciarich). Images are often stored on removable Zip or Flip disks and these items may be the key to proving intent and more severe crimes such as manufacture and distribution. For instance, a disk may contain files useful for decrypting the suspect's data or it may become evident that the suspect used removable disks to swap files with local cohorts.

The severity of the crime and the category of cybercrime will largely determine how much digital evidence is collected. When dealing with computer hardware as contraband or evidence (e.g. component theft), the technical and legal issues are not complex, just get the hardware. Additionally, no sophisticated seizure process or analysis of items will be necessary unless the hardware was used to commit a crime. When the computer is an instrumentality used to disseminate child pornography or commit online fraud, greater care is required to preserve the contents of the computer. In homicide and child pornography cases, it is often reasonable to seize everything that might contain digital evidence. However, even in a homicide or child pornography investigation, the other uses of the computers should be considered. If a business depends on a computer that was collected in its entirety when only a few files were required, the digital investigator could be required to pay compensation for the business lost.

CASE EXAMPLE (STEVE JACKSON GAMES, 1990):

On March 1, 1990 US federal agents searched the premises and computers of the Steve Jackson Games company for evidence relating to a hacker group that called itself the Legion of Doom. Steve Jackson Games designed and published role-playing games based on fictional ways of breaking into computer systems. They also ran a Bulletin Board System called Illuminati to provide support and private e-mail services to their customers. In addition to seizing computers and everything that looked like it was related to a computer, the federal agents confiscated all copies of a book that was under development at Steve Jackson Games. No charges were ever brought against Steve Jackson Games or anyone else as a result of this raid, but Steve Jackson Games did suffer significant losses. After several unsuccessful attempts to recover the seized items, Steve Jackson Games decided to sue the Secret Service and the individual agents for the wrongful raid of their business. During the trial, it was determined that Secret Service personnel/delegates had read and deleted private e-mail that had not yet been delivered to its intended recipients (the Secret Service denied this until it was proven). Steve Jackson Games dropped the charges against the individual agents to speed up the trial and the court ruled that the government had violated the Electronic Communications Privacy Act (ECPA) and the Privacy Protection Act (PPA). The court awarded Steve Jackson Games $51,040 in damages, $195,000 in attorneys' fees and $57,000 in costs.

9.4.1 COLLECTING AND PRESERVING HARDWARE

Although the focus of this chapter is on the data stored on computers, a discussion of hardware is necessary to ensure that the evidence it contains is preserved properly. When dealing with hardware as contraband, instrumentality, or evidence, it is usually necessary to collect computer equipment. Additionally, if a given piece of hardware contains a large amount of information relating to a case, it can be argued that it is necessary to collect the hardware.

There are two competing factors to consider when collecting hardware. On the one hand, to avoid leaving any evidence behind, a digital investigator might want to take every piece of equipment found. On the other hand, a digital investigator might want to take only what is essential to conserve time, effort and resources and to reduce the risk of being sued for disrupting a person's life or business more than absolutely necessary. Some computers are critical for running institutions like hospitals and taking such a computer could endanger life. Additionally, sometimes it simply is not feasible to collect hardware because of its size or quantity.

> It is simply unacceptable to suggest that any item connected to the target device is automatically seizable. In an era of increased networking, this kind of approach can lead to absurd results. In a networked environment, the computer that contains the relevant evidence may be connected to hundreds of computers in a local-area network (LAN) spread throughout a floor, building, or university campus. That LAN may also be connected to a global-area network (GAN) such as the Internet. Taken to its logical extreme, the "take it because it's connected" theory means that in any given case, thousands of machines around the world can be seized because the target machine shares the Internet. (Guidelines, Department of Justice 1994)

If it is determined that some hardware should be collected but there is no compelling need to collect everything in sight, the most sensible approach is to employ the *independent component doctrine.* The independent component doctrine states that digital investigators should only collect hardware "for which they can articulate an independent basis for search or seizure (i.e. the component itself is contraband, an instrumentality, or evidence)" (Department of Justice 1994). Also, digital investigators should collect hardware that is necessary for the basic input and output of the computer components that are being seized. For instance, rather than collecting hard drives as independent components, it is generally prudent to collect the entire chassis that the hard drives are connected to in case it is needed to access them. BIOS translation or hard drive controller incompatibilities can prevent another system from reading regular IDE hard disks containing evidence, making it necessary to connect the hard drives to the system that originally contained them. If a computer system must remain in place but it is necessary to take the original hard drive, a reasonable compromise is to duplicate the hard drive, restoring the contents onto a similar hard drive that can be placed in the computer, and take the original into evidence.

If digital investigators decide to collect an entire computer, the collection of all of its peripheral hardware like printers and tape drives should be considered. It is especially important to collect peripheral hardware related to the type of digital evidence one would expect to find in the computer. When looking for images, any nearby digital cameras, videocassette recorders, film digitization equipment, and graphic software disks and documentation should be collected. The reasoning behind seizing these peripherals is that it might have to be proved that the suspect created the evidence and did not just download it from the Internet. It can sometimes be demonstrated that a particular scanner was used to digitize a given image. Any software installation disks and documentation associated with the computer should also be collected. This makes it easier to deal with any problems that arise during the examination stage. For example, if documents created using a certain version of Microsoft Word are collected, but the installation disks are not, it might not be possible to open the documents without that version of Microsoft Word. Additionally, if the suspect owns a book describing how to use encryption software, this may be an indication that the suspect used encryption and other concealment technology.

Printouts and papers that could be associated with the computer should be collected. Printouts can contain information that has been changed or deleted from the computer. Notes and scraps of paper that could contain Internet dial-up telephone numbers, account information, e-mail addresses, etc. should be collected. Although it is often overlooked, the garbage often contains very useful evidence. A well-known forensic scientist once joked that whenever he returns home after his family has gone to bed, he does not bother waking his wife to learn what happened during the day, he just checks the garbage.

When a computer is to be moved, specially prepared floppy disks should be put in the disk drives to prevent the system from accidentally booting from the hard drive and to protect the drives during transit. Evidence tape should be put around the main components of the computer in such a way that any attempt to open the casing or use the computer will be evident. Taping the computer will not only help to preserve the chain of evidence but will also warn people not to use the computer. Loose hard drives should be placed in anti-static or paper bags and sealed with evidence tape. Additionally, digital investigators should write the date and their initials on each piece of evidence and evidence tape.

Any hardware and storage media collected must be preserved carefully. Preservation also involves a secure, anti-static environment such as a climate-controlled room with floor to ceiling solid construction to prevent unauthorized entry. Computers and storage media must be protected from dirt, fluids, humidity, impact, excessive heat and cold, strong magnetic fields, and static electricity. According to the US Federal Guidelines for Searching and Seizing Computers discussed in Chapter 2, safe ranges for most magnetic media are 50–90°F and 20–80% humidity. There are many anecdotes about computer experts who religiously backed up important information carefully, but then, destroyed the backups by inadvertently exposing them to (or storing them in) unsuitable conditions. Leaving disks in a hot car, a damp warehouse, or near a strong magnetic field can result in complete loss of data, so care should be taken. Fortunately, there are equally many stories about recovery of digital evidence despite criminals' attempts to destroy it, so not all hope is lost when faced with damaged digital evidence.

Another difficult decision when collecting hardware is whether to turn the computer off immediately or leave it running and collect volatile data from RAM. Most law enforcement training programs recommend turning all computers off immediately in all situations. For instance, the *Good Practice Guide for Computer Based Evidence*, by the Association of Chiefs of Police in the United Kingdom advises digital investigators to unplug the power cable from the computer rather than from the wall plate or using the power switch. This precaution anticipates the possibility that a computer's power switch is rigged to set off explosives or destroy evidence. Additionally, removing power abruptly rather than shutting the system normally may preserve evidence such as a swap file that would be cleared during the normal shutdown process.[2]

Although caution often saves lives, there are many situations in which such extremes can do damage. For example, abruptly turning off a large, multiple user systems attached to a network can destroy evidence, disrupt many people's lives, and even damage the computer itself. Therefore, careful attention must be given to this crucial stage of the collection process. The *Good Practice Guide for Computer Based Evidence* renders a strong opinion in this matter:

> It is accepted that the action of switching off the computer may mean that a small amount of evidence may be unrecoverable if it has not been saved to the memory but the integrity of the evidence already present will be retained.

However, this approach is questionable when dealing with systems that have gigabytes of RAM or the data in volatile memory are important to the investigation. For example, if digital investigators notice a suspect at a computer typing a warning message to an accomplice, that message is stored in RAM and will be lost if the computer is unplugged. A photograph of the screen is certainly helpful but it may also be desirable to collect the actual data. Saving data in RAM onto an external disk is a safe approach whereas

[2]The guide does not mention the need to remove the computer's casing to examine the internals of the computer. A computer's casing should be removed to unplug power cables from hard drives, seat all cards properly, ensure that the computer does not contain explosives, and note any anomalies inside the computer like an extra disconnected hard drive.

printing may overwrite evidence by creating spool files on the evidentiary system. When investigating computer intrusions, it is usually desirable to capture information related to active processes and network connections that are stored in RAM. Active network connections can also be important in traditional investigations such as homicides. Ultimately, the digital investigator must decide if there is useful evidence in volatile memory and how to obtain that information with minimal impact on the system.

Preview (Chapter 19): Examining RAM – It may be possible to collect the necessary information by running programs from (and saving the data) to an external device. Specialized utilities like netstat, fport, and handle can be used to display information about network connections and processes on Windows machines. If this approach is taken, every action must be documented copiously along with the time and MD5 value of command output.

9.4.2 COLLECTING AND PRESERVING DIGITAL EVIDENCE

When dealing with digital evidence (information as contraband, instrumentality, or evidence) the focus is on the contents of the computer as opposed to the hardware. There are two options when collecting digital evidence from a computer: just copying the information needed, or copying everything. If a quick lead is needed or only a portion of the digital evidence on the computer is of interest (e.g. a log file), it is more practical to search the computer immediately and just take the information required. However, if there is an abundance of evidence on the computer, it often makes sense to copy the entire contents and examine it carefully at leisure.

The approach of just taking what is needed has the advantage of being easier, faster, and less expensive than copying the entire contents. For instance, in some cases it may be sufficient to only collect active files and not deleted data, in which case a normal backup of the system might suffice. However, if only a few files are collected from a system, there is a risk that digital evidence will be overlooked or damaged during the collection and preservation process.

Preview (Chapter 19): Computer intruders have developed collections of programs, commonly called *rootkits*, to replace key system components and hide the fact that a computer has been broken into. Until recently, rootkits were only developed for UNIX systems but are now being developed for Windows NT. Using trusted copies of system commands can circumvent most rootkits, but additional precautions are required when dealing with more sophisticated computer criminals.

CASE EXAMPLE

A group of computer intruders gained unauthorized access to an IRIX server and used it to store stolen materials, including several credit card databases stolen from e-commerce Web sites. A system administrator made copies of the stolen materials along with log files and other items left by the intruders. The system administrator combined all of the files into a large compressed archive and transferred the archive, via the network, to a system with a CD-ROM burner. Unfortunately, the compressed archive file became corrupted in transit but this was not realized until the investigators attempted to open the archive at a later date. By this time, the original files had been deleted from the IRIX system. It was possible to recover some data from the archive file but not enough to build a solid case.

There is also a risk that the system has been modified to conceal or destroy evidence (e.g. using a rootkit) and valuable evidence might be missed. For instance, if digital investigators need log files from a computer, there may be additional deleted logs in unallocated space that could be useful. When collecting only a few files from a system, it is still necessary to document the collection process thoroughly and chronicle the files in their original state. For instance, obtain a full listing of all files on the disk with associated characteristics such as full path names, date–time stamps, sizes, and MD5 values.

Given the risks of only collecting a few files, in most cases, it is advisable to acquire the full contents of the disk because digital investigators rarely know exactly what the disk contains. Before copying data from a disk, it is advisable to calculate the MD5 value of the original disk – this hash value can be compared with copies to demonstrate that they are identical. When collecting the entire contents of a computer, a bitstream copy of the digital evidence is usually desirable (a.k.a. forensic image, exact duplicate copy).

A bitstream copy duplicates everything in a cluster, including anything that is in the slack space and other areas of the disk outside of the file system's reach, whereas other methods of copying a file only duplicate the file and leave the slack space behind (Figure 9.3). Therefore, digital evidence will be lost if a bitstream copy is not made. Of course, this is only a concern if slack space contains important information. If a file contains evidence and the adjacent slack space is not required, a simple file copy will suffice.

The majority of tools can interpret bitstream copies created using EnCase and UNIX `dd`, making them the *de facto* standards. Safeback is another common file format that is used mainly in law enforcement agencies. EnCase and Safeback embed additional information in their files to provide integrity checks. There is one empirical law of digital evidence collection that should always be remembered:

> **Empirical Law of Digital Evidence Collection and Preservation**: If you only make one copy of digital evidence, that evidence will be damaged or completely lost.

Therefore, always make at least two copies of digital evidence and check to make certain that at least one of the copies was successful and can be accessed on another computer. In light of the fact that evidence acquisition tools have had problems that cause them not to copy some data under certain circumstances, it is advisable to make bitstream copies of a disk with two or

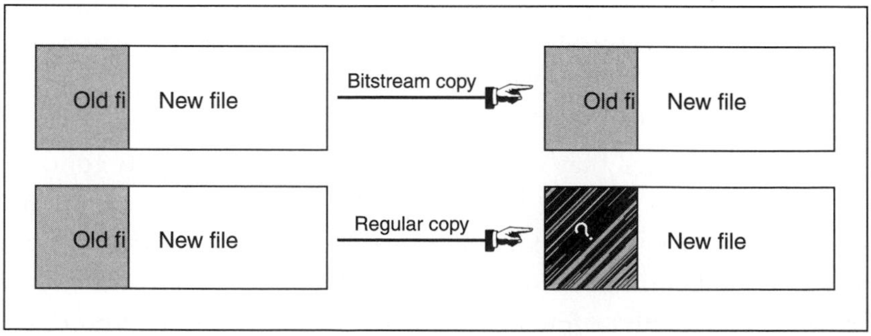

Figure 9.3

Comparing bitstream copying to regular copying.

more tools. For instance, one copy of a hard drive might be made using dd and a second using EnCase. Also, it is imperative that digital evidence is saved onto completely clean disks. If digital evidence is copied onto a disk that already has data on it, that old data could remain in the slack space, commingling with and polluting the evidence. Therefore, it is a good practice to sanitize any disk before using it to collect evidence. To sanitize a disk, use a file wipe program to write a specific pattern on the drive (e.g. 00000000) and verify that this pattern was written to all sectors of the drive. Also document the drive's serial number and the date of sanitization. In addition to preventing digital evidence transfer, sanitizing collection media shows professionalism.[3]

[3]If evidence from multiple sources is being stored on a single collection drive, create a unique directory structure for each source to avoid overwriting files collected previously by oneself or others.

As a rule, computers used to store and analyze digital evidence should not be connected to the public Internet. There is a risk that individuals on the Internet will gain unauthorized access to evidence.

Preview (Chapters 10 and 11): An Evidence Acquisition Boot Disk enables examiners to determine which computers contain evidence by booting the system, previewing it, and searching for keywords. It is also possible to use this method to collect evidence via cables (parallel and network).

Whether all available digital evidence or just a portion is collected, the task is to get the evidence from the computer with the least amount of alteration. One approach is to bypass the operating system on the computer that contains evidence using a specially prepared boot disk and make a bitstream copy of the hard drive as described in Chapters 10 and 11.

In certain situations, it may not be possible or desirable to boot the suspect's computer from a floppy disk. The next best alternative is to remove the hard drive(s) from the suspect computer and move them to an evidence collection system for processing.[4] Although removing a disk from a computer and placing it in an evidence collection system requires more knowledge of computers than booting from a trusted diskette, it has several advantages. First, it might be difficult or impossible to boot the system from an evidence acquisition boot disk (e.g. no floppy/CD drive, BIOS password set). Second, the evidence collection software that is generally available requires a DOS boot disk – this will not work with Apple or Sun systems. Third, it is easier to develop an evidence collection procedure that involves a known evidence collection system than many unknown systems.

[4]Handle hard drives with great care. Touching parts of the drive with fingertips that have static electricity buildup can damage the drive. Roughly removing or inserting the data cable can break pins. Although such damage may be repairable, the cost and time required to repair the drive may be prohibitive.

There are several ways to make a bitstream copy of a hard drive. Hardware duplication devices such as those made by Intelligent Computer Solutions[5] and Logicube[6] are useful for copying data from one IDE or SCSI drive to another. This is useful for preserving the original drive by minimizing the number of times it is copied. However, it is still necessary to examine the evidence on the drive by connecting it to an examination system with hardware and software optimized to support the forensic process (e.g. manual BIOS configuration, drive bays). Additionally, adapters are required to accommodate the many different kinds of storage devices. Even within the SCSI family, there are different types of interfaces. In one case, a Sun Sparc 5 system contained evidence on two hard drives with 80-pin Single

[5]http://www.ics-iq.com

[6]http://www.logicube.com

Connector Attachment (SCA 80) SCSI interfaces. An adapter was obtained from Blackbox[7] that enabled the SCA 80 drives to be plugged into a generic 50-pin SCSI card and power cable. Adapter cables for connecting both SCSI and IDE laptop hard drives to a standard computer are also available.

[7] http://www.blackbox.com

Remember that it is often possible to ask the system owner or administrator for assistance. If data is protected or encrypted, a system owner or administrator might be able to help gain access to it. It is usually safe to allow a system administrator to operate a computer while assisting the digital investigator. However, a suspect must never be allowed to operate a computer. Instead, the suspect should be asked to provide the information required.

The advantages and disadvantages of the three collection options are summarized in Table 9.2.

Table 9.2

Advantages and disadvantages of the three collection options described in Section 9.4.2.

COLLECTION METHOD	RELEVANT CYBERCRIME CATEGORIES	ADVANTAGES	DISADVANTAGES
Collect hardware	■ Hardware as fruits of crime ■ Hardware as instrumentality ■ Hardware as evidence ■ Hardware contains large amount of digital evidence	■ Requires little technical expertise ■ The method is relatively simple and less open to criticism ■ Hardware can be examined later in a controlled environment ■ Hardware is available for others to examine at a later date (opponents, other examiners, using new techniques)	■ Risk damaging the equipment in transit ■ Risk not being able to boot (BIOS password) ■ Risk not being able to access all evidence on drive (e.g. encrypted file system) ■ Risk destroying evidence (contents of RAM) ■ Risk liability for unnecessary disruption of business ■ Develop a bad reputation for heavy-handedness
Collect all digital evidence, leave hardware	■ Information as fruits of crime ■ Information as instrumentality ■ Information as evidence	■ Digital evidence can be examined later in a controlled environment ■ Working with a copy prevents damage of original evidence ■ Minimize the risk of damaging hardware and disrupting business	■ Requires equipment and technical expertise ■ Risk not being able to boot (BIOS password) ■ Risk not being able to access all evidence on drive (e.g. encrypted file system) ■ Risk missing evidence (Protected Area) ■ Risk destroying evidence (contents of RAM) ■ Time consuming ■ Methods are more open to criticism than collecting hardware because more can go wrong
Only collect the digital evidence that you need	■ Information as fruits of crime ■ Information as instrumentality ■ Information as evidence	■ Allows for a range of expertise ■ Can ask for help from system admin/owner ■ Quick and inexpensive ■ Avoid risks and liabilities of collecting hardware	■ Can miss or destroy evidence (e.g. rootkit) ■ Methods are most open to criticism because more can go wrong than collecting all of the evidence

9.5 EXAMINATION AND ANALYSIS

Recall that an examination involves preparing digital evidence to facilitate the analysis stage. The nature and extent of a digital evidence examination depends on the known circumstances of the crime and the constraints placed on the digital investigator. If a computer is the fruit or instrumentality of a crime, the digital investigators will focus on the hardware. If the crime involves contraband information, the digital investigators will look for anything that relates to that information, including the hardware containing it and used to produce it. If information on a computer is evidence and the digital investigators know what they are looking for, it might be possible to extract the evidence needed quite quickly.

In some instances, digital investigators are required to perform an onsite examination under time constraints. For instance, if the investigation is covert or the storage medium is too large to collect in its entirety, an examination may have to be performed on premises. Swift examinations are also necessary in exigent circumstances, for example, when there is a fear that another crime is about to be committed or a perpetrator is getting away. In other situations a lengthy, in-depth examination is required in a controlled environment.

In any case, the forensic examination and subsequent analysis should preserve the integrity of the digital evidence and should be repeatable and free from distortion or bias.

9.5.1 FILTERING/REDUCTION

Before delving into the details of digital evidence analysis, a brief discussion of data reduction is warranted. With the decreasing cost of data storage and increasing volume of commercial files in operating system and application software, digital investigators can be overwhelmed easily by the sheer number of files contained on even one hard drive or backup tape. Accordingly, examiners need procedures (such as the one based on the guidelines in Chapter 24) to focus in on potentially useful data. The process of filtering out irrelevant, confidential or privileged data includes:

- Eliminating valid system files and other known entities that have no relevance to the investigation.
- Focusing an investigation on the most probable user-created data.
- Managing redundant files, which is particularly useful when dealing with backup tapes.
- Identifying discrepancies between digital evidence examination tools, such as missed files and MD5 calculation errors.

Less methodical data reduction techniques, such as searching for specific keywords or extracting only certain file types, may not only miss important clues but can still leave the examiners floundering in a sea of superfluous data. In short, careful data reduction generally enables a more efficient and thorough digital evidence examination.

9.5.2 CLASS/INDIVIDUAL CHARACTERISTICS AND EVALUATION OF SOURCE

Two fundamental questions that need to be addressed when examining a piece of digital evidence are what is it (classification/identification) and where did it come from (evaluation of source). The process of identification involves classifying digital objects based on similar characteristics, called *class characteristics.*

> An item is classified when it can be placed into a class of items with similar characteristics. For example, firearms are classified according to caliber and rifling characteristics and shoes are classified according to their size and pattern. (Inman and Rudin 1997)

For instance, Europol and other cooperating law enforcement agencies can compare characteristics of child pornography found in one case with a database of images seized in past investigations. Using this system, similar segments of fabric and other patterns in photographs can be found, potentially providing digital investigators with additional evidence that can help determine where the photograph was taken or help identify the offender or victim.

As another example of the usefulness of class characteristics, to determine if a file with a ".doc" extension is a Microsoft Word or WordPerfect document, it is necessary to examine the header, footer, and other class characteristics of the file. Similarly, there are different types of graphics files (e.g. JPEG, GIF, TIFF) making it possible to be specific when classifying them as shown in Table 9.3.

Such class characteristics are useful for locating fragments of digital objects on a disk. For instance, searching an entire hard drive for all occurrences of class characteristics like "JFIF" is a more thorough way to search for JPEG images than simply looking at the file system level for files with a ".jpg" file extension. In addition to finding fragments of deleted images in unallocated space, searching for class characteristics will identify JPEG files that

Table 9.3

Header of a JPEG file viewed in hexadecimal (left) and ASCII (right) showing the signature "JFIF".

```
FFD8FFE0 00104A46 49460001 02010048   | ╪ α  ..JF IF.. ...H |  16
00480000 FFED0ECA 50686F74 6F73686F   | .H..  φ.╧ Phot osho |  32
7020332E 30003842 494D03E9 00000000   | p 3. 0.8B IM.Θ .... |  48
00780000 00010048 00480000 000002F4   | .x.. ...H .H.. ...⌠ |  64
0240FFEE FFEE0306 02520000 052803FC   | .@ ε  ε.. .R.. .(.ⁿ |  80
0000072B BAD00000 00000000 0030072B   | ...+ ║⊥.. .... .0.+ |  96
BE400000 00010000 00010000 FFFF072B   | ╛@.. .... ....   .+ | 112
```

have been renamed with a ".doc" extension to hide them from the unwary digital investigator.

There are hundreds of thousands of unique file formats, making it impossible to be familiar with every variation of every kind of digital evidence.[8] File classification tools such as the UNIX file command store class characteristics for various file types (referred to as magic numbers in UNIX) in magic files. However, when the file type is unknown, it becomes necessary to research file formats and compare unknown items with known samples. Searching the Internet for class characteristics of an unknown file is one approach to finding similar items.

If the meaning or significance of a class characteristic is not clear, it may be necessary to experiment. For instance, some applications embed data in image files such as the "Photoshop 3.0.8B" in Table 9.3. Asserting that a defendant manufactured this image because the defendant's computer has this version of Photoshop installed may not be correct. Does this class characteristic indicate that Photoshop 3.0.8B was used to create the image or simply used to modify an existing image? To answer this question, it is necessary to perform empirical experiments – creating and modifying images using Photoshop and comparing them with the image in question.

When digital evidence is found on a disk, it is not safe to assume that the data originated there. It is possible that the file was copied from another system or downloaded from the Internet. For instance, class characteristics of a JPEG file found on a hard drive are shown in Figure 9.4 using ACDSee,[9]

[8] Specifications for many file formats are available at http://www.wotsit.org/

[9] http://www.acdsystems.com

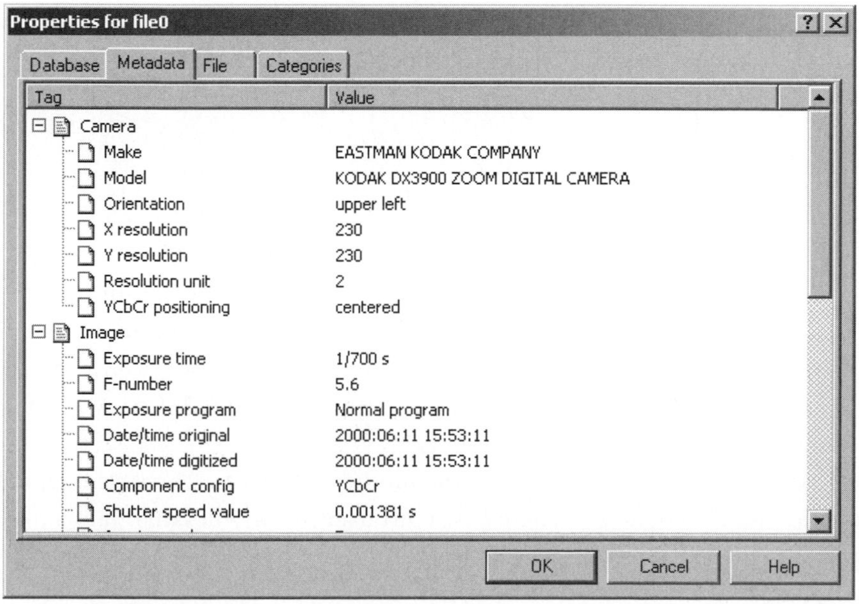

Figure 9.4

Additional class characteristics of EXIF file displayed using ACDSee. The date and time embedded in this file (15:53 on 06/11/2000) is inaccurate because the camera's clock was not set to the correct time, emphasizing the importance of documenting system time when collecting any kind of computerized device.

indicating that the JPG file was created using a Kodak DX3900 digital camera. This information should prompt digital investigators to look for the associated camera as an additional source of evidence.

Using class characteristics such as those in Figure 9.4, one can assert that the evidence is consistent with a given camera. With enough class characteristics associating a piece of evidence with a specific computer, it can be argued that a preponderance of evidence indicates that this computer was involved.

To understand how similar files from different computer systems can contain different class characteristics, compare the ASCII characters in a file created on a Windows system with one created on UNIX.

On a computer running Windows 2000:

```
C:\>echo The suspect's name is John > windowsfile
C:\>od -c windowsfile
0000000   T  h  e     s  u  s  p  e  c  t  '  s     n  a
0000020   m  e     i  s     J  o  h  n     \r  \n
0000035
C:\>md5sum windowsfile
c52f34e4a6ef3dce4a7a4c573122a039 windowsfile
```

On a computer running UNIX:

```
$ echo The suspect\'s name is John > unixfile
$ od -c unixfile
0000000 T  h  e     s  u  s  p  e  c  t  '  s     n  a
0000020 m  e     i  s     J  o  h  n\n
0000033
$ md5sum unixfile
0dc789ca62a3799abca7f1199f7c6d8c unixfile
```

The difference between these two files is caused by the different ways that Windows and UNIX represent an End Of Line (EOL). Windows represents an end of line using a carriage return and line feed (x0D0A = \r \n), whereas UNIX just uses a line feed character (x0A = \n = ASCII 10). Macintosh computers just use a carriage return (x0D = \r = ASCII 13).

Netscape history databases provide another example of how class characteristics can vary between systems. Web browser history files maintain a list of recently visited Web sites and are useful for determining when or how often certain sites were visited, and may even contain private information such as

SYSTEM (FILE NAME)	HEADER
Windows (netscape.hst)	00 06 15 61 00 00 00 02 00 00 **04 D2** 00 00 10 00
Linux (history.dat)	00 06 15 61 00 00 00 02 00 00 **04 D2** 00 00 10 00
Solaris (history.dat)	00 06 15 61 00 00 00 02 00 00 **10 E1** 00 00 10 00
Macintosh (Netscape History)	00 06 15 61 00 00 00 02 00 00 **10 E1** 00 00 10 00

Table 9.4

Headers of Netscape history databases from different systems.

passwords to certain sites. The first line of Netscape history files from four systems are shown in hexadecimal form in Table 9.4.

To understand the differences between the headers in Table 9.4, we need to research the file format. Netscape history databases are in Berkeley Database (DB) version 1.85 format. Searching the Sleepycat Web site leads to details about the database format in the magic file that is used to interface with the UNIX file command.[10] The relevant segment of the Berkeley DB magic file is shown here:

```
0 long 0x00061561 Berkeley DB
>4 long >2 1.86
>4 long <3 1.85
>0 long 0x00061561 (Hash,
>4 long 2 version 2,
>4 long 3 version 3,
>8 long 0x000004D2 little-endian)
>8 long 0x000010E1 native byte-order)
```

The last two lines explain the difference between the Netscape history files. Intel systems such as the one running Windows and Linux in this example are little endian whereas Macintosh and most UNIX systems are big endian. Therefore, if a Netscape history database found on a Windows system contains the 10E1 character, this is inconsistent and it is likely that the file originated from a Macintosh or UNIX computer. Interestingly, older versions of Netscape used an undocumented variation of Berkeley DB on the Windows platform that has the distinctive first line "00 06 15 61 00 00 00 02 00 00 **04 B3** 00 00 10 00".

When evaluating the source of a piece of digital evidence, a forensic examiner is essentially being asked to compare items to determine if they are the same as each other or if they came from the same source. The aim in this process is to compare the items, characteristic by characteristic, until the examiner is satisfied that they are sufficiently alike to conclude that they are related to one another. Ultimately, this comes down to probabilities. What is the probability of two similar items occurring independently? Archaeologists have been dealing with this question for centuries.

[10] *http://www.sleepycat.com/docs/ref/install/magic.s5.be.txt*

> In studying relationships, it is necessary to base conclusions on more than a single artifact or trait. Similarities between assemblages are more significant than isolated trait similarities. For example, two dry caves a hundred miles apart may yield arrowheads of the same kind, sandals and basketry woven by the same technique, and similar simple wooden objects like drills used for making fire. Such similarity in pattern may be convincing evidence of relationship, even though the individual objects are simple in manufacture and so widely used that they would be of little significance taken individually. (Meighan 1966)

Constellations of similar characteristics are relevant in evaluating the relationship between digital evidence and its source. The more characteristics an item and potential source have in common, the more likely it is that they are related. The type of object must also be taken into account, since simple objects have a higher probability of occurring in more than one place independently whereas complex items have a lower possibility. Also, the method of manufacture of a piece of digital evidence can indicate skill level of creator (e.g. a computer program written in C++ versus in Visual Basic).

For example, in computer intrusion investigations, it is ultimately necessary to determine if items on the suspect's computer originated from the compromised system and if items on the compromised system originated from the suspect's computer. In one case, the intruder's Windows computer contained a list of the compromised UNIX machines with associated usernames and passwords (some associated sniffer logs were also found on the suspect's disk), and hacking tools that had been found on the compromised systems. Most of the individual hacking tools did not originate from any of the machines involved – they were common programs that could be downloaded from the Internet. However, the suspect had inserted his nickname into some of the programs and had used one of the compromised systems to compress the tools into a TAR file. In addition to preserving the particular directory and subdirectory structure on the compromised system, the TAR file preserved the associated username – one of the accounts that the intruder had stolen (see Table 9.5).

Additionally, the TAR file on both systems had the same MD5 value, indicating that they were identical. In isolation, each characteristic might not establish a solid relationship between the evidence and its source, but in combination the link could be seen clearly. Similarly, a Postscript file generated on a UNIX system when a document was printed may contain the full path name of the file, the username that printed the file, along with the date and time the document was printed.

It is useful to formalize the different ways that a piece of evidence can be related to a source. The relationships described in Table 9.6 are not mutually exclusive.

```
% hexdump    -C tools.tar
00000000     74 6f 6f 6c 73 2f 00 00     00 00 00 00 00 00 00 00     |tools/..........|
00000010     00 00 00 00 00 00 00 00     00 00 00 00 00 00 00 00     |................|
*
00000060     00 00 00 00 30 30 34 30     37 35 35 00 30 30 32 36     |....0040755.0026|
00000070     32 31 31 00 30 30 30 30     31 35 31 00 30 30 30 30     |211.0000151.0000|
00000080     30 30 30 30 30 30 30 00     30 37 33 34 36 30 31 31     |0000000.07346011|
00000090     35 32 30 00 30 30 31 32     31 31 37 00 35 00 00 00     |520.0012117.5...|
000000a0     00 00 00 00 00 00 00 00     00 00 00 00 00 00 00 00     |................|
*
00000100     00 75 73 74 61 72 00 30     30 6b 6e 6f 77 00 00 00     |.ustar.00know...|
00000110     00 00 00 00 00 00 00 00     00 00 00 00 00 00 00 00     |................|
00000120     00 00 00 00 00 00 00 00     00 67 72 70 31 33 00 00     |.........grp13..|
00000130     00 00 00 00 00 00 00 00     00 00 00 00 00 00 00 00     |................|
00000140     00 00 00 00 00 00 00 00     00 30 30 30 30 32 34 37     |.........0000247|
00000150     00 30 30 30 30 30 30 33     00 00 00 00 00 00 00 00     |.0000003........|
```

Table 9.5

User account (know) and group (grp13) information preserved in a TAR file.

RELATIONSHIP	DESCRIPTION	EXAMPLES
Production	Source produced the evidence	Compressed TAR files created on a given UNIX computer. Images created on a given digital camera
Segment	Source is split into parts and parts of the whole are scattered	Fragments of a Word document found in unallocated space that are related to an intact version on the disk
Alteration	Source is an agent or process that alters or modifies the evidence	Photoshop used to change images. Programs used to delete log entries or change date–time stamps of files
Location	Source is a point in space	Digital photograph shows a portion of a bedroom or neighborhood. Evidence contains an IP address

Table 9.6

Relationships between evidence and its source.

Of course, differences will often exist between apparently similar items, whether it is a different date–time stamp of a file, slightly different data in a document, or a difference between cookie file entries from the same Web site.

> … total agreement between evidence and exemplar is not to be expected; some differences will be seen even if the objects are from the same source or the product of the same process. It is experience that guides the forensic scientist in distinguishing between a truly significant difference and a difference that is likely to have occurred as an expression of natural variation. But forensic scientists universally hold that in a comparison process, differences between evidence and exemplar should be explicable. There should be some rational basis to explain away the differences that are observed, or else this value of the match is significantly diminished. (Thornton 1997)

The concept of a significant difference is important because it can be just such a difference that distinguishes an object from all other similar objects, that is, it may be an *individual characteristic*. Although such characteristics are rarer than class characteristics, it is important to keep in mind that digital evidence may contain a unique characteristic that individualizes it, that is, links it to a particular source with a high degree of probability. Some individual characteristics are created at random – a digitized photograph may contain a line that is consistent with a scratch on the glass of a given flatbed scanner. Similarly, a floppy drive may create a unique pattern in the magnetic media when it writes data to the disk, enabling digital investigators to determine if digital evidence was saved using a given drive. Other individual characteristics are created purposefully for later identification (e.g. an identification number associated with a computer). These unique characteristics of a piece of digital evidence can be used to link cases, generate suspects and associate a crime with a specific computer.

For instance, files created using Office 97 for Windows and Office 98 for Macintosh contain a Global Unique Identifier (GUID) that may be associated with a specific computer. As an example, one Word 97 document created on a computer with the Ethernet address 00-10-4B-DE-FC-E9 contained the following:

```
_PID_GUID‰AN{2083B360-E6EF-11D2-9DC8-00104BDEFCE9}
```

and another document created on the same computer contained the following:

```
_PID_GUID‰AN{CC79EA90-E6EE-11D2-9DC8-00104BDEFCE9}
```

Notice the unique Ethernet address at the end of each line. To see this line the document must be viewed using a program that does not interpret the word processor commands (e.g. a simple text viewer). However, the GUID will not contain an address if the computer does not have a network interface card. Instead, a number is randomly generated when Microsoft Office is installed. Also, it is not safe to assume that a file was created on a given machine simply based on an address in the GUID. For instance, the GUID value in an Excel spreadsheet may change when the document is modified using a different computer, indicating where the file was last modified as opposed to where it was originally created.

So, additional examination is required to determine the precise relationship between a Microsoft Office file and its source (production, alteration, or inconclusive). Notably, Office documents contain other details that can be useful for evaluation of source such as printer names, directory locations, creator, and creation/modification date–time stamps.

CASE EXAMPLE

In 1999, a virus/worm called Melissa hit the Internet. Melissa traveled in a Microsoft Word document that was attached to an e-mail message. This virus/worm propagated so quickly that it overloaded many e-mail servers, and forced several large organizations to shut down their e-mail servers to prevent further damage. It was widely reported that David Smith, the individual who created the virus/worm, was tracked down with the help of a feature of Microsoft Office.

Although some individuals claimed that they tracked down the author of the Melissa virus using the network interface card in the GUID of infected documents, the New Jersey State Police actually apprehended David Smith using information obtained from AOL. The security department at AOL noticed that a stolen account was used to post the virus/worm an Internet newsgroup and that David Smith had connected to AOL through his local Internet service provider, i.e. using the "Bring your own provider" feature. However, before investigators could use this connection to locate Smith, he had realized the severity of his crime and thrown his computer in a dumpster. Although Smith confessed to the crime, his computer was never retrieved so the network interface card could not be compared with GUID information (Geraghty, M. e-mail communication).

9.5.3 DATA RECOVERY/SALVAGE

In general, when a file is deleted, the data it contained actually remain on a disk for a time and can be recovered. The details of recovering and reconstructing digital evidence depends on the kind of data, its condition, the operating system being run, the type of the hardware and software, and their configurations. These details are described in later chapters but some aspects that are common to all situations are presented here.

When a deleted file is partially overwritten, part of it may be found in slack space and/or in unallocated space. It may be possible to extract and reconstitute such fragments to view them in their near original state. Such recovery is easier for file types that have more human readable components, such as Microsoft Word documents, because an individual can often infer the order and importance of each component. Finding and reconstituting file fragments can be more difficult when the header information has been overwritten but it may still be possible to repair the damage. For instance, if the header of a Word document is overwritten, the remaining fragment can be compared with other documents to determine how much of the header was lost. A suitable piece of another document's header can then be grafted onto the fragment to enable Microsoft Word to recognize and display the file. This can be more difficult with image and audio/visual files since the header contains important information such as image height and width, color information, and other information needed to display the image. Therefore, grafting a header from another file may result in odd hybrids but can give a sense of the original file as shown in Figure 9.5.

Figure 9.5
Fragments of an overwritten JPEG file partially reconstituted by grafting a new header onto the file.

There are also binary files on a computer that contain a large amount of information. For example, many operating systems and computer programs use swap files to store information temporarily while it is not being used. For instance, Windows NT uses a file named "pagefile.sys," and UNIX uses dedicated swap partitions (areas on a disk or entire disks) to store information temporarily. Hibernation files are another fruitful source of data because they contain all of the information necessary to restore the previous session. It is conceivably possible to reconstruct the full session using this data but this is difficult in practice.

Additionally, data is stored in binary form by many programs including e-mail programs, compression applications, and word processing programs. For instance, Netscape history databases mentioned earlier contain deleted entries that can be recovered. Similarly, Microsoft Outlook stores e-mail in a file that requires special processing to read and deleted e-mails may still be present in the Outlook binary file. Microsoft Office documents can contain images and other media that may be of interest in an investigation. Furthermore, binary files can contain hidden data placed there by offenders or for legitimate purposes. Some museums place digital watermarks in images of their artwork to help them determine if someone has taken or used a picture without permission.

Encryption presents a significant challenge in the recovery stage of a digital evidence examination. Encryption software like PGP is becoming more commonplace, allowing criminals to scramble incriminating evidence using very secure encoding schemes, making it unreadable. The three main approaches to getting around encryption programs like PGP are to find the encrypted data in unencrypted form, obtain the passphrase protecting the private key, or guess the passphrase. Digital evidence examiners might be able to find passphrases or unencrypted versions of data in unallocated space or swap files. Alternatively, digital investigators might be able to obtain a decryption passphrase by searching the area surrounding a system for slips of paper containing the passphrase, interviewing the suspect, or surreptitiously monitoring the suspect's computer use. The Password Recovery Toolkit and Forensic Toolkit can be combined systematically to test keywords found on a disk to determine if they are the passphrase. The Password Recovery Toolkit can also be configured to use various dictionaries and customized suspect profiles in an effort to guess the passphrase. Other techniques and tools for performing these operations are discussed in later chapters.

In addition to being technically involved, recovering encrypted data can be challenging from a legal viewpoint.

> Stored data must be retrieved in such a way as to ensure that its provenance can be proved in court, and handled in such a way as to maintain the 'chain of evidence'. Decryption of stored data must therefore take place in accordance with best practice on computer forensic evidence. In general, this may require access to the decryption key rather than the plain text (otherwise doubt might be cast in court on the authenticity of the plain text) (Encryption and Law Enforcement, UK Cabinet)

In light of this issue, England enacted the Regulation Investigatory Power Act (RIPA), requiring individuals to disclose their encryption keys on demand or face a 2-year sentence. However, such penalties are insignificant to some offenders, particularly when disclosing their encryption key would result in public disgrace and a longer sentence. In one case involving child pornography and exploitation, the suspect was uncooperative and digital investigators resorted to guessing his PGP passphrase, a time-consuming process that has a low chance of success. The investigators were unable to guess the suspect's passphrase before he committed suicide (citation). In the United States, it is difficult to compel defendants to disclose encryption keys because this is viewed as self-incrimination and is protected under the Fifth Amendment. However, such refusals reflect badly on defendants and a clever attorney can sometimes use this to their advantage, either in arranging a plea bargain or convincing a jury to assume the worst.

Although it may be feasible to obtain an encryption passphrase by monitoring the suspect's computer use, this approach is invasive and can raise privacy issues. For instance, in United States v. Scarfo, the defense argued that the FBI violated wiretap statutes when they installed a key logger system on Scarfo's computer. Although full details of the monitoring system were protected under the Classified Information Procedures Act, court records indicate that the system only captured keystrokes while the computer was not connected to the Internet via the modem. This explanation satisfied the court during an in camera, *ex parte* hearing but most key loggers do not function in this manner and this technique is of limited effect when a computer is continuously connected to the Internet or when the suspect writes e-mail offline and only connects to the Internet to send the messages. The court addressed this concern by comparing key logging to searching a closet or file cabinet.

> That the KLS (Key Logging System) certainly recorded keystrokes typed into Scarfo's keyboard other than the searched-for passphrase is of no consequence. This does not, as Scarfo argues, convert the limited search for the passphrase into a general exploratory search. During many lawful searches, police officers may not know the exact nature of the incriminating evidence sought until they stumble upon it. Just like searches for incriminating documents in a closet or file cabinet, it is true that during a search for a passphrase "some innocuous [items] will be at least cursorily perused in order to determine whether they are among those [items] to be seized."
> (United States v. Scarfo)

Even when data on a disk is deleted and overwritten, a "shadow" of the data might remain as shown in Figure 8.3. These shadow data are a result of the minor imprecision that naturally occurs when data are being written on a disk. The arm that writes data onto a disk has to swing to the correct place, and it is never perfectly accurate. Skiing provides a good analogy. When you ski down a snowy slope, your skis make a unique set of curving tracks. When people ski down behind you, they destroy part of your tracks when they ski over them but they leave small segments.

A similar thing happens when data is overwritten on a disk – only some parts of the data are overwritten leaving other portions untouched. A disk can be examined for shadow data in a lab with advanced equipment (e.g. scanning probe microscopes, magnetic force microscopes) and the recovered fragments can be pieced together to reconstruct parts of the original digital data.

9.6 RECONSTRUCTION

As discussed in Chapter 5, investigative reconstruction leads to a more complete picture of a crime – what happened, who caused the events when, where, how, and why. The three fundamental types of reconstruction – temporal, relational, and functional – are discussed in the following sections.

9.6.1 FUNCTIONAL ANALYSIS

In an investigation, there are several purposes to assessing how a computer system functioned:

- To determine if the individual or computer was capable of performing actions necessary to commit the crime.
- To gain a better understanding of a piece of digital evidence or the crime as a whole.
- To prove that digital evidence was tampered with.
- To gain insight into an offender's intent and motives. For instance, was a purposeful action required to cause the damage to the system or could it have been accidental?
- To determine the proper working of the system during the relevant time period. This relates to authenticating and determining how much weight to give digital evidence as described in Chapter 7.

For example, a log file generated by a suspect's Eudora e-mail client appears to support his claim that he was checking e-mail from his home computer when the crime was committed across town. However, Eudora was configured to save his password and automatically check for new messages every 15 minutes. Therefore, the Eudora log file does not support the suspect's alibi as was originally thought.

CASE EXAMPLE (GREATER MANCHESTER 1974–1998):
Harold Shipman, a doctor in England, killed hundreds of his patients over several decades. To conceal his activities, Shipman regularly deleted and altered patient records in his Microdoc medical database. Digital investigator, John Ashley, studied the database software and found that it maintained an audit trail of changes. This audit trail showed discrepancies, including dates of altered records that helped demonstrate Shipman's intent and guilt. Interestingly, during the trial, Shipman claimed that he was aware of the Microdoc audit trail feature and that he knew how to deceive the system by changing the internal date of the computer. (Baker 2000)

As another example of how functional details can be important, consider illegal materials found on a computer that appear to have been downloaded from the Internet. The digital investigator calculated that 4,000 Mbytes of data were placed on the system in 6 minutes. However, the Internet connection speed is 10 Mbps, which has a theoretical maximum transfer rate of 75 Mbytes per minute (10 Mbits/second × 60 seconds ÷ 8 bits/byte). Therefore, the materials could not have come from the Internet and must have been placed on the system in some other way. Similarly, before asserting that an individual intentionally created a given file on a computer, it is advisable to consider alternative ways that the data may have been placed on the system.

CASE EXAMPLE

Files containing images of young girls (a.k.a. lolita material) were found on a work computer and their locations and creation times implicated a specific employee. The employee denied all knowledge of the materials and further investigation found that an adult pornographic Web site that the employee visited had created the files by exploiting a vulnerability in Internet Explorer.

It may be necessary to experiment with a program to determine how it functions and understand the meaning of data it creates. In one case, the offender claimed that he could not remember the password protecting his encryption key because he had changed it recently. By experimenting with the same encryption program on a test system, the digital evidence examiner observed that changing the password updated the modification date–time stamp of the file containing the encryption key. An examination of the file containing the suspect's encryption key indicated that it had not been altered recently as the suspect claimed. Faced with this information, the suspect admitted that he had lied about changing the password.

CASE EXAMPLE (GERMANY 1989):

Michael Peri, an electronic signals analyst in the military intelligence section stationed near the East German border was convicted of, and subsequently pled guilty to, providing the East German government with US government secrets stored on a laptop computer. Peri would not divulge what information he had given the East Germans and it was necessary to analyze the laptop and diskettes for evidence of espionage.

> … some investigators might think all that was needed was to copy the diskettes and hard drive, look at any documents or free/slack space for any classified documents and, if so, charge Peri with espionage. However, the charge of espionage requires proof that such information was transmitted to a foreign power, not just its presence. (Flusche 2001)

Two files associated with printing from a word processing application called MultiMate had been modified while Peri was in East Germany with the laptop. One of these files contained a reference to a type of printer that was not present in the US military unit in question. The second file, named "wpque.sys," contained a reference to a classified document found on one of the diskettes. By testing the functionality of MultiMate on an identical laptop to determine the significance of these two files, the examiners were able to demonstrate that a secret document had been printed while Peri was in East Germany with the laptop.

Applying the pattern of file changes from the testing to the two MultiMate system files in the root directory would show that on February 22, 1989, at about 11:52 A.M. (adjusting for the one-hour time difference with the laptop), someone initiated a change to the program MultiMate to change its printer designation to a LaserJet A, and then 51 minutes later, used the printer to print out a document with the partial name NEXB.DOC.

Interestingly, in this case the laptop was dusted for fingerprints. Although none were found on the keyboard and case, indicating that it had been wiped to destroy fingerprint evidence, a thumbprint was found on one bootable diskette found in the laptop's floppy drive and several fingerprints (not Peri's) were found on the screen, possibly where someone pointed to data being displayed.

In addition to testing individual programs, it is often desirable to see how the entire system functioned and was configured. For instance, when investigating computer intrusions, it is often necessary to examine a rootkit using a clone of the compromised system to understand fully how the rootkit functions and what evidence it may have destroyed or concealed. To perform this type of functional analysis without altering the original evidence, digital evidence examiners create a clone of the original system by restoring the contents of the hard drive to a new drive.

9.6.2 RELATIONAL ANALYSIS

In an effort to identify relationships between suspects, victim, and crime scene, it can be useful to create nodes that represent places they have been, e-mail and IP addresses used, financial transactions, telephone numbers called, etc. and determine if there are noteworthy connections between these nodes. For instance, in large-scale fraud investigation, representing fund transfers by drawing lines between individuals and organizations can reveal the most active entities in the fraud. Similarly, depicting e-mail messages sent and received by a suspect can help investigators spot likely cohorts by the large numbers of messages exchanged.

CASE EXAMPLE

A woman receives a threatening e-mail message and investigators track it back to a particular apartment. The man in the apartment appears to be cooperative and investigators cannot find any related digital evidence on his computer or any connection between him and the victim. However, by relational analysis of all e-mails on his computer and on the victim's computer, investigators determine that they both know one person in common: the woman's ex-boyfriend. A follow-up interview with the man reveals that the ex-boyfriend had been staying at the apartment when the message was sent. An examination of the ex-boyfriend's Web mail account reveals that he sent the threatening message.

In an intrusion investigation, drawing connections between computers on a relational diagram can provide an overview of the crime and can help locate sources of digital evidence that were previously overlooked.

[11]http://www.xanalys.com
[12]http://www.i2.co.uk
[13]http://www.netmap.com

Link analysis tools such as Watson,[11] The Analyst's Notebook,[12] and NetMap[13] provide a graphical interface to a database containing details gathered during an investigation.

9.6.3 TEMPORAL ANALYSIS

When investigating a crime, it is usually desirable to know the time and sequence of events. Fortunately, in addition to storing, retrieving, manipulating, and transmitting data, computers keep copious account of time. For instance, most operating systems keep track of the creation, last modification and access times of files and folders. These date–time stamps can be very useful in determining what occurred on a computer. In intellectual property theft investigations, date–time stamps of files can show how long it took the intruder to locate the desired information on a system. A minimal amount of searching indicates knowledge of where the data was located whereas a prolonged search indicates less knowledge. In a child pornography investigation, the suspect claimed that his wife put pornography on his personal computer without his knowledge during a bitter breakup to reflect poorly on him in the custody battle over their children. However, date–time stamps of the files indicated that they were placed on his system while his estranged wife was out of the country visiting family. Also, the suspect's computer contained remnants of e-mail and other online activities, indicating that he was using the computer at the time.

In addition to file date–time stamps, some individual applications embed date–time information within files or create log files or databases showing times of various activities on the computer, such as recently visited Web pages. Various locations of date–time information are presented in later chapters. All of these times can be skewed and even rendered useless, however, if their context is not documented. Therefore, when investigating a crime that involves computers, it is important to pay particular attention to the current date and time, any discrepancy between the actual time and the system time, the time zone of the computer clock, and the time stamps on individual digital objects.

Note that any errors in the setting of the system clock would be evident in e-mail messages sent from the system. If the system clock were several hours slow, it would place an incorrect date–time stamp in outgoing e-mail message headers. This can cause great confusion when trying to reconstruct events since it can give the impression that an individual was aware the content of an e-mail before the message was sent. For instance, if an e-mail message contains a link to a Web page but the browser history shows that the individual accessed the Web page a day before the message appears to have been sent, this can cause confusion. Looking at the e-mail header will show correct date–time stamps from servers that handled the message while it was being delivered.

CASE EXAMPLE

In a homicide investigation, one suspect claimed that he was out of town at the time of the crime. Although his computer suffered from a Y2K bug that rendered the date–time stamps on his computer useless, e-mail messages sent and received by the suspect showed that he was at home when the murder occurred, contrary to his original statement. Caught in a lie, the suspect admitted to the crime.

The simple act of creating a timeline of when files were created, accessed, and modified can result in a surprising amount of information. Creating a timeline of events can help an investigator identify patterns and gaps, shedding light on a crime and leading to other sources of evidence. For instance, Table 9.7 shows a timeline of a missing woman's activities on the days preceding her disappearance as reconstructed from her computer. This chronological sequencing of events helped investigators determine that the victim had traveled to Virginia to have a BDSM encounter with a man she met online. When investigators searched the man's home, they found the missing woman's body.

Table 9.7

Timeline of activities on victim's computer show e-mail correspondences, online chat sessions, deleted files, Web searching for maps, and online travel plans.

DATE	ACTIVITY
Day 1	Bondage/Sadomasochistic (BDSM) Web sites viewed, probably by missing individual
Day 2	Hotmail e-mail correspondences of a sexual/BDSM nature with unknown individual, IP address indicates Virginia. At around the same time as Hotmail is checked, Web pages from BDSM sites visited.
Day 3	Logs of online chat sessions show conversation of a sexual/BDSM nature with unknown individual, IP address indicates Virginia
Day 4	Driving directions obtained from Mapquest, address of destination in Virginia
Day 4	Files deleted
Day 4	No activity after 8 P.M.

Representing temporal information in different ways can highlight patterns. For instance, Figure 9.6 shows a histogram of date–time stamps from a computer used by shift workers in a company. One employee is suspected of

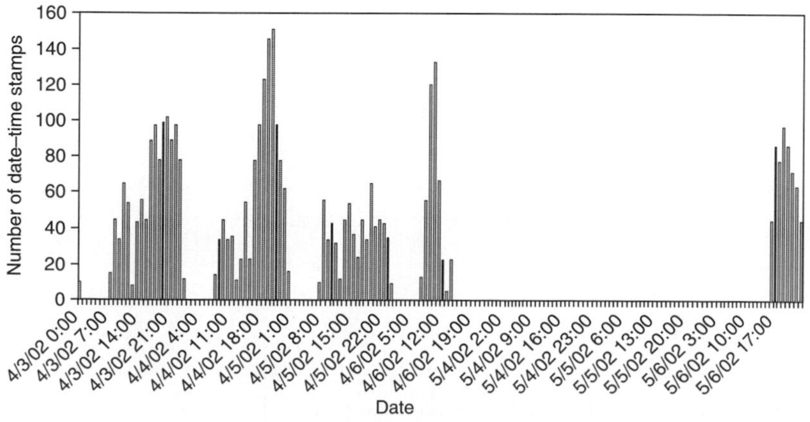

Figure 9.6

Histogram of date–time stamps (created and last modified) showing gaps during suspect's shifts.

viewing obscene and possibly illegal materials during his midnight to 8 A.M. shift but the date–time stamps place the activities on the previous shift (4 P.M. to midnight), implicating his coworker.

The gaps in Figure 9.6 suggest that the computer was not used during the suspect's shift but it is known from his access of network resources from the computer that he was using the computer at these times, indicating that the suspect regularly changed the system clock at the beginning of his shift. Interestingly, in one instance the suspect appears to have accidentally changed the month setting of the clock in addition to the time, creating 8 hours of "fill" on May 6 after 1600 hours, probably corresponding to a gap during his shift on April 6, supporting the hypothesis that he tampered with the system clock. Additionally, an automated backup process that was initiated by a central server contacting the computer in question every night at 0200 hours appeared in the Windows NT Application Event Log 8 hours earlier, supporting the theory that the clock had been altered.

The spike in Figure 9.6 on the morning of April 6 corresponds to the discovery of the obscene materials. The employee who discovered the material caused this flurry of activity because he used the computer to contact his supervisor, installed software on the computer in an effort to show his supervisor the materials, and performed other actions on the system that may have destroyed digital evidence. The supervisor viewed the materials and contacted investigators – the computer was only shutdown after the digital investigators arrived to examine the system.

Another approach to analyzing date–time information is using a grid to accentuate patterns in which events occurred. Table 9.8 shows e-mail sent by the head of a criminal group over several months to other members of the group. Communication about a criminal plan began in mid-June, dropped off in early July, and picked up again as the September 11 deadline approached.

Digital investigators should seek new ways to represent visually temporal information to help them recognize patterns. Plotting times on concentric circles or a spiral may cause certain patterns to stand out (Figure 9.7).

One question that arises when dealing with computers is: how important is accurate time? It has been argued that since computers can represent time

Email Address	Sun, Jun 16	Fri, Jun 21	Sun, Jun 23	Wed, Jun 26	Sat, Jun 29	Sun, Jun 30	Thu, Jul 11	Fri, Jul 26	Mon, Jul 29	Fri, Aug 2	Wed, Aug 14	Thu, Aug 15	Thu, Aug 29	Sun, Sep 8	Wed, Sep 11
member1	xx				x	x							xxx	xx	x
member2	xx		x	x			x		x		x	x	x	x	x
member3	xx	x	x	x			x	x		xxx			x	x	x

Table 9.8 Grid showing e-mail message sent by a suspect over several months to several members of a criminal group.

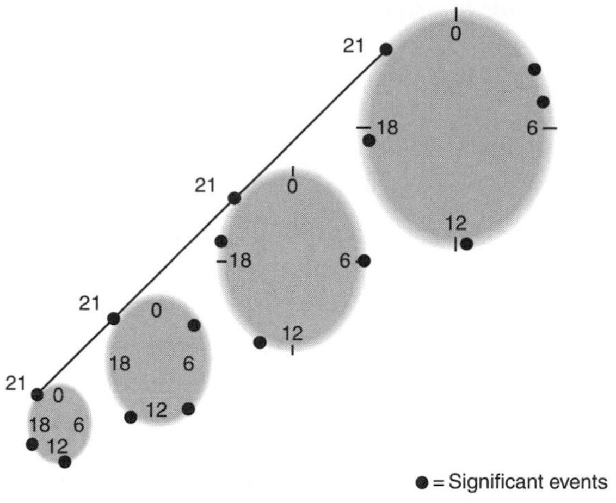

Figure 9.7

Conceptual image of 24-hour clocks with MAC times for several days with a line connecting significant events on sequential days.

to within a few milliseconds, all time-related information from computers should be this accurate. In some instances, when trying to distinguish between events that occurred in the same second, this degree of accuracy may be warranted. However, in most cases, differences in seconds are unimportant and it may even be sufficient to have times that are accurate to within a few minutes. Requiring millisecond accuracy in all situations is neither necessary nor desirable since it would create an insurmountable hurdle for most investigations involving computers.

9.6.4 DIGITAL STRATIGRAPHY

When time markers are obliterated, more imaginative approaches are required to get a sense of when data was created. Concepts from other fields can be translated into the digital realm to develop new analysis techniques such as *digital stratigraphy*.

Stratigraphy is the scientific study of layers (a.k.a. strata) in geology and archaeology with the aim of determining the origin, composition, distribution, and time frame of each stratum. Applying this concept to data stored on a disk can be fruitful in some investigations. For instance, when the creation time of a document is at issue, an examination of how data are positioned and overlaid on the disk may give a sense of when the document was created. If part of one document is found to be overwritten by another document, there is a good chance that the overwritten document was created first. This concept was applied in an extortion case to demonstrate that the suspect had created a document before leaving for holiday.

> During the investigation of an alleged blackmail attempt, a number of fragments of deleted material were recovered from a computer belonging to Mr S. These fragments when subjected to an analysis procedure provided a recognized sequence of revisions

> and changes to the blackmail letter over a period of time. Mr S had been on holiday for two weeks and although admitting that he had written a similar letter, he suggested that the letter had been modified on his computer by someone else during his absence. It was not possible to ascribe a reliable date or time to all of the fragments and in any case computer dates and times indicate only the setting of the internal clock and may have no relevance to real world dates and times.
>
> It happened however, that one of the fragments was in what is known as the "slack space" of another file (the owning file). The significance of this is that it is technically possible to show that the contents of slack space must have existed on the machine before the creation of the owning file. In this case the owning file was a letter to Mr S's bank manager and the date marking on the file was two days before Mr S went on his holiday. The bank manager was able to confirm receipt of the letter a day after the indicated date. Thus it could be shown that that fragment of the blackmail letter together with all previous fragments existed on the computer at least two days before the holiday. It will be seen that the content of the letter was immaterial except insofar as it enabled the bank manager to identify it unequivocally. (Bates 1999)

Notably, when a Microsoft Office document is being edited, data that are cut may still exist in the document or associated temporary files on disk enabling digital investigators to deduce that certain data were created prior to the last modified time of the document.

Windows date–time information exists in MS Word files, directory entries, cookie files, Internet-related files, NT Event logs, and may other files. UNIX has date–time information in various system logs and Internet-related files. Once deleted, these files form an underlying layer of time related data upon which newer files are saved. Examining slack space for time related data is challenging since systems store time in various formats. A useful tool for converting computer representations of time is the forensic date and time decoder[14] shown in Figure 9.8.

[14] http://www.digital-detective.co.uk

Keep in mind that there is more to digital stratigraphy than examining the time frame of layers. Useful conclusions may be reached based on the position of data on a disk (e.g. scattered versus concentrated), the origin of various

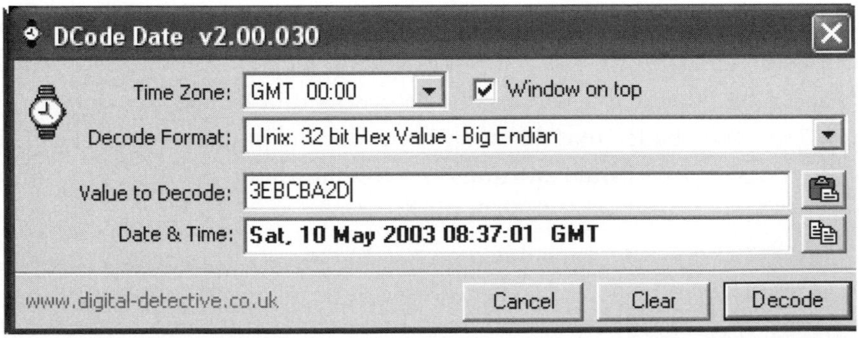

Figure 9.8

Forensic Date & Time Decoder. These times are generally GMT and must be adjusted for time zones.

fragments (e.g. from one source versus many sources), or the composition of the data. For instance, if two pieces of a file are located in clusters on either side of a large, contiguous file, it is likely that the fragmented file was created after the contiguous file. Similarly, proximity of data in swap files may indicate synchronicity but additional research must be performed before this assertion can be made.

As another example, a computer that is running a Linux operating system may have a large number of Microsoft Windows operating system files in unallocated space that contain information specific to the hardware of the machine (e.g. address of the Ethernet card), indicating that the machine was running Microsoft Windows before Linux was installed. The reason for this phenomenon is that formatting and repartitioning a disk does not overwrite all of the data on the disk. Therefore, when a new operating system is installed, it creates a new file structure on the disk and overwrites some data from the previous operating system but much of the previous data still exists in unallocated space.

As more is learned about how different systems store data, other applications to digital stratigraphy will be developed.

9.7 REPORTING

The last stage of a digital evidence examination is to integrate all findings and conclusions into a final report that conveys the findings to others and that the examiner may have to present in court. Writing a report is one of the most important stages of the process because it is the only view that others have of the entire process. Unless findings are communicated clearly in writing, others are unlikely to appreciate their significance. A well-rendered report that clearly outlines the examiner's findings can convince the opposition to settle out of court, while a weakly rendered report can fuel the opposition to proceed to trial. Assumptions and lack of foundation in evidence result in a weak report. Therefore, it is important to build solid arguments by providing all supporting evidence and demonstrating that the explanation provided is the most reasonable one.

Whenever possible, support assertions with multiple independent sources of evidence and include all relevant evidence along with the report since it may be necessary in court to refer to the supporting evidence when explaining findings in the report. Clearly state how and where all evidence was found to help decision-makers to interpret the report and to enable another competent examiner to verify results. Presenting alternative scenarios and demonstrating why they are less reasonable and less consistent with the evidence can help strengthen key conclusions. Explaining why other explanations are unlikely or impossible demonstrates that the scientific method was applied – that an effort was made to disprove the given conclusion but that it withstood critical scrutiny. If there is no evidence to support an alternative scenario, state whether it is more likely that relevant evidence was missed or simply not present. If digital evidence was altered after it was collected, it is crucial to mention this in the report, explaining the cause of the alterations and weighing their impact on the case (e.g. negligible, severe).

A sample report structure is provided here:

- Introduction: case number, who requested the report and what was sought, who the wrote report, when, and what was found.
- Evidence Summary: summarize what evidence was examined and when, MD5 values, laboratory submission numbers, when and where the evidence was obtained, from whom and its condition (note signs of damage or tampering).
- Examination Summary: summarize tools used to perform the examination, how important data were recovered (e.g. decryption, undeletion), and how irrelevant files were eliminated (see Chapter 24).
- File System Examination: inventory of important files, directories, and recovered data that are relevant to the investigation with important characteristics such as path names, date–time stamps, MD5 values, and physical sector location on disk. Note any unusual absences of data.
- Analysis: describe and interpret temporal, functional, and relational analysis and other analyses performed such as evaluation of source and digital stratigraphy.
- Conclusions: summary of conclusions should follow logically from previous sections in the report and should reference supporting evidence.
- Glossary of Terms: explanations of technical terms used in the report.
- Appendix of Supporting Exhibits: digital evidence used to reach conclusions, clearly numbered for ease of reference.

In addition to presenting the facts in a case, digital investigators are generally expected to interpret the digital evidence in the final report. Interpretation involves opinion and every opinion rendered by an investigator has a statistical basis. Therefore, in a written report, the investigator should clearly indicate the level of certainty he/she has in each conclusion and piece of evidence to help the court assess what weight to give them. The C-Scale (Certainty Scale) described in Chapter 7 provides a method for conveying certainty when referring to digital evidence and qualify conclusions appropriately. Some digital investigators use a less formal system of degrees of likelihood that can be used in both the affirmative and negative sense: (1) Almost definitely, (2) Most probably, (3) Probably, (4) Very possibly, and (5) Possibly.

When determining the certainty level of a given piece of digital evidence it may be important to consider the context. For instance, many Macintosh computers are unauthenticated and allow any user to change the system clock, making it more difficult for digital investigators to have confidence in the date–time stamps and to attribute activities to an individual. Computers that were not handled properly causing evidence to be altered or destroyed, make it more difficult to make strong assertions about the evidence they contain. Additionally, a wily offender may arrange evidence to misdirect digital investigators and the certainty of the evidence is reduced if there is no corroborating data from multiple independent sources.

In addition to a final, full-blown, technical report, digital investigators may be required to write reports for less technical decision-makers. For instance, managers in an organization may need to know what transpired to help them determine the best course of action. The public relations department may need details to relay to shareholders. Attorneys may need a summary report to help them focus on key aspects of the case and develop search or arrest warrants or interview and trial strategy. A measure of hard work and creativity is required to create clear, non-technical representations of important aspects in a case such as timelines, relational reconstructions, and functional analyses. However, the effort required to generate such representations is necessary to give attorneys, juries, and other decision-makers the best chance of understanding important details and making informed decisions.

9.8 SUMMARY

This chapter presents concepts from forensic science and computer science that can be used to process and analyze digital evidence stored on a computer. The Forensic Science concepts described in this chapter are applicable to any investigation and are applied to specific operating systems and computer networks in later chapters. Although this chapter focuses on information, it also provides some suggestions for dealing with hardware as contraband, fruits of crime, instrumentality, and evidence.

Computer technology is evolving rapidly but the fundamental components and operations are relatively static. A central processing unit starts the basic input and output system, which performs a power-on self test and loads an operating system from a disk. The process of collecting, documenting and preserving evidence also remains fairly static, making it possible to develop standard operating procedures (SOP) to avoid gross mistakes.

CASE EXAMPLE

A system administrator of a large organization was the key suspect in a homicide. The suspect claimed that he was at work at the time and so the police asked his employer to help them verify his alibi. Coincidentally, this organization occasionally trains law enforcement personnel to investigate computer crimes and was eager to help in the investigation. The organization worked with police to assemble an investigative team that seized the employee's computers – both from his home and his office – as well as backup tapes of a server the employee administered. All of the evidence was placed in a room to which only members of the team had access. These initial stages were reasonably well documented but the reconstruction process was a disaster. The investigators made so many omissions and mistakes that one computer expert, after reading the investigator's logs, suggested that the fundamental mistake was that the investigators locked all of the smart people out of the room. The investigators in this case were either unaware of their lack of knowledge or were unwilling to admit it.

This case demonstrates how critical it is for digital investigators to realize their limitations and seek help when necessary. As a result of the investigators' omissions and mistakes, the suspect's alibi could not be corroborated. Digital evidence to support the suspect's alibi was found later but not by the investigators. If the investigators had sought expert assistance to deal with the large amount of digital evidence they might have quickly confirmed the suspect's alibi rather than putting him through years of investigation and leaving the murderer to go free.

Given the variety of systems and situations, it is difficult to create procedures that anticipate all eventualities. Additionally, writing down exactly how something should be done limits the individual's ability to make intelligent decisions and gives attorneys opportunity to criticize such intelligent decisions because they were not part of a SOP. Therefore, an SOP should contain general descriptions of important steps and should be used as a memory aid rather than a rigid guide.

Digital investigators must be capable of going beyond procedures, applying the concepts presented in this chapter to new situations. Comparing items to discern class characteristics or determine where they originated is a fundamental task in forensic analysis. On their own, class characteristics may not be particularly illuminating, but in combination they can help direct an investigation, eliminate suspects, or create a break in a theory. Evaluation of source often requires extensive searching of surroundings, examination of similar objects, and comparative research. Evaluating the source of digital evidence is particularly important when trying to prove that an individual manufactured child pornography, created a computer virus, or stole a piece of intellectual property. In the case of child pornography, class characteristics can indicate that one image was created on the defendant's digital camera while another image was a photograph that was digitized using his neighbor's flatbed scanner.

Performing temporal, functional, and relational analyses of digital evidence is necessary to recreate a complete picture of a crime. Combining the results of such analyses into a full investigative reconstruction can help investigators understand the crime and the offender as detailed in Chapter 5. As the final stage, reporting is one of the most important activities and should be given the time and attention it deserves. Without a clearly written report, it is difficult for decision makers to understand the results of a digital evidence examination and impairs their ability to reach a verdict based on the truth.

REFERENCES

Baker R. (2000) "Harold Shipman's Medical Practice 1974–1998", Department of Health Audit Report, (Available online at http://www.doh.gov.uk/hshipmanpractice/shipman.pdf)

Bates J. (1999) "Judicial Review relating to Search Warrants – Discussion Paper", International Journal of Forensic Computing (Available online at http://www.forensiccomputing.com/archives/judicial.html)

Flusche K. J. (2001) "Computer Forensic Case Study: Espionage, Part 1 Just Finding the File is Not Enough!", Information Systems Security, March/April 2001, Auerbach

Mattei M., Blawie J. F. and Russell A. (2000) Connecticut Law Enforcement Guidelines for Computer Systems and Data Search and Seizure, State of Connecticut Department of Public Safety and Division of Criminal Justice.

Meighan C. W. (1966) Archaeology: an Introduction, p. 18, San Francisco: Chandler Publishing Company.

Thornton J. I. (1997) "The General Assumptions and Rationale of Forensic Identification", for David L. Faigman, David H. Kaye, Michael J. Saks, & Joseph Sanders, Editors, Modern Scientific Evidence: The Law and Science of Expert Testimony, Volume 2, St. Paul, MN: West Publishing Company.

United States Department of Justice (2001) "Electronic Crime Scene Investigation: A Guide for First Responders", National Institute of Justice, NCJ 187736 (Available online at http://www.ncjrs.org/pdffiles1/nij/187736.pdf).

CASES

Honeywell v. Rand (1973) District Court, Minnesota, 4th division, Civil Action Number 4-67 CIV. 138 (Available online at http://www.cs.iastate.edu/jva/court-papers/).

United States v. Carey (1998) Appeals Court, 10th Circuit, Case Number 98-3077 (Available online at http://laws.findlaw.com/10th/983077.html)

## CHAPTER 10

FORENSIC EXAMINATION OF WINDOWS SYSTEMS

In addition to being familiar with the tools and techniques for acquiring and examining digital evidence from a computer running Microsoft Windows, digtal investigators should develop a familiarity with the underlying operating systems, files systems, and applications.

Understanding file systems helps appreciate how information is arranged, giving insight into where it can be hidden on a Windows system and how it can be recovered and analyzed. An understanding of Windows NT accounts, file access controls, and general security is also necessary to answer questions like: Who had access to the system and files it contained? Was it possible for an outsider to gain unauthorized access to the system from the Internet? Similarly, it is necessary to understand components such as Active Directory to locate and interpret digital evidence relating to systems that are part of a Windows 2000 domain.

Digital investigators must also keep abreast with new developments in this area such as ".NET" framework. The ".NET" framework can be thought of as an operating system within an operating system. It is an execution environment, similar in concept to Java, that is designed to run on post-Windows 95 operating systems (Windows 98/ME/NT/2000/XP) and provide a common environment for programs. This enables programmers to write applications in their preferred language (e.g. Visual Basic, C++, Perl) and compile them for the ".NET" environment, providing greater flexibility and functionality. A program compiled and linked to run in the ".NET" Framework environment has a new EXE or DLL format that can only be executed on a system that contains the framework. The ".NET" framework is optimized for network activities and enhances the capabilities of the operating system it is running on – making it easier to develop network applications.

Given the variety of Windows operating systems and applications, it is not possible to describe or even identify every possible source of information that might be useful in an investigation. Furthermore, each case is different, requiring digital investigators to explore and research components. The following sections provide examples of important aspects of Windows

Digital Evidence and Computer Crime Second Edition
ISBN: 0-12-163104-4

systems with the expectation that the reader will carefully consider each area more closely to find new ways to extract information from them using the techniques covered in the previous chapter.

10.1 WINDOWS EVIDENCE ACQUISITION BOOT DISK

Whether copying evidence from a disk, previewing a system to verify that a crime occurred, or performing a keyword search to determine if the computer contains useful evidence, the computer's operating system should be bypassed to avoid altering evidence, and to avoid any tricks or traps that an advanced user might have set up. As described in Chapter 8, most computers store their operating system on a hard drive, and this operating system can be bypassed using a boot disk. However, extra precautions are required to write protect the drive and ensure that the digital evidence is not altered while it is being processed.

The first step in creating a Windows Evidence Acquisition Boot Disk is to modify the "command.com" and "io.sys" system files to prevent it from accessing any system components on the evidentiary drive. The second step is to delete the "drvspace.bin" file because it attempts to open compressed volumes. A detailed description of this process along with a sample script is available in Larson (2002). Alternatively, Windows systems can be booted using a Linux floppy disk or CD-ROM such as FIRE described in the next chapter.

Until recently, the most common approach to write protecting a hard disk was using software. Recall from Chapter 8 that operating systems write data to hard disks through a computer's BIOS (basic input and output system). Specifically, there are a group of BIOS functions collectively named "INT13h" that control disk access (e.g. read, write, format). A carefully constructed program such as such as PDBlock[1] can intercept calls to these INT13h functions, thus preventing write access to a hard drive. This software approach to write protecting a hard disk is not always successful because of the variations between systems. A more reliable alternative is to connect a piece of hardware to the hard drive that blocks the signals that would cause the disk to be modified. These hardware write blockers have some limitations, preventing access to certain types of disks.

[1] *http://www.digitalintel.com/pdblock.htm*

There are two notable nuances to using a Windows Evidence Acquisition Boot Disk. When devices such as a Zip drive or Ethernet card are being used to transfer data to a collection disk, the necessary drivers must be stored on and loaded from the boot disk. For instance, Ethernet drivers are needed when using a tool like EnCase to preview or acquire evidence via a network cable. Also, because MS-DOS does not support NTFS, it is not possible to save evidence files to an NTFS drive when using a Windows Evidence Acquisition

Boot Disk. Using FAT32 on collection disks allows for large evidence files to be saved. Boot disks should be virus checked before use to avoid damaging the computer and the digital evidence that it contains.

10.2 FILE SYSTEMS

The simplest Windows file systems to understand are the FAT (file allocation table) file systems: FAT12, FAT16, and FAT32. To locate data on a volume, these file systems use directories and a FAT. The root directory (e.g. C:\) is at a pre-specified location on the volume so that the operating system knows where to find it (recall Figures 8.5 and 10.1). This directory contains a list of files and subdirectories on a floppy diskette with their associated properties as shown here through Norton Disk Editor.[2]

[2]This floppy diskette is referenced in a case example later in this chapter. A bitstream copy of this disk is available on the Web site associated with this book (http://www.disclosedigital.com/decc2/).

[] Disk Editor

Object Edit Link View Info Tools Help More>

Name	.Ext	ID	Size	Date	Time	Cluster	76 A R S H D V
Sector 19							
SALES		Vol	0	4-13-03	3:36 pm	0	A - - - - V
ix.doc		LFN				0	- R S H - V
skiways-getaf		LFN				0	- R S H - V
SKIWAY~1	DOC	File	21504	5-13-03	11:58 am	184	A - - - - -
todo.txt		LFN				0	- R S H - V
TODO	TXT	File	122	5-13-03	12:40 pm	226	A - - - - -
t		LFN				0	- R S H - V
newaddress.tx		LFN				0	- R S H - V
NEWADD~1	TXT	File	122	5-13-03	12:42 pm	227	A - - - - -
greenfield.do		Del LFN				0	- R S H - V
σREENF~1	DOC	Erased	19968	5-08-03	2:34 pm	275	A - - - - -
april		Del LFN				0	- R S H - V
σPRIL		Erased	0	5-08-03	2:41 pm	157	- - - - D -
contacts.xls		LFN				0	- R S H - V
CONTACTS	XLS	File	16896	2-18-01	12:49 pm	314	A R - - - -
		Unused directory entry					
Sector 20							
		Unused directory entry					
		Unused directory entry					

Root Directory — Sector 19

A:\ — Offset 0, hex 0

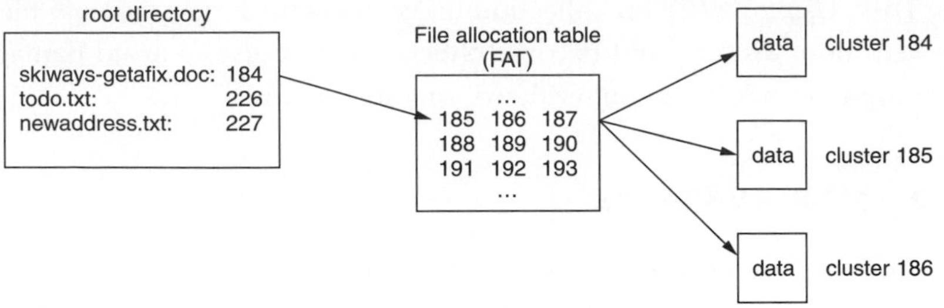

Figure 10.1

Root directory (skyways-getafix.doc, starts in cluster 184) → FAT→ data in clusters 184–225 (42 clusters×512 bytes/clusters = 21504 bytes).

This view of the FAT shows the last modified date and time of each file. The last accessed data and the creation date and time can be displayed by selecting the "More→" menu.[3] Notably, FAT file systems do not record the last accessed time, only the last accessed date. Listing the contents of a volume using the dir command displays some of this information but does not show the starting cluster – a critical component from the file system perspective. In addition to indicating where the file begins, the starting cluster directs the operating system to the appropriate entry in the FAT. The FAT can be thought of as list with one entry for each cluster in a volume. Each entry in the FAT indicates what the associated cluster is being used for. The following output from Norton Disk Editor shows a file allocation table from the same floppy diskette.

[3]FAT represents time since January 1, 1980 and NTFS represents times as the number of 100-nanosecond intervals since January 1, 1601 00:00:00 UTC.

[ ] Disk Editor

Object Edit Link View Info Tools Help

0	0	0	0	0	0	0	0
185	186	187	188	189	190	191	192
193	194	195	196	197	198	199	200
201	202	203	204	205	206	207	208
209	210	211	212	213	214	215	216
217	218	219	220	221	222	223	224
225	<EOF>	<EOF>	<EOF>	0	0	0	0
0	0	0	0	0	0	0	0
0	0	0	0	0	0	0	0
0	0	0	0	0	0	0	0
0	0	0	0	0	0	0	0
0	0	0	0	0	0	0	0
0	0	0	0	0	0	0	0
0	0	0	0	0	0	0	0
0	0	0	0	0	0	0	0
0	0	0	0	0	0	0	0
0	0	0	0	0	0	0	0
0	0	315	316	317	318	319	320

321	322	323	324	325	326	327	328
329	330	331	332	333	334	335	336
337	338	339	340	341			
FAT (1st Copy)						Sector 1	
Drive A:						Cluster 184, hex B8	

Clusters containing a zero are those free for allocation (e.g. when a file is deleted, the corresponding entry in the FAT is set to zero). If a FAT entry is greater than zero, this is the number of the next cluster for a given file or directory. For instance, the root directory indicates that file "skyways-getafix.doc" begins at cluster 184. The associated FAT entry for cluster 184, shown in bold, indicates that the file is continued in cluster 185. The FAT entry for cluster 185 indicates that the file is continued in cluster 186, and so on (like links in a chain) until the end-of-file (EOF) marker in cluster 225 is reached. In this example, Cluster 226 relates to a different file ("todo.txt") that occupies only one cluster and therefore does not need to reference any other clusters and simply contains an EOF.

Subdirectories are just a special type of file containing information such as names, attributes, dates, times, sizes, and the first cluster of each file on the system. For instance, before the directory named "april" on the floppy diskette was deleted and overwritten, it occupied cluster 157 and contained the following:

```
2E202020 20202020 20202010 00343675 | .               .    .46u
A82EA82E 00003675 A82E9D00 00000000 | ¿.¿.   ..6u   ¿.¥.   ....
2E2E2020 20202020 20202010 00343675 | ..              .    .46u
A82EA82E 00003675 A82E0000 00000000 | ¿.¿.   ..6u   ¿...   ....
E573006B 00690077 0061000F 002C7900 | ρs.k   .i.w   .a..   .,y.
73002E00 64006F00 63000000 0000FFFF | s...   d.o.   c...   ..
E54B4957 41595320 444F4320 002A8373 | ρKIW   AYS    DOC    .*âs
A82EA82E 00001448 8E2E7600 004E0000 | ¿.¿.   ...H   Ä.v.   .N..
E567006C 006F0062 0061000F 00236C00 | ρg.1   .o.b   .a..   .#l.
63006F00 6D002E00 64000000 6F006300 | c.o.   m...   d...   o.c.
E54C4F42 414C7E31 444F4320 00A97B73 | ρLOB   AL~1   DOC    .⌐{s
A82EA82E 00002848 8E2E0200 004E0000 | ¿.¿.   ..(H   Ä...   .N..
E5680061 006E0064 0072000F 00156900 | ρh.a   .n.d   .r..   ..i.
67006800 74002E00 64000000 6F006300 | g.h.   t...   d...   o.c.
E5414E44 52497E31 444F4320 00618173 | ρAND   RI~1   DOC    .aüs
A82EA82E 00000648 8E2E4F00 004E0000 | ¿.¿.   ...H   Ä.O.   .N..
E565006E 00670069 006E000F 001D7500 | ρe.n   .g.i   .n..   ..u.
69007400 79002E00 64000000 6F006300 | i.t.   y...   d...   o.c.
E54E4749 4E557E31 444F4320 00A17D73 | ρNGI   NU~1   DOC    .í}s
A82EA82E 00005047 8E2E2900 004C0000 | ¿.¿.   ..PG   Ä.).   .L..
```

This translates to the following directory listing with four deleted files:

Name	Created	Written	Accessed	Size	Cluster
.	05/08/03 02:41:44PM	05/08/03 02:41:44PM	05/08/03	0	157
..	05/08/03 02:41:44PM	05/08/03 02:41:44PM	05/08/03	0	0
σskiways.doc	03/19/80 12:03:50AM	03/03/80 12:03:30AM	01/14/80	4294901760	6553600
σKIWAYS.DOC	05/08/03 02:28:06PM	04/14/03 09:00:40AM	05/08/03	19968	118
σglobalcom.doc	03/03/80 12:03:24AM	03/04/80 12:01:28AM	03/15/80	6488175	7143424
σLOBAL~1.DOC	05/08/03 02:27:54PM	04/14/03 09:01:16AM	05/08/03	19968	2
σhandbright.doc	03/07/80 12:03:18AM	03/04/80 12:01:28AM	03/08/80	6488175	7602176
σANDRI~1.DOC	05/08/03 02:28:02PM	04/14/03 09:00:12AM	05/08/03	19968	79
σenginuity.doc	03/09/80 12:03:42AM	03/04/80 12:01:28AM	03/20/80	6488175	7929856
σNGINU~1.DOC	05/08/03 02:27:58PM	04/14/03 08:58:32AM	05/08/03	19456	41

When an individual instructs a computer to open a file in a subdirectory (e.g. "C:\april\handbright.doc"), the operating system goes to the root directory, determines which cluster contains the desired subdirectory (cluster 157 for "april"), and uses the directory information in that cluster to determine the starting cluster of the desired file (cluster 79 for "handbright.doc"). If the file is larger than one cluster, the operating system refers to FAT for the next cluster for this file. The entire file is read by repeating this "chaining" process until an EOF marker is reached.

FAT12 uses 12-bit fields for each entry in the FAT and is mainly used on floppy diskettes. FAT16 uses 16-bit fields to identify a particular cluster in the FAT and there must be fewer than 65,525 clusters on a FAT16 volume. This is why larger clusters are needed on larger volumes – a 1 Gbyte volume can be fully utilized with 65,525 16 kB clusters (32 sectors per cluster) whereas a 2 Gbyte volume requires clusters that are twice as big; that is, 65,525 32 kB clusters (64 sectors per cluster). FAT32 was created to deal with larger hard drives by using 28-bit fields in the FAT (4 bits of the 32-bit fields are "reserved"). FAT32 also makes better use of space, by using smaller cluster sizes than FAT16 – this can be a disadvantage for investigators because it can reduce the amount of slack space.[4]

[4]FAT16 file systems in Windows 95 and later versions support long file names, storing the long names using Unicode format in special entries in the parent directory. For more detailed discussion see Sammes and Jenkinson (2000, pp. 164–165).

NTFS is significantly different from FAT file systems, storing information in a Master File Table (MFT), supporting larger disks more efficiently (resulting in less slack space), and providing file and directory level security using Access Control Lists (ACLs), and more. The MFT is a list of records that contains most of the information needed to locate data on the disk. Records in the MFT contain the created, last modified, and last accessed dates and times. Directories are treated much like any other file in NTFS but are called *index entries* and store directory entries in a B-Tree to accelerate access and facilitate resorting when entries are deleted. Instead of using ASCII to represent data such as file and folder names, NTFS uses an encoding scheme called *Unicode.* This difference must be taken into account when performing text searches.

NTFS creates MFT entries as they are needed. However, recovering deleted files in NTFS can be complicated by the fact that unused entries in the MFT are reused before new ones are created. Therefore, when a file is deleted, the next file that is created may overwrite the MFT entry for the deleted file. However, if many files are created and then deleted, causing the MFT to grow, those entries will remain indefinitely since new files will reuse earlier entries in the MFT. Another feature of NTFS that makes it more difficult to recover a deleted file is that it keeps directory entries sorted by name. When a file is deleted, a resorting process occurs that may overwrite the deleted directory entry with entries lower down in the directory, breaking a crucial link between the file name and the data on disk.

NTFS is a journaling file system, retaining a record of file system operations that can be used to repair any damage caused by a system crash. There are currently no tools available for interpreting the journal file (called "$Logfile") on NTFS to determine what changes were made. This is a potential rich source of information from a forensic standpoint that will certainly be exploited in the future. For more detailed discussion of NTFS, see the *Handbook of Computer Crime Investigation,* Chapter 7 (Sheldon 2002).

10.3 OVERVIEW OF DIGITAL EVIDENCE PROCESSING TOOLS

Prior to making a bitstream copy of a disk, it may be necessary to perform a keyword search to determine if there is relevant digital evidence on the system. This is particularly useful when looking for specific items on a large number of systems. The most efficient approach to searching many computers is to boot them using an evidence acquisition boot disk and run a disk search utility from the DOS prompt. EnCase,[5] DiskSearch Pro,[6] and Linux have this keyword search capability. Once a system with useful evidence has been identified, a full bitstream copy can be made for further examination.

[5]*http://www.encase.com*
[6]*http://www.forensics-intl.com*

As noted in Chapter 2, Safeback[7] was one of the earliest DOS-based tools for copying digital evidence sector-by-sector. Booting from a floppy disk, Safeback can make an exact copy of a drive in a way that preserves its integrity. Since then, several tools have been developed to acquire evidence from a disk, including EnCase, Forensic Toolkit[8], SnapBack DatArrest,[9] and Byte Back.[10] Rather than calculating integrity checks of acquired data separately, EnCase and Safeback store acquired data along with integrity checks

[7]*http://www.sydex.com*

[8]*http://www.accessdata.com*
[9]*http://www.cdp.com*
[10]*http://www.toolsthatwork.com*

at regular intervals throughout their evidence files. Some believe that these proprietary formats are not an exact copy of the disk even though they contain all of the information from the disk. The crux of this argument is that data is arranged differently in the proprietary file format than on the original disk (Scott 2003). Others believe that this proprietary format is better because it maintains integrity checks throughout the file, enabling digital investigators to identify what portion of data is creating a problem if there is a problem. Whichever approach is used, courts are generally satisfied provided the evidence can be authenticated as described in Chapter 7. Also, since it is advisable to make two copies using different tools, one copy can be made in a proprietary format and the other using the *de facto* dd standard.

Most of these tools can either use information from the BIOS, or bypass the BIOS and access the disk directly to ensure that no false information in the BIOS causes a partial acquisition. Some of these tools contained bugs that prevented them from acquiring all of the data on some drives. For this reason, it is important to compare the amount of data that were copied with the size of the drive (Cylinders × Heads × Sectors per track) as described in Chapter 8.

Once digital evidence has been acquired, there are two main approaches to viewing digital evidence: physically and logically. The physical view involves examining the raw data stored on disk using a disk editor such as Norton DiskEdit or WinHex. Data are generally shown in two forms in a disk viewer: in hexadecimal form on the left and in plain text on the right. The advantage of DiskEdit is that it can run from a bootable floppy disk but WinHex has more examination and analysis capabilities such as recovering all slack or unallocated space, and comparing files to find any differences. For instance, Figure 10.2 shows WinHex being used to compare two seemingly identical Microsoft Word documents created at different times to locate internal date–time stamps discussed later in this chapter.

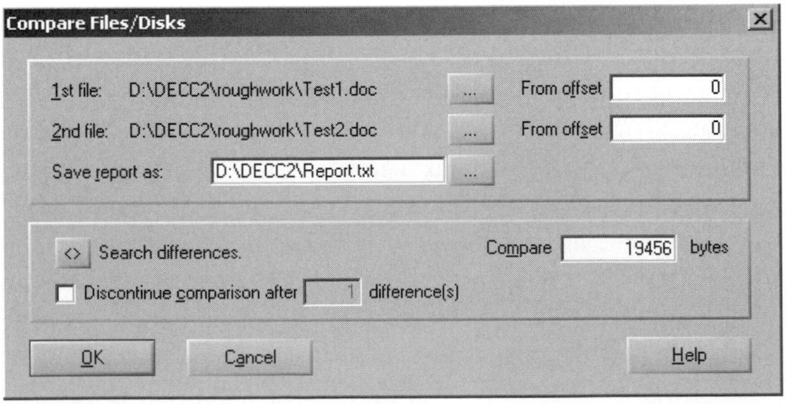

Figure 10.2

WinHex "File Manager Compare" feature.

The logical view involves examining data on a disk as it is represented by the file system. In the past digital investigators used Norton Commander (Figure 10.3) to view the file structure on a drive. Viewing the file system in this way facilitates certain types of analysis but does not show underlying information that is visible using a disk editor. Also, Norton Commander displays limited file information such as name, size, modification time, and attributes.[11]

Each of the above methods of viewing a disk has limitations. For instance, when searching for a keyword, a physical sector-by-sector search will not find occurrences of the keyword that are broken across non-adjacent sectors (the sectors that comprise a file do not have to be adjacent). On the other hand, a physical examination gives access to areas of the disk that are not represented by the file system such as file slack and unallocated space. Integrated tools like EnCase and Forensic Toolkit (FTK[12]) on Windows, and The Sleuth Kit[13] on UNIX combine both of these and other features into a single tool, enabling an examiner to view a disk physically and logically. EnCase and FTK have many other capabilities that facilitate examine of digital evidence, some of which are demonstrated later in this chapter. It is critical to realize that any tool that represents data on a disk can contain bugs that misinterpret data. Therefore, verify important results using multiple tools.

[11] *Some file viewing programs alter last accessed date–time stamps and should not be used on the original disk.*

[12] *http://www.accessdata.com*

[13] *http://www.sleuthkit.org*

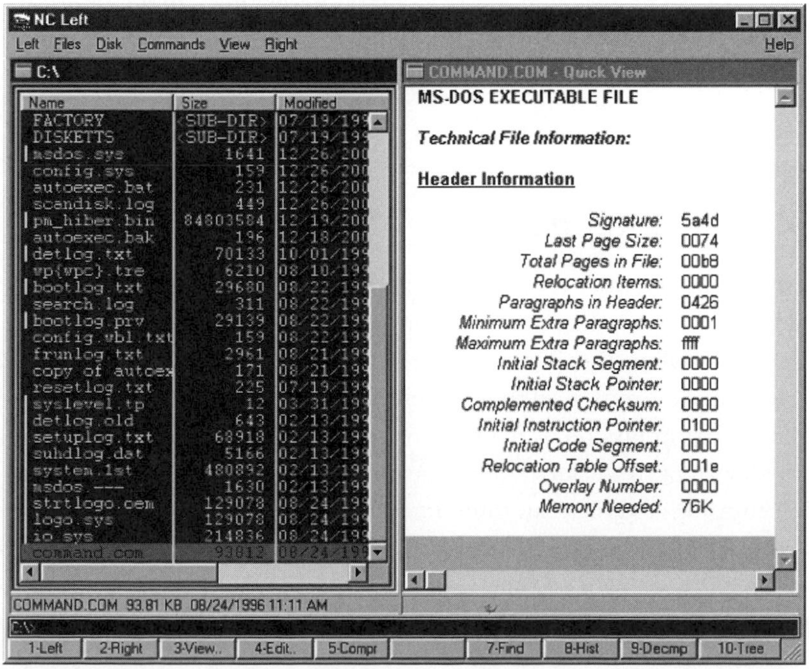

Figure 10.3
Norton Commander.

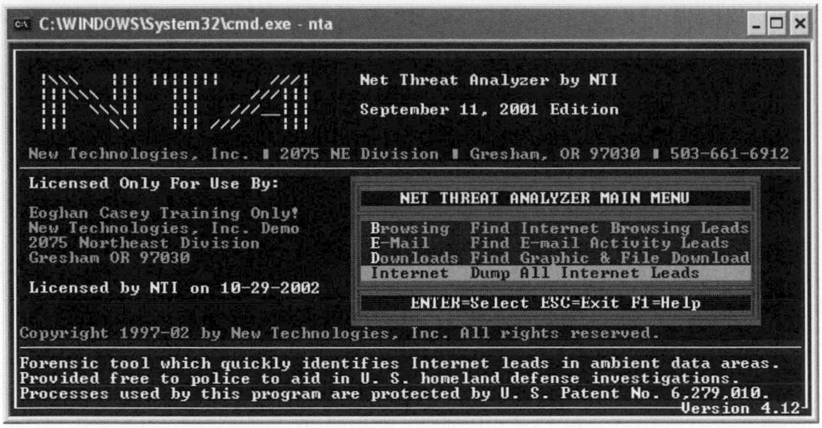

Figure 10.4

NTI Net Threat Analyzer.

Other tools exist to facilitate specialized tasks during examination such as Maresware utilities described in Chapter 24. Also, Net Threat Analyzer from NTI (www.secure-data.com) will search a binary file such as unallocated space or a swap file for Internet-related data such as e-mail addresses and Web pages (Figure 10.4).

No single tool is suitable for all purposes and it is advisable to verify important findings with multiple tools to ensure that all findings are accurate. In some cases, it is advisable to verify results at the lowest level using a disk editor. There is still some debate regarding the best approach to examine digital evidence – using tools from the command line or through a Graphical User Interface. Provided the forensic principles outlined in Chapter 9 are abided by, it does not matter if the tool has a Windows interface or must be run from the command line.

10.4 DATA RECOVERY

Although automated tools are necessary to perform routine forensic examination tasks efficiently, it is important to understand the underlying process to explain them in court or perform them manually in situations where the tools are not suitable. There are two main forms of data recovery in FAT file systems: recovering deleted data from unallocated space and recovering data from slack space.[14]

Recently deleted files can sometimes be recovered from unallocated space by reconnecting links in the chain as described in Section 10.2. For instance, to recover the deleted file named "greenfield.doc" on the aforementioned floppy diskette it is necessary to modify its entry in the root directory, replacing the sigma ("σ") with an underscore ("_") as shown here. The sigma is

[14] *A full discussion of recovering lost or hidden partitions is beyond the scope of this text. EnCase, gpart and testdisk (see Chapter 11) can be used to recover partitions on disks with incorrect or damaged partition tables.*

used on FAT file systems to indicate that a file is deleted. Notably, this recovery process must be performed on a copy of the evidentiary disk because it requires the examiner to alter data on the disk.

Name	.Ext	ID	Size	Date	Time	Cluster	76 A R S H D V
_REENF~1	DOC	Erased	19968	5-08-03	2:34 pm	275	A - - - - -

Then it is necessary to observe that the file begins at cluster 275 and its size is the equivalent of 39 clusters (19,968 bytes ÷ 512 bytes/cluster = 39 clusters). Assuming that all of these clusters are contiguous, the FAT can be modified to reconstruct the chain as shown here in bold.

[] Disk Editor

Object Edit Link View Info Tools Help

0	0	0	0	0	0	0	0
185	186	187	188	189	190	191	192
193	194	195	196	197	198	199	200
201	202	203	204	205	206	207	208
209	210	211	212	213	214	215	216
217	218	219	220	221	222	223	224
225	<EOF>	<EOF>	<EOF>	0	0	0	0
0	0	0	0	0	0	0	0
0	0	0	0	0	0	0	0
0	0	0	0	0	0	0	0
0	0	0	0	0	0	0	0
0	0	0	0	0	0	0	0
0	0	0	**276**	**277**	**278**	**279**	**280**
281	**282**	**283**	**284**	**285**	**286**	**287**	**288**
289	**290**	291	292	**293**	**294**	**295**	**296**
297	**298**	299	300	**301**	**302**	**303**	**304**
305	**306**	307	308	**309**	**310**	**311**	**312**
313	**<EOF>**	315	316	317	318	319	320
321	322	323	324	325	326	327	328
329	330	331	332	333	334	335	336
337	338	339	340	341			

FAT (1st Copy) Sector1

Drive A: Cluster 275, hex 113

Recovering fragmented files is more difficult. When a file was not stored in contiguous clusters, it requires more effort and experience to locate the next link in the chain.

Recovering deleted directories is beyond the scope of this book. The process involves searching unallocated space for an unique pattern associated with directories and is covered in the *Handbook of Computer Crime Investigation*, Chapter 7, pp. 143–145 (Sheldon 2002). EnCase has a useful utility for automatically recovering all deleted directories (and the files they contained) on a FAT volume.

10.4.1 WINDOWS-BASED RECOVERY TOOLS

The recovery process described above is time consuming and must be performed on a working copy of the original disk. More sophisticated examination tools like EnCase and FTK can use a bitstream copy of a disk to display a virtual reconstruction of the file system, including deleted files, without actually modifying the FAT. All of these tools recover files on FAT systems in the most rudimentary way, assuming that all clusters in a file are sequential. Therefore, in more complex situations, when files are fragmented, it is necessary to recover files manually. Windows-based tools like EnCase and FTK can also be used to recover deleted files on NTFS volumes.

10.4.2 UNIX-BASED RECOVERY TOOLS

Linux can be used to perform basic examinations of FAT and NTFS file systems as described in Chapter 11. Also, tools like fatback,[15] The Sleuth Kit, and SMART[16] can also be used to recover deleted files from FAT file systems. For instance, the following shows fatback being used to recover a deleted file from a FAT formatted floppy diskette.

[15]http://sourceforge.net/projects/biatchux/
[16]http://www.asrdata.com

```
examiner1% fatback -l biotechx.log hunter-floppy.dd
Parsing file system.
/ (Done)
fatback> dir
Sun Apr   13   15:36:52   2003       0   SALES
Sun May   13   11:58:10   2003   21504   SKIWAY~1.DOC    skiways-
getafix.doc
Sun May   13   12:40:48   2003     122   TODO.TXT        todo.txt
Sun May   13   12:42:18   2003     122   NEWADD~1.TXT    newaddress.txt
Sun May    8   14:34:16   2003   19968   ?REENF~1.DOC    greenfield.do
Sun May    8   14:41:44   2003       0   ?PRIL/          april
Sun Feb   18   12:49:16   2001   16896   CONTACTS.XLS    contacts.xls
fatback> copy greenfield.do /e1/biotechx/recovered
```

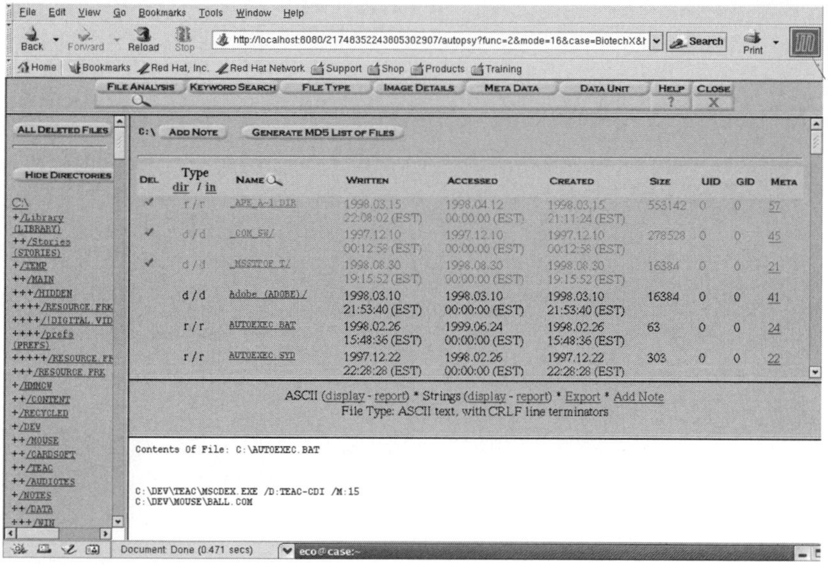

Figure 10.5

The Sleuth Kit and Autopsy Forensic Browser being used to examine a FAT file system (checkmarks indicate files are deleted).

The contents of a deleted file named "greenfield.doc" was extracted and saved into a file named "/e1/biotechx/recovered" using the **copy** command within **fatback**. The Sleuth Kit, combined with the Autopsy Forensic Browser, can be used to examine FAT file systems through a Web browser interface and recover deleted files as shown in Figure 10.5.

The Sleuth Kit and Autopsy Forensic Browser enable examiners to examine data at the logical and physical level and can also be used to recover files from NTFS file systems. The Sleuth Kit can also be used to recover slack space from FAT and NTFS systems using "dls –s". Although SMART has a feature to recover deleted files from FAT and NTFS volumes, it does not display them in a logical view, making the recovery process somewhat cumbersome.

10.4.3 FILE CARVING WITH WINDOWS

Another approach to recovering deleted files is to search unallocated space, swap files, and other digital objects for class characteristics such as file headers and footers. Conceptually, this process is like carving files out of the blob-like amalgam of data in unallocated space. File carving tools such as DataLifter (Figure 10.6) and Ontrack's Easy-Recovery Pro (Figure 10.7) can recover many types of files including graphics, word processing, and executable files. Also, user defined files can be carved using WinHex "File Recovery by Type" feature or EnCase E-scripts available from the EnScript library.[17] Specialized tools like NTI's Graphics Image File Extractor also exist to extract specific types of files. Some of these tools can extract images from other files such as images stored in Word documents.

[17] *http://www.encase.com/support/escript_library.shtml*

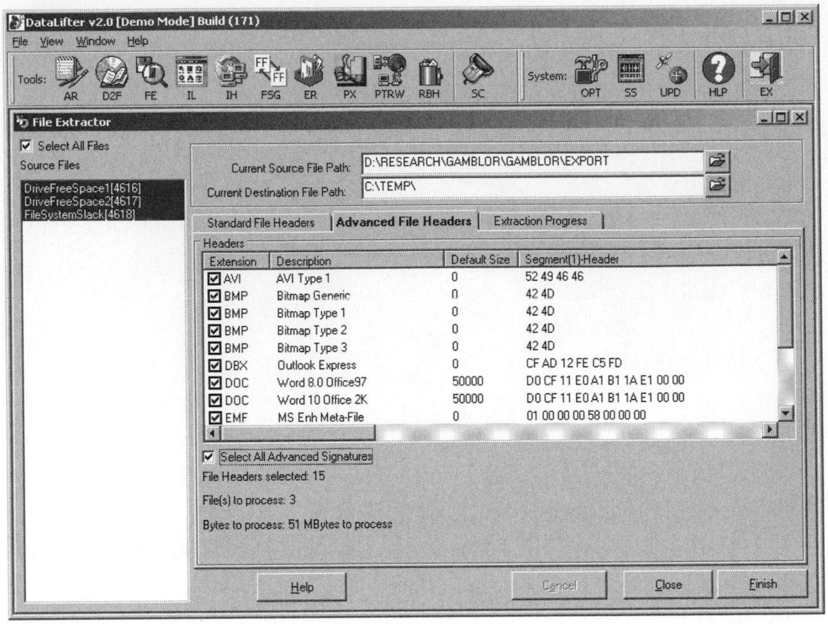

Figure 10.6

DataLifter being used to carve files from two blobs of unallocated space and one blob of file slack from a system.

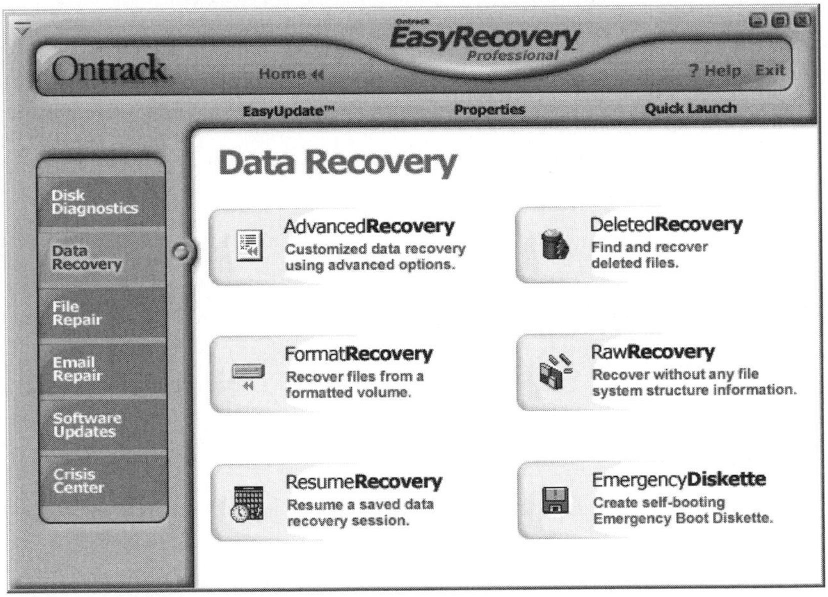

Figure 10.7

Easy-Recovery Pro from Ontrack.

One of the main limitations of these tools is that they generally rely on files having intact headers. Therefore, when file headers have been obliterated, it may be necessary to search for other class characteristics of the desired files and piece fragments together manually. Even when it is not possible to piece recovered fragments together, it may be possible to extract useful information from them. For instance, cluster 37 of the aforementioned floppy disk contains a Word document fragment with Windows date–time stamps from April 14, 2003, at around 0800 hours, shown here in bold.

```
52006F00  6F007400  20004500  6E007400  |  R.o. o.t.  .E. n.t.  |  16
72007900  00000000  00000000  00000000  |  r.y. .... .... ....  |  32
00000000  00000000  00000000  00000000  |  .... .... .... ....  |  48
00000000  00000000  00000000  00000000  |  .... .... .... ....  |  64
16000501  FFFFFFFF  FFFFFFFF  03000000  |  ....           ....  |  80
06090200  00000000  C0000000  00000046  |  .... .... └... ...F  |  96
00000000  4095D28D  8502C301  007F3AEF  |  .... @ò╥ì à.├. .⌂:∩  |  112
8502C301  25000000  80000000  00000000  |  à.├. %... Ç... ....  |  128
31005400  61006200  6C006500  00000000  |  1.T. a.b. l.e. ....  |  144
00000000  00000000  00000000  00000000  |  .... .... .... ....  |  160
00000000  00000000  00000000  00000000  |  .... .... .... ....  |  176
00000000  00000000  00000000  00000000  |  .... .... .... ....  |  192
0E000201  FFFFFFFF  05000000  FFFFFFFF  |  ....      ....       |  208
00000000  00000000  00000000  00000000  |  .... .... .... ....  |  224
00000000  00000000  00000000  00000000  |  .... .... .... ....  |  240
00000000  09000000  00100000  00000000  |  .... .... .... ....  |  256
57006F00  72006400  44006F00  63007500  |  W.o. r.d. D.o. c.u.  |  272
6D006500  6E007400  00000000  00000000  |  m.e. n.t. .... ....  |  288
```

Slack space contains fragments of data that can be recovered but that can rarely be reconstituted into complete files. However, if a small file overwrote a large one, it may be possible to recover the majority of the overwritten file from slack space. It is easiest to recover textual data from slack space because it is recognizable to the human eye. Figure 10.8 shows remnants of a shopping cart on CD Universe in slack space.

Interestingly, the slack space shown in Figure 10.8 is associated with a deleted file that was recovered.

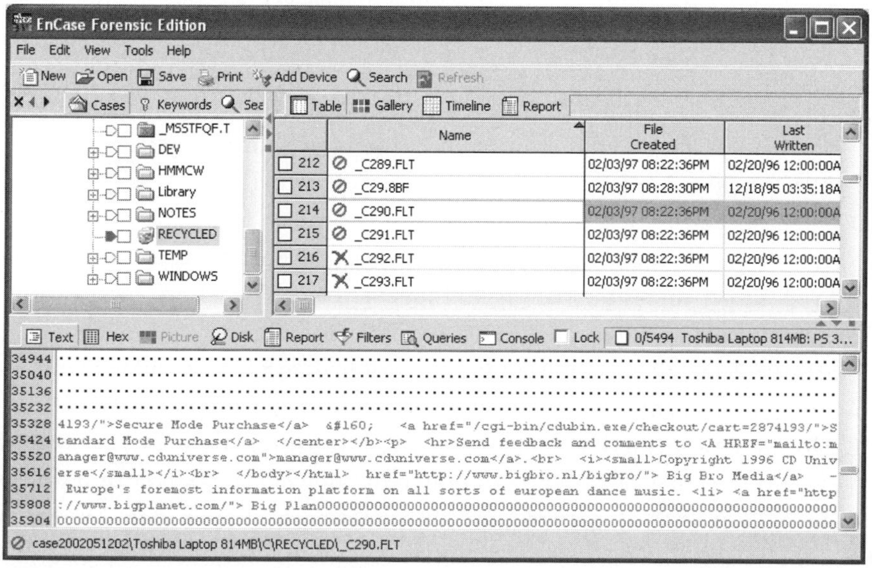

Figure 10.8

File slack of a recovered file viewed using EnCase.

10.4.4 DEALING WITH PASSWORD PROTECTION AND ENCRYPTION

It is generally acceptable, and usually desirable, for digital investigators to overcome password protection or encryption on a computer they are processing. In some instances, it is possible to use a hexadecimal editor like Winhex to simply remove the password within a file. There are also many specialized tools that can bypass or recover passwords of various files. For instance, NTI sells a collection of password recovery tools that they have validated.[18] Other companies such as Lostpassword.com[19] sell password bypassing programs. Free, unvalidated tools are available from Russian Password Crackers[20] and other Web sites.

[18] *http://www.forensics-intl.com/breakers.html*

[19] *http://lostpassword.com/*

[20] *http://www.password-crackers.com/crack.html*

When performing a functional reconstruction using a restored clone of a Windows NT/2000/XP system, it may be necessary to bypass the logon password using a program like ntpasswd,[21] or ERD Commander.[22] In situations where the actual password is needed, tools like LC4[23] (formerly L0pht Crack) are available that attempt to guess passwords in Windows NT password files. Other tools are available to recover passwords for Windows 95/98 systems, dial-up accounts, and other programs that store passwords in PWL files.

[21] *http://home.eunet.no/~pnordahl/ntpasswd/*

[22] *http://www.winternals.com/products/repairandrecovery/erdcommander2002.asp*

[23] *http://www.atstake.com/research/lc/*

The most powerful and versatile password recovery programs currently available are PRTK and DNA from Access Data. The Password Recovery Toolkit can recover passwords from many file types and is useful for dealing with encrypted data. Also, it is possible for a DNA network to try every key in less time by combining the power of several computers. Access Data's Distributed Network Attack (DNA) application can brute force Adobe Acrobat and Microsoft Word/Excel files that are encrypted with 40-bit encryption. Using a cluster of approximately 100 off-the-shelf desktop computers and the necessary software, it is possible to try every possible 40-bit key in 5 days. For example, the *Wall Street Journal* was able to decrypt files found on an Al Qaeda computer that were encrypted using the 40-bit export version of Windows NT Encrypting File System (Usborne 2002).

However, Microsoft Windows EFS generally uses 128-bit keys (Microsoft 2001) and because each additional bit doubles the number of possibilities to try, a brute force search quickly becomes too expensive for most organizations or simply infeasible, taking million of years. Therefore, before brute force methods are attempted, some exploration should be performed to determine if the files contain valuable evidence and if the evidence can be obtained in any other way. It may be possible to locate unencrypted versions of data in unallocated space, swap files, and other areas of the system. Alternatively, it may be possible to obtain an alternative decryption key. For instance, Encrypted Magic Folders[24] advises users to create a recovery disk in case they forget their password. In one investigation, finding this disk

[24] *http://www.pc-magic.com/*

enabled the digital evidence examiner to decrypt data that none of the above mentioned tools could recover.

Similarly, when EFS is used, Windows automatically assigns an encryption recovery agent that can decrypt messages when the original encryption key is unavailable (Microsoft 1999). In Windows 2000, the built-in Administrator account is the default recovery agent (an organization can override the default by assigning a domain-wide recovery agent provided the system is part of the organization's Windows 2000 domain). Notably, prior to Windows XP, EFS private keys were weakly protected and it was possible to gain access to encrypted data by replacing the associated NT logon password with a known value using a tool like ntpasswd and logging into the system with the new password.

10.5 LOG FILES

Attribution is a major goal and log files can record which account was used to access a system at a given time. User accounts allow two forms of access to computers: interactive login and access to shared resources. Both forms of access can significantly expand the pool of suspects in an investigation. If illegal materials are found on a computer, individuals with legitimate access to the computer are the obvious suspects. However, there is the possibility that someone gained unauthorized access to the computer and stored illegal materials on the disk. Similarly, if secret information is stolen from a computer system or a computer is used to commit a crime, it is possible that someone gained unauthorized access to the computer.

Windows NT/2000/XP store log files in the "%systemroot%\system32\config\" directory (most commonly "c:\winnt\system32\config\") (Table 10.1).

System log files can contain the information about user accounts that were used to commit a crime and can show that a user account might have been stolen. The Application and System event logs also contain information about user activities on a system. Additionally, NT Event Logs can be correlated with file system traces to determine what occurred while a given account was logged in. Unfortunately, Windows 95/98 do not have logs of this kind and, on Windows NT, most logging options are disabled by default, so if a system was not configured to keep more detailed logs prior to an incident, much of the information that could have been gathered will be lost.

Table 10.1

Windows NT Event Logs.

FILE	DESCRIPTION
Appevent.evt	Contains a log of application usage
Secevent.evt	Records activities that have security implications such as logins
Sysevent.evt	Notes system events such as shutdowns

Since it is usually desirable to search and sort log files during an investigation, the type of graphical user interface to log files can be a hindrance. Several utilities exist that will process log files from Windows NT and 2000. The most basic utility is dumpel from the Windows NT and Windows 2000 Resource Kits. Be aware that it is often necessary to extract Event Message Files from a system to obtain complete and accurate information from the event logs on that system. A detailed procedure for examining NT event logs is provided in the *Handbook of Computer Crime Investigation*, Chapter 9 (Casey *et al.* 2002, pp. 225–228).

10.6 FILE SYSTEM TRACES

An individual's actions on a computer leave many traces that digital investigators can use to glean what occurred on the system. For instance, when a file is downloaded from the Internet, the date–time stamps of this file represent when the file was placed on the computer. If this file is subsequently accessed, moved, or modified, the date–time stamps may be altered to reflect these actions. Understanding how date–time stamps of files are updated under different circumstances can enable digital investigators to infer the associated actions. A summary of common actions and the associated date–time stamp changes on FAT and NTFS file systems is provided in Table 10.2.

Because moving a file within a volume does not change file times, the original (deleted) directory entry for the file is identical to the new directory entry, enabling forensic examiners to determine where files were moved from as long as the original directory entry exists. Also evident from Table 10.2, when a file is copied within a volume or moved from a hard drive to external media like a floppy diskette, the created and last accessed date–time stamps of the new file are updated but the last modified date–time stamp remains the same, resulting in a last modified time prior to the creation time. When digital evidence examiners encounter this counterintuitive situation for the first time, they sometimes assume that concealment behavior is at work such as system clock changes.

Table 10.2

Date–time stamp behavior on FAT and NTFS file systems.

ACTION	LAST MODIFIED DATE–TIME	LAST ACCESSED DATE–TIME	CREATED DATE–TIME
File moved within a volume	Unchanged	Unchanged	Unchanged
File moved across volumes	Unchanged	Updated	Updated
File copied (destination file)	Unchanged	Updated	Updated

When a file with these counterintuitive date–time stamps is found, indicating that it was copied from somewhere else, it may be possible to locate the original file by searching all available storage media for files with the same MD5 hash value, the same creation time, and/or the same name. However, this date–time stamp phenomenon also occurs when a file is downloaded from certain types of file servers on the Internet. For instance, when a file is copied from a network shared on a remote Windows system, the "creation" date-time stamp is updated to the local system time but the last written date–time stamp is not. The same thing occurs when a file is downloaded from a remote UNIX machine using the file transfer feature of Secure Shell (SSH). Notably, this does not apply to all servers (e.g. FTP). So, if the file was downloaded from a file server on the Internet, it may not be feasible to find the original file but it may still explain the counterintuitive date–time stamps. Finding the original file is useful for addressing the argument that someone on the Internet uploaded the file to the defendant's computer without his knowledge via NetBIOS.[25] Although this is a weak argument unless there is evidence to support unauthorized access, it is useful to have evidence that the defendant had knowledge of the files on the system. For more detailed discussion of examining moved and copied files, see the *Handbook of Computer Crime Investigation*, Chapter 7, pp. 140–142 (Sheldon 2002).

[25]NetBIOS/SMB, also called Common Internet File System (CIFS), is used by Windows to share resources over networks such as printers and portions of a disk.

Notably, the last accessed and modified date–time stamps of the parent directory listing (".") are updated when files are moved out of and copied into the directory because the entries in the associated directory files are being added to and deleted. Similarly, when a file is deleted from a directory, the last modified and accessed date–time stamps of the parent directory listing are updated.

Microsoft Office documents retain quite a bit of information called *metadata*, including the location where a file was stored on disk, the printer, and the original creation date and time. These metadata can be useful for locating file fragments that were generated while documents were being edited. Additionally, the date–time stamps embedded in the file can be useful for temporal analysis. Printing also creates useful artifacts on Windows file systems. Rather than sending data directly to the printer, computer systems can store print jobs on disk temporarily and send them to the printer as it becomes available. In this way, the application being used to print is not tied up while the job is printing. Windows 95/98 stores information relating to printed files in C:\Windows\Spool\Printers and Windows NT/2000 stores them in C:\WinNT\System32\Spool\Printers. These files can contain the name (or URL) of the printed file, application used to print, printer name, file owner, and even the raw data of the print job in. Also, since these files are created when the associated item is printed, the date–time stamps on these files indicate when it was printed. When printing in EMF mode, the associated spool file (0020.SPL) contains names of temporary files that were created during the printing process as shown here:

```
Microsoft Word-
Document2.LPT1:.STP...........FTM.%..... C:\WINDOWS\TEMP\~EMF115D.TMP.ENP.
.........STP\
..............FTM.%........C:\WINDOWS\TEMP\~EMF1639.TMP.ENP ......STP........FTM.%.\
..C:\WINDOWS\TEMP\~EMF1646.TMP.ENP.....STP............FTM.% ...
C:\WINDOWS\TEMP\~\
EMF164D.TMP.ENP .....STP..........FTM.% ... C:\WINDOWS\TEMP\~EMF1742.TMP.ENP ...\
.STP...............FTM.% ... C:\WINDOWS\TEMP\~EMF1749.TMP.ENP .... STP ............... FT\
M.% ... C:\WINDOWS\TEMP\~EMF1410.TMP.ENP ..STP .....FTM.% ... C:\WINDOWS\TE\
MP\~EMF1407.TMP.ENP................END
```

These temporary enhanced metafiles essentially contain an image of segments of the printed document. Some of these EMF files may have been overwritten but those that still exist on disk can be opened with a suitable viewer to see what was printed. These copies can be useful if the original file is modified, encrypted, or non-existent, as in above example "Document2" was never saved.

A detailed case example is provided here to demonstrate how some of the many traces created by activities on Windows systems can be useful in an investigation. The floppy disk referenced in the File System section is used in the following case example:

CASE EXAMPLE

A company called "BioTechX" believes that an ex-employee, Henry Hunter, stole proprietary information and is using it to acquire their best customers by selling the same product for less money. In addition to stealing thousands of tablets of their primary product "BioFixIt," the company believes that Hunter stole test results relating to BioFixIt and is sending their best customers letters offering the same product at a reduced price. Hunter claims that he did not steal any information and that he is selling a product named "Getafix" created by his new company, BioFix, to individuals he met at conferences and trade shows.

An examination of the Windows 95 computer Hunter used when he worked at BioTechX has the following traces from the day he left the organization (May 12, 2003), indicating that he accessed three files containing BioFixIt test data.

Name	File Created
C:\WINDOWS\Recent\s072602.txt.lnk	05/12/03 11:36:38AM
C:\WINDOWS\Recent\s062602.txt.lnk	05/12/03 11:27:32AM
C:\WINDOWS\Recent\s052302.txt.lnk	05/12/03 11:25:08AM

File system traces from May 8 indicate that Hunter accessed the company customer list and created and printed letters to customers. Although this activity was part of his job, it demonstrated that Hunter had access to customer names and addresses. During the examination, it was noted that this computer had Ethernet address 00-60-97-ED-DC-2E and its system clock was 11 minutes fast.

With this evidence of probable cause, investigators obtained a search warrant to search Hunter's home computer and associated media. Of greatest interest was the floppy diskette containing the following (deleted entries marked with a "*"):

Name	File Created	Last Written
newaddress.txt	05/13/03 12:42:16PM	05/13/03 12:42:18PM
todo.txt	05/13/03 12:37:54PM	05/13/03 12:40:48PM
skiways-getafix.doc	05/13/03 12:32:00PM	05/13/03 11:58:10AM
contacts.xls	05/08/03 02:43:14PM	02/18/01 12:49:16PM
*greenfield.do	05/08/03 02:43:00PM	05/08/03 02:34:16PM
*april	05/08/03 02:41:44PM	05/08/03 02:41:44PM

Notably, the MD5 value and date–time stamps of contacts.xls indicate that it was copied from the BioTechX computer that Hunter used. Hunter claimed that he had not realized "contacts.xls" was on the floppy and denied using the information it contained after he left BioTechX. However, a copy of this file was found on his computer in a directory named "sales" with date–time stamps showing that it had been created on May 13, 2003.

A closer examination of the floppy disk uncovered remnants of the allegedly stolen BioFixIt test data. However, it was not immediately apparent when the test data were placed on the floppy disk and Hunter claimed that they were there since 2002 when they were originally given to him. Looking at disk clusters adjacent to the test data showed the following:

Clusters 42: Partially overwritten Word document fragment from BioTechX computer used by Hunter, created on April 14, 2003.

Cluster 184: Word document "skyways-getafix.doc" from Hunter's home computer, created on May 14, 2003.

> The fact that the test data had partially overwritten a Word document created on April 14, 2003, and was partially overwritten by a Word document created on May 14, 2003, strongly suggests that the test data were placed on the floppy diskette between these dates, not in 2002 as Hunter claims.

Be aware that date–time stamps can be affected by external influences. For instance, files extracted from a compressed Zip archive can retain the date–time stamps from the system where they originated. Also, file date–time stamps can be changed to any value using a simple program such as touch.pl.[26] Therefore, it is important to look for other data on the system or network to corroborate these date–time stamps.

[26] *http://patriot.net/~carvdawg/perl.html*

10.7 REGISTRY

Windows systems use the Registry to store system configuration and usage details in what are called "keys." Registry files (a.k.a. hives) on Windows 95 and 98 systems are located in the Windows installation directory and are named "system.dat" and "user.dat." The Registry on Windows NT/2000/XP is comprised of several hive files located in "%systemroot%\system32\config" and a hive file named "ntuser.dat" for each user account.

Registry files recovered from an evidentiary system can be viewed using the Windows NT **regedt32** command on an examination system using the Load Hive option on the Registry menu. Registry files can also be viewed using third-party applications like EnCase or Resplendent Registrar.[27] The values in some Registry keys are stored in hexadecimal format but can be converted to ASCII and saved to a text file using the "Save Subtree As" File menu option of **regedt32**. For instance, the following Registry key shows the names of files that were played recently using Windows MediaPlayer ("< sid >" is substituted for security identifier of the user on the system):

[27] *http://www.resplendence.com*

```
Key Name: HKEY_USERS\<sid>\Software\Microsoft\MediaPlayer\Player\
RecentURLList
          Class Name:        <NO CLASS>
          Last Write Time:   5/9/2003 – 1:48 PM
          Value 0
            Name:            URL0
            Type:            REG_SZ
            Data:            H:\porn\movie1.avi

          Value 1
            Name:            URL1
            Type:            REG_SZ
            Data:            H:\porn\movie2.avi
```

The Registry values in this example referenced files on an external, removable hard drive that was not attached to the system when it was collected. Upon finding these references in the Registry, investigators sought and found the external hard drive. Similar Registry keys exist for other programs and for different file extensions as shown here:

Preview (Chapter 19): Trojan horse programs such as SubSeven and Back Orifice use Registry keys (and other mechanisms) to persist on a system after it is rebooted. The programs give an individual to have full remote control of a computer. Although AntiVirus programs can detect many Trojans in their default state, intruders can modify the programs to avoid detection.

```
Key Name: HKEY_USER\<sid>\Software\Microsoft\Windows\CurrentVersion\
Explorer\ComDlg32\OpenSaveMRU\zip
        Class Name:         Shell
        Last Write Time:    5/9/2003 – 1:17 PM
        Value 0
        Name:               a
        Type:               REG_SZ
        Data:               H:\porn\bodyshots1.zip
        <cut for brevity>
        Value 9
        Name:               j
        Type:               REG_SZ
        Data:               H:\porn\bodyshots2.zip
```

As the name suggests, the "Last Write Time" value indicates when a value in the Registry key was altered or added.

Some keys protect the data they contain, encoding them using a simple cipher such as the one shown here:

```
Key Name:
HKEY_USER\<sid>\Software\Microsoft\Windows\CurrentVersion\Explorer\
UserAssist\{5E6AB780-7743-11CF-A12B-00AA004AE837}\Count
    Class Name:         <NO CLASS>
    Last Write Time:    9/11/2002 – 9:28 AM

    Value 1
    Name: HRZR_EHACNGU:T:\sebfg\sebfg.ong

    Value 2
    Name: HRZR_EHACNGU:T:\rapnfr3.rkr
```

The first entry refers to "g:\frost\frost.bat" and the second entry refers to "g:\encase3.exe".

10.8 INTERNET TRACES

Accessing the Internet leaves a wide variety of information on a computer including Web sites, contents viewed, and newsgroups accessed. For instance, Windows systems maintain a record of accounts that are used to connect to the Internet as shown in Figure 10.9.[28]

Additionally, some Windows systems maintain a log of when the modem was used (e.g. ModemLog.txt) and some Internet dial-up services maintain a detailed log of connections such as the AT&T/IBM Global Network Dialer "Connection Log.txt" and "Message Log.txt" files shown here:[29]

[28]The Internet Account Manager section in the registry often contains default accounts that were not added by the user, such as the Bigfoot and Infospace accounts in Figure 10.9.

[29]The AT&T/IBM Global Network Dialer creates other logs containing useful information, such as ErrorLog.txt and ARLOG.TXT. File names and contents may differ in different versions of the dialer software.

```
--------------------------------------------------------------------------------
Dialer Connection Log
--------------------------------------------------------------------------------
2000/01/12   15:22:39   usinet janedoe dialed 06-3365-3946
2000/01/12   15:41:48   Disconnected after 00:19:04
2000/01/12   17:03:10   ----------------------------------------
2000/01/12   17:03:10   usinet janedoe dialed 06-3365-3946
2000/02/29   23:05:34   ----------------------------------------
2000/02/29   23:05:34   usinet janedoe dialed 06-3365-3946
2000/02/29   23:09:26   Disconnected after 00:03:49
2000/04/18   20:53:09   ----------------------------------------
2000/04/18   20:53:09   usinet janedoe dialed 06-3365-3946
2000/04/18   20:58:17   Disconnected after 00:05:08
--------------------------------------------------------------------------------
Dialer Message Log
--------------------------------------------------------------------------------
The date is Tuesday, February 29, 2000.
The time is 11:04:56 PM.
<cut for brevity>
Modem is 3Com (3C562D-3C563D) EL III LAN+336 Modem PC Card.
Modem log file truncated.
Set up Dial-Up Networking entry IBM Global Network.
Login profile is "johndoe".
The login ID is login.Internet.usinet.johndoe.
Connecting with the IBM Global Network entry.
Opened c:\windows\ModemLog.txt.
RAS dial connect state is 0 (0).
RAS dial connect state is 1 (0).
Initializing the serial port...
Initializing the modem and dialing 06-3365-3946...
<cut for brevity>
02-29-2000 23:05:21.65 – Recv: <cr><lf>CONNECT 115200<cr><lf>
Modem-to-modem speed is 115200 bps.
02-29-2000 23:05:21.65 – Interpreted response: Connect
Setting up the network link...
02-29-2000   23:05:21.65 – Connection established at 31200bps.
02-29-2000   23:05:21.65 – Error-control on.
02-29-2000   23:05:21.65 – Data compression on.
RAS dial connect state is 14 (0).
RAS dial connect state is 8192 (0).
Local IP address is 139.92.104.85.
Gateway IP address is 152.158.45.46.
<cut for brevity>
```

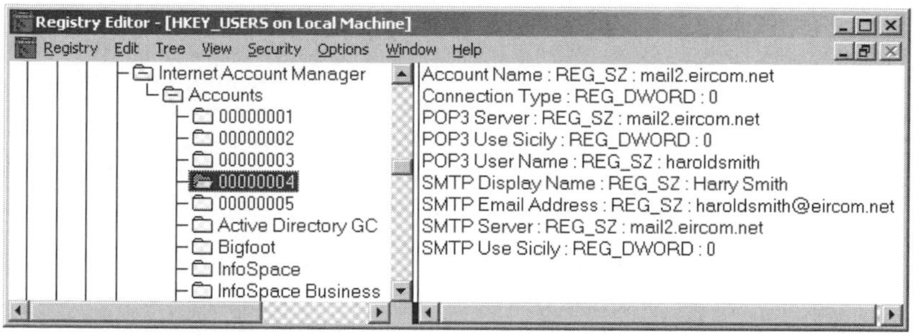

Figure 10.9
Internet Account Manager.

10.8.1 WEB BROWSING

When an individual first views a Web page the browser caches the page and associated elements such as images on disk – the creation and modification times are the same time as the page was viewed. When the same site is accessed in the future, the cached file is accessed. The number of times that a given page was visited is recorded in some Web browser history databases. Look for all information related to downloaded files (e.g. in registry, on external media, etc.) to get a better sense of how they were placed on the computer and what was done with them afterwards. Any other activities that were going on at the time the files were being placed on the computer and viewed/manipulated may give a clue as to who was performing the actions.

As mentioned in Chapter 9, Netscape maintains a database of Web sites visited in a file named "netscape.hst."[30] Entries that have been marked as deleted by Netscape can be recovered using programs such an EnCase E-Script as shown here:

[30] The date–time stamps in these history databases are obtained from the local computer, not the remote server.

# of Times Visited, First Accessed, Last Accessed, Link

4, 01/13/02 05:13:54PM, 01/13/02 06:04:37PM **DELETED** http://www.paraben-foren

3, 01/13/02 06:08:32PM, 01/13/02 06:08:43PM, **DELETED** DE101: Introduction to Digital Evidence

6, 01/13/02 05:51:56PM, 01/13/02 06:11:32PM, **DELETED** http://accessdata.com/images2000/ftk_scrn_graph_sm.jpg

3, 01/13/02, 06:08:32PM 01/13/02 06:08:43PM, **DELETED** DE101: Introduction to Digital Evidence

Internet Explorer maintains similar information in files named "index.dat." These databases can contain a wealth of information including sites accessed and search engine details. Some open source utilities have been developed to extract information from "index.dat" files and other files.[31]

[31] UNIX versions available at http://odessa.sourceforge.net/ and Windows versions available at http://www.foundstone.com

CASE EXAMPLE

Prosecutors upgraded the charge against Robert Durall, 40, to first-degree murder based on what they described as evidence of premeditation found on his office computer. He had been charged with second-degree murder. A co-worker

> told police he had discovered a number of temporary files on Durall's office computer that showed he had used Internet search engines to find Web sites with key words including "kill + spouse," "accidental + deaths," smothering, poison, homicides and murder, according to court documents. A plus sign tells the search engine to only pull up sites that use both terms as key words. (September 4, 1998, Associated Press)

It can be tedious to examine each entry in a Web browser history file but the results are often worth the effort. To facilitate analysis, attempt to group them by time or web site to help interpretation but do not assume that an entry implies intent to view page. Some Web sites redirect browsers to different locations and even make unauthorized changes to a system (Microsoft 2002).

> In addition to caching files on disk, Web browsers cache a small amount of files in memory. For instance, files being held in memory by Netscape can be listed using about:image-cache and about:memory-cache. To view a list of files cached on disk by Netscape, use about:cache and to list the global history, use about:global.

Web browsers also store temporary files in a cache folder to enable quicker access to frequently visited pages. Cache folders contain fragments of pages that were recently viewed, including images and text. Recent versions of Internet Explorer maintain information about these files in another index.dat database and earlier versions used files named MM256.DAT and MM2048.DAT. Netscape maintains this information in a Berkeley DB file named fat.db. Interestingly, Mozilla maintains a file named "_CACHE_001_" that shows HTTP responses containing the current date and time according to the Web sever clock which may be more accurate than the local system clock.

Even after these temporary files are deleted, they can be recovered to reveal a significant amount of information such as Web-based e-mail (e.g. Hushmail.com), purchases (e.g. E-bay.com, Amazon.com), financial transactions (e.g. online banking, Paypal.com), travel itineraries (e.g. Expedia.com), and information from private databases.

Some Web sites keep track of an individual's visits and interests by placing information in cookie files associated with the Web browser. For example, Amazon.com uses cookie files to keep track of the purchases and get a better idea of an individual's interests, enabling them to recommend other books that may be of interest. Netscape stores cookies in a cookies.txt file and Internet Explorer maintains cookies in the Windows\Cookies directory, along with an associated index.dat file (Handbook). Each cookie entry contains information that may be useful in an investigation. For instance, Figure 10.10 depicts the contents of a cookie file created by Mapquest, showing recent searches that may be useful when trying to determine where an individual went.

Notably, the presence of a cookie does not necessarily prove that an individual intentionally accessed a given Web site. For instance, some advertisements on Web pages use cookies, creating references to the advertised site even though the user did not actually view Web pages on that site. Also, in some situations, a Web browser may be automatically redirected to multiple

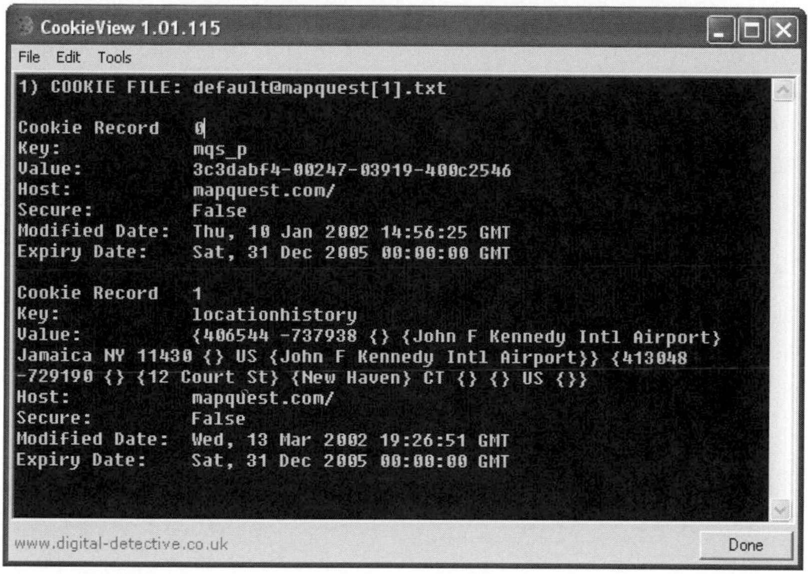

Figure 10.10

A cookie created by MS Internet Explorer showing recent Mapquest searches viewed using CookieView (http://www.digitaldetective.co.uk)

sites, creating files in disk cache and entries in the history database even though the user did not intend to visit any of the sites.

10.8.2 USENET ACCESS

As well as storing all of the URLs that have been accessed, Web browsers with Usenet readers keep a record of which Usenet newsgroups have been accessed. For instance, Netscape's newsreader stored information in a file with a ".rc" extension. MS Internet News stores quite a bit of information about newsgroup activities in the News directory. You will find this News directory where you installed MS Internet News (the default directory is C:/Program Files/Internet Mail and News/user/).

The following contents of a "news.rc" file shows newsgroups that were subscribed to and which messages were read:

```
alt.binaries.cracks! 1-271871,271884,271887,271915,271992
alt.binaries.hacking.utilities! 1–8905,8912,8921,8924,8926,8929,8930,8932
alt.binaries.hacking.computers! 1-1651,1653,1659
alt.binaries.mp3! 1–5441,5443,5445
alt.teens.advice: 1–4244, 4256, 4257
```

The exclamation point after the name of the newsgroup indicates that the user was once subscribed to that newsgroup but has since unsubscribed. A colon after the name indicates that the user is currently subscribed to that newsgroup (e.g. alt.teens.advice). The numbers are reference numbers that a news server uses to keep track of which articles have been downloaded and read. The first range of numbers on each line refer to old messages – the news server will only deliver newer messages. The remaining numbers tell you which articles were read the last time the user looked at the newsgroup. For instance, the last time the user looked at alt.teens.advice, he read two messages. You could look in his newsreader to determine which messages they were – the reference number is contained in the Xref: line of the header (e.g. Xref: news.server.com alt.teens.advice:4256). It is important to realize that these reference numbers are unique to the server used, they do not refer to all of Usenet.

This information can help investigators narrow their search of Usenet to a selection of groups.

10.8.3 E-MAIL

E-mail clients often contain messages that have been sent from and received at a given computer. While Netscape and Eudora store e-mail in plain text files, Outlook, Outlook Express, and AOL use proprietary formats that require special tools to read. Even when e-mail is stored in a non-proprietary format, it is necessary to decode MIME encoded message attachments.

Figure 10.11 shows FTK being used to view a file containing e-mail with Word document attachments. FTK can interpret a variety of proprietary formats, including Outlook. EnCase can also interpret some of these proprietary formats using the View File Structure feature. Another approach to viewing proprietary formats, such as America Online (AOL), is to restore them to a disk and view them via the AOL client. In some cases it is possible to recover messages that have been deleted but have not been purged from

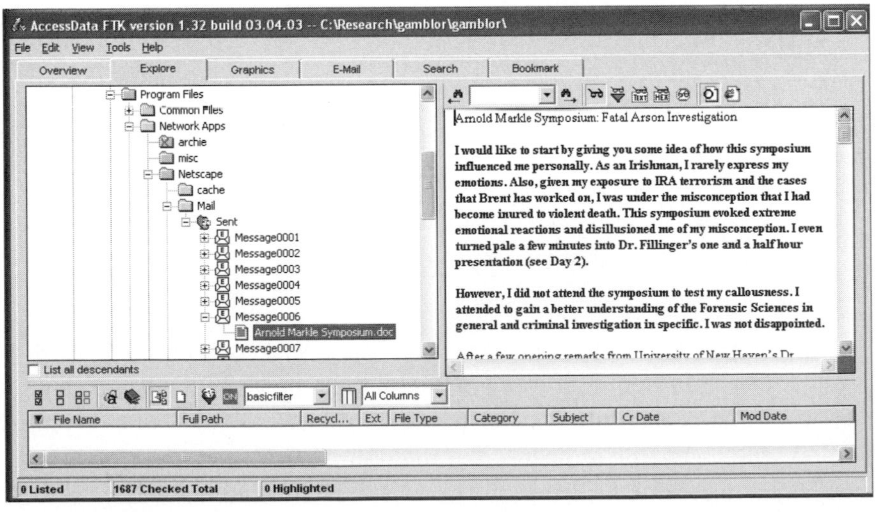

Figure 10.11

FTK showing Word document as e-mail attachments (base 64 encoded).

e-mail files. For additional details about recovering and examining e-mail from Microsoft Exchange server, see Chapter 9 of the *Handbook of Computer Crime Investigation.*

10.8.4 OTHER APPLICATIONS

Yahoo Pager, AOL IM, and other Instant Messenger programs do not retain archives of messages by default but may be configured to log chat sessions. Peer-to-peer file sharing programs may retain a list of hosts that were contacted or files that have been accessed but give very limited information besides this. IRC and other online chat clients may retain more logs but only if the user saves them. Therefore, remnants of these more transient Internet activities are more likely to be found in swap space and other areas of the hard disk. Therefore, the best chance of obtaining information relating to these applications is to search portions of the hard drive where data may have been stored temporarily or to monitor network traffic from the individual's machine while these programs are in use.

10.8.5 NETWORK STORAGE

An important component of any forensic examination is identifying any remote locations where digital evidence may be found. A victim might maintain a Web site or an offender may transfer incriminating data to another computer on the Internet or a home or corporate network. One of the most common remote storage locations is an individual's Internet Service Provider (ISP). In addition to storing e-mail, some ISPs give their customers storage space for Web pages and other data. Files can be transferred to these remote systems using programs such as FTP, Secure CRT, and Secure Shell (SSH). So, in addition to looking for information about Internet accounts in the registry as mentioned earlier, search for traces of file transfer applications.

For instance, WS-FTP creates small log files each time it is used to transfer files, showing file locations, FTP server names, and times of transfer. Secure CRT and SSH can be configured to maintain individual configuration files for each computer that a user connects to frequently. A list of systems that have been accessed may also be available if the user opted to save a copy of each server's public encryption key. Other programs use the Registry to record the names or IP addresses of remote systems that have been accessed. For instance, the Telnet program on some Windows systems maintains a list of recently accessed systems as shown in Figure 10.12. This can also be useful in computer intrusion investigations – showing a connection between the intruder's computer and the compromised systems.

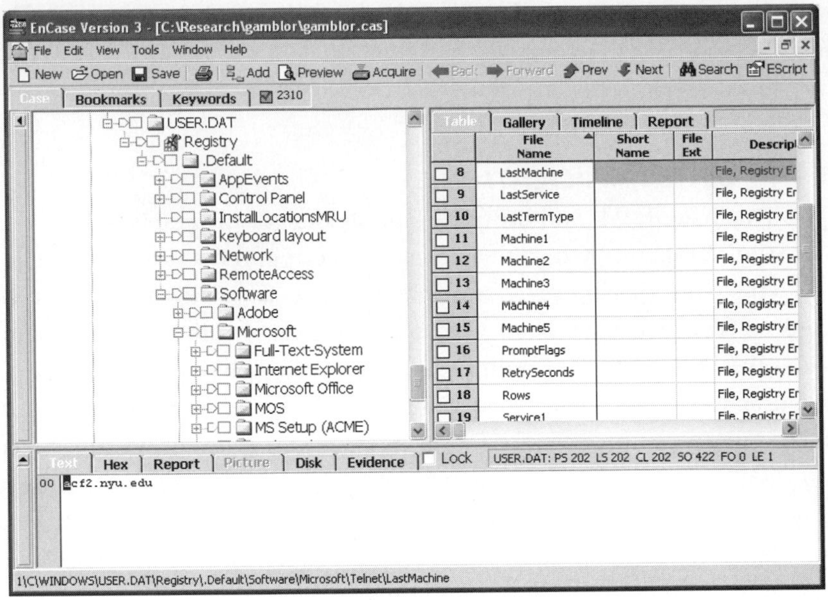

Figure 10.12

Registry showing remote systems recently accessed using Telnet.

Another common form of remote storage is a shared network drive. Most Windows machines can make all or part of their hard drives available on a network. Many organizations use Windows file servers to provide their users with this type of file storage space. Home users also use this network file sharing capability to transfer data between computers rather than using removable media as shown in Figure 10.13.

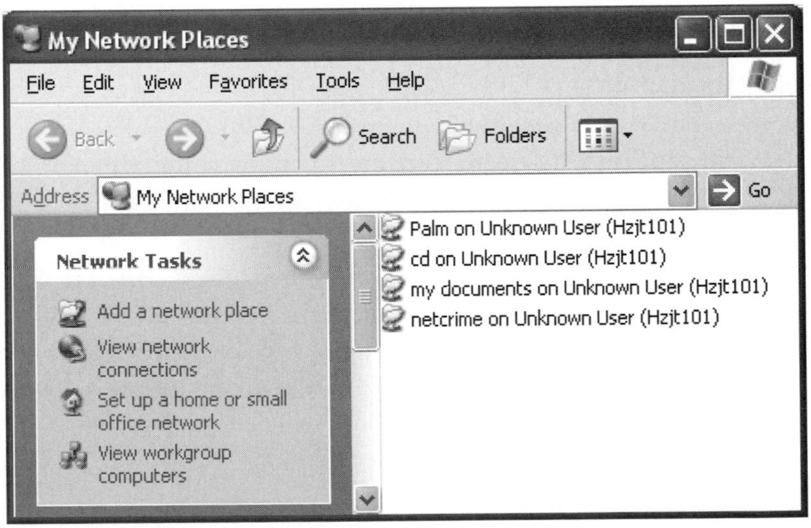

Figure 10.13

Network Neighbourhood on a Windows XP computer connected to a home network.

A list of active network shares can be found in the HKEY-USERS/<sid> /Network/ Registry key as shown in Figure 10.14. Notably, an ability to mount a network share does not necessarily imply that the account could access data

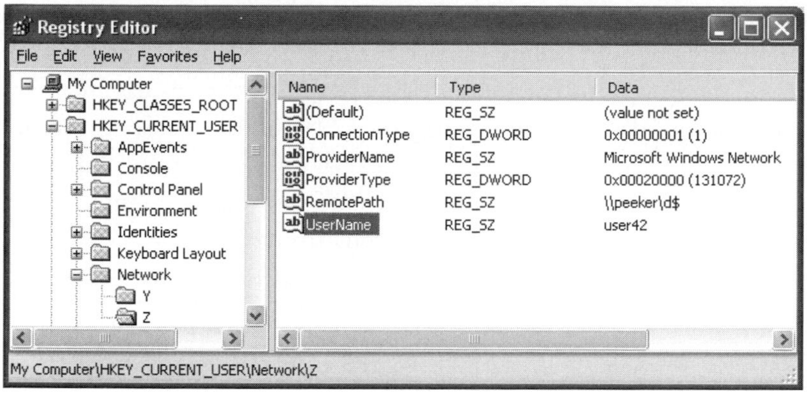

Figure 10.14

Active network file shares.

on that drive. Therefore, examine access control lists to determine if the account could write to or even read from a given network share.

Remnants of network file sharing can also be found in various Registry keys under "HKEY_Users\< sid >\Software\Microsoft\Windows\CurrentVersion\Explorer\". Some of the Explorer subkeys that may contain relevant entries are: RecentDocs, RecentDocs\Nethood, MountPoints, StreamMRU, and RunMRU. The data in these registry keys may be in hexadecimal form that can be converted manually or automatically using the "Save Subtree As" feature of the Registry Editor in Windows NT/2000 (regedt32). Additionally, in some cases it may be fruitful to search for remnants of network file shares scattered around the system (e.g. in registry slack, user.dmp, swap, unallocated space) using a grep expression like "\\\\[A – Z] + \\[A – Z] +."

This is by no means a definitive guide for locating remote storage locations. There are many other remote storage options, including free disk space (e.g. www.freedrive.com, www.filesanywhere.com), the Briefcase feature on Yahoo!, and compromised systems used by intruders to squirrel away files. Most remote storage options require users to enter passwords. It is not advisable for digital investigators to access these remote storage locations without proper authorization, even if they know the password. For instance, a computer may be configured automatically to connect to a remote file storage area. Although it may be possible to access the associated data over the network, doing so might alter evidence and exceed the scope of a search warrant.

10.9 PROGRAM ANALYSIS

When performing a functional reconstruction of a system or application to gain a better understanding of associated digital evidence, it is often desirable to perform empirical testing. For instance, when investigating a computer intrusion, it may be useful to analyze a malicious program (e.g. SubSeven) to see what sorts of evidence it leaves behind on a system. When investigating an online casino, it can be useful to understand more about the inner workings of any gambling programs they distribute to ensure that they do not disclose the investigator's identity or expose the computer in a dangerous manner. The three primary approaches to analyzing a program are to: (a) examine the source code; (b) view the program in compiled form; and (c) run the program in a test environment.

The approach of examining source code was used in United States v. Hersh after digital evidence examiners were unable to decrypt files that they believed contained child pornography.

> ...encrypted files found on a high-capacity Zip disk. The images on the Zip disk had been encrypted by software known as F-Secure, which was found on Hersh's computer. When agents could not break the encryption code, they obtained a partial source code from the manufacturer that allowed them to interpret information on the file print outs. The Zip disk contained 1,090 computer files, each identified in the directory by a unique file name, such as "sfuckmo2," "naked31," "boydoggy," "dvsex01, dvsex02, dvsex03," etc., that was consistent with names of child pornography files. The list of encrypted files was compared with a government database of child pornography. Agents compared the 1,090 files on Hersh's Zip disk with the database and matched 120 file names. Twenty-two of those had the same number of pre-encryption computer bytes as the pre-encrypted version of the files on Hersh's Zip disk. (Unites States v. Hersh)

Based on these findings, the court was convinced that the encrypted files contained child pornography.

The remainder of this section focuses on simple methods of running a program in a test environment. A convenient approach to creating a test environment for program analysis is to use VMWare,[32] a program that runs one operating system in a window on another operating system, creating a virtual machine. For instance, Windows 2000 could be installed and run in a virtual machine using VMWare on Windows XP. The supporting operating system, in this case Windows XP, is protected from any actions taken in the Windows 2000 virtual machine. Similarly, Linux can be installed and run in a VMWare virtual machine on Windows.

[32] *http://www.vmware.com*

Once a suitable test environment has been created, it is advisable to create a baseline of the system. By comparing this baseline to the system after the program of interest has been executed will reveal what changes the program made to the system, including file creation, system file alteration, and Registry modifications. For instance, changes to the Registry can be viewed by comparing it against a baseline using Regsnap.[33] Similarly, Tripwire[34] can be used to create a file system baseline and show alterations after the program of interest has been executed. File system activity can also be reconstructed after the act using the Windows search feature or using one of the digital evidence

[33] *http://www.webdon.com*
[34] *http://www.tripwire.com*

analysis tools mentioned earlier. Alternatively, Registry and file system activities can be observed in realtime using Regmon and Filemon.[35]

In some cases it may be desirable to observe processes and network traffic related to a given program. Details about processes and network connections can be observed using various tools from Sysinternals.com. More details about processes can be seen using Strace for NT.[36] Network traffic can be captured and analyzed using the tools and techniques described in Chapter 16.

[35] *http://www.sysinternals.com*

[36] *http://razor.bindview.com/tools/*

10.10 SUMMARY

Microsoft is continually developing new systems that bring new sources of digital evidence. Although the next generation of Microsoft file systems will be significantly different from its predecessors, many of the existing systems will continue to be sources of digital evidence. Therefore, an understanding of existing file systems and artifacts is a necessary component of a digital evidence examiner's training. Additionally, there will be similarities between new systems and their predecessors and certain features will remain the same. An understanding of existing systems will make it easier for digital evidence examiners to become familiar with new systems.

REFERENCES

Casey E., Larson T. and Long, H. M. (2002) Network Analysis, Casey, E. (editor) *Handbook of Computer Crime Investigation*, pp. 225–228, London: Academic Press.

Larson T. (2002) "Evidence Acquisition Boot Disk" (Available online at http://www.disclosedigital.com/eabd.html).

Microsoft (1999) "Back Up the Recovery Agent Encrypting File System Private Key in Windows 2000", Microsoft KB Q241201 (Available online at http://support.microsoft.com/default.aspx?scid=kb;[LN];Q241201)

Microsoft (2000) "FAT32 File System Specification", (Available online at http://www.microsoft.com/hwdev/hardware/fatgen.asp)

Microsoft (2001) "Detailed Explanation of FAT Boot Sector", MS KB140418 (Available online at http://support.microsoft.com/default.aspx?scid=KB;en-us;q140418)

Microsoft (2001) "Maximum Partition Size Using FAT16 File System" (Available online at http://support.microsoft.com/default.aspx?scid=kb;EN-US;118335)

Microsoft (2001) "General Information about Microsoft Office XP Encryption", Microsoft KB Article Q290112 (Available online at http://support.microsoft.com/default.aspx?scid=kb;en-us;Q290112)

Microsoft (2002) "Description of the FAT32 File System" (Available online at http://support.microsoft.com/default.aspx?scid = kb;EN-US;154997).

Microsoft (2002) "Limitations of FAT32 File System" (Available online at http://support.microsoft.com/default.aspx?scid = KB;en-us;q184006).

Sammes T. and Jenkinson B. (2000) *Forensic Computing: A Practitioner's Guide*, London: Springer.

Scott M. (2003) "Independent Review of Common Forensic Imaging Tools", Memphis Technology Group Report (Available at http://mtgroup.com/papers/ForensicImagingTools.pdf)

Sheldon B. (2001) Forensic Analysis of Windows Systems, Casey, E. (editor) *Handbook of Computer Crime Investigation*, London: Academic Press

Usborne D. (2002) "Has an old computer revealed that Reid toured world searching out new targets for al-Qa'ida?", UK Independent (Available online at http://www.independent.co.uk/story.jsp?story=114885)

CASES

United State v. Hersh (2001) Appeals Court, 11th Circuit, Case Number 00-14592 (Available online at: http://laws.lp.findlaw.com/11th/0014592opn.html)

## CHAPTER 11

FORENSIC EXAMINATION OF UNIX SYSTEMS

Over the past three decades many different types of UNIX have developed, resulting in commercial systems such as Solaris, AIX, and HP-UX as well as free operating systems like Linux, OpenBSD, and FreeBSD. UNIX operating systems are generally designed to be powerful, stable, and networked, creating an ideal platform for critical components of the Internet and smaller networks. As a result, many e-commerce Web sites, corporate financial databases, and other likely targets of criminal abuse run on UNIX systems. In addition to being a common source of digital evidence, Linux systems provide an excellent platform for forensic examination with tools for acquiring and examining digital evidence.

Although UNIX systems may seem complex, this is largely due to the fact that most of the information about the system is available for review. For instance, configuration and log files are often in plain text, allowing examiners to review quickly important aspects of a system. Additionally, individuals have easy access to the underlying source code, enabling a deeper understanding of the operating system. The openness of UNIX operating systems presents both opportunities and challenges for digital evidence examiners. For instance, this openness allows offenders to modify the system to conceal or destroy evidence. Conversely, this openness can make it easier to find evidence and examiners can compare evidence with the original source code to find any modifications.

Given the variety of UNIX operating systems and applications, it is not possible to describe or even identify every possible source of information that might be useful in an investigation. This chapter concentrates on Linux – one of the many varieties of UNIX. Furthermore, each case is different, requiring digital evidence examiners to explore and research components. The following sections provide examples of important aspects of UNIX systems with the expectation that the reader will carefully consider each area more closely to find new ways to extract information from them using the techniques covered in Chapter 9.

Digital Evidence and Computer Crime Second Edition
ISBN: 0-12-163104-4

11.1 UNIX EVIDENCE ACQUISITION BOOT DISK

Because UNIX can be instructed to access drives in read-only mode, conceivably any bootable CD-ROM or floppy diskette containing a UNIX operating system can serve as an evidence acquisition boot disk. However, one boot disk will not work with all UNIX systems because different types of UNIX systems typically have different kinds of hardware that are not compatible with each other. One boot disk is needed to boot a Solaris running on Sun Sparc-based hardware while another is needed to boot an Intel-based system running Linux. One boot disk might not even be sufficient for all Intel-based systems running Linux, since it may not have all of the necessary drivers to access all devices (e.g. Firewire drives, Ethernet cards).

CASE EXAMPLE

A Sun Ultrasparc, Enterprise 3500 system contained evidence on a 9 GB Seagate ST-19171FC Fibre Channel FC-AL, Dual Port (Barracuda 9) hard drive. Because of the unusual interface on this hard disk, it was not feasible to connect it to the available evidence collection system. Therefore, it was necessary to boot the Enterprise server from a Solaris CD-ROM and make a bitstream copy of its hard drives to a sanitized external SCSI drive using the dd command.

Notably, an evidence acquisition boot disk with Linux for Intel-based systems can be used to boot and access a Windows computer. For instance, FIRE (fire.dmzs.com) is a bootable Linux CD-ROM that can be used to acquire evidence from Intel-based systems. Like EnCase, FIRE enables remote previewing of a system via a network cable as shown in Figure 11.1.

Figure 11.1

Remote view of a Windows system using FIRE with its VNC connection feature.

Although UNIX systems can reliably mount most hard drives in read-only mode, there is still a possibility that it could make changes on an evidentiary device so some examiners use a hardware write-blocker as a precaution.

11.2 FILE SYSTEMS

There are many different UNIX file systems including UFS (UNIX File System) and ext2 (Extended File System 2) that have similar structures. Although directories play a role in UNIX file systems, they are much simpler than their Windows counterparts, only containing a list of filenames and their associated inode (index node) numbers. Every file has an associated entry in the inode table, identified by the inode number, which contains all information about the file, apart from its name. As shown in Figure 11.2, the contents of an inode include date–time stamps, the number of bytes in the file, and which clusters (a.k.a. blocks) on the disk contain the data.

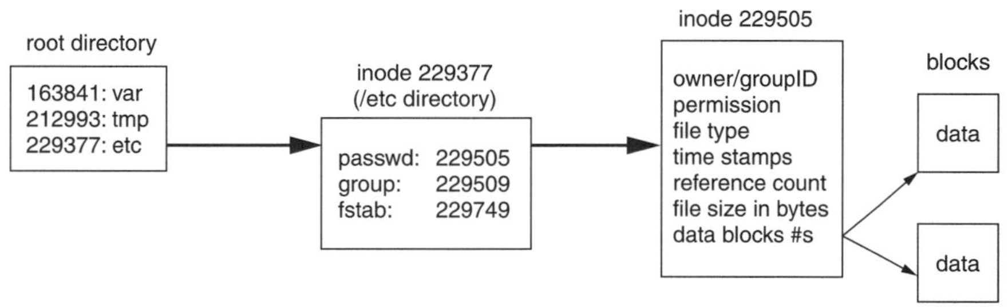

Figure 11.2

Conceptual representation of a directory and inode where the file types include regular, directory, symbolic link, and socket.

As shown in Figure 11.3, UNIX file systems break each partition into *block groups*, each with its own inodes and data blocks. Compartmentalizing data in this way prevents catastrophic file system damage because there is no single point of failure. If the disk area containing one block group is

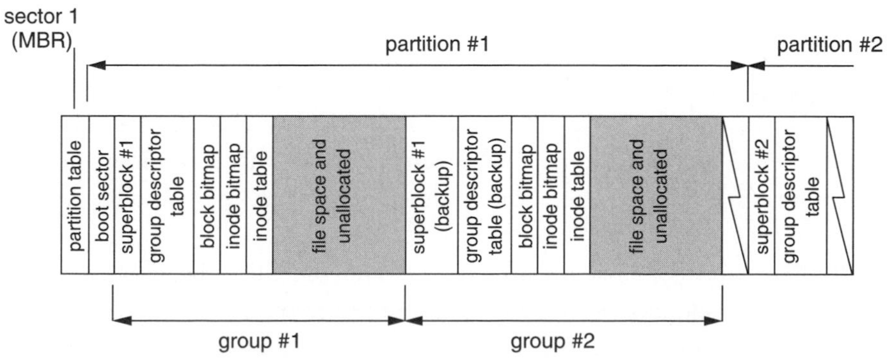

Figure 11.3

Overview of UNIX file systems.

damaged, only the data in that group are impacted, leaving data in the other groups intact.[1]

[1]Block groups are sometimes called cylinder groups because they are comprised of one or more consecutive disk cylinders.

In addition to containing data, each block group contains duplicates of critical file system components; that is, the superblock and group descriptors, to facilitate recovery if the primary copy is damaged. The superblock contains information about the file system such as block size, number of blocks per block group, the last time the file system was mounted, last time it was written to, and the sector of the root directory's inode.[2]

[2]The root directory is always associated with inode number 2 and is denoted by a "/". The first inode is generally used to keep track of bad blocks.

As the name suggests, group descriptors contain the most important information for each block group including the location of the inode table (a list of inodes and their locations). Group descriptors for all of the block groups are duplicated in each block group in case of file system corruption. So, if the primary group descriptor in any block group is damaged, a backup copy of the group descriptor can be used to repair the damage. If the inode table itself is damaged, it becomes more difficult to reconstruct the files in that block group.

Applying the library card catalog analogy from Chapter 8 to UNIX file systems, imagine a library with several divisions, each with its own books and associated card catalog. If an absent minded librarian loses his list of the locations of each division's card catalog, he can obtain an identical list from any other division. However, if the card catalog in one division is damaged or destroyed, this information is not duplicated anywhere, making it more difficult to find books in that division. Fortunately, because of the compartmentalization, damage to one division's card catalog does not adversely effect other divisions.

To summarize, when a system is commanded to access a file such as "/etc/passwd," it first looks in the superblock for the sector of inode number 2 to find the root directory as shown in Figure 11.4. The system then reads the root directory until it finds the entry for "etc" with its associated inode number (inode 0x00038001 = 229377 in Figure 11.4), reads the data blocks referenced by inode 229377 until it finds the entry for "passwd," and accesses the associated inode to identify the data blocks occupied by the password file (Figure 11.5).

If a file contains more data than can be referenced by the direct blocks field in its inode, additional indirect blocks are used to store this information. In other words, the indirect blocks contain lists of data blocks that contain the file. Even larger files may require additional indirection, in which case indirect blocks will contain lists of more (secondary or 2x) indirect blocks that in turn contain lists of data blocks that contain the file. Some file systems even allow for a third level of indirection as noted in Figure 11.5.

As shown in Figure 11.5, Linux maintains a date–time stamp of when each file was deleted. In this instance, the file has not been deleted and the value is set to a default value. This is zero from a UNIX standpoint since it represents time in epoch time, the number of seconds since January 1, 1970 00:00:00 UTC.

When a file is deleted on a UNIX system, the file's directory entry is hidden from view and the system notes that the associated inode is available for reuse. The file's directory entry, inode, and data remain on the disk

```
                          lde v2.6.0 : ext2 : /dev/hdd2
Inode:          2 (0x00000002)  Block:          0 (0x00000000)

0x00000002: drwxr-xr-x  21      4096 .
0x00000002: drwxr-xr-x  21      4096 ..
0x0000000B: drwxr-xr-x   2     16384 lost+found
0x00008001: drwxr-xr-x   2      4096 boot
0x00010001: drwxr-xr-x  17     77824 dev
0x00020001: drwxr-xr-x   2      4096 proc
0x0000000C: -rw-r--r--   1         0 .autofsck
0x00028001: drwxr-xr-x  17      4096 var
0x00034001: drwxrwxrwt   8      4096 tmp
0x00038001: drwxr-xr-x  49      4096 etc
0x00048001: drwxr-xr-x  15      4096 usr
0x00598003: drwxr-xr-x   2      4096 bin
0x00640003: drwxr-xr-x   3      4096 home
0x0064C003: drwxr-xr-x   2      4096 initrd
0x00650003: drwxr-xr-x   7      4096 lib
0x00660003: drwxr-xr-x   4      4096 mnt
0x0066C003: drwxr-xr-x   2      4096 opt
0x00670003: drwxr-x---   7      4096 root
0x0067C003: drwxr-xr-x   2      4096 sbin
0x0044C04C: drwxr-xr-x   2      4096 misc
0x000E0021: drwxr-xr-x   4      4096 e1
```

Figure 11.4

Contents of the root directory's inode, interpreted as a directory using Linux Disk Editor. http://lde.sourceforge.net

```
                          lde v2.6.0 : ext2 : /dev/hdd2
Inode:      229505 (0x00038081)  Block:          0 (0x00000000)

-rw-r--r--   1 root      root          1186  Tue Sep 24 08:57:40 2002

TYPE: regular file  LINKS:   1              DIRECT BLOCKS=0x000703F9
MODE: \0644         FLAGS: \10
UID: 00000(root)    GID: 00000(root)
SIZE: 1186          SIZE(BLKS): 8

ACCESS TIME:        Tue Nov 26 11:10:18 2002
CREATION TIME:      Tue Sep 24 08:57:40 2002
MODIFICATION TIME:  Tue Sep 24 08:57:40 2002
DELETION TIME:      Wed Dec 31 19:00:00 1969


                                                 INDIRECT BLOCK=
                                                 2x INDIRECT BLOCK=
                                                 3x INDIRECT BLOCK=
```

Figure 11.5

inode for /etc/passwd

until they are overwritten. Some systems such Solaris, ext3, and newer versions of ext2 remove the inode number in the directory, thus breaking the link between directory entries and inodes, making it more difficult to recover deleted files. Also, some systems like HP-UX delete directory entries completely, making file recovery even more difficult. Furthermore, newer file systems also break the link between the inode and the sectors that contained the data, thus removing all file system references to the data.

UNIX ctime is not equivalent to NTFS creation time (NTFS record modified time is closer). File modifications do not change the ctime. The difference between a change (ctime) and a modification (mtime) in UNIX is the difference between altering the label of a package and altering its contents (Peek *et al.* 1997). A change alters a file's inode whereas a modification alters the contents of the file. For instance, when someone changes permissions on a file it is a change, whereas when someone adds to the contents of a file it is a modification.

The ext3 Linux file system is similar to ext2 but adds journaling capabilities to facilitate file system recovery and repair after a system crash. As with NTFS, there are currently no tools available for interpreting the journal file on ext3 to determine what changes were made. This is a potential rich source of information from a forensic standpoint that will certainly be exploited in the future.

11.3 OVERVIEW OF DIGITAL EVIDENCE PROCESSING TOOLS

Linux has several features that make it ideal as a digital evidence acquisition and examination system. Linux contains many useful utilities that are designed to work together – the output of one tool can be fed into another tool easily. This ability to pipe (represented by a vertical bar "|") output from one program into another creates great flexibility. For instance, after sanitizing a disk (dd if=/dev/zero of=/dev/fd0; sync), the following command combination can be used to verify that all sectors are filled with zeros:

```
dd if=/dev/hda | xxd | grep -v "0000 0000 0000 0000 0000 0000 0000 0000"
```

This command looks for anything that is non-zero and should return nothing provided the disk has been properly sanitized. Also, Linux supports many file system types and can be used to examine media from UNIX, Windows, Macintosh, and other more arcane systems. Linux also permits direct access to devices, making it easier to acquire data from damaged media and bypass copy protection on certain memory cards. Furthermore, Linux is open source, creating a large technical support base and allowing digital evidence examiners to verify and augment its operation.

Prior to making a bitstream copy of a disk, it may be necessary to perform a keyword search to determine if there is relevant digital evidence on the system. This is particularly useful when looking for specific items on a large number of systems. The most efficient approach to searching many computers is to boot them using an evidence acquisition boot disk and run a disk search utility from the UNIX prompt. The grep command on Linux provides this keyword search capability. Once a system with useful evidence has been identified, a full bitstream copy can be made.

The mainstay of acquiring digital evidence using UNIX is the dd command. The simplest example is using dd to make a bitstream copy of a floppy disk: "dd if=/dev/fd0 of=floppycopy.dd." The dd command has many options, allowing the user to specify the block size of the evidentiary drive and to save segments of a bitstream copy in multiple files (e.g. to fit on compact disks). The output of dd can be saved in a file as shown above, or put directly onto a blank hard drive to create a clone, or can be sent through a network connection to a remote collection system using netcat. In addition to copying disks, the dd command can be used to perform analysis such as classifying data on storage media as described in the *Handbook of Computer Crime Investigation*, Chapter 8 (Seglem *et al.* 2001).

There are some nuances to copying a UNIX disk in this way that are worth mentioning. By default, dd assumes that each sector on a disk is 512 bytes. Copying large disks in 512 byte segments is inefficient and may cause confusion when copying tapes with interblock gaps. Also, when UNIX creates a file system on a disk, it takes into account disk geometry (recall cylinder/block groups). Therefore, if the two disks have even a slightly different geometry, a computer may not be able to find and boot the operating system from the new hard disk because it will be in a slightly different location on the disk. However, although the new disk will not be bootable, it will still be mountable and can be examined using another UNIX system.

When dealing with hard drives that have multiple partitions, it is advisable to make a bitstream copy of the entire disk first and then extract individual partitions later as needed (Carrier 2003a).[3] In this way, a complete copy of the original drive is preserved. Also, before making a bitstream copy, in addition to calculating the MD5 value of the drive, it is useful to document the hard drive that is being copied. To obtain information about a hard drive and the partitions on the drive, use the following commands on Linux:

```
examiner1% grep hd /var/log/dmesg
    ide0: BM-DMA at 0xa890–0xa897,   BIOS settings:   hda:DMA, hdb:pio
    ide1: BM-DMA at 0xa898–0xa89f,   BIOS settings:   hdc:pio, hdd:pio
hda: HITACHI_DK23DA-20, ATA DISK drive
hda: 39070080 sectors (20004 MB) w/2048KiB Cache,     CHS=2584/240/63,
UDMA(100)
hda: hda1 hda2 hda3 hda4 < hda5 >
examiner1% /sbin/hdparm -I /dev/hda
/dev/hda:
ATA device, with non-removable media
        Model Number:           HITACHI_DK23DA-20
        Serial Number:          14RM3D
        Firmware Revision:      00J2A0F3
Standards:
        Used: ATA/ATAPI-5 T13 1321D revision 3
        Supported: 5 4 3 2 & some of 6
Configuration:
        Logical         max     current
        cylinders       16383   16383
        heads           16      16
        sectors/track   63      63
        --
```

[3] *Some versions of UNIX, including BSD, have different partition tables than Linux and Windows, requiring a different approach to extracting partitions (Carrier 2003b).*

```
        CHS current addressable sectors:    16514064
        LBA user addressable sectors:       39070080
        device size with M = 1024*1024:        19077 MBytes
        device size with M = 1000*1000:        20003 MBytes   (20 GB)
Capabilities:
<cut for brevity>
examiner1% /sbin/sfdisk -l -uS /dev/hda
Disk /dev/hda: 2584 cylinders, 240 heads, 63 sectors/track
Units = sectors of 512 bytes, counting from 0
 Device    Boot     Start        End   #sectors  Id  System
/dev/hda1   *          63     211679     211617  83  Linux
/dev/hda2          211680   20684159   20472480  83  Linux
/dev/hda3        20684160   22317119    1632960  82  Linux swap
/dev/hda4        22317120   39070079   16752960   f  Win95 Ext'd (LBA)
/dev/hda5        22317183   39070079   16752897  83  Linux
```

It is also important to calculate the message digest value of data on the disk for later comparison. Linux provides message digest utilities such as md5sum and sha1sum that can be used to verify the integrity of digital evidence. The following combination of commands uses dd to extract data from a floppy disk and feed it to md5sum to calculate the MD5 value of the disk:

```
examiner1% dd if=/dev/fd0 bs=512 | md=sum
2880+0 records in
2880+0 records out
de3af39674f76d1eb2d652543c536a32 -
```

This MD5 value can be compared with that of the evidence after it is collected as shown here:

```
examiner1% dd if=/dev/fd0 of=hunter-floppy.dd bs=512
2880+0 records in
2880+0 records out
examiner1% md5sum hunter-floppy.dd
de3af39674f76d1eb2d652543c536a32 hunter-floppy.dd
```

The DCFL created an enhanced version called dcfl-dd[4] that can calculate MD5 values of data at regular intervals during the copying process.

Once a bitstream copy has been created, it can by "mounted" for examination. Linux provides a loopback interface that allows access to a file as if it were a disk, enabling digital evidence examiners to work on a copy as if it were the original, including accessing the file system and performing

[4] *http://sourceforge.net/projects/biatchux/*

searches. For instance, the following commands mount a bitstream copy (readonly, via a loopback device) to generate a list of files with their MD5 values and a list of all files modified in the past day.

```
examiner1% date
Tue May 13  18:01:50 EDT 2003
examiner1% mount -o ro,loop –t vfat hunter-floppy.dd /e1/case2/exhibit1
examiner1% find /e1/case2/exhibit1 -type f -exec md5sum { } \;
bca6aa0863902c44206dc3f09ccde765  skiways-getafix.doc
adcbb2fe3bcdeb62addf4ea27f15ac7c  todo.txt
d787d1699ae3c3a81fe94a9482038176  newaddress.txt
9064112159ad06c597ccfa7e70f4ec44  contacts.xls
examiner1% find /e1/case2/exhibit1  -mtime 0  –ls
6 21 -rwxr-xr-x 1 root root 21504 May 13 11:58 skiways-getafix.doc
7 0  -rwxr-xr-x 1 root root 122   May 13 12:40 todo.txt
8 0  -rwxr-xr-x 1 root root 122   May 13 12:42 newaddress.txt
```

Some forms of examination can be performed on the evidence file itself as opposed to mounting the file system. For instance, the evidence file can be viewed using a hexidecimal viewer like **xxd** or can be searched for keywords using **strings** or **grep** as shown here:

```
examiner1% strings hunter-floppy.dd | grep sales
Write additional Getafix sales letters
examiner1% cat biotechx-keywords
patient
GUID
examiner1% grep -aibf biotechx-keywords hunter-floppy.dd
30573:_PID_GUIDäAN 443A4AC0–6E57–11D7–865E-006097EDDC2Eþÿÿÿ
37959:patient#   infected   cellcount
62023:patient#   infected   cellcount
86603:patient#   infected   cellcount
125313: _PID_GUIDäAN {D2D244A2–0FE4–11D0–9B61–00AA003CF91Aþÿÿÿ
150373: _PID_GUIDäAN {443A4AC0–6E57–11D7–865E-006097EDDC2Eþÿÿÿ
170341: _PID_GUIDäAN {443A4AC0–6E57–11D7–865E-006097EDDC2Eþÿÿÿ
```

However, this approach to examining a disk is severely limited because it does not indicate which files contained the keywords.

Additionally, utilities for Linux are available from Maresware such as **hashl** and **catalog** for listing message digest values and date–time stamps of files, **hexdumpl** for viewing digital evidence in hexadecimal and ASCII form, and

strsrch for finding keywords. The output of **hexdumpl** is slightly different from **xxd**, showing the byte offset in decimal rather than hexadecimal.

```
examiner1% hexdumpl netscape.hst
00000000 00000000 00000000 E8217A3D I ....  ....  ....  Φ ! z =  I 4352
E8217A3D 01000000 01000000 536F7572 I Φ!z=  ....  ....  Sour     I 4368
6365466F 7267652E 6E65743A 2050726F I ceFo  rge.  net:  Pro      I 4384
6A656374 2046696C 656C6973 74006874 I ject  Fil   elis  t.ht     I 4400
74703A2F 2F736F75 72636566 6F726765 I tp:/  /sou  rcef  orge     I 4416
2E6E6574 2F70726F 6A656374 2F73686F I .net  /pro  ject  /sho     I 4432
7766696C 65732E70 68703F67 726F7570 I wfil  es.p  hp?g  roup     I 4448
5F69643D 31333935 36267265 6C656173 I _id=  1395  6&re  leas     I 4464
655F6964 3D343530 313900E4 217A3DA6 I e_id  =450  19.∑ ! z=ª    I 4480
217A3D03 00000001 00000053 6F757263 I !z=.  ....  ...S  ourc     I 4496
65466F72 67652E6E 65743A20 50726F6A I eFor  ge.n  et:   Proj     I 4512
65637420 496E666F 202D204C 696E7578 I ect   Info  – L   inux     I 4528
204E5446 53206669 6C652073 79737465 I NTF   S fi  le s  yste     I 4544
6D207375 70706F72 74006874 74703A2F I m su  ppor  t.ht  tp:/     I 4560
2F736F75 72636566 6F726765 2E6E6574 I /sou  rcef  orge  .net     I 4576
2F70726F 6A656374 732F6C69 6E75782D I /pro  ject  s/li  nux-     I 4592
6E746673 2F00C221 7A3DA721 7A3D0700 I ntfs  /· T!  z=°!  z=..     I 4608
00000000 00000068 7474703A 2F2F7366 I ....  ...h  ttp:  //sf     I 4624
6164732E 6F73646E 2E636F6D 2F62616E I ads.  osdn  .com  /ban     I 4640
6E65722F 73666F73 30303231 656E2E67 I ner/  sfos  0021  en.g     I 4656
69663F31 30333134 31333838 33009621 I if?1  0314  1388  3.û!     I 4672
7A3D9621 7A3D0100 00000100 0000536F I z=û!  z=..  ....  ..So     I 4688
75726365 466F7267 652E6E65 743A2057 I urce  Forg  e.ne  t: W     I 4704
656C636F 6D650068 7474703A 2F2F736F I elco  me.h  ttp:  //so     I 4720
examiner1% xxd netscape.hst
00010f0: 0000 0000 0000 0000 0000 0000 e821 7a3d  ..............!z=
0001100: e821 7a3d 0100 0000 0100 0000 536f 7572  .!z=........Sour
0001110: 6365 466f 7267 652e 6e65 743a 2050 726f  ceForge.net:Pro
0001120: 6a65 6374 2046 696c 656c 6973 7400 6874  ject Filelist.ht
0001130: 7470 3a2f 2f73 6f75 7263 6566 6f72 6765  tp://sourceforge
0001140: 2e6e 6574 2f70 726f 6a65 6374 2f73 686f  .net/project/sho
0001150: 7766 696c 6573 2e70 6870 3f67 726f 7570  wfiles.php?group
0001160: 5f69 643d 3133 3935 3626 7265 6c65 6173  _id=13956&releas
0001170: 655f 6964 3d34 3530 3139 00e4 217a 3da6  e_id=45019..!z=.
0001180: 217a 3d03 0000 0001 0000 0053 6f75 7263  !=5........Sourc
0001190: 6546 6f72 6765 2e6e 6574 3a20 5072 6f6a  eForge.net: Proj
00011a0: 6563 7420 496e 666f 202d 204c 696e 7578  ect Info – Linux
00011b0: 204e 5446 5320 6669 6c65 2073 7973 7465   NTFS file syste
00011c0: 6d20 7375 7070 6f72 7400 6874 7470 3a2f  m support.http:/
00011d0: 2f73 6f75 7263 6566 6f72 6765 2e6e 6574  /sourceforge.net
00011e0: 2f70 726f 6a65 6374 732f 6c69 6e75 782d  /projects/linux-
00011f0: 6e74 6673 2f00 c221 7a3d a721 7a3d 0700  ntfs/..! z=. !z=..
0001200: 0000 0000 0000 0068 7474 703a 2f2f 7366  .......http://sf
0001210: 6164 732e 6f73 646e 2e63 6f6d 2f62 616e  ads.osdn.com/ban
0001220: 6e65 722f 7366 6f73 3030 3231 656e 2e67  ner/sfos0021en.g
0001230: 6966 3f31 3033 3134 3133 3838 3300 9621  if?1031413883..!
0001240: 7a3d 9621 7a3d 0100 0000 0100 0000 536f  z5.!z=........So
0001250: 7572 6365 466f 7267 652e 6e65 743a 2057  urceForge.net: W
0001260: 656c 636f 6d65 0068 7474 703a 2f2f 736f  elcome.http://so
```

SAMPLE COMMAND	DESCRIPTION
ils -r /dev/hda1	List inodes of deleted files on partition 1 on drive hda
icat /dev/hda1 2	Show the contents of inode 2 on partition 1 on drive hda
unrm /dev/hda1 > unallocated	Extract unallocated space from partition 1 on drive hda
mactime -R –d /e1/case2/exhibit3 12/13/2002	Generate a chronological list of MAC times of files in the /e1/case2/exhibit3 directory and all subdirectories between December 13, 2002, and the present time

Table 11.1

Utilities from The Coroner's Toolkit being used to access a hard drive directly, illustrating the previewing capabilities of many UNIX-based tools.

More advanced examination can be performed using a collection of utilities called The Coroner's Toolkit (TCT).[5] A few example commands with explanations of their function are provided in Table 11.1. These tools can be used on a bitstream copy of a disk or to access a hard drive directly as shown in Table 11.1. Be aware that these tools currently support some UNIX file systems (e.g. UFS, ext2) but not FAT or NTFS. The Grave Robber component of TCT collects data from RAM in a systematic manner as discussed in Chapter 19.

[5] *http://www.porcupine.org/forensics/*

As an example, the second inode can be viewed in hexadecimal form as shown below and compared with Figure 11.4. Note that the inode numbers shown here in bold are little-endian, so inode 229,377 corresponding to the "etc" directory mentioned earlier (hex value "x00 x03 x80 x01") is represented as "x01 x80 x03 x00."

```
examiner1% icat /dev/hdc2 2 | xxd
0000000: 0200 0000 0c00 0102 2e00 0000 0200 0000   ................
0000010: 0c00 0202 2e2e 0000 0b00 0000 1400 0a02   ................
0000020: 6c6f 7374 2b66 6f75 6e64 0000 0180 0000   lost1found......
0000030: 0c00 0402 626f 6f74 0100 0100 0c00 0302   ....boot........
0000040: 6465 7600 0100 0200 0c00 0402 7072 6f63   dev.........proc
0000050: 0c00 0000 1c00 0901 2e61 7574 6f66 7363   .........autofsc
0000060: 6b74 6573 742d 6669 6c65 6d67 0180 0200   ktest-filemg....
0000070: 0c00 0302 7661 7200 0140 0300 0c00 0302   ....var..@......
0000080: 746d 7000 0180 0300 0c00 0302 6574 6300   tmp.........etc.
0000090: 0180 0400 0c00 0302 7573 7200 0380 5900   ........usr...Y.
00000a0: 0c00 0302 6269 6e00 0300 6400 0c00 0402   ....bin...d.....
00000b0: 686f 6d65 03c0 6400 1000 0602 696e 6974   home..d.....init
00000c0: 7264 0000 0300 6500 0c00 0302 6c69 6200   rd....e.....lib.
00000d0: 0300 6600 0c00 0302 6d6e 7400 03c0 6600   ..f.....mnt...f.
00000e0: 0c00 0302 6f70 7400 0300 6700 0c00 0402   ....opt...g.....
00000f0: 726f 6f74 03c0 6700 0c00 0402 7362 696e   root..g.....sbin
0000100: 4cc0 4400 0c00 0402 6d69 7363 2100 0e00   L.D.....misc!...
0000110: 0c00 0202 6531 6c74 ba00 4300 e80e 0502   ....e1lt..C.....
```

The Sleuth Kit[6] (previously TASK) extends TCT to support FAT and NTFS file systems and provides several other powerful utilities.

[6]The Sleuth Kit and the Autopsy Forensic Browser are available at http://www.sleuthkit.org

The **istat** command in The Sleuth Kit can be used to examine specific inodes as shown here. Note that the deletion time is only shown for deleted files. Similar information about regular files can be obtained using the standard Linux **stat** command.

```
examiner1% istat -f linux-ext2 ext2-bitstream.dd 2
inode: 2
Allocated
Group: 0
uid / gid: 0 / 0
mode: drwxr-xr-x
size: 4096
num of links: 21

Inode Times:
Accessed:         Tue  Nov  26  04:03:19  2002
File Modified:    Mon  Nov  25  20:39:17  2002
Inode Modified:   Mon  Nov  25  20:39:17  2002

Direct Blocks:
519
```

The Sleuth Kit can be combined with the Autopsy Forensic Browser to provide different views of data through a Web browser interface (Figure 11.6).

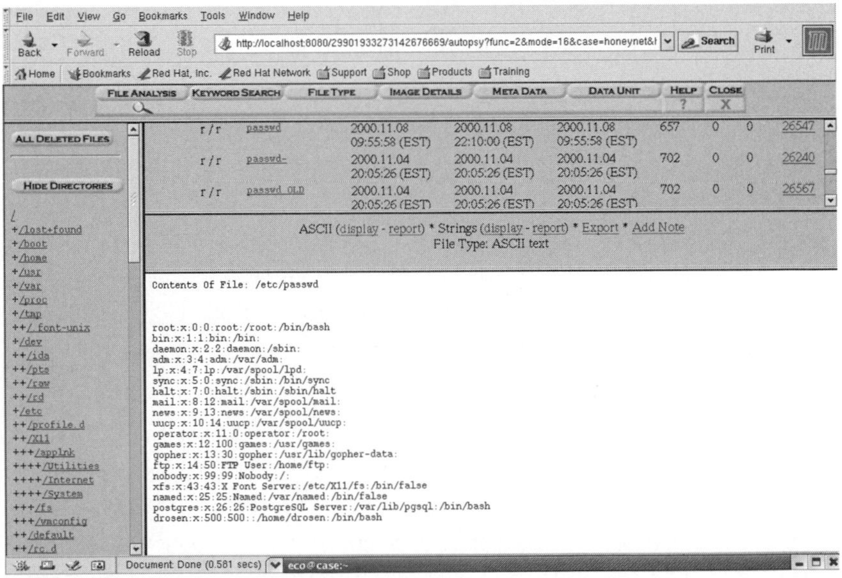

Figure 11.6

Viewing a Linux system using The Sleuth Kit and Autopsy Forensic Browser.

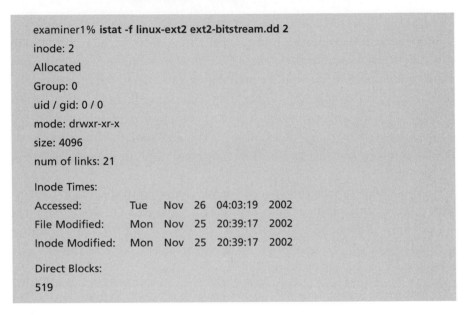

Given the large number of utilities available and the infinite adaptability of Linux, its power as a forensic examination platform is limited only by one's knowledge of the system. Although some Windows-based tools can be used to examine ext2, ext3, and UFS file systems, most do not facilitate examination of inodes and other attributes distinctive to UNIX file systems. Therefore, as mentioned in the previous chapter, no single tool should be relied upon solely. Use tools for their strengths and validate results from one tool by checking them with another.

11.4 DATA RECOVERY

Unlike Windows and Macintosh file systems, UNIX does not have file slack space. When UNIX creates a new file, it writes the remainder of the block with zeros and sets them as unallocated. Therefore, it is not possible to recover deleted data from slack space on UNIX systems. Some tools, such as **testdisk**[7] and **gpart**[8] are available for recovering deleted partitions on UNIX and Windows systems. There are only a few tools, such as **tarfix, fixcpio, tarx,** and **tar_aids** for repairing damages files on Unix.

[7] *http://www.cgsecurity.org/testdisk.html*

[8] *http://www.stud.uni-hannover.de/user/76201/gpart/*

11.4.1 UNIX-BASED TOOLS

One approach to recovering deleted files on UNIX systems is to search for inodes and recover the associated data. For instance, a list of all deleted inodes obtained from a Linux system using **ils** is shown here:

```
examiner1% ils -f linux-ext2 /e1/case2/ext2-bitstream.dd | more
class|host|device|start_time
ils|case|ext2-bitstream.dd|1054082181
st_ino|st_alloc|st_uid|st_gid|st_mtime|st_atime|st_ctime|st_dtime|st_mode|st_nli
nk|st_size|st_block0|st_block1
1 | a | 0 | 0 | 973385730|973385730|973385730|0|0|0|0|0|0
24|f|500|500|973695537|973695537|973695537|973695537|40700|0|0|308|0
25|f|500|500|954365144|973695521|973695537|973695537|100600|0|28587|309|310
26|f|500|500|954365144|973695521|973695537|973695537|100600|0|340|338|0
2049|f|500|500|973695537|973695537|973695537|973695537|40700|0|0|848|0
2050|f|500|500|953943572|973695536|973695537|973695537|100600|0|4178|8490|8491
2051|f|500|500|960098764|973695521|973695537|973695537|100600|0|52345|8495|8496
2052|f|500|500|953943572|973695537|973695537|973695537|100600|0|4860|8548|8549
2053|f|500|500|959130680|973695521|973695537|973695537|100600|0|28961|8553|8554
2054|f|500|500|959130680|973695521|973695537|973695537|100600|0|87647|8583|8584
2055|f|500|500|961959437|973695521|973695537|973695537|100600|0|30799|8670|8671
```

```
2056|f|500|500|959130680|973695521|973695537|973695537|100600|0|50176|8702|8703
2057|f|500|500|953943572|973695537|973695537|973695537|100600|0|21700|8752|8753
2058|f|500|500|959130680|973695521|973695537|973695537|100600|0|22865|8775|8776
2059|f|500|500|959130680|973695521|973695537|973695537|100600|0|14584|8799|8800
2060|f|500|500|953943572|973695521|973695537|973695537|100600|0|12276|8815|8816
2061|f|500|500|959130680|973695521|973695537|973695537|100600|0|10840|8827|8828
2062|f|500|500|959130680|973695521|973695537|973695537|100600|0| 26027|8838|8839
```

Once the inode number of a deleted file is known, the contents of the file can be accessed using **icat** provided the data still exist as shown here for inode 2054 in the previous list (in bold):

```
examiner1% icat -f linux-ext2 ext2-bitstream.dd 2054
/*
        dcc.c — handles:
        activity on a dcc socket
        disconnect on a dcc socket
        ...and that's it! (but it's a LOT)
        dprintf'ized, 27oct95
*/
/*
        This file is part of the eggdrop source code
        copyright (c) 1997 Robey Pointer
        and is distributed according to the GNU general public license.
        For full details, read the top of 'main.c' or the file called
        COPYING that was distributed with this code.
*/
#if HAVE_CONFIG_H
#include <config.h>
```

[9] *http://lde.sourceforge.net/lde_use.html*
[10] *http://recover.sourceforge.net/linux/recover/*

The Linux Disk Editor,[9] Recover,[10] and **debugfs** (Widdowson and Ferlito 2001; Buckeye and Liston 2002) use this approach to recover deleted files on ext2 file systems. The SMART tool also uses this approach to recover deleted files (Figure 11.7).

However, recall that many UNIX file systems remove references from inodes to the sectors that contain the data, breaking the connection between the inode and the data on disk. This fact is evident in the following list of

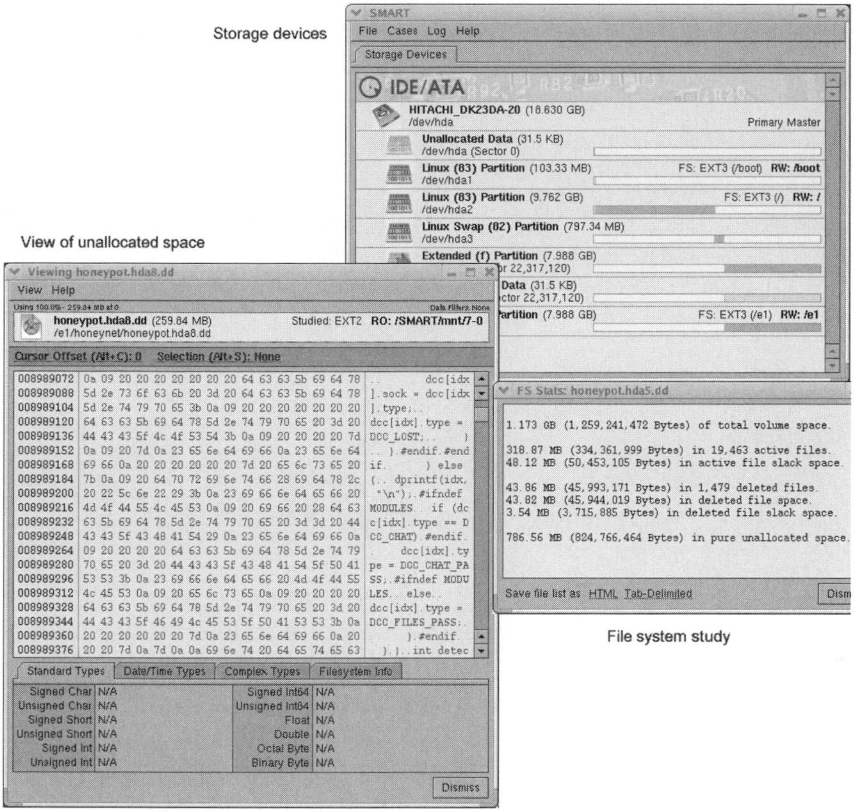

Figure 11.7

SMART file recovery process saves deleted files onto the examination system for further analysis using other tools.

deleted inodes from a Solaris system – all of the starting blocks (the first sector that contained data for each file) are set to zero:

```
examiner1% ils -r -f solaris /e1/case2/ufs-bitstream.dd
class|host|device|start_time
ils|legolas|/e1/morgue/ufs-bitstream.dd|1039101486
st_ino|st_alloc|st_uid|st_gid|st_mtime|st_atime|st_ctime|st_mode|st_nlink|st_size|
st_block0|st_block1
213|f|0|1|1038427233|1038427233|1038427243|0|0|0|0|0
3946|f|0|0|987886669|987886669|987886690|0|0|0|0|0
7698|f|0|60001|987893332|987893332|987893332|0|0|0|0|0
11509|f|0|60001|987893332|987893332|987893332|0|0|0|0|0
15105|f|0|60001|987893332|987893332|987893332|0|0|0|0|0
15260|f|0|0|987886816|987886816|987886830|0|0|0|0|0
15261|f|0|0|987886821|987886821|987886830|0|0|0|0|0
15264|f|0|0|987886449|987886449|987886457|0|0|0|0|0
```

```
15265|f|0|0|987886449|987886449|987886457|0|0|0|0|0
22816|f|0|0|1038421634|1038421621|1038421634|0|0|0|0|0
22817|f|0|0|987893848|987887279|987893848|0|0|0|0|0
34164|f|0|60001|987893333|987893332|987893354|0|0|0|0|0
45493|f|0|0|1038421571|1038421571|1038421634|0|0|0|0|0
45494|f|0|0|1038421571|1038421571|1038421634|0|0|0|0|0
53039|f|0|60001|987893333|987887277|987893354|0|0|0|0|0
56784|f|0|0|987886929|987886922|987886935|0|0|0|0|0
56787|f|0|0|987886930|987886929|987886935|0|0|0|0|0
56788|f|0|0|987886903|987886903|987886917|0|0|0|0|0
60579|f|0|0|987886609|987886609|987886620|0|0|0|0|0
60580|f|0|0|987886601|987886601|987886620|0|0|0|0|0
64394|f|0|1|1038425953|1038425939|1038425983|0|0|0|0|0
64395|f|0|1|1038421500|1038421498|1038421506|0|0|0|0|0
```

Another approach to recovering deleted files is to look in directories for deleted entries, provided they exist.[11] For instance, The Sleuth Kit uses this method to generate a list of deleted files and directories on an ext2 file system using **fls** as shown here:

[11] Recall that Solaris and ext3 clears the inode number in deleted directory entries and HP-UX deletes the entire entry, eliminating this method as a possibility.

```
examiner1% fls -d -r -f linux-ext2 /dev/hdd2
-/- * 0: boot/
-/- * 4(realloc): boot/
-/- * 0: boot/P
-/- * 0: boot/
-/- * 0: boot/
b/- * 0: dev/ataraid/d9p9;3d905a83
b/- * 0: dev/cciss/c7d9p9;3d905a83
c/- * 0: dev/compaq/cpqrid;3d905a83
c/- * 0: dev/dri/card3;3d905a83
b/- * 0: dev/i2o/hdz9;3d905a83
b/- * 0: dev/ida/c7d9p9;3d905a83
c/- * 0: dev/inet/udp;3d905a83
d/d * 933895(realloc): dev/input
c/c * 66319(realloc): dev/ip2ipl0
l/l * 66318(realloc): dev/ip
c/c * 66323(realloc): dev/ip2stat0
c/c * 66320(realloc): dev/ip2ipl1
```

```
c/c * 66321(realloc): dev/ip2ipl2
c/c * 66322(realloc): dev/ip2ipl3
d/d * 983047(realloc): dev/logicalco
-/- * 3355443: dev/
<cut for brevity>
```

The Autopsy Forensic Browser combines these two approaches to list all deleted directory entries that were referencing a given inode (labeled "Pointed to by file") as shown here for inode 3817585 on an ext2 file system:

```
node: 3817585
Pointed to by file:
/tmp/makewhatis3JoBa0 (deleted)
/root/.netscape/cache/1A/cache3DDC0D5A01A20AD  (deleted)
/root/.netscape/cache/1A/cache3DD5997A1200A22   (deleted)
File Type: empty
Details:
Not Allocated
Group: 233
uid / gid: 0 / 0
mode: drwx———
size: 0
num of links: 0
Inode Times:
Accessed: Mon Nov 25 19:08:29 2002
File Modified: Mon Nov 25 19:08:29 2002
Inode Modified: Mon Nov 25 19:08:29 2002
Deleted: Mon Nov 25 19:08:29 2002
Direct Blocks:
```

It is worth reiterating that these tools are not limited to examining UNIX file systems – they can be used to recover files from FAT and NTFS systems.

11.4.2 WINDOWS-BASED TOOLS

Although EnCase recovers some deleted files on ext2 file systems, placing them all in a "Lost Files" area, it does not reference data using inode numbers and does not currently recover deleted directory entries as described earlier in this next section. However, some Windows-based tools do facilitate

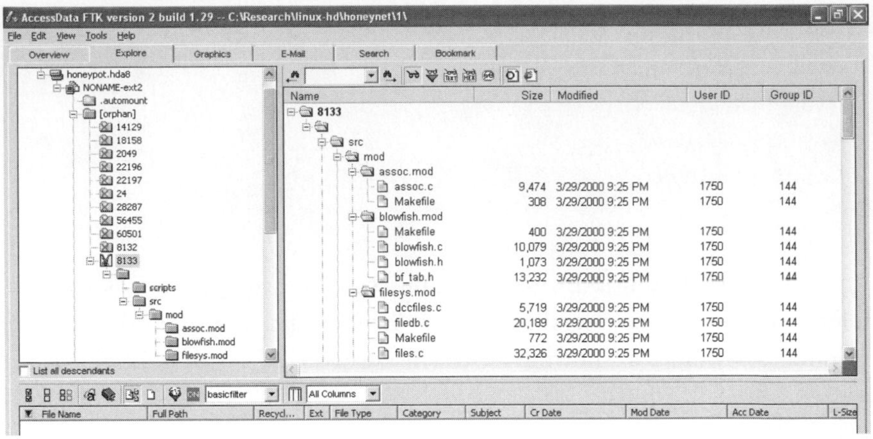

Figure 11.8

FTK used to view ext2 file system in the file "honeynet.hda8.dd," available from http://www.honeynet.org/challenge/.

certain forms of examination that are not readily available in Linux-based tools. As an example, Forensic Toolkit recovers deleted files and folders from ext2 file systems into an area called "[orphan]," organizing and displaying the recovered data in a way that facilitates examination. For instance, as shown in Figure 11.8, FTK uses inode numbers to reference recovered items and provides convenient representations of recovered files such as the deleted TAR file.

11.4.3 FILE CARVING WITH UNIX

Deleted data can also be recovered using class characteristics. For instance, foremost[12] can be used to carve files from any digital object such as an evidence file, unallocated space, or a swap file. The following output shows foremost recovering files from a bitstream copy of a floppy disk:

[12] http://foremost.sourceforge.net

```
examiner1% foremost -o carved-foremost -v floppycopy.dd
foremost version 0.62
Written by Kris Kendall and Jesse Kornblum.

Using output directory: /e1/carved-foremost
Verbose mode on
Using configuration file: foremost.conf
Opening /e1/case2/floppycopy.dd.
Total file size is 1474560 bytes

/e1/case2/floppycopy.dd: 100.0% done (1.4 MB read)
A doc was found at: 17408
Wrote file /e1/case2/carved-foremost/00000000.doc — Success
A doc was found at: 37888
```

```
Wrote file /e1/case2/carved-foremost/00000001.doc -- Success
A jpg was found at: 76800
Wrote file /e1/case2/carved-foremost/00000002.jpg --Success
A jpg was found at: 77230
Wrote file /e1/case2/carved-foremost/00000003.jpg --Success
A jpg was found at: 543232
Wrote file /e1/case2/carved-foremost/00000004.jpg -- Success
A gif was found at: 990208
Wrote file /e1/case2/carved-foremost/00000005.gif --Success
A jpg was found at: 1308160
Wrote file /e1/case2/carved-foremost/00000006.jpg -- Success
Foremost is done.
```

This tool can be instructed to search for any type of file by adding the appropriate header and footer information to its configuration file "foremost.conf." If a file is fragmented, this and other carving methods will only find the first portion of the file since other fragments will not contain the signature header.

Another approach to recovering data is implemented in Lazarus from TCT. Lazarus automatically classifies digital data in the following way:

1 Read a chunk of data (default 1k).
2 Determine if the chunk is text or binary data:
   (a) If text, attempt to classify it based on its contents (e.g. html).
   (b) If binary, attempt to classify it using the UNIX file command.
3 If chunk was successfully classified, compare it with the previous chunk:
   (a) If they are the same class, assume they are in the same file.
   (b) If they are not of the same class, assume they are in different files.
4 If chunk was not successfully classified, compare it with the previous chunk:
   (a) If they are the same type (binary or text), assume they are in the same file.
   (b) If they are different types (binary or text), assume they are in different files.

As with other file carving tools, one of the operative assumptions in this approach is that computers make an effort to save files in contiguous sectors. In this way, Lazarus provides some structure to data on a disk and attempts to reconstruct files fragments in contiguous chunks as shown in Figure 11.9.

Note that this simple but clever method uses the concepts of comparison and classification described in Chapter 9.

Although certain aspects of UNIX file systems make data recovery more difficult, the use of block groups in UNIX file systems can facilitate

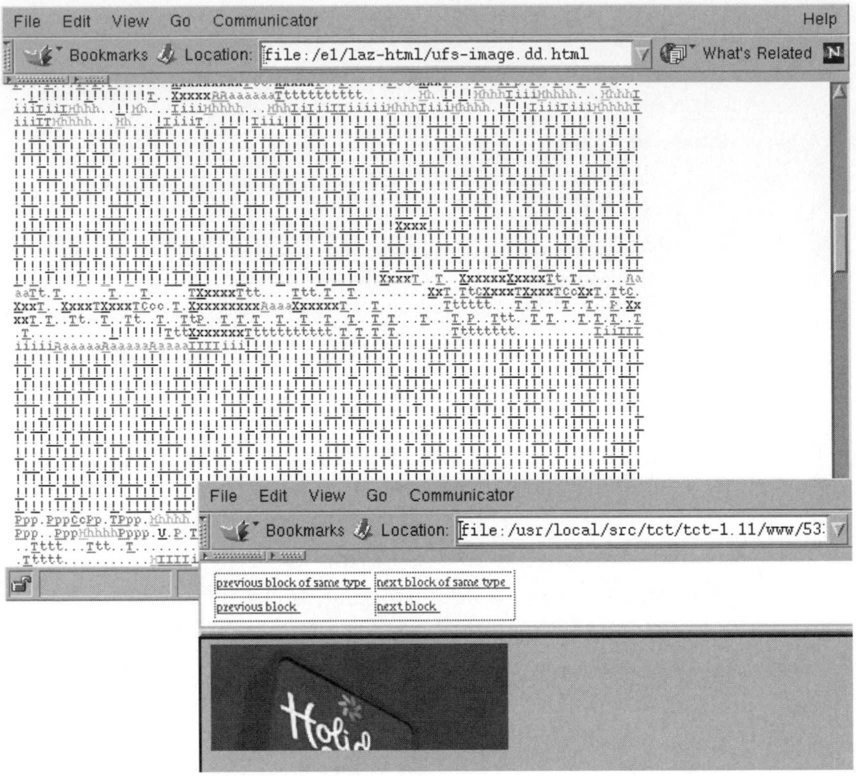

Figure 11.9

Lazarus from the Coroner's Toolkit used to classify data on a disk and recover deleted data such as the partial image shown here.

data recovery because it causes clustering of data on the disk. For instance, all log files in the directory "/var/log" (but not necessarily its subdirectories like "/var/log/argus") will be stored in the same block group. So, rather than searching all unallocated space on the disk for deleted log entries, digital evidence examiners can focus on unallocated space of that block group. For instance, on one Linux system, the "/var/log" directory has inode number 502952 [Figure 11.10(a)] in block group 31 [Figure 11.10(b)].

The "Image Details" screen in the Autopsy Forensic Browser gives the following information about block group 31:

Group: 31:
Inode Range: 502945 – 519168
Block Range: 1015808 – 1048575
Data bitmap: 1015808 – 1015808
Inode bitmap: 1015809 – 1015809
Inode Table: 1015812 – 1016318
Data Blocks: 1015810 – 1015811, 1016319 – 1048575

(a)

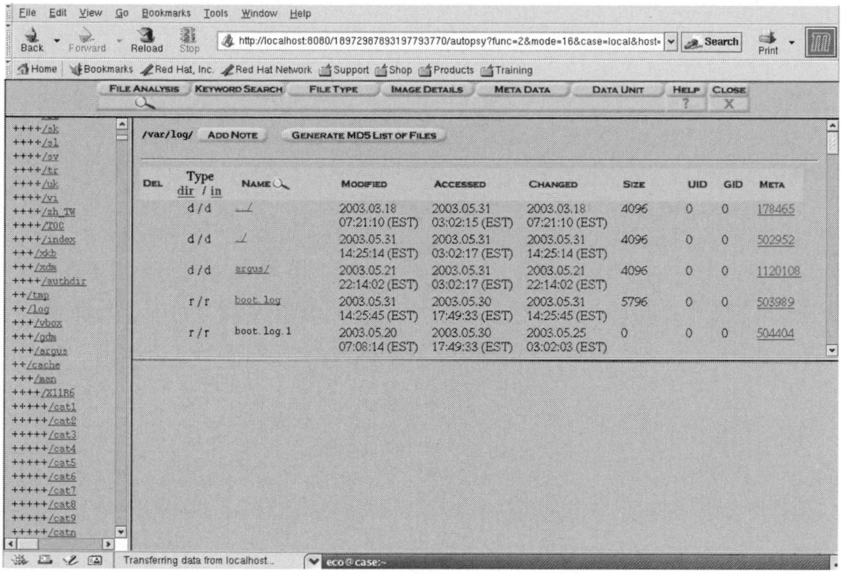

(b)

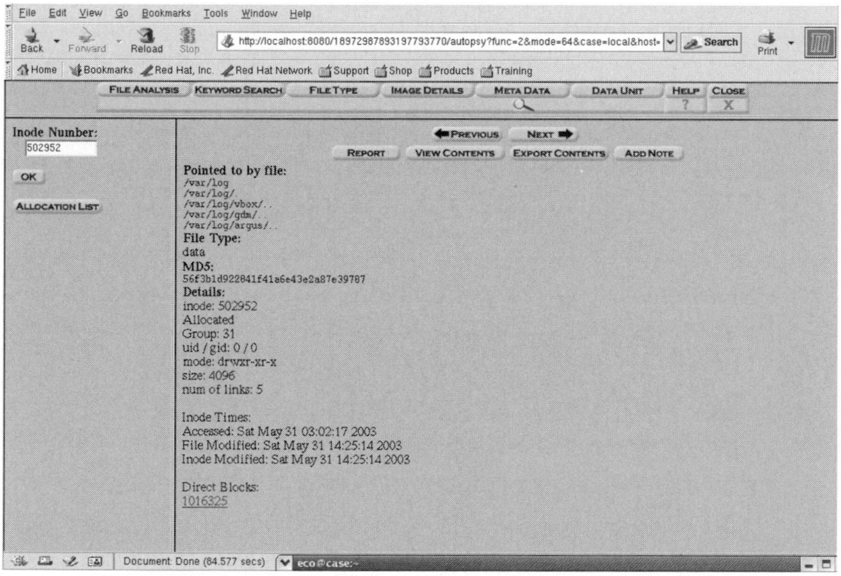

Figure 11.10

The Sleuth Kit showing (a) /var/log directory with inode number 502952 (b) information relating to inode number 502952, including the associated block group 31, which can also be obtained using the istat command.

The unallocated sectors for just this portion of the disk can be extracted using **dls** in the Sleuth Kit and then searched for information of interest as shown here:

```
examiner1% dls -f linux-ext2 /dev/hda2 1016319-1048575 > /e1/block31-
unallocated
examiner1% strings block31-unalloc | grep "Apr 3"
Apr 3 09:54:45 case sshd[792]: Server listening on 0.0.0.0 port 22.
Apr 3 09:55:14 case xinetd[806]: START: sgi_fam pid=1118 from=<no address>
Apr 3 10:20:20 case sshd[165]: Could not reverse map address 192.168.0.3.
Apr 3 10:20:25 case sshd[165]: Failed password for jay from 192.168.0.3 port
                               1176 ssh2
Apr 3 10:20:29 case sshd[165]: Accepted password for jay from 192.168.0.3 port
                               1176 ssh2
Apr 3 10:45:05 case sshd[282]: Could not reverse map address 192.168.0.3.
Apr 3 10:45:09 case sshd[282]: Accepted password for jay from 192.168.0.3 port
                               1177 ssh2
Apr 3 13:23:37 case sshd[765]: Server listening on 0.0.0.0 port 22.
Apr 3 13:24:07 case xinetd[779]: START: sgi_fam pid=1013 from=<no address>
Apr 3 13:47:16 case sshd[117]: Could not reverse map address 192.168.0.5.
Apr 3 13:47:21 case sshd[117]: Failed password for moe from 192.168.0.5 port
                               1553 ssh2
Apr 3 13:47:26 case sshd[117]: Failed password for moe from 192.168.0.5 port
                               1553 ssh2
Apr 3 13:47:30 case sshd[117]: Accepted password for moe from 192.168.0.5
                               port 1553 ssh2
Apr 3 13:47:32 case sshd[119]: subsystem request for sftp
```

However, when searching for log files or other digital evidence, keep in mind that swap space may also contain useful data.

11.4.4 DEALING WITH PASSWORD PROTECTION AND ENCRYPTION

Although a collection of UNIX systems, called a "Beowulf cluster", can be used to attempt to break weak encryption, this approach is rarely effective against strong encryption like PGP. When strong encryption is involved, it is usually necessary to take advantages of weaknesses in the implementation of the encryption program. For instance, files on UNIX machines can be encrypted using the crypt utility as shown here.

```
% crypt -key 'guessme' < plaintext> ciphertext
```

However, if the plaintext file is simply deleted rather than wiped, it may be possible to recover this copy from the hard disk. Furthermore, if the plaintext file was stored in memory, swapped to disk, or backed up to external media, it may be possible to retrieve some or all of these data. Another obvious weakness of the crypt command is the secret key. If an easy to remember key such as "guessme" is used, it may be possible for someone to guess it and gain access to the encrypted data. If a difficult to remember key is used, it may be necessary for the user to write it down in a location that can be referenced the next time the data are decrypted, potentially exposing it to others.

When performing a functional reconstruction using a restored clone of a UNIX system, it may be possible to bypass the logon password by booting into single user mode and manually altering the password file. In situations where the actual password is needed, tools like Crack and Jack the Ripper are available that attempt to guess password entries in UNIX password files.

11.5 LOG FILES

UNIX systems have a variety logs that can be useful in an investigation. Logons and logoffs, or any event on a UNIX computer for that matter, can create entries in one or more system log files. An entry may be made in the lastlog file that can be interpreted using the lastlog command, and in the wtmp and utmp databases that can be interpreted using the last command. The degree of detail in these logs varies depending on how logging is configured. UNIX systems can even be configured to record the commands that each user account executed using process accounting (pacct files are accessed using lastcomm) or the Basic Security Module (BSM) on Solaris. Additionally, servers running on UNIX machines may have logs that can be useful for reconstructing events and tracking down offenders as discussed in Part 3 of this text.

11.6 FILE SYSTEM TRACES

Any activity can make an impression on a UNIX file systems, like footprints in snow. Applications can leave remnants on disk either directly in temporary files or indirectly through swap space. For instance, printing creates spool files (usually in /var/spool/lpd) and other applications create temporary files in /tmp and other areas. A TAR file can bring date–time stamps and userids from other systems. Some UNIX systems have a "/proc" file system with information relating to processes running in memory that can be useful for gaining a more complete picture of what was occurring on a system as discussed in Chapter 19.

The simple act of accessing and manipulating files alters their date–time stamps and this information can be correlated with log file entries to gain a better understanding of which user account was involved. For instance, mactime (in TCT and The Sleuth Kit) can use a time range from a wtmp log to generate a chronological list of MAC times for that period as shown here:

```
# last
eco        pts/3      66-65-113-65.nyc Sun Oct 20 23:45 - 01:08     (00:23)
# mactime  -b body -l "Sun Oct 20 23:45 - 01:08  (05:23)"
Oct 20 02 23:45:42       452 .a. -rw ------ root    root  /etc/pam.d/sshd
Oct 20 02 23:45:47       124 .a. -rw-r--r-- eco     eco   /home/eco/.bashrc
                         191 .a. -rw-r--r-- eco     eco   /home/eco/. bash_profile
Oct 20 02 23:47:30     75428 .a. -r-xr-xr-x root    bin   /usr/bin/ftp
Oct 20 02 23:55:24  22433792 mac -rw-r--r-- eco     eco   /home/eco/secret.pgp
```

These MAC times suggests that the FTP client was used to download a file named "secret.pgp," demonstrating that an understanding of how date–time stamps of files are updated under different circumstances can help digital investigators reconstruct the associated events. Process accounting and command history logs may contain information to corroborate this theory.

A summary of common actions and the associated date–time stamp changes on UNIX is provided in Table 11.2. Unlike Windows, this behavior is clearly documented in UNIX manual pages (see man fstat).

When a file is added to or moved out of a directory, the inode change time of the directory listing (".") is updated as well as the last modified and accessed times. One implication of this behavior is that, when a file is deleted on a UNIX system, the ctime of its parent directory is updated. This time can be correlated with the ctime of deleted inodes (and deletion time on ext2/ext3) to get a sense of which file may have been deleted from the directory as shown later in this section.[13]

Because deleted inodes are not accessible to the file system, deleting a file has the effect of preserving its inode until it is reused. Therefore, when an intruder gains unauthorized access to a UNIX system, installs tools, and

[13]UNIX dates are generally in GMT and may need to be adjusted using the time zone specified in the TZ environment variable.

Table 11.2

Date–time stamp behavior on UNIX.

ACTION	LAST MODIFIED DATE–TIME	LAST ACCESSED DATE–TIME	INODE CHANGE DATE–TIME
File moved within a volume	Unchanged	Unchanged	Updated
File copied (destination file)	Updated	Updated	Updated

deletes files, the inodes of deleted files may be recovered long after the intrusion even if the data are not recoverable. For instance, the following shows **ils** and **mactime** from The Sleuth Kit being used to create a chronological list of modification, access, and creation (MAC) times from deleted files on a Solaris system:

```
examiner1% ils -m -f solaris ufs-image.dd | mactime 4/1/2001
Apr 21 01 16:54:09 0  0 ma. ---------- root  root   <ufs-bitstream.dd-dead-15265>
                      0 ma. ---------- root  root   <ufs-bitstream.dd-dead-15264>
Apr 21 01 16:54:17    0 ..c ---------- root  root   <ufs-bitstream.dd-dead-15265>
                      0 ..c ---------- root  root   <ufs-bitstream.dd-dead-15264.s
Apr 21 01 16:56:41    0 ma. ---------- root  root   <ufs-bitstream.dd-dead-60580>
Apr 21 01 16:56:49    0 ma. ---------- root  root   <ufs-bitstream.dd-dead-60579>
Apr 21 01 16:57:00    0 ..c ---------- root  root   <ufs-bitstream.dd-dead-60579>
                      0 ..c ---------- root  root   <ufs-bitstream.dd-dead-60580>
Apr 21 01 16:57:49    0 ma. ---------- root  root   <ufs-bitstream.dd-dead-3946>
Apr 21 01 16:58:10    0 ..c ---------- root  root   <ufs-bitstream.dd-dead-3946>
Apr 21 01 17:00:16    0 ma. ---------- root  root   <ufs-bitstream.dd-dead-15260>
Apr 21 01 17:00:21    0 ma. ---------- root  root   <ufs-bitstream.dd-dead-15261>
Apr 21 01 17:00:30    0 ..c ---------- root  root   <ufs-bitstream.dd-dead-15261>
                      0 ..c ---------- root  root   <ufs-bitstream.dd-dead-15260>
Apr 21 01 17:01:43    0 ma. ---------- root  root   <ufs-bitstream.dd-dead-56788>
Apr 21 01 17:01:57    0 ..c ---------- root  root   <ufs-bitstream.dd-dead-56788>
Apr 21 01 17:02:02    0 .a. ---------- root  root   <ufs-bitstream.dd-dead-56784>
Apr 21 01 17:02:09    0 m.. ---------- root  root   <ufs-bitstream.dd-dead-56784>
                      0 .a. ---------- root  root   <ufs-bitstream.dd-dead-56787>
Apr 21 01 17:02:10    0 m.. ---------- root  root   <ufs-bitstream.dd-dead-56787>
Apr 21 01 17:02:15    0 ..c ---------- root  root   <ufs-bitstream.dd-dead-56787>
                      0 ..c ---------- root  root   <ufs-bitstream.dd-dead-56784>
Apr 21 01 17:07:57    0 .a. ---------- root  60001  <ufs-bitstream.dd-dead-53039>
Apr 21 01 17:07:59    0 .a. ---------- root  root   <ufs-bitstream.dd-dead-22817>
Apr 21 01 18:48:52    0 mac ---------- root  60001  <ufs-bitstream.dd-dead-15105>
                      0 mac ---------- root  60001  <ufs-bitstream.dd-dead-11509>
                      0 mac ---------- root  60001  <ufs-bitstream.dd-dead-7698>
                      0 .a. ---------- root  60001  <ufs-bitstream.dd-dead-34164>
Apr 21 01 18:48:53    0 m.. ---------- root  60001  <ufs-bitstream.dd-dead-53039>
                      0 m.. ---------- root  60001  <ufs-bitstream.dd-dead-34164>
Apr 21 01 18:49:14    0 ..c ---------- root  60001  <ufs-bitstream.dd-dead-53039>
                      0 ..c ---------- root  60001  <ufs-bitstream.dd-dead-34164>
Apr 21 01 18:57:28    0 m.c ---------- root  root   <ufs-bitstream.dd-dead-22817>
```

```
Nov 27 02 13:24:58    0 . a . ---------- root   bin    <ufs-bitstream.dd-dead-64395>
Nov 27 02 13:25:00    0 m . . ---------- root   bin    <ufs-bitstream.dd-dead-64395>
Nov 27 02 13:25:06    0 . . c ---------- root   bin    <ufs-bitstream.dd-dead-64395>
Nov 27 02 13:26:11    0 ma.   ---------- root   root   <ufs-bitstream.dd-dead-45494>
                      0 ma.   ---------- root   root   <ufs-bitstream.dd-dead-45493>
Nov 27 02 13:27:01    0 . a . ---------- root   root   <ufs-bitstream.dd-dead-22816>
Nov 27 02 13:27:14    0 . . c ---------- root   root   <ufs-bitstream.dd-dead-45494>
                      0 . . c ---------- root   root   <ufs-bitstream.dd-dead-45493>
                      0 m.c   ---------- root   root   <ufs-bitstream.dd-dead-22816>
Nov 27 02 14:38:59    0 . a . ---------- root   bin    <ufs-bitstream.dd-dead-64394>
Nov 27 02 14:39:13    0 m . . ---------- root   bin    <ufs-bitstream.dd-dead-64394>
Nov 27 02 14:39:43    0 . . c ---------- root   bin    <ufs-bitstream.dd-dead-64394>
Nov 27 02 15:00:33    0 ma.   ---------- root   bin    <ufs-bitstream.dd-dead-213>
Nov 27 02 15:00:43    0 . . c ---------- root   bin    <ufs-bitstream.dd-dead-213>
```

The resulting output shows two periods of high activity (April 21, 2001, and November 27, 2002) when a number of files were deleted corresponding with an intruder's activities. The fls utility provides additional information for this time period, showing which directories were modified, accessed, and changed. Combining these data gives digital investigators a sense of where the intruder was operating.

```
% fls -m / -f solaris ufs-image.dd | mactime 4/1/2001
Sat Apr 21 2001 15:45:28   8192  mac    -/drwx---------   0 0   3       /lost+found
Sat Apr 21 2001 15:47:10    512  mac    -/drwxr-xr-x      0 0   3776    /usr
Sat Apr 21 2001 15:51:57    512  . a .  -/drwxrwxr-x      0 3   34006   /opt
                              9  m . c  -/lrwxrwxrwx      0 0   14      /bin ->./usr/bin
                            512  mac    -/drwxrwxr-x      0 3   30225   /mnt
                            512  mac    -/drwxr-xr-x      0 3   37777   /proc
                            512  . a .  -/drwxrwxrwt      3 3   45326   /tmp
                            512  . a .  -/drwxr-xr-x      0 3   64208   /kernel
                              9  m . c  -/lrwxrwxrwx      0 0   20      /lib -> ./usr/lib
Sat Apr 21 2001 15:53:25    512  mac    -/drwxr-xr-x      0 3   18906   /platform
Sat Apr 21 2001 16:32:18    512  mac    -/drwxrwxr-x      0 3   19012   /home
Sat Apr 21 2001 16:35:59    512  m . c  -/drwxrwxr-x      0 3   34006   /opt
Sat Apr 21 2001 16:45:56    512  m . c  -/drwxrwxr-x      0 3   18898   /devices
Sat Apr 21 2001 16:52:58    512  m . c  -/drwxr-xr-x      0 3   64208   /kernel
Sat Apr 21 2001 16:53:00    512  . a .  -/drwxrwxr-x      0 3   41556   /sbin
Sat Apr 21 2001 16:53:01    512  m . c  -/drwxrwxr-x      0 3   41556   /sbin
Sat Apr 21 2001 16:57:54    512  . a .  -/drwxr-xr-x      0 3   7552    /var
```

Sat Apr 21 2001 17:04:30	512	. a .	-/drwxrwxr-x	0	3	18898	/devices
Sat Apr 21 2001 17:07:26	512	mac	-/dr-xr-xr-x	0	0	53030	/xfn
	512	mac	-/dr-xr-xr-x	0	0	30398	/net
Sat Apr 21 2001 17:07:35	1032	. a .	-/-rw----------	0	0	87	/.cpr_config
Sat Apr 21 2001 17:07:40	512	mac	-/drwxr-xr-x	0	0	53037	/vol
Sat Apr 21 2001 17:07:47	512	m . c	-/drwxr-xr-x	0	3	7552	/var
Sat Apr 21 2001 17:07:52	512	. a .	-/drwxr-xr-x	0	60001	53038	/cdrom
Sat Apr 21 2001 18:48:53	512	m.c	-/drwxr-xr-x	0	60001	53038	/cdrom
Sat Apr 21 2001 20:22:41	512	m.c	-/drwxrwxr-x	0	3	128	/export
Sat Apr 21 2001 20:22:42	512	. a .	-/drwxrwxr-x	0	3	128	/export
Sun Apr 22 2001 22:11:02	804520	m.c	-/-rw----------	0	0	211	/core
Sun Apr 22 2001 22:12:32	804520	. a .	-/-rw----------	0	0	211	/core
Wed Nov 27 2002 13:26:11	512	m.c	-/drwxrwxrwt	3	3	45326	/tmp
Wed Nov 27 2002 13:26:21	3072	. a .	-/drwxr-xr-x	0	3	49090	/etc
Wed Nov 27 2002 13:26:32	1032	m.c	-/-rw----------	0	0	87	/.cpr_config
Wed Nov 27 2002 13:26:34	3072	m.c	-/drwxr-xr-x	0	3	49090	/etc
	3584	m.c	-/drwxrwxr-x	0	3	18896	/dev
Wed Nov 27 2002 13:26:37	3584	. a .	-/drwxrwxr-x	0	3	18896	/dev
Wed Nov 27 2002 14:57:03	9	. a .	-/lrwxrwxrwx	0	0	14	/bin -> ./usr/bin
	9	. a .	-/lrwxrwxrwx	0	0	20	/lib -> ./usr/lib

Digital investigators can focus on these periods of high activity, looking for related log files and other data that may help them determine what occurred. When dealing with large amounts of these sorts of data, plotting date–time stamps in a histogram can be useful, showing spikes corresponding to periods of high of activity. For instance, creating a histogram of MAC times using the following command, results in Figure 11.11:

```
% ils -m -f linux-ext2 honeynet.hda8.dd | mactime -d -i /e1/honeynet/hda8.idx.txt
```

Figure 11.11 shows a high number of deleted inodes on November 8, corresponding to the intruder's activities.

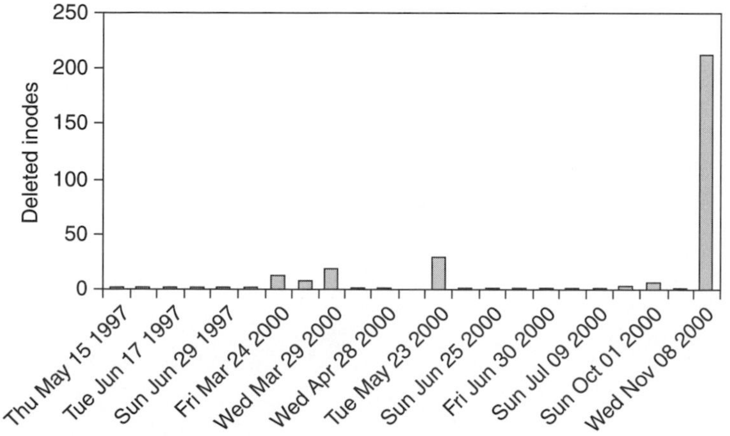

Figure 11.11

A histogram of deleted inodes from a compromised machine showing a spike on November 8 as a result of an intruder's activities.

11.7 INTERNET TRACES

UNIX was specifically designed with networking in mind and has many applications for accessing the Internet. Most of these utilities do not keep logs, but may leave subtle traces of activities in swap space or temporary files as discussed in the previous section. However, some Internet applications create records of activities such as Web resources accessed and e-mails sent and received.

11.7.1 WEB BROWSING

On UNIX, Netscape stores a history of Web sites that were accessed in a Berkeley DB file called "history.dat," and stores information about cache files in a Berkeley DB file called "index.db." These files can be processed using the db_dump185 utility from the Berkeley DB software package as shown here:[14]

[14] *http://www.sleepycat.com*

Times are shown in bold here for clarification and can be converted and adjusted for the time zone. For instance, the above data represents the following:

```
# db_dump185 history.dat
format=bytevalue
type=hash
h_ffactor=60
db_lorder=1234
db_pagesize=4096
HEADER=END
687474703a2f2f72722e73616e732e6f72672f61756469742f6e65746361742e6874
6d00
5a18e53d5a18e53d010000000000000000
687474703a2f2f72722e73616e732e6f72672f61756469742f7472616e7370617265
6e742e67696600
5a18e53d5a18e53d080000000000000000
687474703a2f2f7777772e6365726961732e7075726475652e6564752f686f6d6573
2f636172726965722f666f72656e736963732f00
ce37e53dd332e53d03000000010000006361727269657223a20436f6d7075746572
0466f72656e7369637300
<cut for brevity>
```

```
URL: http://rr.sans.org/audit/netcat.htm
Date Accessed: Wed Nov 27 14:09:14 2002 (GMT -0500)
Accessed: 1

URL: http://rr.sans.org/audit/transparent.gif
Date Accessed: Wed Nov 27 14:09:14 2002 (GMT -0500)
Accessed: 8

URL: http://www.cerias.purdue.edu/homes/carrier/forensics/
Last Accessed: Wed Nov 27 16:23:26 2002 (GMT -0500)
First Accessed: Wed Nov 27 16:02:11 2002 (GMT -0500)
Accessed: 3
```

In this instance, the first and last visited times are equal but the "transparent.gif" file was accessed eight (8) times because it is referenced in the "netcat.htm" page eight (8) times. However, the db_dump185 utility does not display entries that have been marked for deletion but still exist in the file. Deleted entries can be seen by viewing the raw data in the format last time visited, first time visited, number of times visited, and URL.

The Netscape cache "index.db" database can also be processed using db_dump185 as shown here:

```
# db_dump185 index.db
format=bytevalue
type=hash
h_ffactor=16
db_lorder=1234
db_pagesize=4096
HEADER=END
3200000026000000687474703a2f2f7777772e676f6f676c652e636f6d2f696d6167
65732f726573312e6769660000000000

a900000005000000fb75b33ddd17e53dff3dfe7fa8060000000000000000001c0000003
1442f636163686533444535313744443031323046433 72e676966000000000000010
00000000000000000000000000000000a000000696d6167652f676966000000000
00000000000a80600000000000000000000000000000000000000000000000000000
000000000000000000000000000000000000000000000000000000000000000000000
0000000

3400000028000000687474703a2f2f7777772e61747374616b652e636f6d2f696d6
16765732f636c6561722e6769660000000000
```

```
ab00000005000000e27d6c3ae417e53d000000003100000000000000001c00000030
342f636163686533444535313745343031423046433372e676966000000000100000
000000000000000000000000000a000000696d6167652f676966000000000000000
000031000000000000000000000000000000000000000000000000000000000000000
000000000000000000000000000000000000000000000000000000000000000000000

420000003600000068747470 3a2f2f7777772e61747374616b652e636f6d2f6e6176
696d616765732f626c616e6b5f73756273656374696f6e2e676966000000000

b900000005000000f87d6c3ae417e53d000000006e0000000000000001c0000003
0342f636163686533444535313745343032363046433372e67696600000000000100
0000000000000000000000000000000a000000696d6167652f676966000000000000
0000000006e000000000000000000000000000000000000000000000000000000000
000000000000000000000000000000000000000000000000000000000000000000000
00000000000000000000000000000000000

3f0000003300000068747470 3a2f2f7777772e6c696e757867617a657474652e636f
6d2f67782f6e61766261722f74616c6b6261636b2e6a706700000000000
```


Obviously, some interpretation is required – the above data represent the following:

URL:	http://www.google.com/images/res1.gif
Content Length:	1704
Content type:	image/gif
Local filename:	1D/cache3DE517DD0120FC7.gif
Last Modified:	Sun Oct 20 23:35:23 2002
Expires:	Sun Jan 17 14:14:07 2038
URL:	http://www.atstake.com/images/clear.gif
Content Length:	49
Content type:	image/gif
Local filename:	04/cache3DE517E401B0FC7.gif
Last Modified:	Mon Jan 22 13:37:22 2001
Expires:	No expiration date sent
URL:	http://www.atstake.com/navimages/blank_subsection.gif
Content Length:	110
Content type:	image/gif
Local filename:	04/cache3DE517E40260FC7.gif
Last Modified:	Mon Jan 22 13:37:44 2001
Expires:	No expiration date sent

The Last Modified date is when the file was changed on the server, not on the local computer.

Other information discussed in Chapter 10 such as cookies and newsgroup access can be found on a UNIX machine. Some UNIX utilities have been developed to extract information from Internet Explorer cookie and "index.dat" files.[15] Information about newsgroups that have been accessed are stored in a file named ".newsrc" that is usually located in the individual's home directory.

[15]http://odessa.sourceforge.net/

11.7.2 E-MAIL

On UNIX systems that receive e-mail, incoming messages are held in "/var/spool/mail" in separate files for each user account until a user accesses them. Outgoing messages are stored temporarily in "/var/spool/ mqueue/ mail" but are generally deleted after they are sent. Incoming and outgoing e-mail messages may also be stored in files under the home directories of each user. UNIX generally stores e-mail in text files, making them easier to process. However, there may be MIME encoded attachments that must be extracted and decoded using utilities like mimencode or mpac.[16]

Although there some UNIX utilities are available for converting Outlook PST files to Linux readable format[17] and other proprietary formats, they are not designed with digital evidence in mind and may not recover deleted messages. Therefore, it is advisable to process proprietary e-mail formats like Outlook and AOL using Windows systems.

[16]http://www.usinglinux.org/converters/

[17]http://www.sourceforge.net/projects/o12mbox/

11.7.3 NETWORK TRACES

UNIX systems are often configured to print, log, and store user data (e.g. files, e-mail, passwords) on remote systems. Therefore, it is vital to look for traces of connections to remote locations on a network and can lead to additional sources of digital evidence. Quickly identifying other likely sources of digital evidence on a network will increase the chances of obtaining the data before they are altered or lost.

As with Windows, individual applications like ncftp retain logs when used to transfer files from remote computers and SSH can store a list of public keys for each host that was accessed in files named "known_hosts." Similarly, ".Xauthority" files contain lists of remote systems that are accessed using X, a method of viewing remote systems via an X windows interface. Also, UNIX system logs can contain information relating to connections to remote systems and the "/etc/hosts" file often contains a list of computers that are communicated with frequently.

Shared network drives are common in UNIX environments. The file system mount table ("/etc/fstab") shows local and remote file systems that are automatically mounted when the system is booted. For instance, the last two lines of an "/etc/fstab" file from a Linux system indicate that user home directories and e-mail are stored on a remote system named central:

```
# cat /etc/fstab
/dev/hda1      /             ext2      defaults            1 1
/dev/hda7      /tmp          ext2      defaults            1 2
/dev/hda5      /usr          ext2      defaults            1 2
/dev/hda6      /var          ext2      defaults            1 2
/dev/hda8      swap          swap      defaults            0 0
/dev/fd0       /mnt/floppy   ext2      user,noauto         0 0
/dev/hdc       /mnt/cdrom    iso9660   user,noauto,ro      0 0
none           /dev/pts      devpts    gid55,mode5620      0 0
none           /proc         proc      defaults            0 0
central:/home/accts    /home/accts                         nfs
bg,hard,intr,rsize=8192,wsize58192
central:/var/spool/mail    /var/spool/mail    nfs
bg,hard,intr,noac,rsize=8192,wsize=8192
```

A list of currently mounted drives, including those not listed in /etc/fstab (e.g. those mounted by individual users) is kept in "/etc/mtab" ("/etc/mnttab" on Solaris 7 and later versions). Similar information is also maintained in /proc/mounts on systems like Linux that maintain a /proc file system. In addition to using NFS, remote network resources on Windows systems can be accessed from UNIX using Samba.[18] Therefore, digital evidence examiners may be able to find remnants of Windows network file shares (e.g. "\\server\resource") and directory listings (e.g. "C:\winnt\system32*.exe").

[18] *http://www.samba.org*

UNIX computers can be configured to send logs to remote systems in the /etc/syslog.conf as shown here:

```
# cat /etc/syslog.conf
*.*                                @remote-server
```

Additionally, the /etc/printcap file is used to send print jobs to remote systems as shown in the following segment:

```
# cat /etc/printcap
lp0llp:\
   : sd=/var/spool/lpd/lp0:\
   : mx#0:\
   : sh:\
   : rm=remote-server:\
   : rp=lp0:\
   : if=/var/spool/lpd/lp0/filter:
```

As mentioned in Chapter 10, it is not advisable for digital investigators to access these remote storage locations without proper authorization. The most effective way to obtain evidence from such systems is to gain physical access to each system, following standard operating procedures to preserve and recover the data.

11.8 SUMMARY

Given the large number of UNIX systems that exist, it is necessary for digital evidence examiners to be familiar with UNIX file systems. Although UNIX may appear to be more complex than Windows, this is largely because many operations involve commands rather than graphical user interface. However, UNIX systems are arguably easier to understand because they are more transparent – these systems' configuration and functions are plainly visible and it is even possible to view the source code of many Unix operating systems and utilities.

Linux is a powerful forensic platform that can be used to examine many file systems, including FAT and NTFS. Tools like The Sleuth Kit and SMART provide a graphical user interface, simplifying the process of performing digital evidence examinations using UNIX systems.

REFERENCES

Buckeye B. and Liston K. (2002) "Recovering Deleted Files in Linux", February 2002, Sysadmin Magazine. (Available online at http://www.samag.com/documents/s=7033/sam0204g/sam0204g.htm).

Carrier B. (2003a) "Splitting The Disk – Part 1", Sleuth Kit Informer, Issue #2, March 15, 2003 (Available online at http://www.sleuthkit.org/informer/sleuthkit-informer-2.html).

Carrier B. (2003b) "Splitting The Disk – Part 2", Sleuth Kit Informer, Issue #5, July 15, 2003 (Available online at http://www.sleuthkit.org/informer/sleuthkit-informer-5.html).

Peek J., O'Reilly T. and Loukides M. (1997) *Unix Powertools*, California: O'Reilly.

Seglem K., Luque M. E. and Murphy S. E. (2001) Forensic Analysis of UNIX Systems, Casey, E. (editor) *Handbook of Computer Crime Investigation*, London: Academic Press.

Widdowson L. and Ferlito J. (2001) "Tales from the Abyss: UNIX File Recovery", January 2001, SysAdmin Magazine (Available online at http://www.samag.com/documents/s=1441/sam0111b/0111b.htm).

## CHAPTER 12

FORENSIC EXAMINATION OF MACINTOSH SYSTEMS

Apple Macintosh systems receive less attention than other systems as a source of digital evidence, probably because there are fewer of them and people are less familiar with them. However, these systems cannot be ignored since criminals use them and the user-friendly graphical user interface does not translate into a user-friendly digital examination. If anything, digital evidence examiners need to dedicate more attention to these systems. More of the newer, colorful, compact Macintosh desktop, and laptop systems are being sold worldwide and the emergence of UNIX-based MacOS X has attracted more technical users who appreciate the power of UNIX and the convenience of the Macintosh interface. There are only a few tools for examining digital evidence on a Macintosh. As a result this chapter provides a necessarily brief introduction to Macintosh systems.

12.1 FILE SYSTEMS

As with other systems, Macintosh stores its partition table in the first sector on disk. The first sector of each volume contains the boot sector and additional details about the volume are stored in the third sector. Like FAT16 and FAT32, the Macintosh HFS and HFS Plus (HFS+) file systems use 16 and 32 bits, respectively, to address clusters on a disk. HFS supports a maximum of 2^{16} (65536) clusters and HFS Plus has a maximum of 2^{32} clusters. The main files comprising HFS are the Catalog and Extents Overflow files. The Catalog file is comparable to a master file table, containing records for each file and folder on the system with attributes such as date–time stamps. HFS represent time as the number of seconds since midnight, January 1, 1904, GMT.

Records in the Catalog file are stored in a balanced tree (B-tree), which is a simple database that enables efficient searching. Each record in the Catalog file has a unique number called *a catalog node ID* (CNID). The Catalog file has four types of records: folders, files, folder threads, and file threads. Although the format of folder and file records varies between HFS and HFS Plus, they

Digital Evidence and Computer Crime Second Edition
ISBN: 0-12-163104-4

contain similar information. Folder records contain the following fields, in addition to some details used by the system.

Record type: 0 × 0100

Name: folder name

Valence: number of files and folders directly contained by this folder

CNID: unique catalog node ID

Creation date: when this folder was created

Modification date: when a file or folder was created or deleted inside this folder, or when a file or folder was moved in or out of this folder

Access date: not maintained by HFS (always set to zero)

Backup date: when this folder was last backed up

File records contain the following fields, in addition to some details used by the system.

Record type: 0 × 0200

Name: file name

CNID: unique catalog node ID

Creation date: when this file was created

Modification date: when a file modified by extending, truncating, or writing either of the forks

Access date: not maintained by HFS (always set to zero)

Backup date: when this file was last backed up

Data fork: information about the location and size of the data fork

Resource fork: Information about the location and size of the resource fork

The attentive reader will notice that folder records do not contain lists of their contents, and files have two storage areas on disk (a.k.a. forks). HFS uses folder and file thread records in the Catalog file to link names with the associated file or folder records using the unique CNID. These file and folder thread records also contain references to parent folders that are used to construct the file system hierarchy and directory listings that most users are familiar with. Files on an HFS volume have two forks: a data fork that stores the contents of a file, and a resource fork with a special data structure for information such as icons and menu items. The first eight clusters of each fork (a.k.a. extents) are listed in each file's Catalog record. Any additional extents are stored in the Extents overflow file, which is also organized as a B-tree.

Figure 12.1(a) and (b) shows a file record in a HFS Catalog file in interpreted form and hexadecimal form, respectively. This file is located under the Trash folder, indicating that it was deleted but the Trash had not been emptied.

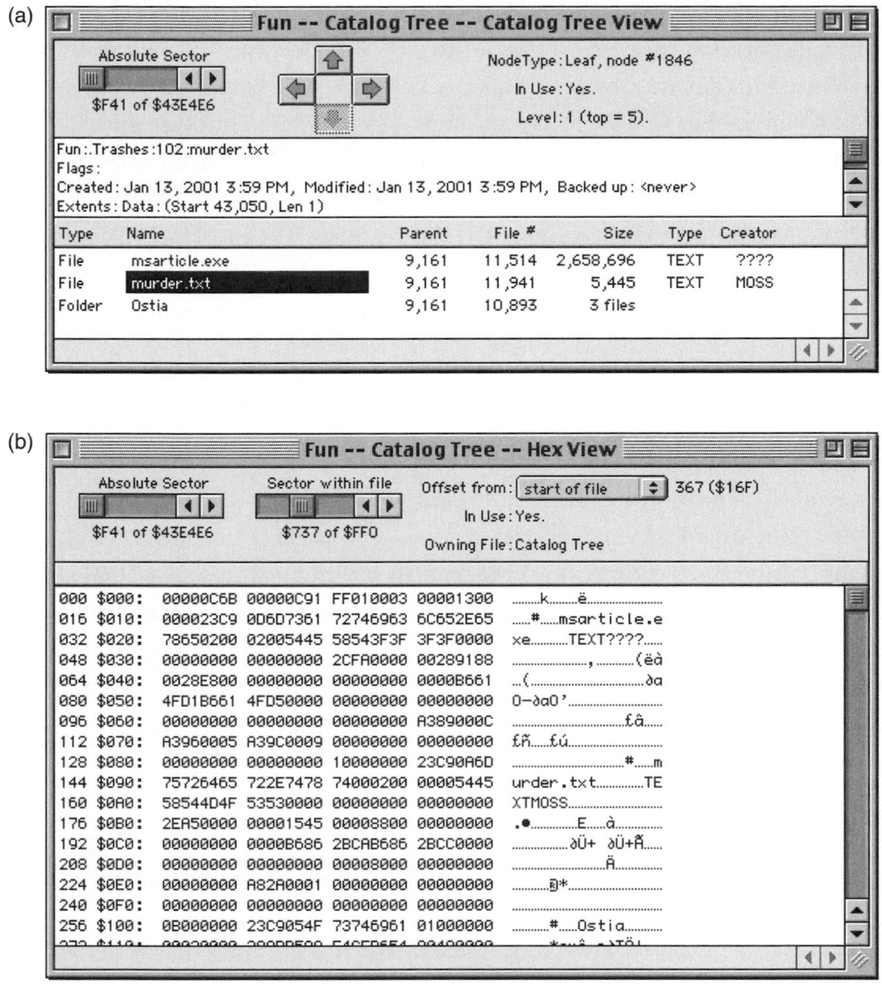

Figure 12.1

(a) File record interpreted using Norton Disk Editor. (b) Same file record in hexadecimal form.

Notice that, rather than relying entirely on file extensions to determine the type of data in a file, HFS stores this information in Catalog records. However, this information can be altered and should not be relied on to classify files.

When a file is moved to the Trash on a Macintosh, it is actually moved to a Trash folder but is not marked as deleted. The file is only marked as deleted when the Trash is emptied but the data remains on disk until it is overwritten. A file is marked as deleted by setting the *key length* value within the associated Catalog database key to zero. Also, when a file is deleted, its Catalog entry may be deleted, removing all references to the data on disk. Because of the complexity of the Catalog file, it is difficult to recover deleted files manually. Fortunately, automated tools exist that scan the Catalog B-tree and find deleted entries.

One significant change in HFS Plus is that it stores file and folder names in Unicode format. As with NTFS, the use of Unicode can have an impact on text searches. Also, be aware that MacOS X is UNIX based and supports the UNIX File System (UFS). Although digital evidence examiners can use many of the lessons from Chapter 11 to examine UFS, there are slight nuances when MacOS X is involved. For instance, MacOS X uses hidden files (e.g. ._filename) to translate the concept of HFS resource forks to UFS. Also, a file named "/etc/.hidden" contains a list of files that MacOS X hides – generally this only references system files but any filename could be hidden in this way.

12.2 OVERVIEW OF DIGITAL EVIDENCE PROCESSING TOOLS

The most common approach to creating a bitstream copy of a hard drive from a Macintosh system is to remove it and connect it to another computer. Although it is possible to boot Macintosh systems using a CD-ROM, this is mainly useful for noting the time of the system clock and copying individual files from the system. If it is necessary to boot a Macintosh using a CD-ROM, hard drives should be disconnected from the system first to avoid accidental alteration. In one case, a system administrator who was helping investigators attempted to boot an iBook using a CD-ROM but mistakenly booted from the hard drive, altering file date–time stamps in the process.

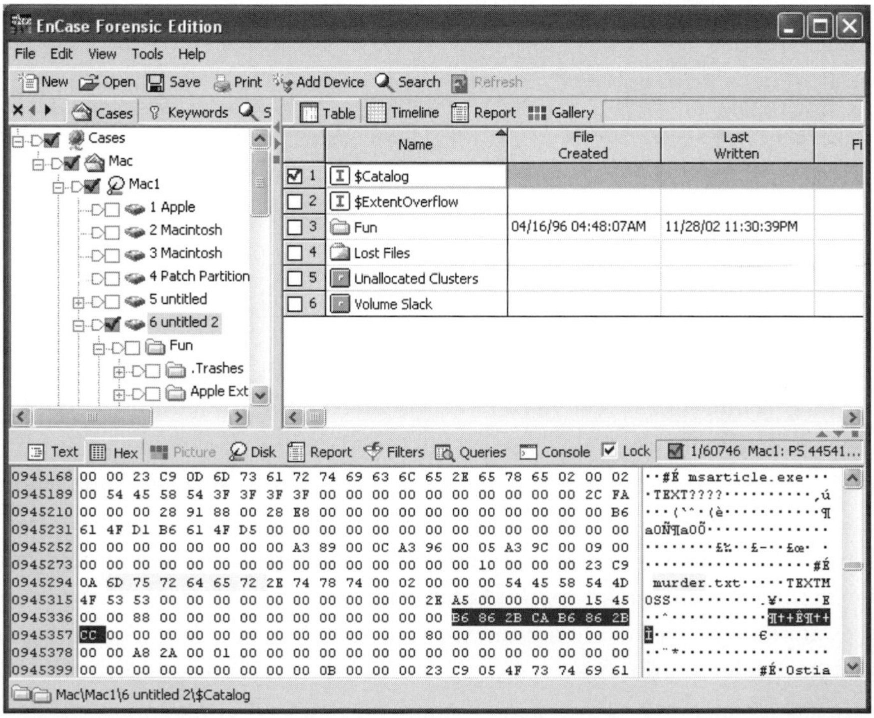

Figure 12.2

HFS viewed in EnCase showing Catalog file record from Figure 12.1.

HFS and HFS Plus can be acquired and examined using MacOS X, Linux, SMART, or EnCase on Windows. Be aware that when MacOS X boots up, it will attempt to mount an evidence disk unless automount is turned off, an eventuality that digital evidence examiners will want to avoid. Figure 12.2 shows the same file as Figure 12.1 viewed using EnCase.

Currently, digital evidence examiners can use The Sleuth Kit on MacOS X to examine NTFS, FAT, UFS, and EXT but not HFS file systems.

There are various utilities for examining special Macintosh files such as Desktop databases discussed later in this chapter. Also, corrupt Catalog files can be repaired using tools such as Disk Warrior[1] or Norton Disk Doctor, recovering files, folders, and related file system details than were not previously visible. To run these tools, it is necessary to create a clone of the original system and perform recovery or other examination operations on the copy.

[1] http://www.alsoft.com

12.3 DATA RECOVERY

One approach to recovering deleted files and folders on Macintosh systems is to make a clone of the evidentiary drive, connect it to a Macintosh system, and use tools like Norton Utilities, Disk Warrior, or ProSoft Data Rescue.[2] In Figure 12.3 all of the deleted files found by Norton Unerase appear to be fully recoverable. Even when a file has a low chance of recoverability, Norton Unerase may be able to perform a full recovery.

[2] http://www.prosoftengineering.com

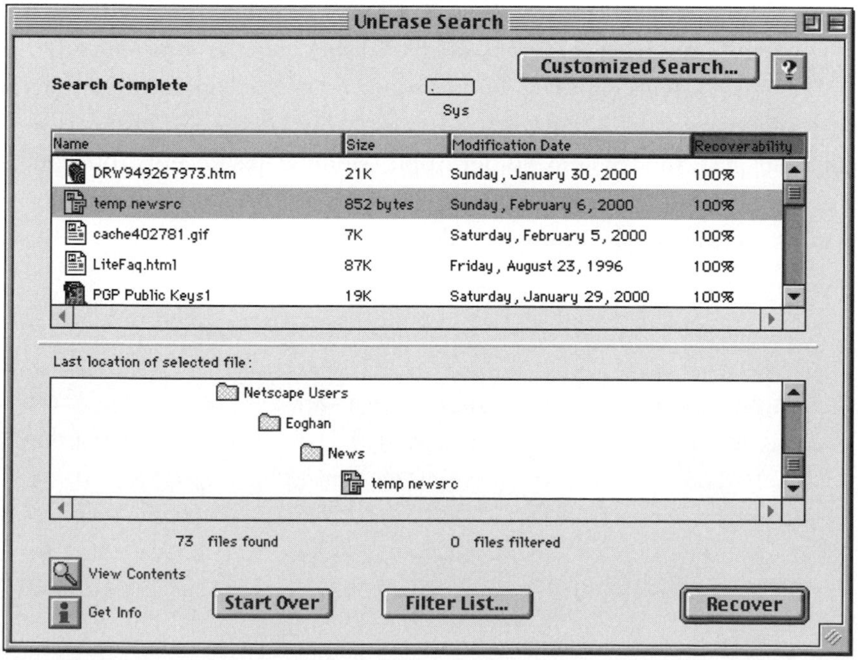

Figure 12.3
Norton Unerase.

It is advisable to try several tools since one may recover more deleted files than others in certain circumstances.

Another approach to recovering deleted files and folder on HFS is to use EnCase. Some deleted files may be listed in the table view and others may be recoverable using E-scripts to carve files from unallocated space. Other carving tools mentioned in previous chapters, such as **foremost** on Linux, can be used to recover files from unallocated space.

12.4 FILE SYSTEM TRACES

When files on HFS are moved or copied, their date–time stamps are not updated – as far as the system is concerned, only the contents of the parent directories have changed. A summary of common actions and the associated date–time stamp changes on MacOS 9 is provided in Table 12.1.

Table 12.1

Date–time stamp behavior on MacOS 9.

ACTION	LAST MODIFIED DATE–TIME	LAST ACCESSED DATE–TIME	CREATED DATE–TIME
Moving files	Unchanged	N/A	Unchanged
Copying files	Unchanged	N/A	Unchanged
Parent directories	Updated	N/A	Unchanged

Macintosh reduces the chances of accidental data loss by maintaining redundant information in the catalog about files and using the Trash folder. The main volume on a Macintosh system has a folder named "Trash" where deleted files are stored in case the user later decides he/she needs the data. All other volumes have folders named ".Trashes" for the same purpose.

Macintosh systems maintain a list of recently accessed applications and files to provide users with easy access to commonly used items. For instance, as the names suggest, the "System Folder:Apple Menu Items:Recent Applications" and "System Folder:Apple Menu Items:Recent Documents" folders list recently accessed applications and files.

Name	File Created	Last Written
APPENDIX-II.doc	01/28/03 03:22:22PM	01/28/03 03:22:22PM
AZ_v_BASS_2001.doc	01/22/03 11:58:57AM	01/22/03 11:58:57AM
CHAPTER3-new.doc	01/28/03 03:21:42PM	01/28/03 03:21:42PM
CHAPTER4.doc	01/28/03 03:22:10PM	01/28/03 03:22:11PM
Chapters 1 & 2.doc	01/28/03 03:20:54PM	01/28/03 03:20:54PM
notes-network.txt	11/20/02 07:25:42PM	11/20/02 07:25:42PM
The Crown v Speyer	12/09/02 10:51:29AM	12/09/02 10:51:29AM

The associated "System Folder:Preferences:Apple Menu Options Prefs" file also contains information about recently accessed files on the system as shown here.

```
7358003A  ECAC0000  01FFFFFB  0000287E  |  sX.:  ∞¼..   .  √   ..(~   | 2,064
B6DA88CA  12546865  2043726F  776E2076  |  ╢┌ê╨  .The   Cro   wn v   | 2,080
20537065  79657272  7265616C  2E646F63  |   Spe  yerr   real  .doc   | 2,096
00000000  000001FF  FFFB0000  098DB852  |  ....  ...    √..   .ì╕R   | 2,112
62230F41  5050454E  4449582D  49492E64  |  b#.A  PPEN   DIX-  I I.d   | 2,128
6F63003A  AED00049  5CA40016  7358003A  |  oc.:  «╨.I   \ñ..  sX.:   | 2,144
ECAC0000  01FFFFFB  0000098D  B85261ED  |  ∞¼..  .  √   ...ì  ╕Raφ   | 2,160
0C434841  50544552  342E646F  632E646F  |  .CHA  PTER   4.do  c.do   | 2,176
636F63F0  0000000C  00167358  00000000  |  coc≡  ....   ..sX  ....   | 2,192
000001FF  FFFB0000  098DB852  61F41043  |  ...    √..   .ì╕R  a⌠.C   | 2,208
48415054  4552332D  6E65772E  646F636F  |  HAPT  ER3-   new.  doco   | 2,224
63B00000  00BF0016  73580000  00000000  |  c░..  ·┐··   sX..  ....   | 2,240
01FFFFFB  0000057E  B6ED3AA8  116E6F74  |  .  √  ...~   ╢φ:¿  .not   | 2,256
65732D6E  6574776F  726B2E74  78745250  |  es-n  etwo   rk.t  xtRP   | 2,272
00B2D950  00167358  00000000  00000000  |  .█┘P  ..sX   ....  ....   | 2,288
FFFB0000  057EB6EB  E4130E6E  6F746573  |   √..  .~╢δ   Σ..n  otes   | 2,304
2D303333  312E7478  746F6348  525000B2  |  -033  1.tx   tocH  RP.█   | 2,320
D9500016  73580000  00000000  01FFFFFB  |  ┘P..  sX..   ....  .  √   | 2,336
0000098D  B852621E  0E415050  454E4449  |  ...ì  ╕Rb.   .APP  ENDI   | 2,352
582D492E  646F63B8  003AAED0  000000BF  |  X-I.  doc╕   .:«╨  ...┐   | 2,368
00167358  003AECAC  000001FF  FFFB0000  |  ..sX  .:∞¼   ...    √..   | 2,384
098DB7E8  EC9C1A43  68617074  65727320  |  .ì╖Φ  ∞£.C   hapt  ers    | 2,400
31202661  6D703B61  6D703B20  322E646F  |  1 &a  mp;a   mp;   2.do   | 2,416
63580000  00000000  01FFFFFB  0000098D  |  cX..  ....   .  √  ...ì   | 2,432
```

CASE EXAMPLE

A suspect's computer was examined but no incriminating digital evidence was found. However, entries relating to PGP in the Recent Applications, suggested that someone may have encrypted or wiped data on the system.

On each volume of a Macintosh system, there is a database in files named "Desktop DB" and "Desktop DF". This Desktop database contains information about activities on the system including programs that were run and files and Web sites that were accessed. These database files can be viewed using a program like Desktop DB Diver. Notably, when viewing applications that were run on the system, the "creation date" in "Deskop DB" files corresponds to the creation date–time stamp of the associated executable, indicating when the application was installed on the system, not when it was first used. Also, when a Web page is saved to disk using Netscape or Internet Explorer, the URL is inserted into a "comments" field of the file. These comments are also stored in the Desktop database and can persist long after the associated file is deleted.

It is instructive to observe the simple case of file system traces on external media such as floppy diskettes and memory cards. When files are saved to a HFS formatted floppy diskette, a Desktop Folder is created to store files that the user wants to appear on the Macintosh Desktop when the floppy is inserted into a system. A number of interesting file system traces are created when files are saved from a Macintosh to a floppy diskette or memory card (e.g. from a digital camera) formatted using FAT. In addition to a folder named "resource.frk" that contains the resource forks of files saved from HFS, Apple's PC Exchange program creates two files named "finder.dat" and "fileid.dat" are created. Using the Sleuth Kit to examine a floppy diskette formatted with FAT and used to store files from a Macintosh. Note that the last accessed times of the files copied from a Macintosh onto a FAT formatted disk are meaningless since the HFS does not maintain access times.

```
examiner1% dd if=/dev/disk3 | md5
2880+0 records in
2880+0 records out
X bytes transferred in Y secs (Z bytes/sec)
d14cbf5e5dccbbbace817409b494c602
examiner1% dd if=/dev/disk3 of=fat-mac-floppy.dd
2880+0 records in
2880+0 records out
X bytes transferred in Y secs (Z bytes/sec)
examiner1% fls -l -f fat12 /morgue/fat-mac-floppy.dd
<note added by author    last written              created                   size>
r/r 3: pubring.pkr          1999.01.05 12:32:14 (EST) 1999.01.05 11:11:06 (EST) 1146
r/r 4: secring.skr          1999.01.05 12:32:14 (EST) 1999.01.05 11:11:12 (EST) 1099
r/r 5: FINDER.DAT           1999.01.28 22:15:30 (EST) 1999.01.28 21:57:36 (EST) 1628
r/r 6: Desktop              1999.01.28 19:57:42 (EST) 1999.01.28 21:57:42 (EST) 0
r/r 7: FILEID.DAT           1999.01.28 20:42:02 (EST) 1999.01.28 21:57:42 (EST) 704
r/r 8: NAV QuickScan        1999.03.18 19:51:52 (EST) 1999.01.28 21:57:36 (EST) 582
d/d 20: RESOURCE.FRK        1999.01.28 21:57:42 (EST) 1999.01.28 21:57:42 (EST) 512
d/d * 25: Desktop Folder    1999.04.03 23:15:08 (EST) 1999.04.03 23:15:08 (EST) 0
d/d * 27: Trash             1999.04.03 23:15:10 (EST) 1999.04.03 23:15:10 (EST) 0
d/d * 34: Temporary Items   1999.04.03 23:15:10 (EST) 1999.04.03 23:15:10 (EST) 0
r/r 37: OpenFolderListDF_   1999.01.28 22:15:30 (EST) 1999.01.28 22:15:30 (EST) 0
```

The "finder.dat" file contains information that Macintosh systems use to organize the files on screen and the "fileid.dat" file contains long file names. Interestingly, a segment of the "finder.dat" file shown here contains date–time stamps (in bold) for files on the disk and some date–time stamps from 1 year prior (April 10, 1998 and June 1, 1998).

```
examiner1% task/bin/icat -f fat12 /morgue/fat-mac-floppy.dd 5 | xxd
<cut for brevity>
0000250: 4944 454e 5449 5459 2020 2084 0b53 4543     IDENTITY ..SEC
0000260: 5249 4e47 2e53 4b52 0000 0793 b154 0793     RING.SKR.....T..
0000270: b198 0084 4c30 5345 4352 494e 5445 5854     .....L0SECRINTEXT
0000280: 646f 7361 0100 0000 0081 0000 0000 0000     dosa...........
0000290: 0000 0000 0000 0000 0000 0002 b2b7 a3d0     ...............
00002a0: b2b7 b6ce 0000 0000 7fff fff0 5345 4352     ................SECR
00002b0: 494e 4720 534b 5284 0b50 5542 5249 4e47     ING SKR..PUBRING
00002c0: 2e50 4b52 0000 0793 b154 0793 b198 0084     .PKR.....T......
00002d0: 4c30 5055 4252 494e 5445 5854 646f 7361     L0PUBRINTEXTdosa
00002e0: 0100 0000 0001 0000 0000 0000 0000 0000     ...............
00002f0: 0000 0000 0000 0002 b2b7 a3ca b2b7 b6ce     ...............
0000300: 0000 0000 7fff ffef 5055 4252 494e 4720     .............PUBRING
0000310: 504b 5284 114e 4156 2051 7569 636b 5363     PKR..NAV QuickSc
<cut for brevity>
```

These "finder.dat" files may contain names and date–time stamps of files deleted from the diskette using a non-Macintosh system that does not update these files. Also, keep in mind that the date–time stamps on the files in "resource.frk" may not be identical to those of the corresponding data fork if changes were made to the data using Windows.

12.5 INTERNET TRACES

Older Macintosh systems were not designed with Internet access in mind and do not retain log files of network activities. More recent versions, such as MacOS 9 and MacOS X, come with Web servers and other Internet servers that have associated log files. On all systems, Internet applications such as Netscape, Internet Explorer, and Eudora create records of activities such as Web resources accessed and e-mail sent and received.

12.5.1 WEB ACTIVITY

On Macintosh systems, Netscape user profiles in "System Folder: Preferences: Netscape:Users" contain files named "Netscape History," and sometimes a second "Netscape History Old" file, which contain a history of Web sites that were accessed. These files are in Berkeley DB format and can be interpreted

Figure 12.4

IE Cache.waf file viewed using WAFInspec.

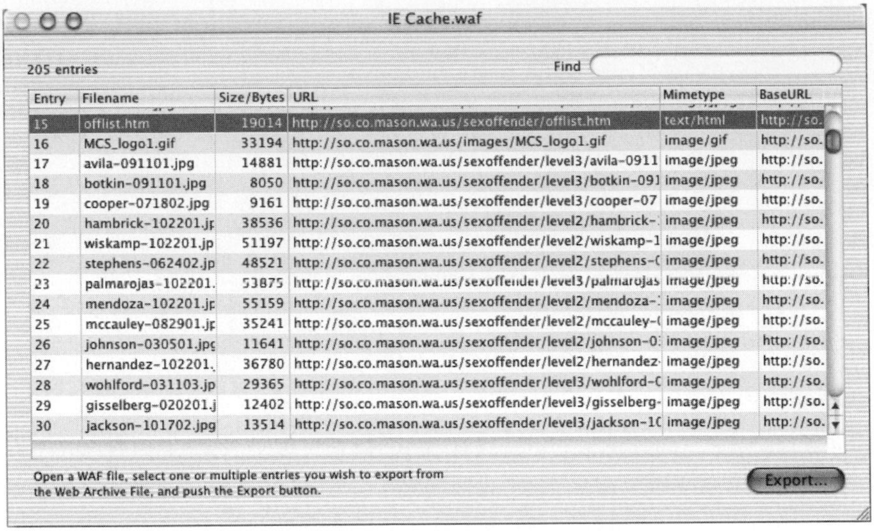

as detailed in previous chapters. Netscape stores cached files in each user's Cache folder along with details such as the associated URL and when they were accessed in Acachelog.txt and Ccachelog files. Each user's cookies are stored in a file named "MagicCookie."

On operating systems prior to MacOS X, Internet Explorer related files are in its installation directory, "System:Explorer:History.html," "System: Preference:Internet Prefs," and "System Preferences:MS Internet Cache: cache.waf." Rather than storing each cached item in separate files, a WAF file organizes cached content and associated information in a single Web Archive Format. MacOS X keeps most Internet Explorer files in each user's home directory under "Library/Preferences/Explorer/," and stores cached data using a Web Archive Format file in "Library/Caches/MS Internet Cache." The contents of these Web Archive Format file can be viewed using WAFInspec[3] on MacOS X (Figure 12.4). The Export function of WAFInspec extracts cached content such as images and HTML pages from these files. Alternatively, Web content can be carved out of the "cache.wav" file.

[3]http://www.executive-computing.de/MacOSX/Applications/Freeware/WAFInspec/

Internet Explorer stores cookie files in different places, depending on the version of the browser: version 2 in "System Folder:Preferences: Explorer: Cookies.txt"; version 3 in "System Folder:Preferences:Internet Preferences"; version 4 in "System Folder:Preferences:MS Preference Panels:Cookies".

Internet Explorer stores Web browser history entries in an HTML file named "History.html" with date–time stamps in UNIX numeric format as shown here (e.g. 1052078766 = Sun, 04 May 4, 2003 15:06:06 – 05:00).

```
<A HREF="http://www.cantenna.com/thankyou.html"
LAST_VISIT="1052078766" ADD_DATE="1052078766"
VISITATION_COUNT="2" OBJECT_TYPE="LINK">Cantenna WiFi Booster

<A HREF="https://www.paypal.com/cgi-bin/webscr?__track=_xclick-flow:
p/xcl/pay/buy-confirm:_xclick-payment-confirm-submit"
ADD_DATE="1052078378" LAST_VISIT="1052078754" VISITATION_COUNT="6"
OBJECT_TYPE5"LINK">PayPal – PayPal Website Payment

<A HREF="https://www.paypal.com/cgi-bin/webscr?__track=_xclick-flow:
p/xcl/pay/buy-index-blank_reg:_xclick-user-submit" ADD_DATE="1052078185"
LAST_VISIT="1052078727" VISITATION_COUNT="5"
OBJECT_TYPE="LINK">PayPal – PayPal Website Payment

<A HREF="http://www.google.com/search?hl=en&lr=&ie=ISO-
8859–1&q=human+poison+herbs"

ADD_DATE="1049641841" LAST_VISIT="1049642467" VISITATION_COUNT="3 "
OBJECT_TYPE="LINK">
```

12.5.2 E-MAIL

Some e-mail applications log details of incoming and outgoing messages, such as the Eudora log shown here.

```
Fri Jan 28 21:44:46 2000
101 1:38.27.0 mail.domain.net 9543
101 1:0.1.7 Sending John Doe, 9:44 PM -0500, What do you think?.
101 1:0.2.51 Succeeded.

Fri Jan 28 21:47:46 2000
102 1:3.0.2 mail.domain.net 9543
102 1:0.1.19 Sending Janet Smith, 9:47 PM -0500, Re: Important Questions.
102 1:0.2.52 Succeeded.

Fri Jan 28 21:52:57 2000
103 1:5.11.47 mail.domain.net 9543
103 1:0.0.58 Sending George Baker, 9:52 PM -0500, Re: Meeting tomorrow.
103 1:0.2.26 Succeeded.

Fri Jan 28 22:03:27 2000
MAIN 8:3.14.4 eco@corpus-delicti.com
MAIN 8:0.0.0 enter the
104 1:0.0.24 mail.domain.net 9543
MAIN 8:0.4.42 Dismissed with 1.
104 1:0.37.29 Sending Sam Rider, 10:03 PM -0500, What I forgot on the phone.
104 1:0.39.10 Succeeded.
```

Although Eudora on any operating system can be configured to log the same type of information, by default, Eudora for Macintosh records more information than Eudora for Windows. Outlook Express stores e-mail under "Documents:Microsoft User:Data:Outlook Express:Identities."

12.5.3 NETWORK STORAGE

MacOS X is Unix based and has many of the same network sharing capabilities described in the previous chapter. Both MacOS 9 and MacOS X maintain a list of recently accessed file servers. MacOS 9 maintains this information in "System Folder:Apple Menu Items:Recent Servers" and MacOS X stores the list under each user's home directory as shown here.

```
[macosx:~/Library/Recent Servers] user13% ls -l
total 0
-rw-r—r— 1 user13 staff 0 Apr 4 13:44 idisk.mac.com-user13
```

The iDisk is a remote file storage service, offered by Apple as part of their ".Mac" program, which is common among Macintosh users and is available from Windows systems as well.

Some third party applications enable file sharing between MacOS 8 and Windows systems on a network. For instance, the DAVE application enables Macintosh systems to communicate using NetBIOS. Although DAVE can be configured to maintain a log of basic activities, such as when a remote host started and stopped a NetBIOS session, the logs have limited use because they do not record the time of events as shown here.

```
Node DARA     started a session on Saturday, December 1, 2001
Node OISIN    started a session on Saturday, December 1, 2001
Node OISIN    stopped a session on Saturday, December 1, 2001
Node PEEKER   started a session on Saturday, December 1, 2001
Node PEEKER   stopped a session on Saturday, December 1, 2001
Node DARA     stopped a session on Saturday, December 1, 2001
```

Older versions of MacOS use AppleTalk to share resources on a network but do not retain logs.

12.6 SUMMARY

Despite their friendly appearance, Macintosh systems are quite complex and powerful. Recovering deleted files manually is a difficult task because of the intricate structure of the Catalog file. Existing tools can be used to perform basic digital evidence examinations of Macintosh systems, including the viewing file structure and recovering deleted data. However, there is a dearth of tools for interpreting special files such as "Desktop DB" and "cache.wav." There is a need for more digital evidence examination tools and research for Macintosh systems. As on other systems, Internet applications on Macintosh systems can keep records of activities. With the emergence of MacOS X and ".Mac," these systems contain more network-related data.

CHAPTER 13

FORENSIC EXAMINATION OF HANDHELD DEVICES

Personal Digital Assistants (PDAs) and mobile telephones are rapidly becoming a necessity for many people around the world, including criminals. Many vice officers and courts consider mobile telephones and pagers as an integral part of drug trafficking and dealing. This connection has motivated some halfway houses and schools to prohibit these devices. Also, parole boards are including these and other electronic devices in the list of items that certain parolees cannot possess.

CASE EXAMPLE (UNITED STATES v. RANDOLPH 2002):
Randolph was convicted for armed robbery, burglary and criminal conspiracy. After his release to a halfway house, Randolph fled the halfway house and did not return. Randolph was later found at his sister's house and, upon entering his room one of the parole officers, Agent Taylor, saw a mobile telephone and pager in plain view. Randolph's possession of these items was in violation of specific conditions of his parole, which provided that he was "not to possess, on your person, property, or residence, any electronic paging devices such as pagers, cell phones, digital phones, etc." When Agent Taylor neared the cell phone, Randolph blurted out, "Oh, that cell phone doesn't even work." Agent Taylor then looked under the bed, which occupied much of the room, and found a 9 mm semiautomatic Llama firearm, which was loaded. Further search of the room uncovered body armor (in violation of Count #2 of Randolph's special conditions of parole) as well as drug paraphernalia, such as a scale and vials.

The defense argued that the search violated Randolph's right to privacy under the Fourth Amendment. However the court held that a warrant was not required to search the room of a fugitive from a halfway house.

> Under the circumstances here – where the officers were dealing with a fugitive who had been convicted of armed robbery – the officers had ample justification to conduct what Agent Taylor described as a "protective sweep" of Randolph's bedroom. In the course of that sweep, once Agent Taylor saw the forbidden cell phone and pager, she had double justification for looking for other contraband and, most seriously, weapons. She found more contraband when she and her colleagues discovered the body armor, and the gun was found in a place where Randolph could well have reached it if he was unhandcuffed as he dressed.

Digital Evidence and Computer Crime Second Edition
ISBN: 0-12-163104-4

Although compact, these handheld devices can contain significant digital evidence including schedules, memos, address books, e-mail messages, passwords, credit card numbers, and other personal information. Some devices, such as Qualcomm's Kyocera models, combine a Palm OS PDA with a mobile telephone to provide a wider range of features and correspondingly more types of digital evidence. Other handheld devices are optimized for data acquisition such as bar code scanning and scientific measurements (e.g. voltage, temperature, acceleration). Furthermore, some PDAs and mobile telephones use Bluetooth and other wireless protocols to communicate with other nearby computers to form proximity networks (impromptu communities).

Many handheld devices can already be used to exchange photographs and access the Internet. As the technology develops, higher data transmission rates will allow individuals to transfer larger files and use handheld devices in much the same way as we currently use laptop systems. This rapid development of mobile computing and communication technology creates opportunities for criminals and investigators alike. This chapter describes the basic operation of handheld devices and presents tools and techniques for acquiring and examining digital evidence on these devices.

Many investigators do not realize that handheld devices can be a valuable source of digital evidence and fail to preserve them as such. It is not unheard of for an investigator to make calls from a victim's mobile telephone, using numbers programmed into the telephone to speed dial family members of the victim. As with any computer, operating a handheld device can destroy existing evidence. Furthermore, digital evidence in handheld devices can be lost completely if its batteries run down and can be overwritten by new data it receives over wireless networks. Therefore, it is advisable to acquire evidence from handheld devices promptly.

This chapter describes the structure and operation of handheld devices, how they structure data, and tools that can be used to process the digital evidence they contain. Notably, handheld devices are just one type of embedded system. A more in-depth treatment of embedded systems, including GSM mobile telephones is provided in the *Handbook of Computer Crime Investigation*, Chapter 11 (Van der Knijff 2001).

13.1 OVERVIEW OF HANDHELD DEVICES

Handheld devices are simple computers with a CPU, memory, batteries, input interfaces such as a keypad or mouthpiece, and output interfaces such as a screen or earpiece. Data in memory are generally the focus of a forensic examination, but some understanding of the input/output components are needed to access these data. In some instances, it may be sufficient manually to operate a device and read information from the display. However, to recover deleted data or perform more advanced examination, specially designed tools are needed to interface with the device. Knowledge of how data is manipulated and stored on handheld devices is sometimes needed to acquire all available digital evidence from handheld devices without altering it and translate it into a human readable form. For instance, placing a Palm OS device on a cradle and HotSyncing it with a computer to obtain information from the device will not copy all data and may even destroy digital evidence.

When learning about handheld devices, it is helpful to consider one type in depth. All handheld devices have many similarities so an understanding of one can be generalized to others. Therefore, this section focuses on one of the most common types of PDA: those running Palm OS.

13.1.1 MEMORY

Handheld devices generally have two types of memory: read only memory (ROM) and random access memory (RAM). The ROM contains the operating system and other software needed for basic functions and RAM is used to store user data and software. As the name suggests, data in ROM cannot be altered but it can retain its contents indefinitely even when it is not being supplied with electrical power, providing a stable platform for critical system components. However, the inability to upgrade critical system software in ROM is inconvenient. To provide greater flexibility, newer devices use programmable ROM that can be modified a limited number of times but will still retain its contents for several years without power. Currently, the most common form of programmable ROM used in PDAs and mobile telephones is called FLASH.

In addition to storing data in RAM, many devices have the ability to save data on removable memory modules similar to those found in a digital camera. Some memory modules are the size of a postage stamp and can store hundreds of megabytes of data (see Figure 13.6).

13.1.2 DATA STORAGE AND MANIPULATION

Handheld devices are designed to make efficient use of their limited amount of memory. For instance, Palm OS divides its memory into partitions called *heaps* and further divides some of these heaps into *chunks* for storing the equivalent of files (called *databases* on Palm OS). Each chunk is referened using a "Local ID," which is basically the position in memory relative to beginning of the memory card it is on.

The two main heaps on Palm OS are the dynamic and storage heaps. The dynamic heap is used for transient data storage and the storage heap is used for long-term storage (e.g. user data). So, when a user enters data into a database on Palm OS, the software uses the dynamic heap for temporary storage while it is running and saves the data in the storage heap for long-term storage. Data in the dynamic heap, such as decrypted data, is overwritten frequently and is completely reinitialized when the device is soft reset.[1]

[1] Inserting a pin into a small hole in the back of the device that contains the reset button causes a soft reset.

Data in the storage heap is less volatile but can be deleted by the user or by the device under certain circumstances. For instance, when Palm OS cannot find enough memory in the storage heap to save a piece of data, it uses a process called heap compaction to rearrange data and expunge deleted records as discussed later in this section. Heap compaction also occurs when a Palm OS device is soft reset. More drastically, a hard reset effectively reformats the memory, recreating empty heaps.[2] Some data may still be recoverable after a hard reset because it only initializes areas that are important and may not clear all memory.

[2] A hard reset can be caused by power loss or by holding down the power key while inserting a pin into a small hole in the back of the device that contains the reset button.

Some devices use the FAT file system to arrange data in memory but Palm OS uses databases. Databases are relatively simple structures that maintain data in records. Each Palm Database (PDB) consists of a *database header*, followed by a table of *record entry headers*, followed by the *records* containing the data as shown in Table 13.1.

Database records can be deleted in several ways on Palm OS. The DmRemoveRecord function is the least complicated and simply removes a record. However, this method of deletion does not allow for synchronization between the device and another computer. To accommodate the need for synchronization, two other deletion methods are available: DmDelete Record and DmArchiveRecord.[3] DmDeleteRecord sets the delete bit in the record and removes the pointer to the data's location in memory. The record is then moved to the end of the database and, during the next synchronization, the corresponding record is deleted from the desktop. The DmArchive Record method, on the other hand, allows a user to archive a deleted record on the desktop during the next synchronization before it is removed from the handheld. Specifically, DmArchiveRecord sets the delete bit in the record but does not free the associated data chunk until the next synchronization.

[3] Other deletion methods exist such as DmRemoveSecretRecords and DmDetachRecord but they are less common.

All of these deletion methods ultimately cause Palm OS to "forget" where the associated data was located in memory. Although data may still exist in this unallocated space, handheld operating systems are quite efficient and unallocated data may only exist for a short time. Some of the digital evidence

Table 13.1

PDB format.

DATABASE HEADER (78 BYTES)		
FIELD	**BYTES**	**VALUE**
DB Name	32	Database name
Attributes	2	e.g. hidden, readonly, copyprevent
Version	2	Application – specific version of the database
Creation time	4	Seconds since 12:00 A.M. on January 1, 1904
Modification time	4	Most recent modification
Backup time	4	Last time database was backed up
Modification number	4	Number of times database has been modified
AppInfo offset	4	Optional application specific information
SortInfo offset	4	Optional application specific information
Type	4	Set by the application/system (e.g. pqa, data)
Creator	4	Application identifier (e.g. clpr)
Unique ID seed	4	Used by PalmOS to create unique record Ids
Next Record list ID	4	Location of 2nd record list with more records
Number of records	2	Number of records in this record list
First Entry	2	
RECORD LIST	**VARIABLE**	**ONE ENTRY PER RECORD WITH THE FOLLOWING FIELDS**
Offset	4	Location of record from the start of database
Record attribute	1	e.g. private (1), modified (4), deleted (8)
Unique ID	3	Unique number for each record
Gap	variable	Empty
Data	variable	Data in database

collection tools described later in this chapter cause a soft reset after acquiring evidence from Palm OS devices, triggering heap compaction that overwrites data in unallocated space.

13.1.3 EXPLORING PALM MEMORY

To learn more about the operation of Palm OS devices and how their data are structured, it is useful to experiment with the Palm Debugger (do not experiment on an evidentiary device). Palm devices must be put in debug mode before they are accessible to the Palm Debugger. This is achieved using a special combination of graffiti symbols (ℓ,..,2).[4] The Palm Debugger can be used in combination with the Palm OS Emulator (POSE) to experiment without actually using a physical device. The emulator can also be used to synchronize with the desktop via TCP/IP, a useful feature for testing and for viewing evidentiary databases in a safe environment.

The following shows commands in the Palm Debugger being used to query a device directly. The **cardinfo** command obtains information about specified memory chips in the device, the **heaplist** command obtains information about the heaps on a specific card.[5]

[4]Handspring devices may require the "up" button to be depressed while writing the "2." Palm OS 4.0 and above require the power on password (if it is set) before putting device in console debug mode.

[5]Card 0 refers to the internal RAM and ROM on a Palm OS device. Additional memory chips or modules are numbered card 1, card 2, etc.

```
> cardinfo 0
Name: PalmCard
Manuf: Palm Computing
Version: 0001
CreationDate: B28C66DF
ROM Size: 00127FFC
RAM Size: 00400000
Free Bytes : 003ED792
Number of heaps: #3
> heaplist 0
index   heapID   heapPtr    size       free       maxFree    flags
-------------------------------------------------------------------
0       0000     00001510   0001EAF0   0001890E   00017ED6   4000
1       0001     0002010E   003DFEF2   003D4E84   003D4D42   4000
2       0002     10C08212   00127DEE   0000C2CA   0000C2C2   4001
```

The memory sizes are represented in hexadecimal format – note the discrepancy between memory sizes reported by **cardinfo** and **heaplist** in this example.

- RAM size (cardinfo): 4,194,304 bytes (x00400000);
- RAM size (heaplist): 4,188,642 bytes (dynamic + storage heaps);
- ROM size (cardinfo): 1,212,412 bytes (x00127FFC);
- ROM size (heaplist): 1,211,886 bytes (x00127DEE).

The reason for this difference is that **cardinfo** provides information about the memory card whereas **heaplist** reports how the card has been divided by Palm OS (some RAM is reserved by the operating system). Notably, although cardinfo reports the correct RAM size, it does not report ROM size accurately – the actual size of the memory chips in this device are larger (2,097,152 bytes). This discrepancy is due to the Palm OS not using the entire ROM chip and most tools calculate this value incorrectly. Currently, only Pilot-link and **pdd** calculate the actual size of ROM correctly by extracting information from the processor directly (Grand 2002).

Although differences in ROM size may not seem significant, from an investigative perspective it is important to keep in mind that FLASH is not ROM. Individuals may use extra space in FLASH to backup important data or hide incriminating evidence. For instance, an individual can store data in FLASH on a Palm OS device using programs such as FlashPro and JackFlash. After installing FlashPro on this device, **heaplist** shows two additional heaps that have been created in FLASH.

> heaplist 0

index	heapID	heapPtr	size	free	maxFree	flags
0	0000	00001510	0001EAF0	000188B0	00017E6C	4000
1	0001	0002010E	003DFEF2	003C0756	003C06EA	4000
2	0002	10C08212	00127DEE	0000C2CA	0000C2C2	4001
3	0003	10D303FA	00001C06	00000000	00000000	4001
4	0004	10D32000	000CE000	00000000	00000000	4001

Interestingly, the Palm OS Emulator will report when such modifications have been made (Figure 13.1).

Ironically, certain features of FLASH can facilitate data recovery, making it easier for forensic examiners to obtain valuable evidence.

> The block structure [of FLASH] has two important implications for forensic investigations. Firstly, these systems are mostly built in such a way that erased files are only marked in the FAT as erased but can still be retrieved. After formatting, the blocks are indeed physically erased and cannot be retrieved. In addition, difference physical versions of one logical file can be present. This occurs when the size of the files is much smaller than the FLASH block size, which makes it more efficient to erase a file only if here is no more free space available. (Van der Knijff 2001, p. 321)

Additionally, data are generally stored in contiguous chunks rather than in a fragmented manner, making it easier to recover entire databases.

The structure of data on Palm OS allows individual to manufacture a database with any properties and import it into a device. For instance, using

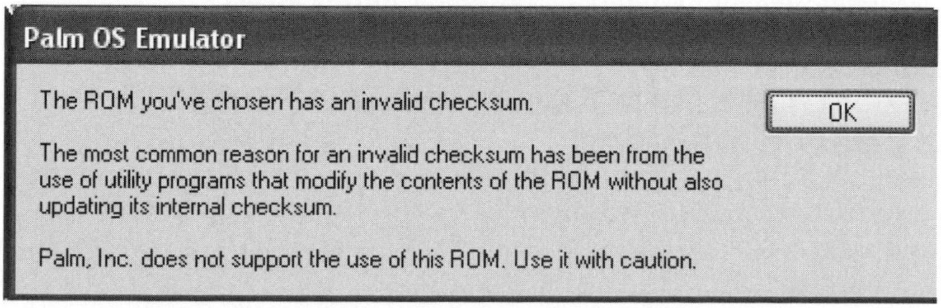
Palm OS Emulator
The ROM you've chosen has an invalid checksum.
OK
The most common reason for an invalid checksum has been from the use of utility programs that modify the contents of the ROM without also updating its internal checksum.
Palm, Inc. does not support the use of this ROM. Use it with caution.

Figure 13.1

Warning message displayed by Palm OS Emulator when loading a copy of ROM that has been modified using FlashPro.

a program called files2pdb.java,[6] a database with a fabricated date–time stamp can be created using the following command.[7]

[6] http://www.righto.com/pilot/pdb.html
[7] The header can be viewed using pilot-file from the Pilot-link package discussed later in this chapter.

```
examiner1% java files2pdb -n TestDB -a 0 -v 0 -t DATA -md 1470137001 \
-cd 1470138004 -bd 1470148004 -c test TestDB MemoDB1 MemoDB2
$ pilot-file -h TestDB
name: "TestDB"
flags: 0x0
version: 0
creation_time: 1950–08–02 07:40:04
modified_time: 1950–08–02 07:23:21
backup_time: 1950–08–02 10:26:44
modification_number: 1
type: 'DATA', creator: 'test'
```

When this database is imported into a device, its last modified date–time stamp is updated but the fabricated creation and backup date–time stamps are not.

```
examiner1% pilot-file -h TestDB
name: "TestDB"
flags: 0x0
version: 0
creation_time: 1950–08–02 07:40:04
modified_time: 2001–09–30 13:58:00
backup_time: 1950–08–02 10:26:44
modification_number: 11
type: 'DATA', creator: 'test'
```

The ability to insert fabricated data may be used by criminals in some cases and should be kept in mind as a possibility during analysis.

13.2 COLLECTION AND EXAMINATION OF HANDHELD DEVICES

In addition to collecting a handheld device itself, it is important to look for associated items that might contain data or help extract data from the device. Removable memory and SIM cards can contain more data than the device itself and interface cables and cradles may be needed to connect the device to an evidence collection system. As with any other computer, document the types of hardware and their serial numbers, taking photographs and notes as appropriate. If a device is on when it is found, leave it on if possible because turning it off may activate password protection, making it more difficult to extract data from the device later. Also, document any information visible on the display including the date and time of the system clock.

Since data in RAM will be lost if it does not receive power, adequately charged batteries are crucial. Even when the device appears to be off, it is consuming some battery power. If there is any indication that the batteries in the device are low, consider replacing them with fully charged batteries. Also, to protect the device against damage or accidental activation, package it in an envelope or bag. Keep in mind that some devices can receive data through wireless networks that might bring new evidence but might overwrite existing data. Therefore, an investigator must make a calculated decision to either prevent or allow the device to receive new data over wireless networks.[8]

[8]Some devices can be reconfigured to prevent communication with the network. Devices that do not have such a feature can be isolated from radio waves by wrapping them in aluminum foil. However, this can cause the battery in the device to deplete more quickly.

After taking precautions to preserve data on the device, examine it for physical damage or suspicious modifications. In most cases, a cursory examination of the exterior of the device will suffice. However, when dealing with a very technically savvy or dangerous offender, some investigative agencies X-ray devices to detect internal damage or modifications. Be aware that a blank display may simply indicate that the screen is damaged and it may still be possible to extract evidence via cable or replace the screen if a manual examination is necessary. A manual examination is sometimes sufficient if investigators only need a particular piece of information from the device. Before performing a manual examination of a device, it is advisable to become familiar with its operation using an identical test device. For this reason, and to enable tool testing and tool development, forensic laboratories that specialize in this type of examination maintain an extensive collection of handheld devices. When performing a manual examination, it is important to record all actions taken with device to enable others to assess whether the examination was performed satisfactorily.

When investigators require all logical files or deleted evidence from a device, special tools are used to acquire and examine this data. Some of these tools are described in the following sections. A more in-depth coverage of tools and techniques for processing evidence on handheld devices and other embedded systems is available in the *Handbook of Computer Crime Investigation*, Chapter 11 (Van der Knijff 2001).

13.2.1 PALM OS

There are several tools for acquiring digital evidence from devices running Palm OS, each with their own advantages and limitations. Some of these tools can only perform full memory dumps, while others can also extract logical databases. Some of these tools can only acquire evidence through a serial connection, while others can also acquire evidence via a USB connection. Some of these tools require the device to be placed in console debug mode, causing a soft reset after collection. Recall from the previous section that a soft rest triggers heap compaction, which overwrites deleted records, so this method only allows one chance to acquire deleted data from the device.

The two most versatile tools for acquiring and examining digital evidence from a Palm OS device are Pilot-link[9] and PDA Seizure. Both of these programs can interpret Palm databases, enabling examiners to view data in their logical form.

[9] *http://www.pilot-link.org*

13.2.1.1 UNIX-BASED TOOLS

Although not specifically designed for evidence processing, the Pilot-link package contains UNIX utilities that can be used to acquire and examine data from a Palm OS device. These utilities include pi-getram and pi-getrom for obtaining memory dumps, pilot-xfer for accessing and copying data logically, and the previously mentioned pilot-file for interpreting and examining logical Palm databases. The following output shows pi-getram in the process of obtaining a physical copy of data in RAM. Unfortunately, pi-getram does not always capture the full contents of RAM, a severe limitation that may be corrected in the future.

```
examiner1% pi-getram /dev/cua1 pda1-evidence
Please insert the Palm in the cradle and press the HotSync button.
Generating pda1-evidence3.1.0.ram
299264 of 2097152 bytes
```

The resulting memory dumps do not provide any structure to the data they contain, leaving it to the examiner to extract any information they can, such as passwords, hidden data, and deleted items. The following output shows pilot-xfer being used to make a logical copy of databases in RAM.

```
examiner1% pilot-xfer -p /dev/cua0 —Illegal —sync pda1-evidence
 Please press the HotSync button now...
Connected...
Synchronizing pda1-evidence/Unsaved Preferences.prc
Synchronizing pda1-evidence/AddressDB.pdb
Synchronizing pda1-evidence/MemoDB.pdb
Synchronizing pda1-evidence/ToDoDB.pdb
Synchronizing pda1-evidence/MailDB.pdb
Synchronizing pda1-evidence/DatebookDB.pdb
Synchronizing pda1-evidence/Saved Preferences.prc
Synchronizing pda1-evidence/NetworkDB.pdb
Synchronizing pda1-evidence/Secret!.prc
Synchronizing pda1-evidence/Secret2.pdb
<cut for brevity>
Synchronizing pda1-evidence/ROM Transfer.prc
Synchronizing pda1-evidence/FlashPro.prc
Synchronizing pda1-evidence/FlashPro Setup.prc
Synchronizing pda1-evidence/FlashPro Uninstall.prc
RAM backup done.
```

The "--Illegal" option instructs pilot-xfer to extract the "Unsaved Preferences" database that contains the password protecting private data on the device. The pilot-xfer utility has several other useful options including "--Flash" to copy not operating system files from FLASH and "--archive" to recover deleted records that have their archive bit set. Making a logical copy of databases does not recover records with their delete bit set, even if the data is still in memory. These deleted data are preserved in a full memory dump.

Notably, the pi-getram and pi-getrom utilities use a HotSync conduit and therefore do not require the device to be in console debug mode. Interestingly, although they use the HotSync feature, these utilities do not change the Last HotSync date on the device. So, although the Pilot-link package was not designed specifically for processing digital evidence, it does not alter data on the device by causing a soft reset or updating the Last HotSync date.

13.2.1.2 WINDOWS-BASED TOOLS

PDA Seizure was specifically designed to collect and examine digital evidence on personal digital assistants and can also be used to extract databases from a Palm device as shown in Figure 13.2. To capture this logical structure, PDA Seizure uses the HotSync feature of Palm OS devices.

Although earlier versions of PDA Seizure changed the Last Hotsync date–time stamp, this has been remedied in version 2.5. By default, extracted databases only include deleted records that have the archive bit set. Other deleted entries are not recovered using the PDA Seizure logical copy feature, even if they still exist in memory. A full memory dump will preserve these deleted data. PDA Seizure can dump the contents of RAM and ROM into a file. Also, PDA Seizure calculates hash values of all digital evidence it collects for future integrity checking and provides several useful features including a report generator and search function.

Aspects of another tool called **pdd** have been incorporated into PDA Seizure. One advantage of **pdd** is that it obtains the actual size of ROM from the CPU rather than relying on the Palm Application Programming Interface (API). This is significant because the Palm API only reports the amount of ROM that the device uses rather than the actual size of the ROM chip, thus misinforming any program that relies on the API for information. Therefore, a tool that relies on the Palm API may miss any data that has been stored in portions of ROM that are not used by Palm OS.

EnCase provides the basic ability to dump RAM and ROM, performing integrity checks and initiating chain of custody as usual. The features of each of these tools are compared in Table 13.2. The RAM and ROM memory sizes detected by each tool are compared in Table 13.3.

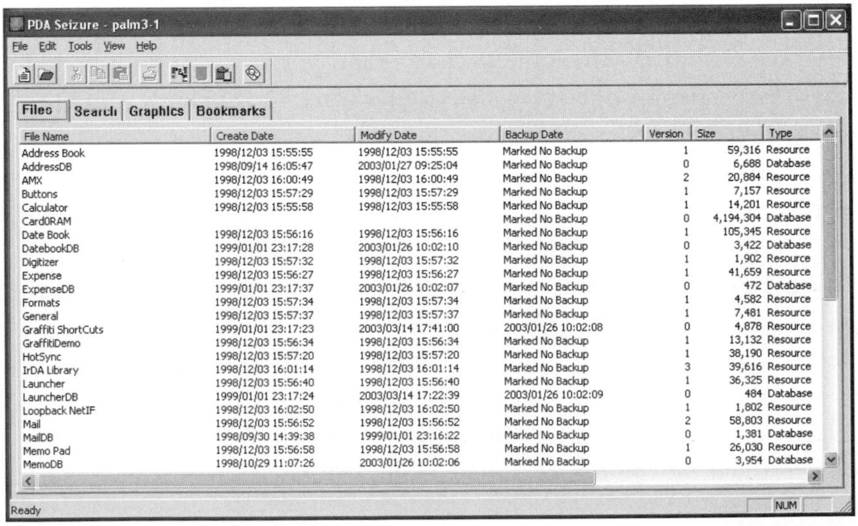
PDA Seizure - palm3-1
File Edit Tools View Help
Files Search Graphics Bookmarks
File Name Create Date Modify Date Backup Date Version Size Type
Address Book 1998/12/03 15:55:55 1998/12/03 15:55:55 Marked No Backup 1 59,316 Resource
AddressDB 1998/09/14 16:05:47 2003/01/27 09:25:04 Marked No Backup 0 6,688 Database
AMX 1998/12/03 16:00:49 1998/12/03 16:00:49 Marked No Backup 2 20,884 Resource
Buttons 1998/12/03 15:57:29 1998/12/03 15:57:29 Marked No Backup 1 7,157 Resource
Calculator 1998/12/03 15:55:58 1998/12/03 15:55:58 Marked No Backup 1 14,201 Resource
Card0RAM Marked No Backup 0 4,194,304 Database
Date Book 1998/12/03 15:56:16 1998/12/03 15:56:16 Marked No Backup 1 105,345 Resource
DatebookDB 1999/01/01 23:17:28 2003/01/26 10:02:10 Marked No Backup 0 3,422 Database
Digitizer 1998/12/03 15:57:32 1998/12/03 15:57:32 Marked No Backup 1 1,902 Resource
Expense 1998/12/03 15:56:27 1998/12/03 15:56:27 Marked No Backup 1 41,659 Resource
ExpenseDB 1999/01/01 23:17:37 2003/01/26 10:02:07 Marked No Backup 0 472 Database
Formats 1998/12/03 15:57:34 1998/12/03 15:57:34 Marked No Backup 1 4,582 Resource
General 1998/12/03 15:57:37 1998/12/03 15:57:37 Marked No Backup 1 7,481 Resource
Graffiti ShortCuts 1999/01/01 23:17:23 2003/03/14 17:41:00 2003/01/26 10:02:08 0 4,878 Resource
GraffitiDemo 1998/12/03 15:56:34 1998/12/03 15:56:34 Marked No Backup 1 13,132 Resource
HotSync 1998/12/03 15:57:20 1998/12/03 15:57:20 Marked No Backup 1 38,190 Resource
IrDA Library 1998/12/03 16:01:14 1998/12/03 16:01:14 Marked No Backup 3 39,616 Resource
Launcher 1998/12/03 15:56:40 1998/12/03 15:56:40 Marked No Backup 1 36,325 Resource
LauncherDB 1999/01/01 23:17:24 2003/03/14 17:22:39 2003/01/26 10:02:09 0 484 Database
Loopback NetIF 1998/12/03 16:02:50 1998/12/03 16:02:50 Marked No Backup 1 1,802 Resource
Mail 1998/12/03 15:56:52 1998/12/03 15:56:52 Marked No Backup 2 58,803 Resource
MailDB 1998/09/30 14:39:38 1999/01/01 23:16:22 Marked No Backup 0 1,381 Database
Memo Pad 1998/12/03 15:56:58 1998/12/03 15:56:58 Marked No Backup 1 26,030 Resource
MemoDB 1998/10/29 11:07:26 2003/01/26 10:02:06 Marked No Backup 0 3,954 Database
Ready
NUM

Figure 13.2

Print screen of PDA Seizure showing logical databases.

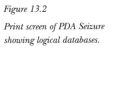

TOOL	MEMORY DUMP	ACTUAL RAM SIZE	ACTUAL ROM SIZE	LOGICAL COPY	SOFT RESET	MD5	USB
EnCase 4	x	x			x	x	x
PDA Seizure 2.5	x	x		x	x	x	x
pdd 1.11	x	x	x		x		
Pilot-link	x		x	x			x

Table 13.2

Feature comparison of tools for processing Palm OS devices.

TOOL	RAM DETECTED (BYTES)	ROM DETECTED (BYTES)
EnCase 4	4194304	1212416
PDA Seizure 2.5	4194304	1572864
pdd 1.11	4194304	2097152
Pilot-link	4063232	2097152

Table 13.3

Memory sizes detected by each tool.

An advantage to having a logical copy of data from a Palm OS device is that databases acquired from the evidentiary system can be exported from PDA Seizure and loaded into the Palm OS Emulator (POSE) where they can be viewed as they were seen by the user (Figure 13.3). This is particularly useful for viewing and presenting evidence because it displays information in a familiar form.

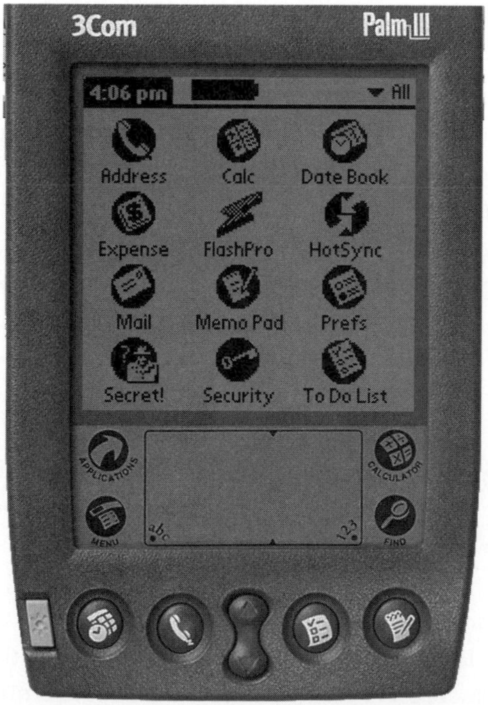

Figure 13.3

Image/data being viewed using Palm OS Emulator (POSE).

To import evidentiary Palm databases into POSE, save it to disk using Pilot-link or PDA Seizure's export features, and simply drag and drop the database into POSE.

13.2.2 WINDOWS CE DEVICES

Like Palm OS devices, Windows CE devices store data in RAM (divided into heaps) and operating system files in FLASH. However, Windows CE arranges data in memory using a different format from Palm OS. Windows CE stores data in an "object store" comprised of a file system, databases (Address book), and a system registry. These devices are capable of accessing the Internet and other remote resources like SQL databases. Therefore, they may contain data relating to an individual's network activities. As with Palm OS, developer kits for Windows CE contain emulators and debugging tools that are useful for learning more about these systems. Additionally, tools like FlashBack enable individuals to save data into Flash memory on Windows CE devices. Currently, PDA Seizure is the only tool for capturing evidence from a Pocket PC and Windows CE devices.

13.2.3 RIM BLACKBERRY

As with Palm OS, the Research in Motion (RIM) Blackberry handheld devices use a database structure with data in separate records. However, RIM devices differ from Palm OS and Windows CE devices in that they store user data in FLASH rather than RAM. As noted earlier, this is an advantage from a data recovery standpoint because data can only be erased from FLASH in 64-kbyte blocks, a time consuming process in computer terms (approximately 5 seconds) that is only performed when absolutely necessary. Therefore, RIM devices simply add new records and new versions of modified records to the end of the database, marking deleted or modified records as old. Only when the file system runs out of space do RIM devices perform the costly process of erasing blocks of FLASH, removing old and deleted records in the process.

RIM devices are designed to provide mobile users with remote access to Internet and corporate systems. In addition to sending and receiving e-mail and text messages, remote storage interfaces enable users to access and save files on a remote server. As a result, these handheld devices often contain e-mail and other network related data of a sensitive nature. Although some RIM devices allow secure, encrypt communication with a remote server, data on the device itself is not automatically encrypted.

Like Palm OS devices, Blackberry provides a debugger (named Programmer) and simulator in their Standard Developer's Kits (SDK) that can be used to access RIM devices. The Program Loader can be used to query the device and make memory dumps, but can cause a soft reset that usually decimates useful information on the device such as logs of recent data transfers and radio towers. Therefore, it is advisable to obtain some of this information by manually examining the device before performing a full memory dump.

13.2.4 MOBILE TELEPHONES

The variety of mobile telephones makes it difficult to develop a single digital evidence examination tool for all of them. Additionally, investigators often only want specific data from the telephone such as recent numbers called and received, and are not interested in recovering deleted data. Furthermore, the large number of phones that exist makes it prohibitively time consuming to perform a lengthy examination of each one. For these reasons, in many cases investigators are required to perform a manual examination of a telephone, reading data from the display and documenting their findings. Digital evidence examiners with specialized tools are usually only employed when deleted data is required or if password protection or encryption must be bypassed.

As a result of the growing popularity of GSM, commercial products have been developed to enable individuals to access their SIM card using their computers. The programs are not designed with digital evidence

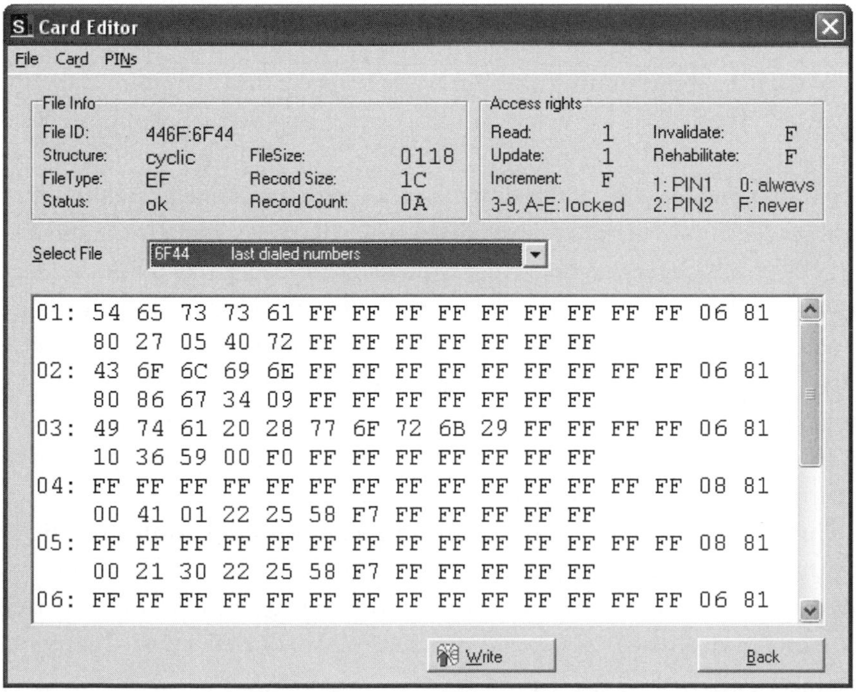

Figure 13.5

A SIM card viewed using Card Editor SIM Manager Pro.

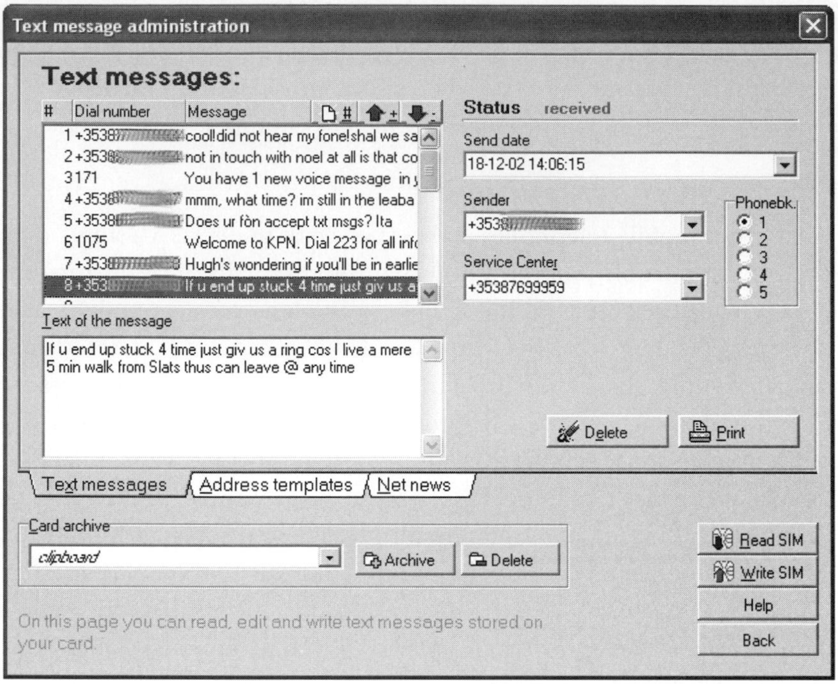

Figure 13.4

Text messages on a SIM card viewed using SIM Manager Pro.

examination in mind but can be useful for collecting the types of information that are obtained during a manual examination. For instance, SIM Manager Pro can be used to read certain data from a GSM SIM card such as text messages as shown in Figure 13.4. This program does not bypass the personal identification number (PIN) that protects the SIM. Therefore, the PIN must be obtained or bypassed in some other way before the SIM can be accessed using this software.

Because this software is not designed with digital evidence examination in mind, it does not write-protect the SIM card. Therefore, when using this tool, great care must be taken not to alter the original data. In fact, SIM Manager Pro is specifically designed to facilitate data entry so that individuals can update their address book via their computer. SIM Manager Pro has a Card Editor that can be used to access and alter certain parts of a SIM card as shown in Figure 13.5. Although it can be useful to view data on the card in uninterrupted form, the Card Editor does not give access to all regions on the card and may not show all data.

The Netherlands Forensic Institute has developed tools specifically for processing digital evidence on mobile telephones, SIM cards, and PDAs, some of which have the ability to recover deleted data. These tools are described in the *Handbook of Computer Crime Investigation*, Chapter 11 (Van der Knijff 2001).

13.3 DEALING WITH PASSWORD PROTECTION AND ENCRYPTION

Palm OS permits users to password protect their device and stores the associated password in encoded form in two places: in the "Unsaved Preferences" database on the device and in a file named "users.dat" file on computers that are used to HotSync the device. Also, if a Palm OS device is on, digital evidence examiners can obtain an encoded version of the password via the InfraRed port using the **notsync**[10] utility on another Palm OS device. Prior to Palm OS 4, these passwords were weakly encoded and could be recovered using **palmcrypt** as shown here.

[10] *http://www.atstake.com/research/tools/password_auditing/*

```
D:\>palmcrypt -d
B8791D707A2359435082DA4E599FBE4BEE675CCE541B346C041B6C55AE81CDF
PalmOS Password Codec
kingpin@atstake.com
@stake Research Labs
http://www.atstake.com/research
August 2000
0x62 0x69 0x72 0x74 0x68 0x64 0x61 0x79 [birthday]
```

It is more difficult to recover data from a Palm OS device that is protected with strong encryption using applications like Secret! and CryptoPad. In such cases, it may be possible to recover data in unencrypted form in the device memory or on the computer used to HotSync the device. Alternatively, it may be possible to obtain or guess the password used to encrypt the data. More advanced tools and techniques for obtaining or guessing passwords from PDAs and mobile telephones are described in the *Handbook of Computer Crime Investigation,* Chapter 11 (Van der Knijff 2001).

13.4 RELATED SOURCES OF DIGITAL EVIDENCE

Data relating to a handheld device can often be found on associated desktop computers and memory modules. For example, when a Palm OS device is synchronized with a desktop computer, data is stored in primary backup files (.dat, *.bak) and archive files (*.dba, *.tda, *.ada). Items that have been erased from the device may still exist on the desktop including e-mail messages and private data. These files are Microsoft Foundation Class (MFC) objects and their format varies depending on the MFC version used. For this reason, tools that are designed to interpret Palm databases may not be able to read these files. To complicate matters, the format of data in Palm memory is not identical to the format of these backup files. Therefore, it may be necessary to interpret meticulously and piece together data in these backup files on the desktop.

13.4.1 REMOVABLE MEDIA

Memory modules are usually formatted with FAT file system and can be treated like any other piece of removable media. For example, some memory cards have a write-protection switch, which should be enabled before the digital evidence acquisition process. Also, like other forms of storage media, some form of drive or adapter is required to provide an interface between the memory module and the digital evidence collection system. Adapters for more types of memory modules are available for desktop and laptop computers (see Figure 13.6).

One complication that can arise with some memory modules is copy protection. This can usually be bypassed using dd on UNIX. Another complication arises when dealing with modules such as GSM SIMs and other smart cards that cannot be accessed using previously mentioned evidence acquisition tools. For instance, Cards4Labs is a tool specifically designed for accessing smart cards of various kinds (Van der Knijff 2001).

13.4.2 NEIGHBORHOOD DATA

Handheld devices often contain remnants of network activity such as e-mail messages and Web clippings obtained using Palm Query Application (PQA). This information can be used to locate related digital evidence on other systems.[11]

For instance, the following portion of RAM dump of a Kyocera device (combination Palm PDA and mobile telephone) contains the number of the telephone and the name of the POP server used to check e-mail. The

[11] *E-mail messages and other information downloaded from the Internet can be transferred onto handheld devices via a desktop computer. Therefore, the presence of such information on a device does not necessarily indicate that the device could access the Internet directly.*

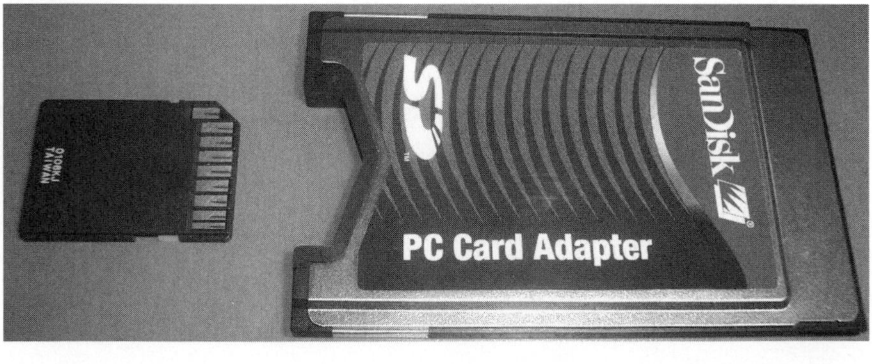

Figure 13.6

A memory module for a Palm OS device along with a PCMCIA interface card. This type of adapter is useful for acquiring digital evidence from memory modules using Windows and Unix based tools such as EnCase and dd.

telephone company may have call records associate with this telephone number and the POP server may have associated logs and e-mail messages.

```
00003230 3300FEC5 B1F42053 5052494E | ..20  3.■┼  ▒[ S  PRIN  | 221,328
54205043 53200028 32303329 20353435 | T PC  S .(  203)   555  | 221,344
2D303733 32000000 00000000 00000000 | -196  6...  ....  ....  | 221,360
00000000 00000000 00000000 00000000 | ....  ....  ....  ....  | 221,376
00000000 00000000 00002832 30332920 | ....  ....  ..(2  03)   | 221,392
3534352D 30373332 00000000 00000000 | 555-  1966  ....  ....  | 221,408
00000000 00000000 00000000 00000000 | ....  ....  ....  ....  | 221,424
00000000 00000000 00000000 0000006C | ....  ....  ....  ...l  | 221,440
AA6D006C AA6D00C0 00C04A75 6E203238 | ¬m.l  ¬m.└  .└Ju  n 28  | 221,456
20323030 31003136 3A32333A 3130004A |  200  1.16  :23:  10.J  | 221,472
756E2032 38203230 30310031 373A3030 | un 2  8 20  01.1  7:00  | 221,488
3A303000 4B543130 31302020 006A043F | :00.  KT10  10    .j.?  | 221,504
C3000257 0000271B 00000000 00000000 | ├..W  ..'.  ....  ....  | 221,520
00000000 00000000 00000000 4C473030 | ....  ....  ....  LG00  | 221,536
30000000 00000000 00000000 00000000 | 0...  ....  ....  ....  | 221,552

4575646F 72612057 65620045 75646F72 | Eudo  ra W  eb.E  udor  | 269,200
61205765 620000C8 00000000 51434F4D | a We  b..╚  ....  QCOM  | 269,216
03200000 00345143 515400D7 00000001 | . ..  .4QC  QT.╫  ....  | 269,232
0000005E 02000CD5 00081700 07000100 | ...^  ...╒  ....  ....  | 269,248
02000000 00000000 00000200 00000474 | ....  ....  ....  ...t  | 269,264
0100010F 64617465 61646472 746F646F | ....  date  addr  todo  | 269,280
6D656D6F 5143444C 73796E63 6C6E6368 | memo  QCDL  sync  lnch  | 269,296
73796E63 01000040 00400040 01002710 | sync  ...@  .@.@  ..'.  | 269,312
75333238 01006C6E 63680000 00780000 | u328  ..ln  ch..  .x..  | 269,328
00560200 0C2E0002 00000000 FF000007 | .V..  ....  ....   ...  | 269,344
00630000 00FF0000 00173FD0 00000000 | .c..  . ..  ..?╨  ....  | 269,360
00000000 0100FFF0 004F0020 0020000A | ....  .. ≡  .O.   . ..  | 269,376
001E001E 001E001E 001E001E 001E001E | ....  ....  ....  ....  | 269,392
001E001E 001E001E 10325476 98BADCFE | ....  ....  .2Tv  ÿ║▄▀  | 269,408
00000000 01000010 02C2124B 0001656F | ....  ....  .┬.K  ..eo  | 269,424
6333002C 0100001E 02C21251 0001656F | c3.,  ....  .┬.Q  ..eo  | 269,440
63332E6D 61696C2E 79616C65 2E656475 | c3.m  ail.  yale  .edu  | 269,456
00000000 002002C2 11F50000 00300000 | ....  . .┬  .┘..  .0..  | 269,472
```

Mobile telephones and Blackberry devices are specifically designed to access wireless networks and may have a substantial amount of neighborhood data.

13.5 SUMMARY

There are a growing number of handheld devices for personal organization and communication, some with access to the Internet. These devices can be an instrumentality of a crime when used to eavesdrop on wireless network traffic. The information they contain can also be an instrumentality of a crime when they are used to steal intellectual property or create and disseminate child pornography. They can also be a source of digital evidence, containing passwords and other useful data, or showing where individuals were at a specific time and with whom they were communicating. In some countries, including Sweden and Japan, it has become routine for investigators to collect handheld devices as evidence. Embedded systems are a challenging source of evidence because the data on them is volatile and different tools are needed to process different devices. Currently tools and training in this area are limited but, given the rapid increase in their use, this is likely to become one of the largest growth areas in the field of digital evidence examination.

REFERENCES

Burnette M. (2002) "Forensic Examination of a RIM (BlackBerry) Wireless Device" (Available online at http://www.rh-law.com/ediscovery/Blackberry.pdf).

Grand J. (2002) Proceedings of the 14th Annual Computer Security Incident Handling Conference, Waikoloa, Hawaii, June 24–28, Forum of Incident Response and Security Teams (Available online at http://www.mindspring.com/~jgrand/pdd/pdd-palm-forensics.pdf and http://www.atstake.com/research/reports/acrobat/pdd_palm_forensics.pdf).

Digital Discovery & e-Evidence, May 2001, "New Rules on Cell Phone Monitoring Systems Fuel Interest in Embedded Systems."

Palm OS Memory Architecture (Available online at http://oasis.palm.com/dev/kb/papers/1145.cfm).

Palm OS Memory and Database Management (Available online at http://oasis.palm.com/dev/kb/papers/2029.cfm).

van der Knijff R. (2001) Embedded System Analysis, Casey, E. (editor) *Handbook of Computer Crime Investigation*, London: Academic Press

CASES

United States v. Randolph (2002) District Court, Eastern District of Pennsylvania, Case Number 02-114 (Available online at http://www.paed.uscourts.gov/documents/opinions/02D0408P.HTM)

PART 3

NETWORKS

## CHAPTER 14

NETWORK BASICS FOR DIGITAL INVESTIGATORS

Until recently, it was sufficient to look at individual computers as isolated objects containing digital evidence. Computing was disk-centered – collecting a computer and several disks would assure collection of all relevant digital evidence. Today, however, computing has become network-centered as more people rely on e-mail, e-commerce, and other network resources. It is no longer adequate to think about computers in isolation since many of them are connected together using various network technologies. Digital investigators examiners must become skilled at following the cybertrail to find related digital evidence on the public Internet, private networks, and other commercial systems. An understanding of the technology involved will enable digital investigators to recognize, collect, preserve, examine and analyze evidence related to crimes involving networks.

When a crime just involves e-mail, an understanding of network protocols is useful but not essential – digital investigators might only require a basic understanding of e-mail to perform an effective investigation. However, most crimes involving networks require digital investigators to be familiar with the underlying technology. Sources of digital evidence on networks include server logs, contents of network devices, and traffic on both wired and wireless networks. An understanding of these technologies is necessary to track down unknown offenders via networks and attribute criminal activity to them. For instance, to investigate computer intrusions effectively, a solid understanding of TCP/IP and the operating system(s) involved is required. At the very least, digital investigators need a basic understanding of networks to interpret digital evidence found on personal computers such as e-mail, Web browser history, and file transfer.

When digital investigators do not have access to a key computer, it is necessary to reconstruct events using only evidence on networks. In a number of cases, sexual predators have persuaded their victims to destroy evidence by removing and disposing of their hard drive before leaving their home to meet the offender. Sources of evidence on the Internet that may reveal whom

Digital Evidence and Computer Crime Second Edition
ISBN: 0-12-163104-4

the victim was communicating with include e-mail and log files on the victim's Internet Service Provider's systems and backup tapes. Additionally, mobile telephone records may help determine whom the victim was communicating with and where he/she went. When a suspect claims that he/she does not have a home computer, credit card billing records, telephone records, and ISP logs may show that the suspect has a home computer and may contain clues of its current whereabouts.

This chapter provides an overview of networks and goes on to describe how these different networks are joined together to form the seemingly homogeneous Internet.[1] This chapter ends with an overview of crimes that occur at different levels of networks. Subsequent chapters go into more detail, discussing network layers.

[1] *The word* internet *is used in lowercase when referring to any connection of dissimilar networks using an internet protocol like TCP/IP. The Internet (capitalized) refers specifically to the global network of interconnected networks.*

14.1 A BRIEF HISTORY OF COMPUTER NETWORKS

As with the electronic computer, the military spurred the creation of computer networks that have developed into the Internet. In 1969, the Advanced Research Projects Agency (ARPA), a part of the Defense Department, began funding companies and universities to develop a communications system to withstand heavy enemy attacks. The primary aim was to enable military installations around the country to communicate even if significant parts of the communications system were destroyed. However, an early memorandum noted that such a system would have additional benefits.

> While highly survivable and reliable communications systems are of primary interest to those in the military concerned with automating command and control functions, the basic notions are also of interest to communications systems planners and designers having need to transmit digital data. (Baran 1964)

By the end of 1969, a primitive network named the ARPANET was in place (Figure 14.1). This network was the foundation of the modern Internet.

In 1991, the World Wide Web (WWW) was released to the general public, making it easier for people to use the Internet. Since then, the Internet has been commercialized and its popularity has grown exponentially. In fact, so many people have been using the Internet that several universities and research organizations decided to set up second, higher speed networks in an effort to bypass the traffic jams on the Internet. One of these high-speed networks is called Abilene.[2]

[2] *http://abilene.internet2.edu*

In a relatively short period, technology has advanced to the point where the lines between computers, televisions, telephones, and print media have been blurred. Many experts in computing and telecommunications agree that, with

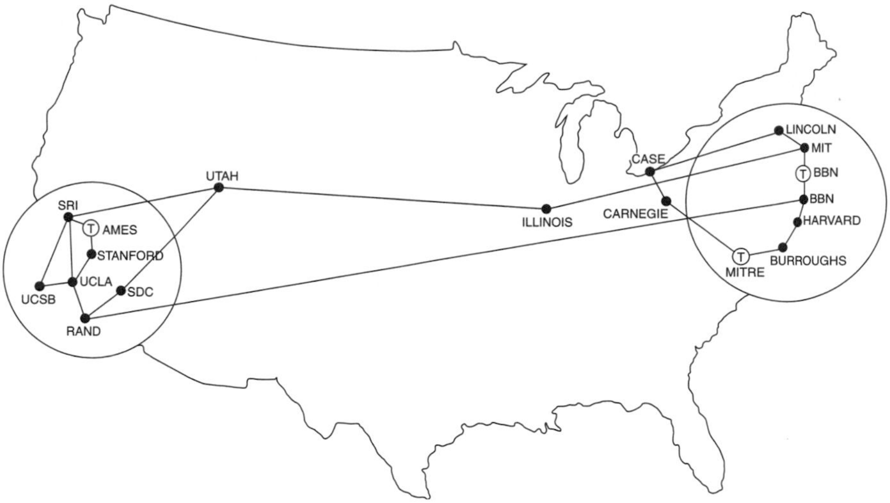

Figure 14.1
Map of ARPANET.

ENIAC	ARPANET	Intel 8080	Mac & IBM PCs	WWW	Internet2
1946	1969	1974	1980s	1991	1999

Figure 14.2
Time line of key events.

this seamlessly integrated global infrastructure in place, the next 5 years of computing and telecommunications will bring more changes than the last 20 years. Already, households and neighborhoods are being connected to networks that enable them to operate, communicate, and collaborate more effectively. This technology enables the owner of a house to control household functions remotely. Conversely, this technology could give criminals access to household appliances. The day approaches when someone from across the world can stage an accident by turning on a gas stove and sparking a toaster to blow up another's house.

14.2 TECHNICAL OVERVIEW OF NETWORKS

A computer connected to a network is generally referred to as a *host*, and uses a modem or network interface card (NIC) to send and receive information over wires or through the air.[3] When more than two hosts are being connected, it is not feasible to link each host directly to every other host – this would result in a ludicrous number of wires terminating at each host. Each time a new host was added to the network, it would have to be wired directly with every other computer. In the past, to avoid this situation, a single network cable was used and

[3]Individuals who are learning about networks for the first time will find that the convenience of using abbreviations and acronyms creates its own difficulties. For instance, the acronym for Media Access Control addresses (MACs) can easily be confused with the abbreviation for Macintosh computers (Macs). The Glossary organizes the terms, abbreviations, and acronyms that are used in this text to assist the reader.

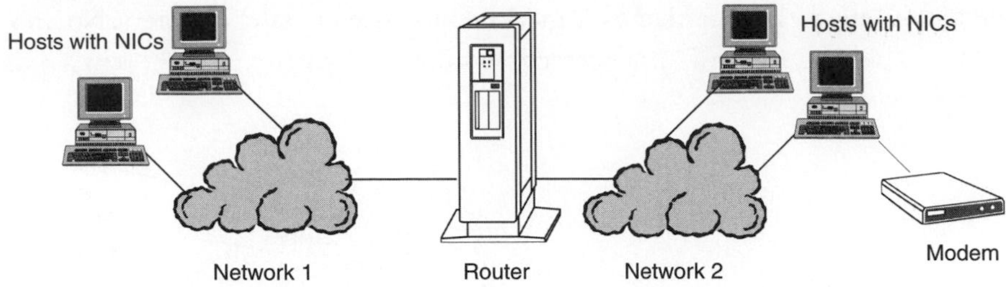

Figure 14.3
Depiction of hosts with NICs connected to a router to form a network.

devices called a *tap* punctured the plastic sheath of the thick cable physically to connect a host to the network. Because this approach was inflexible and difficult to maintain, devices called *hubs* (a.k.a. concentrators) were developed to simulate this single network cable configuration – instead of using taps each host is connected to the hub using a thin cable. To increase network security and efficiency, hubs are being replaced by *switches* that perform a similar function but direct data to their intended destination rather than broadcasting them to all hosts on the network, thus inhibiting one host from eavesdropping on the network traffic of all neighboring hosts. Techniques have been developed to enable eavesdropping on switched networks, undermining the security provided by these devices (Snipe 2000; Convery 2002).

Computers connected to the global Internet communicate using a set of protocols collectively called TCP/IP (Transport Control Protocol/Internet Protocol). As detailed in the next section, the Internet comprises many individual networks. TCP/IP is essentially the common language that enables hosts on these individual, often dissimilar networks to communicate. Each TCP connection (a.k.a. *TCP stream)* is bi-directional: one *flow* for receiving data and a second *flow* for sending data. A tool like Argus[4] can monitor network traffic and maintain logs for later analysis such as the two NetBIOS connections shown here:

[4]http://www.qosient.com/argus/

Date	Time	Proto	Source	Destination
20 May 03	07:11:18	tcp	192.168.0.5.1029 →	192.168.0.2.netbios-ssn
20 May 03	07:12:24	tcp	192.168.0.5.1030 →	192.168.0.3.netbios-ssn

Hosts that are connected to two or more of these networks and direct traffic between them are called *routers.* Routers are a crucial component of computer networks, essentially directing data to the correct place. Although almost any host can be used as a router, most networks use custom-made routers like those produced by Cisco and Juniper. Routers can direct data from one network to another, filter unwanted traffic, and keep logs that can be an excellent source of digital evidence. In addition to system logs, some routers can generate more detailed NetFlow logs, similar to Argus logs, discussed in later chapters. Notably, NetFlow displays individual, unidirectional flows as shown here, whereas Argus displays bi-directional streams:

Start	End	SrcIPaddress	SrcP	DstIPaddress	DstP	Proto
0520.07:11	0520.07:12	192.168.0.5	1029	192.168.0.2	139	6
0520.07:11	0520.07:12	192.168.0.2	139	192.168.0.5	1029	6
0520.07:12	0520.07:13	192.168.0.5	1030	192.168.0.3	139	6
0520.07:12	0520.07:13	192.168.0.3	139	192.168.0.5	1030	6

Because of their importance, routers are at high risk of attack and computer intruders target routers to eavesdrop on traffic and disrupt or gain access to networks.

Firewalls are similar to routers in that they direct traffic from one network to another. However, these security devices are designed to block traffic by default and must be configured to permit traffic that meets certain criteria. Firewalls can keep detailed logs of successful and unsuccessful attempts to reach the hosts that it protects and can be a useful source of digital evidence.

The services that networks enable, such as sending and receiving e-mails, rely on the client–server model. Telnet provides a clear example of client–server communication, enabling remote users to log into a server and execute commands. For example, the following shows a telnet connection from a Windows client to a UNIX server (192.168.0.9) and some resulting log file entries:

```
C:\> telnet 192.168.0.101

Standard telnet does not encrypt traffic, exposing your password and data to
network sniffers. A more secure alternative to telnet is Secure Shell (SSH),
available at www.ssh.org.

login: eoc3
Password: ********
Last login: Thu Apr 3 15:50:33 from 192.168.0.5

WARNING: To protect the system from unauthorized use and to ensure that
the system is functioning properly, activities on this system are monitored and
recorded and subject to audit. Use of this system is expressed consent to such
monitoring and recording. Any unauthorized access or use of this Automated
Information System is prohibited and could be subject to criminal and civil
penalties.
```

Preview (Chapter 17): For the most part, every host on the Internet is assigned a unique number, called an Internet Protocol (IP) address, to distinguish it from other hosts. Before information is sent through the Internet, it is addressed using the IP address of the destination host, much like an envelope is addressed before it is submitted to a postal system. Routers use these IP addresses to direct information through the Internet to its destination. If the sender requires confirmation that the destination host has received a transmission, the Transport Control Protocol (TCP) will perform this task, resending information when necessary. Be aware that TCP performs other functions, such as breaking information into packets and that there are other protocols in the TCP/IP family such as the User Datagram Protocol (UDP), the Internet Control Message Protocol (ICMP), and the Address Resolution Protocol (ARP). It is also worth noting that TCP/IP enables other protocols like Simple Mail Transfer Protocol (SMTP) and Hypertext Transfer Protocol (HTTP) to transmit e-mail and Web pages, respectively.

```
oisin% grep telnet /var/log/messages

Apr 3 15:50:33 oisin inetd[178]: [ID 317013 daemon.notice] telnet[373] from
192.168.0.5 2523

Apr 4 15:59:23 oisin inetd[178]: [ID 317013 daemon.notice] telnet[432] from
192.168.0.5 2531

oisin% last
eoc3   pts/6   192.168.0.5        Fri   Apr  4   15:59     still   logged in
eoc3   pts/2   192.168.0.5        Thu   Apr  3   15:50  -  16:06   (00:16)
ftp    ftp     ACBC4D0B.ipt.aol   Tue   Apr  1   14:41  -  13:04   (8+22:22)
```

This example also demonstrates the need to correlate log files to obtain a more complete picture of what occurred on a system. The associated syslog entry on the server shows the time of the connection and the IP address of the client. However, the syslog entries in this example do not indicate which account was used to make the connection and how long the connection lasted. This information is stored in the wtmp log, accessed here using the last command, showing which user account was used to connect at the time but does not indicate that telnet was used as the connection method.[5]

[5]Some systems record the username and logout time in syslog. However, neither syslogs nor wtmp indicate what activities occurred on the system during the login session – this would require an analysis of MAC times on the file system and process accounting logs or BSM audit records if they exist. Additionally, routers, firewalls, intrusion detection systems, and other network monitoring devices could provide corroborating data.

In the past, a server was viewed as a powerful computer that could provide a service to many smaller computers called clients, much like a law firm provides services to its clients. Some servers allow anyone to access their resources without restrictions (e.g. Web servers) while others (e.g. e-mail servers) only allow access to authorized individuals, usually requiring a user identifier and password. With the increased power and capacity of personal computers, the distinction between clients and servers has blurred. Today, any host can be made into a server by installing software that allows other hosts to access it over a network. This approach is commonly called *peer-to-peer networking* (P2P) to differentiate between this type of file sharing and the traditional client–server model, and was popularized by programs like Napster and Kazaa.

Peer-to-peer networking has been taken one step further by wireless technology that uses radio frequency, infrared, lasers, and microwaves to carry data. For instance, Bluetooth enables computers, personal digital assistants, mobile phones, and household appliances like televisions to communicate with each other. In essence, when a Bluetooth-enabled device is turned on, it attempts to communicate with other devices in its vicinity to create what is commonly called an *ad hoc network* or *piconet.*

Many components of networked systems contain information about the activities of the people who use them. Table 14.1 summarizes some of the information that different network components may have.

INTERNET ACTIVITY	LOGS	ACTIVE STATE DATA
PPP dial-up	TACACS/RADIUS	Terminal server memory
Firewall/router	syslog/NetFlow	show cons
Host logon	Wtmp/NT Event Log	utmp/nbtstat -c
Web server	Access log	netstat -an
E-mail server	messages/syslog	Mail spool
FTP server	xferlog	netstat -an
IRC server	Server/boot logs	netstat -an
Wireless LAN	Device logs	Device memory query
Mobile phone	Call records	Location/conversations

Table 14.1

Examples of log files and active state data relating to various networked systems.

14.3 NETWORK TECHNOLOGIES

Beneath the apparently consistent facade of TCP/IP is a collection of dissimilar network technologies. It is these network technologies that enable multiple hosts to share a single transmission medium such as a wire or the air. When hosts are sharing a transmission medium only one host can use the medium at any given time. This is analogous to a polite conversation between people in which one person talks and the other listens. If two hosts were allowed to use the transmission medium at the same time, they would interfere with each other.

The easiest way to understand network basics is to imagine someone setting up a network. For instance, suppose "Barbara the Bookie" wants to create an online betting site like World Sports Exchange[6] or World Gaming.[7] Once Barbara the Bookie has decided where to incorporate (e.g. England), where to establish operations (e.g. Antigua), and purchased computer equipment, she must select a network technology to connect the Antiguan servers physically. Seven network technologies: ARCNET, Ethernet, FDDI, ATM, IEEE 802.11 (wireless), cellular, and satellite are briefly described here.

14.3.1 ATTACHED RESOURCE COMPUTER NETWORK (ARCNET)

ARCNET was one of the earliest network technologies and the latest version (ARCNET Plus) can transmit data at twenty megabytes per second (20 Mbps).[8] ARCNET uses coaxial cables, similar to the ones used for cable television, to connect the Network Interface Card (NIC) in each host to a central hub. If a single host is damaged or turned off, others on the network can still communicate with each other through the hub. However, if the hub is damaged or turned off, none of the hosts will be able to communicate with each other.

ARCNET uses a method called "token passing" to coordinate communication between each of the hosts connected to the central hub (Figure 14.4). Basically, a token is sent around on the network and when a host wants to

[6]WSEX (http://www.wsex.com) founder Jay Cohen was convicted of violating the US Wire Communications Act by illegally using interstate telephone lines to take online wagers. More specifically, Cohen had accepted sports bets from New Yorkers via the WSEX gambling site in Antigua. In 2001, Starnet Communications International, a subsidiary of World Gaming, Inc., pled guilty to violating Section 202 (1) b of the Canadian criminal code by having a machine in Canada for gambling or betting (http://laws.justice.gc.ca/en/c-46/39421.html). World Gaming has since moved their systems to Antigua and is incorporated in England.

[7]http://www.worldgaming.com

[8]ARCNET Plus is an enhanced version of ARCNET that has the ability to use TCP/IP.

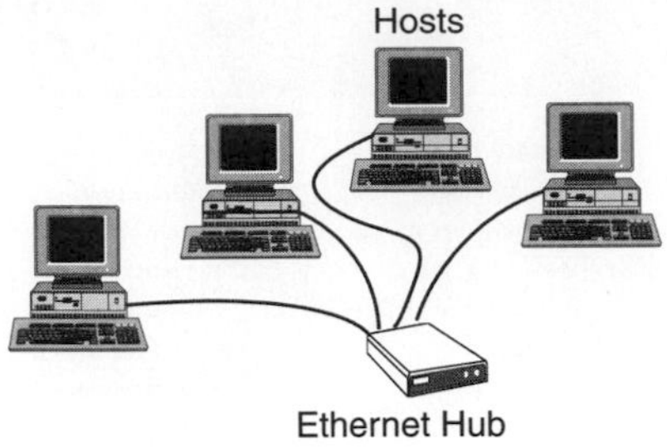

Figure 14.4

Hosts connected to a central hub (star typology).

send data it waits for the token, takes the token, and starts to transmit. When that host has finished transmitting, it relinquishes the token, passing it on to the other hosts on the network thus allowing other hosts to communicate.

14.3.2 ETHERNET

Ethernet has gone through several stage of development and has become one of the most widely used network technologies because it is relatively fast and inexpensive. One of the most recent forms of Ethernet uses wires similar to regular telephone cords. These wires are used to connect the NIC in each host to a central hub or switch that essentially makes the hosts think that they are connected by a single wire.

Instead of token passing, Ethernet uses Carrier Sense Multiple Access with Collision Detection (CSMA/CD) to coordinate communication. Although CSMA/CD is a mouthful, the concept is straightforward. Hosts using Ethernet are like people making polite conversation at a dinner party. At a polite dinner party, if two people start to speak at the same time, they both stop for a moment, one starts to talk again while the other waits. Similarly, when two hosts using Ethernet start to transmit data at the same time, they both sense that the other host is transmitting and they both stop for a random period of time before transmitting again. Ethernet is described in more detail in Chapter 16.

14.3.3 FIBER DISTRIBUTED DATA INTERFACE (FDDI)

As the name suggests, FDDI uses fiber optic cables to transmit data by encoding it in pulses of light. This type of network is expensive but fast, transmitting data at 100 Mbps. Like ARCNET, FDDI uses the token passing technique but instead of using a central hub, hosts on an FDDI network are connected together to form a closed circuit (Figure 14.5). Data travel around this circuit through every host until it reaches its destination. Normally, data only travel in one direction around this circuit. However, if one of the hosts on an FDDI network detects that it cannot communicate with its neighbor, it uses a second, emergency ring to send data around the ring in the opposite direction. In this way, a temporary ring of communication is established until the faulty host can communicate again.

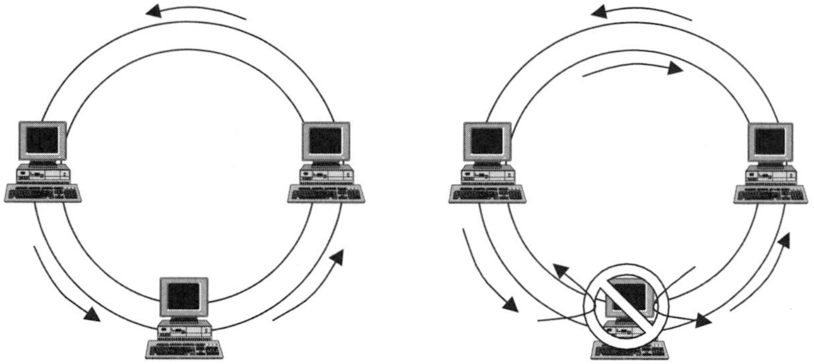

Figure 14.5

Normal FDDI communication versus backup communication when a host is down (double ring topology).

14.3.4 ASYNCHRONOUS TRANSFER MODE (ATM)

ATM uses fiber optic cables and specialized equipment (ATM switches) to enable computers to communicate at very high rates (Gbits per second). Telecommunications companies developed this technology to accommodate concurrent transmission of video, voice, and data. Although it is very expensive, ATM is becoming more widely used.

ATM uses technology similar to telephone systems to establish a connection between two hosts. Computers are connected to a central ATM switch and these switches can be connected to form a larger network. One host contacts the central switch when it wants to communicate with another host. The switch contacts the other host and then establishes a connection between them.

In Chapter 16, ATM is briefly compared with Ethernet to highlight their similarities and differences and describe how they both can be useful as a source of digital evidence.

14.3.5 IEEE 802.11 (WIRELESS)

Unlike the previously summarized network technologies, computers connected using one of the IEEE 802.11 standards do not require wires; they transmit data through the air using radio signals (Figure 14.6). Currently, the two most widely used standards are 802.11a and 802.11b, which use the 2.4 and 5 GHz spectrums, respectively. The 802.11g standard is also becoming popular because of its increased speed and backwards compatibly with 802.11b. *Access points* containing a radio transmitter and receiver form the core of these wireless networks, enabling computers, personal digital assistants, and other devices with a compatible wireless NIC to communicate with each other. In addition to being a conduit for wireless devices, these access points are generally connected to a wired network like an Ethernet network to enable communication with wired devices and the Internet.

The main limitations of 802.11 networks are distance, speed, and interference. A computer must be within a certain distance of an access point to achieve reliable connectivity and even then, data are only transmitted at

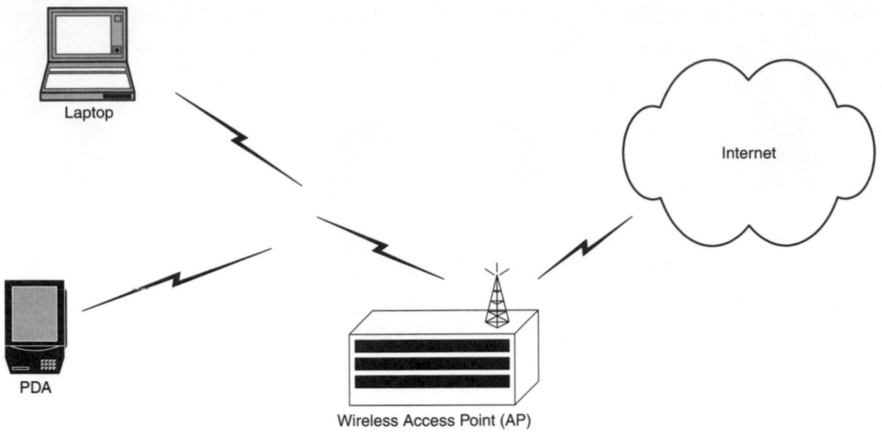

Figure 14.6
Wireless IEEE 802.11 network with a PDA and PC connected to an AP. Also shown is the AP connected to the Internet.

theoretical maximums of 11 and 54 Mbps for 802.11b and 802.11a networks, respectively. Any obstacles between the computer and access point that block radio waves will degrade or prevent connectivity.[9]

[9] *IEEE 802.11a networks interfere with other devices in Europe, making them ineffective. For this reason, the European HiperLAN2 standard was developed for higher speed wireless access.*

Some businesses and hobbyists have intentionally created 802.11 networks for anyone to use. Passers-by can configure their computers to connect to these public wireless networks and access the Internet. Some organizations and home users have unintentionally configured their wireless network insecurely allowing anyone to access them. The emergence of these public and insecure wireless networks has led to a trend called *war driving* – people drive around neighborhoods and business districts with computers configured to locate 802.11 networks. Some individuals will use insecure networks to gain unauthorized access to an organization's network and can even monitor wireless network traffic. Others simply notify other war drivers of the wireless networks they have found either by marking a nearby surface with a symbol that describes the network (called *war chalking*) or by posting them on the Internet.

14.3.6 CELLULAR NETWORKS

Cellular data networks are becoming widely available and increasingly popular. Organizations that depend on mobility (e.g. airlines, package delivery companies) have equipped their employees with hand-held devices that communicate over cellular networks. Cellular networks enable computers to connect to the Internet using a cellular telephone in much the same way as a modem is used to connect using telephone wires. Cellular networks are made up of cell sites that enable individuals within a certain geographical area to place and receive calls. Cell sites are connected to central computers (switches) that process and route calls and keep logs that can be used for billing, maintenance, and investigations. Although cellular networks are primarily used as circuit-switched networks (making direct connections between telephones) they can also function as packet-switched network (making virtual circuits between computers). To function as a packet-switched network, additional equipment is required that extracts packets of data from the wireless network and routes them to their destination.

Most digital cellular networks use Frequency Division Multiple Access (FDMA), Code Division Multiple Access (CDMA), Time Division Multiple Access (TDMA), or a combination of these technologies to transmit data via radio waves. These technologies enable several mobile telephones to share a single communications channel on a mobile telephone network (e.g. AMPS, GSM) by dividing the channel into several time slots, and assigning each telephone its own slot. To enable cellular devices to communicate with other hosts on the Internet, some cellular networks use a protocol Cellular Digital Packet Data (CDPD).[10] However, CDPD has been largely replaced with the higher speed General Packet Radio Service (GPRS) – part of GSM technology that uses a combination of TDMA and FDMA and has Internet Protocol capabilities.

Cellular technology is developing rapidly and the next evolution of GSM (called third generation or 3G) is emerging, providing higher data transmission rates and thus enabling more multimedia services such as music and video. The increasing functionality in cellular network technology is creating new opportunities for criminals and investigators. To understand the potential for investigators, a summary of mobile telephones is provided here. More information about digital evidence on wireless networks and devices is available in Chapter 10 of the *Handbook of Computer Crime Investigation* (Clarke and Gibbs 2001).

Mobile telephones have two numbers that uniquely identify them – an Electronic Serial Number (ESN), and a telephone number or Mobile Identification Number (MIN). When a mobile telephone is manufactured, its microchip is programmed with a unique ESN and when the telephone is given to a subscriber it is assigned a telephone number that people use to call the subscriber. These numbers are used by telephone companies to direct calls to the correct mobile telephone and are used by investigators to locate the phone. Special electronic tracking equipment enables investigators to lock onto an ESN/MIN pair and track it to a general geographical area. Within a given geographical area, triangulation can be used to pinpoint the cellular telephone. Investigators require the assistance of cellular telephone companies to perform this type of tracking.[11]

Most mobile telephone companies maintain communication with all of their mobile telephones at all times even when the telephone is not in use (the telephone must be turned on). This constant communication is used to notify subscribers of voice mail and can be used to track a cellular telephone even when it is not being used to make calls. For instance, the position data relating to a murder victim's mobile telephone can be compared with that of

[10]A CDPD network uses a network technology called Digital Sense Multiple Access with Collision Detection (DSMA/CD) that works just like CSMA/CD. Although it is possible to eavesdrop on a cellular network, CDPD uses encryption to conceal data in transit.

[11]If criminals can obtain an ESN and MIN, they can reprogram a cellular telephone to mimic someone else's telephone. Any calls made from the criminal's telephone will be billed to the valid subscriber. Additionally, it becomes harder to capture criminals when they change the ESN/MIN in their phones. This became such a problem in the late 1990's that most cellular telephones companies use encryption to protect the ESN and MIN of their telephones.

a suspect's to determine if they were in the same vicinity at the time of the crime. In one case, a kidnap victim's mobile telephone was used in real time to track and intercept the car she was being transported in. In several cases offenders have stolen the victim's mobile telephone and in one case the offender apparently called the victim's mother to taunt her. In another case, a victim saw the offender make calls from the crime scene using a mobile telephone. Although the offender was not apprehended in this case, digital evidence did exist on a telephone company's systems that could have been used to generate a short list of suspects. Some cellular telephones even have Global Positioning System (GPS) features that can be used to locate the device quite precisely.

In addition to tracking, cellular telephone companies can provide investigators with call details, toll records, and wiretaps. This information can be used to determine the calling patterns and even the specific activities of a criminal.

14.3.7 SATELLITE NETWORKS

Satellites are becoming more widely used to convey Internet traffic around the globe. Some networks simply use satellite dishes, called Very Small Aperture Terminals (VSATs), to beam communications from the ground to a satellite overhead, which transmits the data to a central location on the ground. As with cellular networks, these VSATs use TDMA, CDMA, and similar technologies to transmit data using radio waves. These networks can support a range of network technologies, including ATM for high speed Internet access. Although some VSATs are portable, they usually only function within a given region or country and they are not as convenient to transport as a cellular telephone.

The Teledesic network is not designed with mobility in mind but aims to provide *Internet-in-the-Sky* access to anywhere in the world such as telecommuters in remote regions or businesses and homes in developing countries that do not have reliable telecommunications infrastructures. Conversely, Mobile Satellite Systems (MMS) like Iridium and Globalstar are designed with mobility in mind, providing global connectivity using mobile telephones. The Iridium Satellite System uses GSM-based technology to transmit data between wireless devices and low earth satellites and can be used to make telephone calls as well as connect to the Internet.

14.4 CONNECTING NETWORKS USING INTERNET PROTOCOLS

Like people who do not speak the same language, two hosts using different network technologies cannot communicate directly. So, a host using FDDI cannot communicate directly with a host using Ethernet. There are two methods of enabling communication between hosts using different network technologies: translators and common languages (Figure 14.7). As with the use

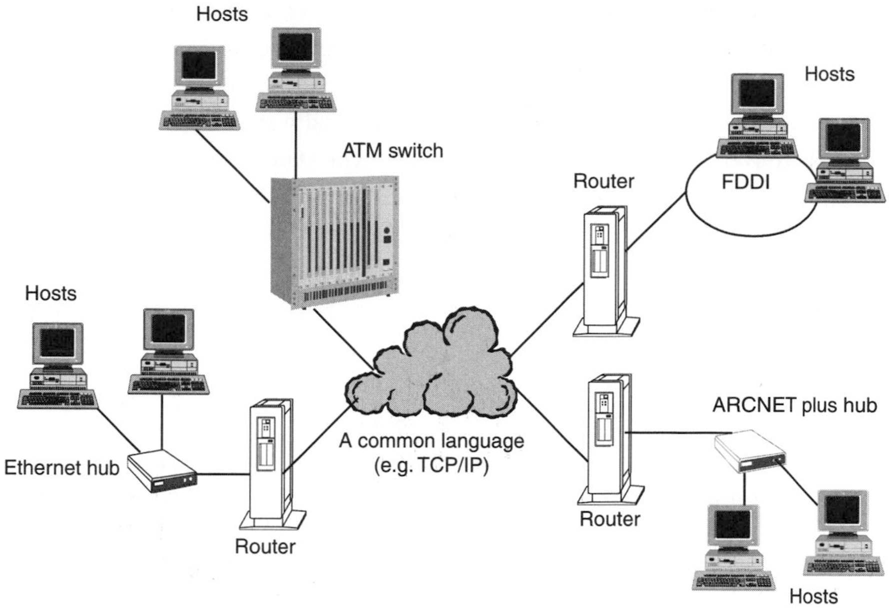

Figure 14.7

Dissimilar networks connected using a common language to form an internet.

of professional translators and common languages like Esperanto, in the computer-networking world there are translators (e.g. translating bridges) and common languages – called internet protocols (e.g. TCP/IP, TP-4/CLNP).

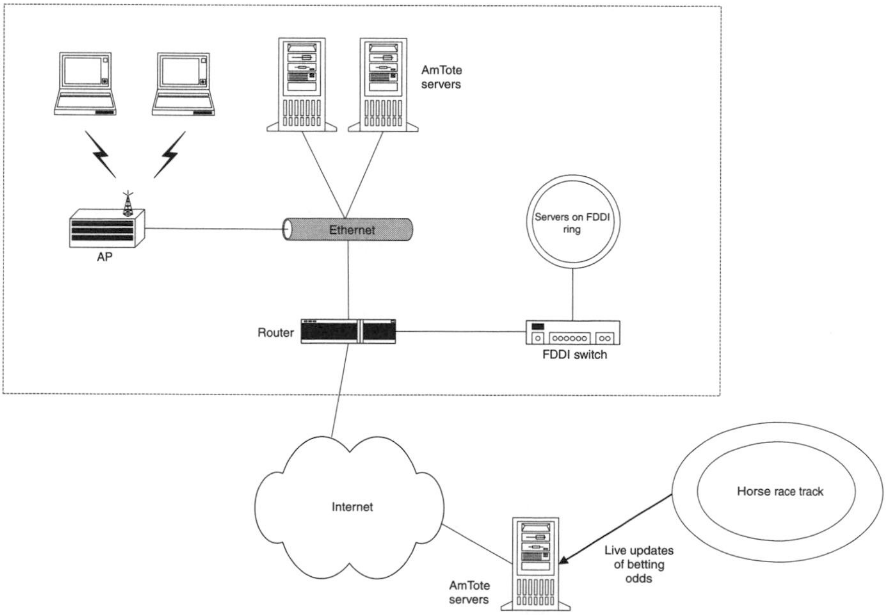

Figure 14.8

Barb the Bookie's Network.

For instance, suppose that Barbara the Bookie decides to connect her servers using FDDI and her workstations using wireless 802.11a technology because it is too difficult to run wires through the concrete walls of the hurricane-proof bunker that houses her network (Figure 14.8). She also

[12] http://www.amtote.com

wants to use AmTote[12] automated totalisator systems that use Ethernet to connect to racetracks and other sports betting venues. Additionally, Barbara the Bookie wants to connect her network to her Internet Service Provider using an ATM link. These networks are essentially speaking different languages. If Barbara just wanted to connect the AmTote systems with her servers on the FDDI network, it might make sense to use a specialized translator to convert from Ethernet to FDDI. However, when connecting many dissimilar networks it is more efficient to join them using devices with the necessary network interface cards and then use a common internet protocol like TCP/IP that every host can understand. This approach is more flexible and scalable, making it easier to modify and expand the network.

Currently, the most widely used internet protocols are the Transport Control Protocol (TCP), the User Datagram Protocol (UDP), and the Internet Protocol (IP). These protocols, along with a few supporting protocols, are collectively referred to as the TCP/IP internet protocol suite – TCP/IP for short. In some respects, TCP/IP is the Internet – currently every host attached to the Internet uses TCP/IP to communicate (Figure 14.9).

To deal with digital evidence on the Internet, digital investigators need a solid understanding of TCP/IP. To understand how TCP/IP works, it is useful to think of it in terms of layers as defined in the Open System Interconnection (OSI) reference model (Figure 14.10). Notably, TCP/IP was developed before the OSI model was formalized and, therefore, does not conform completely to the model. However, there are enough areas of similarity to discuss TCP/IP in terms of the OSI model. A layer model is useful to digital investigators because it provides a framework for understanding evidence, the operation of the technology, how data are created and transported on networks, and associated error, uncertainty, and loss. Examining each layer helps digital investigators develop a mental model of where evidence can be found on networks and how to collect and examine that

Figure 14.9

Conceptual depiction of TCP/IP with arrows indicating communication between modules.

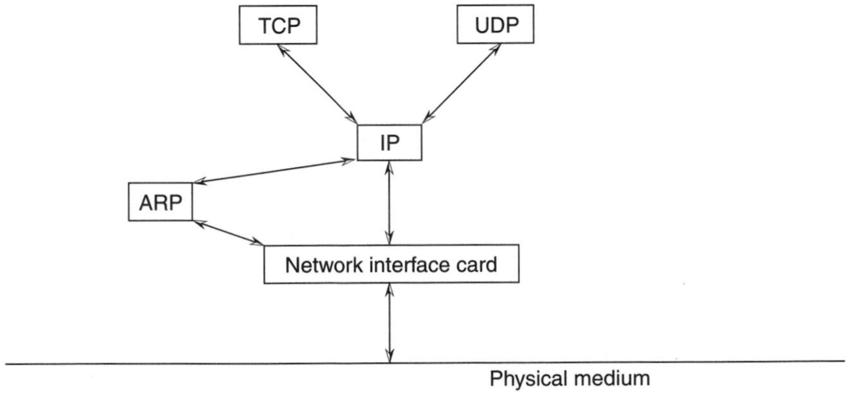

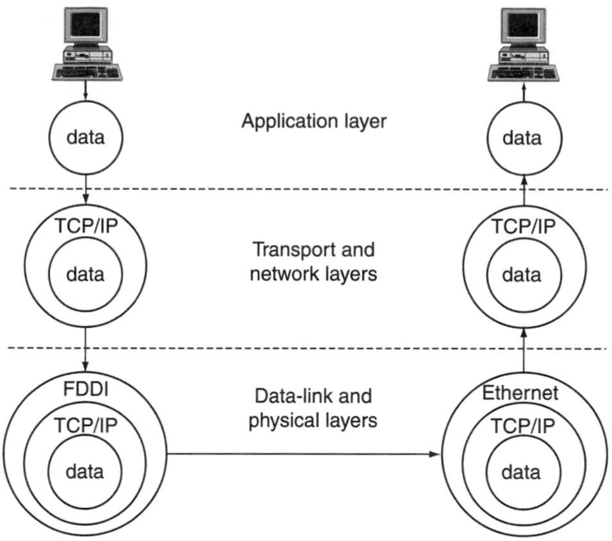

Figure 14.10

A simplified depiction of the Open System Interconnection layers showing where TCP/IP fits.

evidence. They can then apply this generalized mental model to specific networks of any kind.

The OSI reference model divides internets into seven layers: the physical, data-link, network, transport, presentation, and application layers. IP and TCP are network and transport layer protocols, respectively.

Each layer of the OSI model performs specific functions and hides the complexity of lower layers. For example, Barbara the Bookie's Wireless and Ethernet networks occupy the lowest layers of the Internet – the Physical and Data-link layers. A common language like TCP/IP at the Network and Transport layers enables hosts on ARCNET Plus, Ethernet, FDDI, ATM, and 802.11 networks to communicate with each other. The Session, Presentation, and Application layers make it easier for humans to use the network – hiding the inner workings of the lower layers. Provided all networks follow this model, they will be able to interconnect with relative ease.

The OSI reference model is described here briefly and is discussed in more detail in subsequent chapters.

14.4.1 PHYSICAL AND DATA-LINK LAYERS (LAYERS 1 AND 2)

The physical layer refers to the actual media that carries data (e.g. telephone wires, fiber optic cables, radio signals, and satellite transmissions). This layer is not concerned with what is being transported, but without it there would be no connection between computers. While the upper layers enable communication between distant computers, the data-link layer enables basic connectivity between computers that are close to each other. For example, when two hosts are connected by a single wire, the data-link layer puts data into a form that can be carried by the wire and processed by the receiving computer. For instance, hosts connected via modems generally use the Point-to-Point Protocol (PPP) to communicate. Hosts connected using network technologies described earlier in this chapter such as Ethernet use their own cards, cables, and protocols to communicate.[13]

The data-link layer has session-like aspects, establishing, maintaining, and terminating point-to-point connections between neighboring machines. Also, the data-link layer uses addresses to direct data but there addresses are only used locally when data are being transmitted between hosts that are not separated by routing equipment.[14] In short, the data-link layer is responsible for local communications between hosts and once routing, large distances, and multiple networks are involved, the network layer takes over. In addition to formatting and transmitting data according to the specifications of the network technology being used (e.g. Ethernet, 802.11, PPP), the data-link layer ensures that data were not damaged during transmission. Without the data-link layer, data would be sent down from the upper layers and would reach a dead end. Computers would not be able to communicate at all.

The physical and data-link layers are a gold mine from a digital evidence perspective. The Media Access Control (MAC) addresses described earlier in this chapter are part of the data-link layer and can be used to identify a specific computer on a network. These addresses are more identifying than network layer addresses (e.g. IP addresses) because they are generally associated with hardware inside the computer (IP addresses can be reassigned to different computers). Switches and other layer 2 network devices may also contain useful information. Additionally, all information traveling over a network passes through the physical layer. Individuals who can access the physical layer have unlimited access to all of the data on the network (unless it is encrypted). Digital investigators can dip into the raw flow of bits traveling over a network and pull out valuable nuggets of digital evidence. Conversely, criminals can access the physical layer and gather any information that interests them.

[13]A hub joins hosts at the physical level whereas a switch joins them at the data-link layer. When computers are connected with a hub it is as though they were connected with a single wire and any one of them can easily eavesdrop on the network traffic of all other connected hosts. Conversely, switches use MAC addresses to direct traffic to just the intended computer, making eavesdropping more difficult.

[14]Some routers can direct traffic between two machines on the same physical network segment using their MAC (layer 2) addresses thus avoiding the delay that would be caused by peeling away the layer two encapsulation to see the IP (layer 3) addresses. Notably, this only works for machines directly connected to the router – data destined for distant hosts must be routed using their IP addresses because the router cannot easily discover their MAC addresses.

CASE EXAMPLE

Someone within an organization configured his/her computer with the CEO's IP address and sent offensive e-mail messages, making it appear that the CEO had sent them. As soon as they were informed of the problem, the computer security department started monitoring network traffic that appeared to come from the CEO's IP address in the hope that they would catch the perpetrator in the act. Unfortunately, word of the investigation leaked out and the perpetrator did not repeat the offense. Fortunately, information gathered from a router early in the investigation showed that the CEO's IP address had been temporarily associated with the MAC address of another computer. This MAC address was used to locate the offending computer, which belonged to a disgruntled member of the software development department. An examination of the computer confirmed that it had been involved and the disgruntled employee had been using it at the time the messages were sent.

14.4.2 NETWORK AND TRANSPORT LAYERS (LAYERS 3 AND 4)

The network layer is responsible for routing information to its destination using addresses, much like a postal service that delivers letters based on the address on the envelope. If a message must pass through a router to get from one place to another, this layer will include appropriate instructions in the message to help the router direct the message properly. The transport layer is responsible for managing the delivery of data and has some features that are similar to the session layer. For example, the transport layer establishes, maintains, manages, and terminates communications between hosts. The transport layer divides large messages into smaller, more manageable parts and keeps track of the parts to ensure that they can be reassembled or retransmitted when necessary. If desired, the transport layer will confirm receipt of data, like a registered mail service that gives the sender a confirmation when the letter reaches its destination. When data are lost in transit, the transport layer will resend it if desired.

These session-like functions exist in both the session and transport layers because one long-lasting session between a client and server can consist of multiple, shorter duration TCP connections that are effectively subsessions. While TCP maintains these subsessions, ensuring that individual packets (a.k.a. datagrams) are delivered, the session layer maintains the overall continuity of the connection, hiding the underlying discontinuities from the user. For instance, when an individual connects to a remote file server and establishes an NFS or NetBIOS session, he/she can come back to this connection several hours later and still access the remote server even though the original TCP connection was terminated long ago and a new TCP connection must be established.

The network and transport layers are ripe with digital evidence. This is largely because these layers play such an important role in internetworking. Addresses on the network layer (e.g. IP addresses) are used to identify hosts and direct information. Technically proficient criminals can alter this addressing and routing information to intercept or misdirect information, break into computers, hide their location (by using someone else's IP address), or

Preview (Chapter 16): It is not especially difficult to access the physical layer and eavesdrop on network traffic. One method of eavesdropping is to gain physical access to network cables and use specially designed eavesdropping equipment. However, it is much easier to gain access to a computer attached to a network and use that host to eavesdrop. With the proper access privileges and software, a curious individual can listen into all traffic on a network. Computer intruders often break into computer systems and run programs called sniffers to gather information. Also, employees can run sniffers on their computers, allowing them to read their co-workers' or employer's e-mail messages, passwords, and anything else that travels over the network.

Preview (Chapter 17): The transport layer is also responsible for keeping track of which application each piece of data is associated with (e.g. part of an e-mail message or Web page). Port numbers are used to help computers determine what application each piece of data is associated with.

just cause general mischief. Conversely, digital investigators can use this addressing information to determine the source of a crime. On Internet Relay Chat (IRC) networks, some criminals shield their IP address, a unique number that identifies the computer being used, to make it more difficult for an investigator to track them down. Another chat network called ICQ purposefully enables their users to hide their IP address to protect their privacy. However, an investigator who is familiar with the network and transport layers can uncover these hidden IP addresses quite easily as described in Chapter 17.

Computer intruders often use programs that access and manipulate the network and transport layers to break into computers. The simple act of gaining unauthorized access to a computer is a crime in most places. However, the serious trouble usually begins after a computer intruder gains access to a host. A malicious intruder might destroy files or use the computer as a jump off point to attack other systems or commit other crimes. There is usually evidence on a computer that can show when an individual has gained unauthorized access. However, clever computer intruders will remove incriminating digital evidence.

It is important to note that many of the activities on the application layer generate log files that contain information associated with the network and transport layers. For example, when an e-mail message is sent or received, the time and the IP address that was used to send the message are often logged in a file. Similarly, when a Web page is viewed, the time and the IP address of the viewer are usually logged. There are many other potential sources of digital evidence relating to the network and transport layers. A clear understanding of these layers can help digital investigators locate and interpret these sources of digital evidence.

14.4.3 SESSION LAYER (LAYER 5)

The session layer coordinates dialog between hosts, establishing, maintaining, managing, and terminating communications. For example, the session layer verifies that the previous instruction sent by an individual has been completed successfully before sending the next instruction. Also, if the connection between two hosts has been lost, the session layer can sometimes reestablish a connection and resume the dialog from the point where it was interrupted.

The clearest implementation of the session layer is Sun's Remote Procedure Call (RPC) system. RPC enables several hosts to operate like a single computer – sharing each other's disks, executing commands on each other's systems, and sharing important system files (e.g. password files). On UNIX, the Network File System (NFS) and Network Information System protocols depend on RPC. Microsoft uses its own RPC system to enable hosts to share resources. Commands like `showmount` on Unix and `nbtstat` on Windows can be used to display information relating to these kinds of sessions provided they are still active. Also, as noted in Chapters 10 and 11 remnants of such sessions can sometimes be found in configuration files and in unallocated space of a hard drive. However, these kinds of sessions are often temporary and it can be difficult to determine later when they were established or used unless an intrusion detection system, such as NetFlow logs, Argus logs, or some other form of network logging mechanism, recorded the activity.

CASE EXAMPLE

An organization feared that a competitor stole intellectual property from one of their Windows file servers but could find no evidence on the system to confirm their suspicions. The Security Event log did show a suspicious remote logon using an Administrator account but the log did not record the intruder's IP address. Also, it was not clear from the Event logs whether the intruder had downloaded the proprietary information. Fortunately, an intrusion detection system had not only recorded the IP address of the intruder but also captured the associated network traffic. This network traffic revealed that the intruder connected from the competitor's network, had used an Administrator account to establish a NetBIOS session with the file server, and had downloaded the proprietary data to a computer.

Given the limited amount of session-related information that persists on computers and networks, it is not covered separately in this text. Instead, digital evidence relating to sessions is presented in the context of other network layers that may record the activity.

14.4.4 PRESENTATION LAYER (LAYER 6)

When necessary, the presentation layer formats and converts data to meet the conventions of the specific computer being used. This reformatting is necessary because not all computers format and present data in the same way. Some computers have different data formats and use different conventions for representing characters (ASCII or EBCDIC). This is analogous to an exclusive restaurant or club that requires men to wear jackets and ties and will provide these items of clothing to those who do not have them to make them "presentable." Without the presentation layer, all computers would have to be designed in exactly the same way to communicate. Rather than design all computers to process data in exactly the same way, presentation layer protocols have been developed to facilitate communication (e.g. OSI's ASN.1 and Sun's XDR). This layer does not have much evidentiary value and will not receive further attention in this text.

14.4.5 APPLICATION LAYER (LAYER 7)

The application layer provides the interface between people and networks, allowing us to exchange e-mail, view Web pages, and utilize many other network services. Without the application layer, we would not be able to access computer networks. Because the application layer is essentially the user interface to computer networks, it is the most widely used layer and so can be awash with evidence of criminal activity. On this layer, e-mail, the Web, Usenet, Chat rooms, and all of the other network applications can facilitate a wide range of crimes. These crimes can include homicide, rape, torture, solicitation of minors, child pornography, stalking, harassment, fraud, espionage, sabotage, theft, privacy violations, and defamation.

It is no secret that there are national and international pedophile rings, so it should be no surprise that these rings use the Internet. Nonetheless, the amount of evidence of child abuse on the Internet and the numbers of pedophile rings using the Internet has astonished the most veteran crime fighters.

CASE EXAMPLE (UNITED STATES v. ROMERO 1999):
Richard Romero was charged with kidnapping a 13-year-old boy with the intent to engage in sexual activity. Romero befriended the boy on the Internet, initially posing as a young boy himself. Romero persuaded the boy to meet him at a Chicago hotel and travel with him to Florida. After the boy's mother alerted police of her son's absence, a taxi driver reported driving Romero and the boy to a bus station and investigators were able to arrest Romero before he and the boy reached their destination. The FBI found child pornography on Romero's computer and evidence to suggest that Romero frequently befriended young boys on the Internet.

In addition to depositing digital evidence on the Internet, recall from Part 2 of this text that many programs leave corresponding traces of network

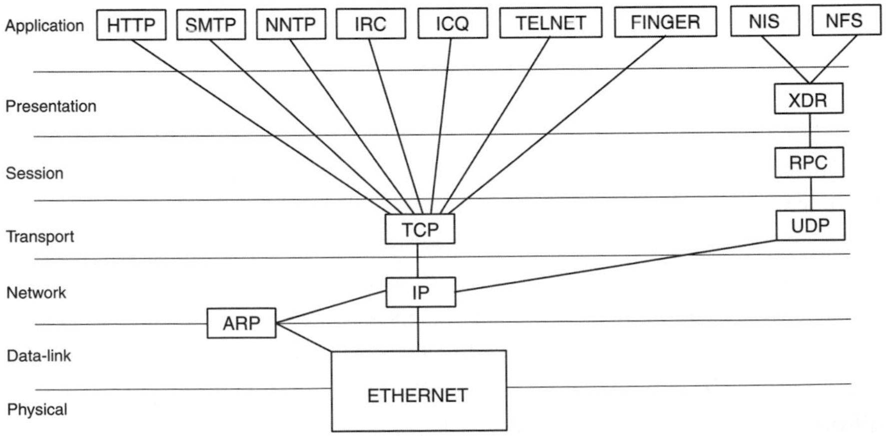

Figure 14.11

Graphical synopsis of the OSI reference model.

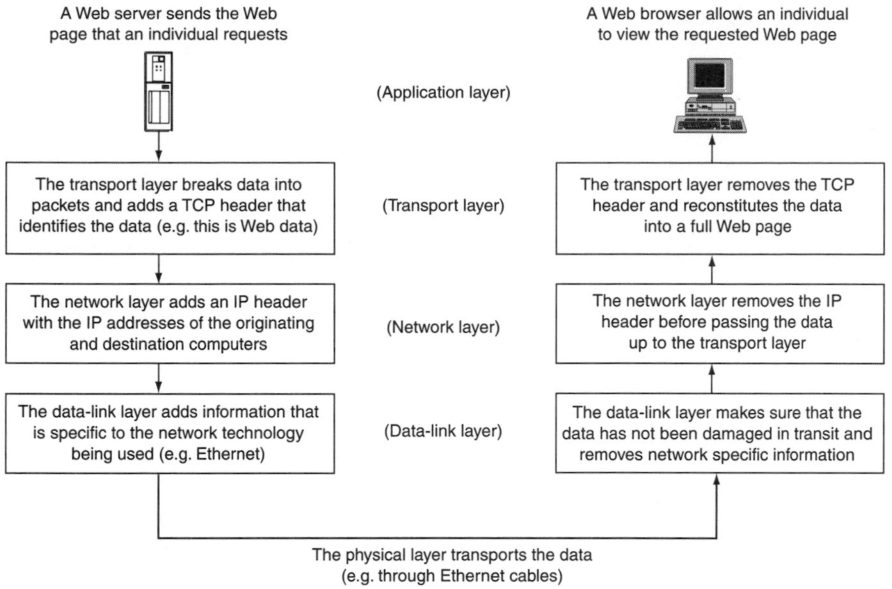

Figure 14.12

How a Web browser accesses the Internet as seen through the OSI model.

activities on personal computers that can point to or be correlated with evidence on the Internet. Web browsers often keep a record of all Web pages visited and temporary copies of materials that were viewed recently. Some e-mail applications retain copies of messages after they are deleted. The process of analyzing common forms of digital evidence on the Internet is covered in Chapter 18.

There are many other Internet applications each with their own investigative and evidentiary challenges and benefits. For example, Hotline Server is a very compact program that enables individuals to turn their personal computers into servers that provide a variety of services including file transfer and chat. Using a Hotline Client, anyone on the Internet can connect directly to a host running the Hotline server to upload or download files. Access to a Hotline Server can be password restricted. This is very similar to a Bulletin Board System (BBS) but is much easier to use. There is currently no reliable way to find Hotline Servers that people want to keep secret – and this makes it more difficult to detect illegal activity. Also, because no central servers are involved, the only evidence of a crime is on the individual computers involved. Fortunately, the Hotline Server can keep a record of every IP address that connects to the server, and every file that is downloaded or uploaded will be noted. This can be a useful source of digital evidence. One should look carefully at every new computer application encountered to determine what kind of digital evidence it can provide as described in Chapter 10.

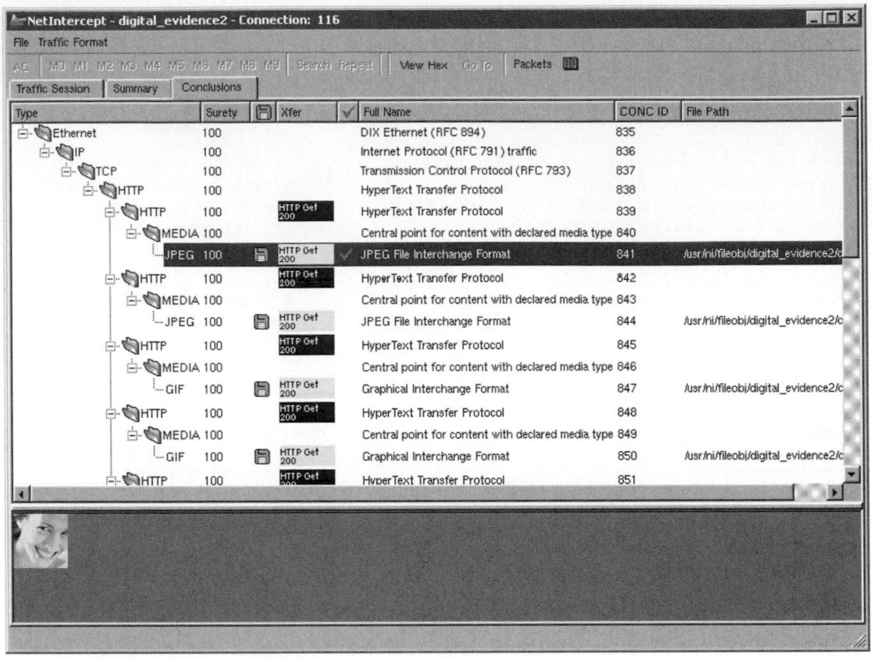

Figure 14.13

NetIntercept (http://www.sandstorm.com) showing components of a Web page both in OSI layers and content recovered from network traffic.

14.4.6 SYNOPSIS OF THE OSI REFERENCE MODEL

Figure 14.11 shows how various things fit into the OSI reference model. We can see how the OSI model applies to the Internet by looking at how a Web browser accesses the Internet (Figure 14.12).

Tools such as NetIntercept can be used to capture network traffic and extract portions for analysis such as the Web page in Figure 14.13. Note that the right section of the screen displays each layer of the Web page traffic from the Ethernet frame (layers 1 and 2), to the IP datagram (layer 3), TCP header (layer 4), HTTP portion (layer 7), and ultimately the contents of the Web page itself.

14.5 SUMMARY

Without an understanding of where information can be found on networks, digital investigators are guaranteed to waste a significant amount of time and are likely to lose valuable digital evidence. Additionally, without an understanding of how networks function, digital investigators will have a harder time making sense of any data they obtain from a network. To address this need, Chapters 16–18 cover three important layers of the OSI model. Chapter 16 details Ethernet and provides guidance for processing digital evidence at the physical and data-link layers. Chapter 17 covers the basics of TCP/IP and describes how digital investigators can process and utilize log files, state tables, and other data relating to the network and transport layers. Chapter 18 discusses the Internet as a source of evidence and addresses key challenges, including anonymity.

REFERENCES

Baran P. (1964) Introduction to Distributed Communications Networks, RM-3420-PR. Santa Barbara, CA: The Rand Corporation (Available online at http://www.rand.org/publications/RM/RM3420/)

Clark D. F. and Gibbs K. E. (2001) Wireless Network Analysis, Casey, E. (editor) *Handbook of Computer Crime Investigation*, London: Academic Press

Convery S. (2002) "Hacking Layer 2: Fun with Ethernet Switches" BlackHat Briefing. (Available online at http://www.blackhat.com/presentations/bh-usa-02/bh-us-02-convery-switches.pdf)

Gauis (2000) "Things to do in Ciscoland when you're dead" Phrack 56 (Available online at http://www.phrack.com/show.php?p=56&a=10)

National Computer Security Association (1997) Internet Security: Professional Reference. Indiana: New Riders Publishing.

Snipe S. (2000) "Why your switched network isn't secure" (Available online at http://www.sans.org/resources/idfaq/switched_network.php)

Sterling B. (1993) A Short History of the Internet. Cornwall, CT: Magazine of Fantasy and Science Fiction. (Available online at http://www.library.yale.edu/div/instruct/internet/history.htm)

CASES

U.S. v. Romero (1999) Appeals Court, 7th Circuit (189 F.3rd. 576), Case number 96 CR 167-1 (Available online at: http://www.laws.lp.findlaw.com/7th/982358.html)

CHAPTER 15

APPLYING FORENSIC SCIENCE TO NETWORKS

Like computers, networks contain digital evidence that can be used to establish that a crime has been committed, determine how a crime was committed, provide investigative leads, reveal links between an offender and victim, disprove or support witness statements, and identify likely suspects. For instance, several hours after the Columbia Space Shuttle crash in 2003, it became evident that a crime was being committed when pieces of the spacecraft were being offered for sale on E-bay. A missing person's e-mail has provided a link between the victim and offender, revealing where she went and who she arranged to meet. Child pornography posted on the Internet has led investigators to victims who were being abused by a family member without the knowledge of other family members, neighbors, or others close to the family. Web proxy logs have been used to demonstrate that an offender took precautions to conceal his illegal activities, shedding doubt on his claims that he did not know what he was doing was wrong. When someone witnesses an unknown offender making a call from his/her mobile phone, it may be possible to obtain records from local base stations for that time period and determine who made calls from the region, thus narrowing the suspect pool.

Processing a hard drive for evidence is a relatively well-defined procedure. When dealing with evidence on a network, however, digital investigators face a number of unpredictable challenges. Data on networked systems are dynamic and volatile, making it difficult to take a snapshot of a network at any given instant. Unlike a single computer, it is rarely feasible to shut a network down because digital investigators often have a responsibility to secure evidence with minimal disruption to business operations that rely on the network. Besides, shutting down a network will result in the destruction of most of the digital evidence it contains. Also, given the diversity of network technologies and components, it is often necessary to apply best evidence collection techniques in unfamiliar contexts.

Additionally, unlike crime in the physical world, a criminal can be several places on a network at any given time. This distribution of criminal activity and associated digital evidence makes it difficult to isolate a crime scene.

Digital Evidence and Computer Crime Second Edition
ISBN: 0-12-163104-4

At the same time, having evidence distributed on many computers can be an advantage in an investigation. The distribution of information makes it difficult to destroy digital evidence. If digital evidence is destroyed on one computer, a copy can often be found on various computers around the network or on backup tapes. Many organizations backup their information regularly and some even store a second copy of all backups in a different location for added protection.

With some adaptation, the methodical approach to processing evidence described in Chapters 4 and 5, and expounded in Chapter 9 can be applied to digital evidence on networks. The initial process of discovery, preparation, and authorization are similar with some added legal and technical complexities. Also, searching for sources of digital evidence on networks requires us to expand the search envelope while maintaining focus and often leads to types of data that require specialized expertise to collect. The general concepts of documentation, collection, and preservation apply to networks but require some adaptation to accommodate different technologies and unique properties of networks.

Although the general analysis techniques described in Chapter 9 (e.g. classification, comparison, individualization) are applicable, analyzing digital evidence from networks often requires specialized knowledge of tools and the underlying network technology. Presenting the resulting findings to non-technical individuals can be challenging but remains one of the most important stages in a forensic examination because an examiner's findings will likely remain unused if they are not understood. This chapter addresses each of these stages in turn, elaborating on how they apply to evidence on computer networks.

15.1 PREPARATION AND AUTHORIZATION

In some cases, digital evidence exists on networks that were not directly involved in a crime and the network administrators are cooperative, often helping digital investigators obtain evidence. Some system administrators even capture useful data routinely to detect and resolve performance and security problems, effectively collecting evidence proactively. However, this proactive evidence gathering might not meet the standards for legal action and digital investigators may need to perform additional steps to preserve this data as evidence. Additionally, there are often more sources of digital evidence on a network than even the system administrators realize. Therefore, to ensure that all relevant data is located, digital investigators must use their understanding of networks in general thoroughly to query system administrators and clearly communicate what types of digital evidence are needed.

It is unreasonable to expect everyone to invest in digital evidence collection systems for their networks. This is the equivalent of expecting all victims of burglary to have an alarm system and surveillance camera in their homes. While victims and incidentally involved parties should not be penalized for lack of preparedness, everyone should be informed of the need to preserve digital evidence after an incident.

CASE EXAMPLE

The alibi of a prime suspect in a homicide case depended on his employer's network. Unfortunately, system administrators who assisted investigators did not know about an administrative console that contained key digital evidence and failed to preserve it promptly. By the time the suspect pointed out the console, it was too late – he was accused of fabricating digital evidence on the console after the fact to support his alibi. If the investigators in this case had not relied on the system administrators' incomplete knowledge of their network, the suspect probably would not be in jail today.

When system administrator cooperation is not forthcoming, digital investigators have to gather intelligence themselves about the target systems before obtaining authorization to seize evidence. For instance, when a Web site is under investigation, it is necessary to determine where the Web servers are located before obtaining authorization to seize the systems. Additionally, it is useful for digital investigators to know what kinds of computers to expect so that they can bring the necessary tools. Digital investigators might also want to copy as much of the material from the Web site as possible prior to the search to demonstrate probable cause or as a precautionary measure.

Collecting digital evidence from a large network requires significant planning, particularly when the administrators are not cooperative. Obtaining information about the target systems prior to the actual search can be a time consuming process.

CASE EXAMPLE

In the investigation of the Starnet online casino, Canadian law enforcement gathered a significant amount of information about the target systems before executing a search warrant. Based on their findings, investigators determined that they needed additional people to assist with the operation and pulled in dozens of agents from the surrounding region. This research and planning enabled them to seize all of the target systems in a matter of minutes.

The process of gathering information about a network can involve reviewing purchase orders, studying security audit reports, scanning the system remotely, and examining e-mail headers, searching the Web, Usenet, DNS, and other Internet resources for revealing details.

> On a practical level, agents may take various approaches to learning about a targeted computer network. In some cases, agents can interview the system administrator of the targeted network (sometimes in an undercover capacity), and obtain all or most of the information the technical specialist needs to plan and execute the search. When this is impossible or dangerous, more piecemeal strategies may prove effective. For example, agents sometimes conduct on-site visits (often undercover) that at least reveal some elements of the hardware involved. A useful source of information for networks connected to the Internet is the Internet itself. It is often possible for members of the

> public to use network queries to determine the operating system, machines, and general layout of a targeted network connected to the Internet (although it may set off alarms at the target network). (USDOJ 2002)

A network vulnerability assessment is a process of identifying weaknesses that could be exploited by computer intruders. Part of this assessment process involves the same tools and techniques used by computer intruders as described in Chapter 19. Tools that gather information by remotely probing computers may cause a firewall or intrusion detection system on the target network to generate an alarm. For instance, if a suspect is using a personal firewall such as Norton Internet Security and Zone Alarms, he/she will receive an alert regarding remote information gathering probes. Additionally, some tools can disrupt systems and should only be used by trained personnel with proper authority. Therefore, before connecting directly to a suspect's system, digital investigators should weigh their need for the information against the risk of alerting the suspect.

This information gathering process is similar to that of network vulnerability assessments, resulting in a list of computers on the network highlighting machines that are likely to contain the most valuable data and summarizing any related information that may be useful for obtaining and analyzing data from the system (Table 15.1).

Before conducting an online investigation, corporate security professionals and law enforcement officers alike should obtain permission to proceed. Even the process of scanning the target system to gather information may create a liability if the target system views this as a malicious attack, particularly if it disrupts their systems. Privacy laws relating to data stored on and transmitted using computers are complex and must be carefully considered to avoid spoiling a case. For instance, a university may not be authorized to probe student or faculty computers for information unless there is a policy that allows such actions under certain circumstances. Law enforcement officers who decide to investigate online child pornography without proper authorization have been accused of illegal activity themselves. Security professionals can only intercept network traffic and review log files without explicit authorization under specific circumstances detailed in privacy legislation. Security professionals can minimize the risk of being criticized for violating a system owner's rights by obtaining written instructions from their attorneys and management. Law enforcement officers can take similar measures to protect themselves legally and professionally.

Once likely sources of digital evidence have been identified, it is often necessary to deploy several groups to preserve everything in a timely manner. Without a clear procedure, there is likelihood that each group will collect evidence differently. Therefore, it is advisable to rehearse likely scenarios and develop a detailed plan with associated checklists, logic diagrams, and customized programs or scripts to maintain consistency and even use two-way radios to maintain communication during the collection process.

As noted in Chapter 3, the difficulty in obtaining authorization to search e-mail, network communications, and other data on networks varies depending on the situation, the country, the type of data, and who is collecting it. In the United States, getting authorization to search recent or unread e-mail is more difficult than old e-mail because of the higher degree of invasiveness. Monitoring network traffic is even more invasive, requiring very strong justification before a court will permit it. In fact, law enforcement may have to demonstrate that they have exhausted all other possibilities before a search warrant will be granted. However, system administrators are permitted to

IP ADDRESS	HOSTNAME	FUNCTION	DIGITAL EVIDENCE	TYPE/VERSION	PRIORITY	NOTES
192.168.1.32	mail.corpX.com	SMTP/POP/IMAP	Suspect's e-mail content, logs, backup tapes, syslogs	Solaris 8	3	Too large to copy entire disk. Just copy e-mail logs
192.168.1.33	dc1.corpX.com	Domain controller	NT Event, IAS, and IIS logs	Windows 2000	3	
192.168.1.34	www.corpX.com	WWW, shell	Web and shell access logs, syslogs, config files	Redhat Linux 8	3	Web access logs in /data /logs
192.168.1.42	ids.corpX.com	Snort IDS	Snort logs and configuration files, syslogs and system config files/details	FreeBSD 5	2	Logs backed up daily to compact disk
192.168.1.45	flow.corpX.com	NetFlow Collectoror	NetFlow logs in raw and text format	Solaris 8	2	Also stored in Oracle database to facilitate searching
192.168.52.23	srv1.corpX.com	File server	Bitstream copy of disk	Windows NT 4	1	
192.168.98.34	wks34.corpX.com	Suspect's Workstation	Bitstream copy of disk	Windows NT 4	1	

Table 15.1

Sample chart created in preparation for acquiring digital evidence from a small corporate network.

monitor traffic on their network when this is necessary to protect the network and data it contains.

When seeking authorization to search a network and digital evidence that may exist in more than one jurisdiction, it is advisable to obtain a search warrant for each location whenever possible.

> When agents can learn prior to the search that some or all of the data described by the warrant is stored remotely from where the agents will execute the search, the best course of action depends upon where the remotely stored data is located. When the data is stored remotely in two or more different places within the United States and its territories, agents should obtain additional warrants for each location where the data resides to ensure compliance with a strict reading of Rule 41(a).
> (USDOJ 2002)

Also, using passwords obtained during investigation to access remote sources of digital evidence usually requires additional authorization. This issue becomes more complex when dealing with different countries. In 2002, legal action was brought against an investigator for gaining remote, unauthorized access to a suspect's computer and collecting evidence over the Internet.

CASE EXAMPLE (SEATTLE 2000):
The FBI successfully prosecuted two Russian computer intruders, Aleksey Ivanov and Gorshkov, for breaking into a number of e-commerce sites in the United States. The FBI lured Ivanov and Gorshkov to the United States for a fictitious job

interview and used Winwhatwhere to capture passwords to the suspects' systems in Russia. Investigators used the passwords to collect incriminating evidence remotely from the suspects' computers. As a result of this action, the Russian government initiated criminal proceedings against one FBI agent for unauthorized access to computers in Russia.

When drawing up an affidavit for a warrant, it is important to specifically mention all desired digital evidence. Without specificity, a search warrant may miss important evidence or might just as easily be overly broad if it authorizes the search and seizure of evidence that is not supported by probable cause. It often helps to speak with the operators of the system involved to determine what types of systems and information they have. If this is not possible, it is generally acceptable to request a range of information provided limiting language is used to specify the crime, the suspects, and relevant time period. It is also recommended to include explicit examples of the records to be seized and indicate that the records may be seized in any form, including digital and paper. An example of such a request is provided here:

> All records associated with the subscriber and account, including screen name(s) and/or account name(s), phone number(s), address(es), credit card numbers used to establish the account, connection records, to include logon dates and times, IP address assigned for each session, origination information for each call, phone number used for access to the system, newsgroups logs, e-mail logs, quantity of local storage provided and percentage utilized (non content information), credit, and billing information for any and all accounts held in the name of John Doe and the address(s) 192.168.12.14, 192.168.12.16, and john.doe@home.com, for the period of (insert date and time covered as nearly as possible and limited to the period of suspected criminal activity). Furthermore, company policy and activities pertaining to the frequency of backup operations and retention periods of information requested herein. The term "records" includes all of the foregoing items of evidence in whatever form and by whatever means they may have been created or stored.

There are two nuances in this example that deserve emphasis. First, e-mail content is not requested, thus avoiding the privacy issues related to stored personal communications, making it easier to obtain a search warrant. Investigators may be able to obtain a significant amount of information quickly and with relative ease by making this clear distinction between subscriber information and the contents of the individual's account. Some organizations, such as E-bay, can even provide law enforcement with certain information about their users (e.g. name, address) without a court order because their user agreement permits such disclosure. Second, note that log files and "origination information for each call" are included in this sample request. The "origination information for each call" generally refers to the fact that some ISPs have Automatic Number Identification (ANI) on their dial-up modem banks, thus enabling digital investigators to trace a connection back to a very specific location (e.g. house, apartment, room).

In large fraud cases in which a network was used to store relevant documents, it might be argued that only the documents were relevant and that investigators should not have be authorized to search log files or other sources of evidence on the network. This argument does not take into account the need for multiple independent sources of digital evidence to corroborate important events and to establish the continuity of offense. Investigators can expect to have their work challenged in court, but can expect reasonable results provided they follow the rules. In one case, the defendant argued that investigators should have been present when a major Internet Service Provider collected digital evidence in response to a search warrant.

CASE EXAMPLE (BACH v. MINNESOTA 2002):
Accused of possessing child pornography, Bach argued that his Fourth Amendment rights were violated because a law enforcement officer was not present when his Internet Service Provider (Yahoo!) collected information relating to his account on their system. Initially, the district court agreed that the warrant was executed outside the presence of a police officer when Yahoo! employees seized e-mail from Yahoo!'s servers in violation of 18 U.S.C. § 3105 and sections 626.13 and 626A.06 of the Minnesota Statutes, and thus the Fourth Amendment.

Sergeant Schaub investigated this incident, discovered that "dlbch15" was Bach and that he had been convicted of criminal sexual conduct in 1996. Eventually, Schaub obtained a state search warrant to retrieve from Yahoo! e-mails between the defendant and possible victims of criminal sexual conduct, as well as the Internet Protocol addresses connected to his account. Both the warrant itself and Schaub's affidavit indicated that the warrant could be faxed to Yahoo! in compliance with section 1524.2 of the California Penal Code. Schaub faxed the signed warrant to Yahoo!. Yahoo! technicians retrieved all of the information from Bach's account at dlbch15@yahoo.com and AM's Yahoo! e-mail account. According to Yahoo!, when executing warrants, technicians do not selectively choose or review the contents of the named account. The information retrieved from Bach and AM's accounts was either loaded onto a zip disc or printed and sent to Schaub. E-mails recovered from Bach's account detail him exchanging pictures with other boys and meeting with them. One e-mail contained a picture of a naked boy. The information retrieved from Yahoo! also included Bach's address, date of birth, telephone number, and other screen names.

Investigators then obtained a search warrant for Bach's house, where they seized a computer, disks, a digital camera, and evidence of child pornography. Based on this information, and the information obtained from Yahoo!, Bach was indicted for possession, transmission, receipt, and manufacturing of child pornography in violation of 18 U.S.C. §§ 2252A(a) (1) and (2), 2252A(a)(5), 2252A(b)(2), 2252(a)(4), 2252(a)(1) and (2), 2252(b)(2), 2251(a) and (d), and 2253(a). Bach moved to suppress the evidence seized from the execution of both warrants. The district court suppressed the information obtained from the warrant executed by Yahoo! (but not the

> information obtained from the subsequent search of his home) because an officer was not present during Yahoo's execution of the first warrant in violation of 18 U.S.C. § 3105 and sections 626.13 and 626A.06 of the Minnesota Statutes, both of which, according to the district court, codify the Fourth Amendment.

Prosecutors appealed this ruling and the court found that Yahoo!'s execution of the search warrant did not violate Bach's Fourth Amendment rights.

Another defendant unsuccessfully appealed on the grounds that information he provided to AOL was private and should not have been made available to investigators (Cox v. Ohio).

15.2 IDENTIFICATION

Recall that the cybertrail is bi-directional. When dealing with a computer as a source of evidence, the crime scene search generally leads to a connected network and ultimately the Internet. Conversely, when digital investigators find digital evidence on the Internet, their search often leads them through a smaller, private network (e.g. ISP, employer, and home networks) to an individual computer. These search areas are depicted in Figure 15.1 with a dashed line between the Internet and the smaller, private network because the division between the two is not always clearly defined. For example, corporate networks often have internal servers that are used to share information within the organization and these servers are sometimes accessible to employees via the Internet.

Given the amount of information that can exist in any of these areas, it is necessary to have a method of quickly locating systems that contain the most useful digital evidence. The first phase is to seek the end-points and intermediate systems such as switches, routers, and proxies. These systems can contain digital evidence that helps establish the continuity of offense and gain a more complete understanding of the crime. For example, log files on an e-mail server used to send harassing e-mail can provide a more complete view of the harasser's activities than a single message. Additionally, intermediate systems like routers and switches may generate detailed logs of network activity, which leads to the second phase. The second phase is to seek log files that provide an overview of activities on the network, such as packet logs from traffic monitoring systems, traffic logs from Argus probes, NetFlow logs from routers, and alert logs from intrusion detection systems. These network level logs are very useful for determining what occurred and which other systems on the network might be involved. For example, when investigating an intrusion into one computer, network level logs may reveal that the same intruder targeted several other systems. The third phase is to look for supporting systems such

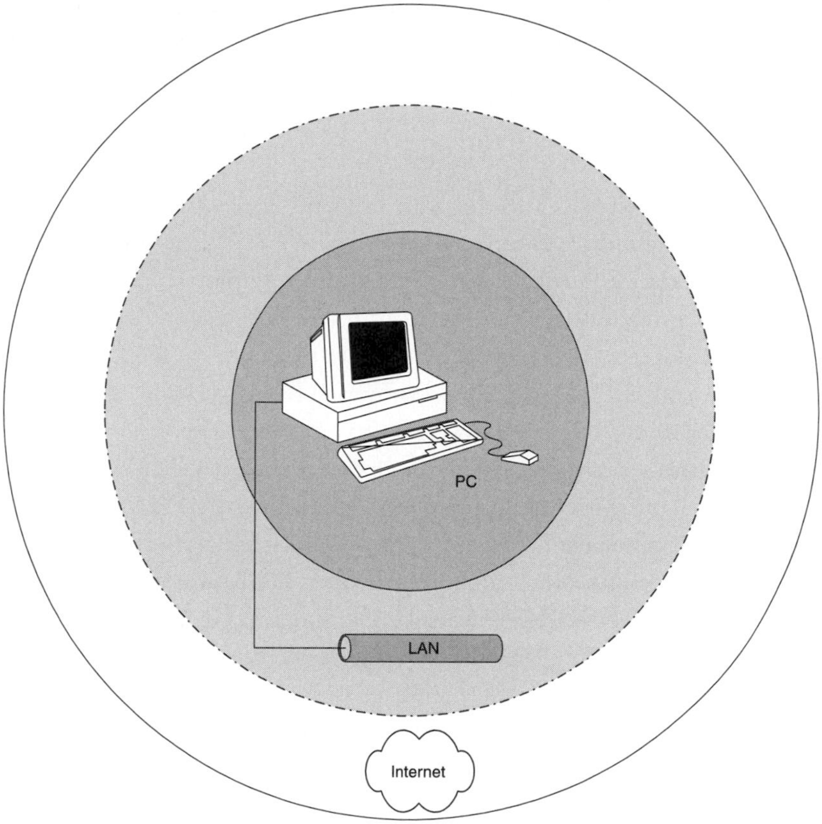

Figure 15.1

Search circles that may contain digital evidence.

as authentication servers and caller-id systems that can help attribute online activities to an individual. In practice, these three phases are conducted simultaneously since, in some instances, the second and third phases may lead to other intermediate system or end-points. This three-phase approach is useful for focusing the search for digital evidence on a network to reconstruct the crime (recall Figure 4.5).

The process of tracking an intruder provides a simple example of following the cybertrail, establishing the continuity of offense, and ultimately apprehending the offender.

CASE EXAMPLE

An investigator examines a compromised machine and determines the source and method of attack. By locating other systems compromised using the same *modus operandi* and by monitoring network traffic to the compromised machines, the investigator determines where the intruder is connecting from. The investigator contacts the ISP, instructs them to preserve the related evidence on their systems, and obtains a search warrant. It transpires that the intruder is using a stolen dial-up account. Fortunately, the ISP has Automatic Number Identification (ANI) information and is able to provide the investigator with the telephone number that the intruder was using to dial into the ISP's modems. This telephone number leads the investigator to the intruder's home. Another search warrant is obtained and the intruder is caught red-handed, logged into compromised systems around the world.

In some cases, a search of an intruder's computer results in more leads and it is necessary to request additional information from telephone companies and ISPs to obtain records to develop a more complete reconstruction of events. For example, all relevant account usage and telephone records can give a more complete view of the intruder's activities.

Preview (Chapter 19): When investigating computer intrusions, it may be desirable to examine a host that is still running to find digital evidence in memory that will be lost when the system is turned off. For instance, active network connections and processes in memory may reveal where the intruder is coming from and what he/she was doing on the system. When performing this type of live host examination, digital evidence should be collected in order of volatility, first preserving data that will change more frequently and then collecting evidence that changes less frequently.

The previous case example demonstrates the time critical nature of this kind of investigation. It may be necessary to analyze evidence immediately to locate other sources of evidence and apprehend an online offender. Having one group collect evidence and another group analyze it immediately is more effective than leaving everything to one individual. However, when an individual is confronted with a choice between collection and analysis, it is best to collect digital evidence carefully first and analyze it later. This issue is complicated when dealing with highly active devices such as routers and dial-up terminal servers because the results of one command often help digital investigators determine what other information to collect from memory, and what command to execute next, requiring simultaneous collection and analysis. This emphasizes the need for standard operating procedures for collecting evidence in such situations. It may not be feasible to have standard operating procedures for all network devices that may be encountered, but the most common ones such as Cisco routers and firewalls can be developed.

The need to correlate multiple sources of evidence and establish continuity of offense to attribute computer intrusions to an individual also applies to other kinds of investigations, including child pornography.

CASE EXAMPLE (UNITED STATES v. HILTON):
The investigator who had examined the defendant's computer was asked to explain his conclusion that pornographic images on the suspect's computer had been downloaded from the Internet. The investigator explained that the files were located in a directory named MIRC (an Internet chat client) and that the date–time stamps of the files coincided with time periods when the defendant was connected to the Internet. The court was satisfied with this explanation and accepted that the files were downloaded from the Internet.

Largely because of the haste required to preserve data on a network and the large amounts of resulting data, digital investigators have made mistakes, implicating the wrong individual. For instance, digital evidence examiners accidentally typed the incorrect time (3:13 P.M. instead of 3:13 A.M.) in a request they sent to AOL, resulting in the wrong subscriber information. In another instance, digital investigators typed the incorrect IP address (192.168.1.45 instead of 192.168.1.54) in a request they sent to Uunet, resulting in the wrong subscriber information. The danger of implicating the wrong individual is compounded when offenders modify digital evidence to misdirect digital investigators. Again, obtaining corroborating evidence from multiple independent sources can mitigate this danger.

Given the expanded search area, potential for mistakes, and wide variety of digital evidence on networks it is necessary to have a methodical approach to searching for evidence on networks. Although it is necessary to follow the cybertrail, connecting the dots to establish the continuity of offense, this is not sufficient to locate sources of evidence that were not directly involved in the commission of a crime but still contain relevant data. For instance, most routers are configured to send their logs to a remote server for permanent storage, making it necessary for investigators to take a slight detour on the cybertrail to collect this useful digital evidence.

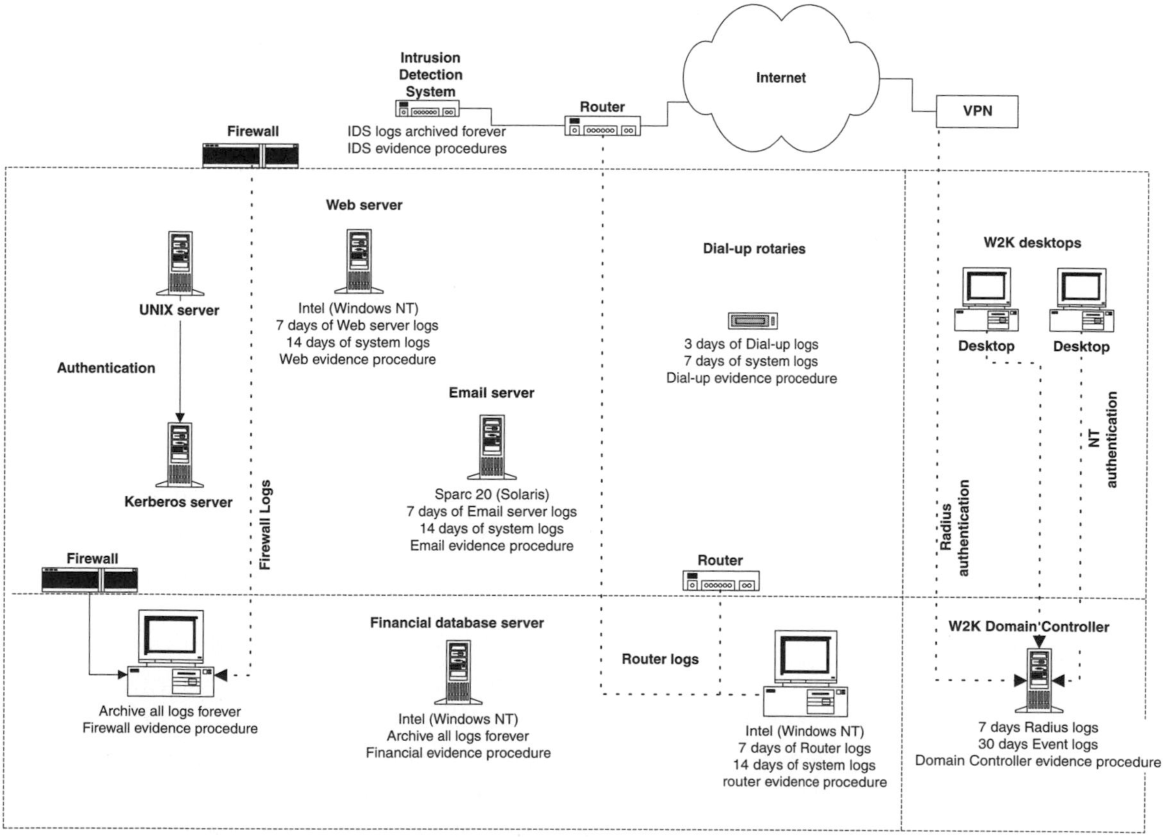

Figure 15.2 Sample digital evidence map.

A graphical depiction of the network and where potential sources of evidence are located – a *digital evidence map* if you will – can greatly facilitate a methodical search. A simplistic digital evidence map is shown in Figure 15.2.

Many organizations have network topology charts showing how the more important network components are connected. Such network charts can be used as a starting point when developing a digital evidence map but digital investigators must be aware that these charts are often outdated (many networks are growing and changing continuously) and are rarely detailed enough for a digital investigator's needs. Therefore, it is important to sit down with the individuals who are familiar with a given network and work with them to develop an accurate, detailed depiction of all relevant systems on a network. Also, information gathered in the preparatory stage of the search (e.g. Table 15.1) can be useful for developing a complete and accurate digital evidence map.

Locating entry points into a network and key servers often leads to the richest sources of digital evidence. Once important servers and network devices are identified, digital investigators can determine what data they retain on disk and in memory, where their logs are stored, and where related configuration files and backups are located.[1] For instance, Cisco firewall and routers are usually configured to send their logs to a remote server for permanent storage and only retain the most recent log entries in memory. However, some information such as the last time the device was rebooted or configured may be stored permanently in memory. Also, system administrators often keep copies of old configuration files and data obtained using administrative and performance monitoring tools that can be useful for determining the past state and operation of network systems.[2]

Before excluding a system as a potential source of evidence, be sure to examine a network component closely before discounting it – important digital evidence can reside in unexpected places. For example, if the routers on a given network only keep logs of anomalies, determine if the anomalies can tell you anything useful. Alternatively, the logs generated by a network component might be of no relevance at all, but the time the network component was last reconfigured could be important. In addition to showing how systems are connected, a digital evidence map should summarize what information can be found at each node on the network, how long the evidence exists, and how it can be obtained (who has the necessary privileges and knowledge to access and collect the evidence). This information enables digital investigators to prioritize, preserving the most volatile, short-lived evidence first (e.g. logs rotated and overwritten once each day).

[1]Keep in mind that additional backup tapes of important systems may be located off-site (e.g. Iron Mountain). Additional time and resources are often required when dealing with backup tapes from large systems (e.g. Tivoli Storage Manager, BrightStor ARCserve Backup) because they use compression and may not have indexes on each tapes, making it more difficult to recover data from them.

[2]Much of this information is obtained through Simple Network Management Protocol (SNMP). If a device has not been queried using SNMP, it can be fruitful to do so before turning the device off.

CASE EXAMPLE

A system administrator who was the prime suspect in a homicide investigation used an IP address that was not officially assigned to him. As a result, searching network logs for traffic from hosts that were officially assigned to him did not result in any useful data, suggesting that the suspect was lying. By the time the error was realized, the network traffic logs had been deleted and overwritten by newer ones and it was not possible to determine if there had been traffic from the unofficial IP address. Use, but do not rely on records that system administrators maintain, and collect full logs.

A digital evidence map might seem like a tedious process with minimal benefits but the effort will pay off the moment you realize that the network contains something you are missing. Without the map, digital investigators might never know that they are missing something or that the network contains what they are missing. Also, rather than shouting "Eureka!" and then running around for hours trying to figure out how to obtain the evidence, you can shout "Eureka!" and run straight to the evidence with the help of your trusty digital evidence map.

15.3 DOCUMENTATION, COLLECTION, AND PRESERVATION

In some instances, it is desirable to preserve digital evidence on a networked system by gaining physical access to the associated computer and making a bitstream copy of the contents using the guidelines provided in Chapter 23. Also, the same procedures are used to preserve loose media and related backup tapes, and collect associated hardware and software needed to read them. The primary differences when dealing with networked systems arise when digital investigators cannot make a bitstream copy of digital evidence.

A bitstream copy may not be viable in some situations because the system cannot be shut down, the hard drive may be too large to copy, or the digital investigator may not have authority to copy the entire drive. Also, digital investigators often rely on large Internet Service Providers to collect evidence from their own systems such as subscriber information. Furthermore, digital investigators may not be able to gain physical access to the system containing evidence, requiring them to collect evidence remotely. Digital investigators also collect digital evidence remotely when there is a strong chance that it will be destroyed before they can reach the machine. For instance, data on the Internet such as Web pages and Usenet messages can be altered or removed at any time and computer intruders often delete log files.

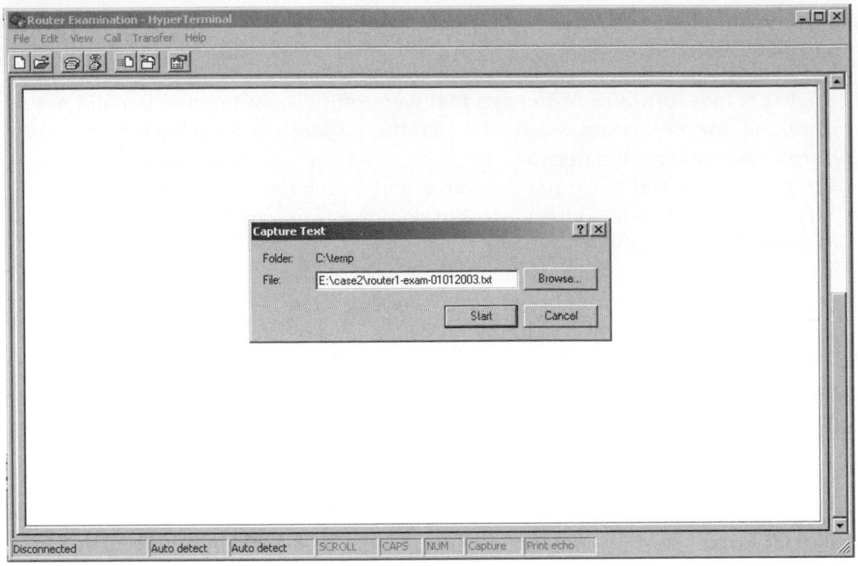

Figure 15.3

HyperTerminal has the capability to record the results of a router examination in a file. The "Capture Text" option is on the "Transfer" menu.

Also, when digital investigators are performing certain tasks, data is only displayed on screen for a moment, making it necessary to preserve the dynamic digital evidence in some way. For example, script on UNIX and the HyperTerminal program available on most Microsoft Windows systems can be used to record the results of an examination of routers, firewalls, and other network devices through a serial cable (Figure 15.3). Also, a second digital investigator observing the collection process can jot down each action and its result while the evidence is being collected. This approach has the added benefit of catching mistakes and making suggestions.

Another example of real time evidence gathering is an IRC chat session in which digital investigators keep a running log of their conversation with a suspect. However, if a significant amount of information is being displayed onscreen it may be desirable to record a visual representation of events. A visual recording can be created using a video camera or a software program that can capture dynamic digital evidence, like a sequence of onscreen events, and can replay them at a later time much like videotape. Notably, these and other programs that are useful for collecting digital evidence do not perform integrity checking and other documentation that can be used to authenticate the data.

In some cases, it is necessary to monitor network traffic in real time to convincingly attribute online activities to an individual and to locate other targets. Many organizations use intrusion detection systems to continuously monitor network traffic and generate alerts when certain patterns occur. Most intrusion detection systems can be configured to capture the network traffic associated with an alert but rarely perform integrity checking on log files or document other system details to help authenticate the data. Therefore, additional measures must be taken to preserve intrusion detection system logs as a source of digital evidence.

When it is not possible to obtain a bitstream copy of digital evidence, digital investigators must creatively employ the principles of preserving digital evidence and establishing chain of custody presented in Chapter 9. For instance, a log file can be preserved by noting the time of the system clock, documenting the file's location and associated metadata (e.g. size, date–time stamps), copying it to a collection disk, calculating its MD5 value, and labeling the collection disk appropriately. If the log is small enough, it can also be printed in paper form, initialed, and dated to provide another form of documentation. Additionally, it is advisable to save a second copy of the log file to a different medium and verify that both copies are readable on another system.

When dealing with network logs, preserving the entire log file rather than individual entries is preferable to only collecting relevant portions because digital investigators may later find that other portions of the log are relevant to the case.

As noted in Chapters 10 and 11, copying a file alters some of its date–time stamps and compressing the files in a TAR or ZIP archive can retain these date–time stamps. However, these archives can become corrupted, making it difficult to extract the original files. Therefore, when collecting individual files from a system, it is advisable to note date–time stamps of files prior to collection, save a copy of the files in an archive to retain their date time stamps, and save copies of the files in uncompressed form to ensure that they are available if the archive is corrupted.

CASE EXAMPLE

In a homicide case, digital investigators collected information from login server relating to the victim's activities but did not collect the entire log file. It was later determined that the offender may have been logged into the server at the same time, allowing them to chat in real time and arrange a meeting an hour later. By the time this was realized, archived copies of the relevant log files had been overwritten (the backup tapes had been reused) and it was not possible to determine who else was accessing the system at the time.

However, some binary log files can only be read using specialized software and just making a copy of the binary file may make analysis more costly and inconvenient. Therefore, in addition to preserving the binary log file, consider saving a copy of the contents in interpreted form. These and other considerations are discussed in more detail in Chapter 17.

A detailed record of the entire collection process should be maintained in digital or written form to help authenticate the resulting copies at a later time. This record should document who collected the evidence, from where, how, when, and why.[3] Given the distributed nature of the Internet and the many potential sources of digital evidence, it can be very challenging to collect even the relatively static digital evidence such as Web pages and Usenet messages. In these simple situations, it may not be possible to obtain the date–time stamps of the associated files on the remote system. Therefore, it is imperative to make every possible effort to document the fact that

[3]These measures help authenticate the log file, but additional information about the system may be needed to determine if the log is complete and accurate. Therefore, if the log file is going to be used in court, make an effort to assess the reliability of the system that created the log file. Additionally, seek evidence from other independent sources that corroborate information in the log file.

evidence was stored on a remote computer, detailing where the original evidence was, when and how it was collected, and by whom. In more complex investigations, it becomes even more challenging to document evidence as it is collected from remote systems.

CASE EXAMPLE

An intruder was caught breaking into a computer system on an organization's network via the Internet. Before disconnecting the system from the network, digital investigators gathered evidence that clearly showed the intruder committing a crime. To achieve the equivalent of a videotape of the crime, digital investigators used a sniffer to monitor network traffic to record all IP packets of the intruder's session. Additionally, they logged into the compromised machine using a client that could keep a log of the session and gathered evidence of the intruder's presence on the system and programs that the intruder was running. In an effort to find related evidence, digital investigators searched neighboring systems (e.g. computers, firewalls, routers, intrusion detection systems) for information relating to the intruder. They found other machines compromised by the same intruder and they connected to those through a backdoor created by the intruder. Because it was not possible to access all of the compromised machines physically and there was a risk that the intruder might destroy evidence on these systems at any moment, digital investigators collected evidence from them remotely. While performing this remote collection, they again used programs that monitored their keystrokes, thus documenting the collection process.

When it is necessary to connect to a computer over a network and collect information about/from the remote system, there are several issues to be aware of, and a few ways to help document the process and demonstrate integrity and authenticity:

- Following a standard operating procedure (reduces mistakes and increases consistency across investigations).
- It is essential to retain a log of actions taken during the collection process and take print screens of important items.
- One must document which server actually contains the data that is being collected because the examiner can be forwarded from one server to a server in another country.
- Calculate the MD5/SHA1 values of all evidence prior to transferring them if possible, and after transferring them from the remote host.
- Consider digitally signing and encrypting the files and saving them to read only media.

In a number of cases, investigators gained remote access to the host that a computer intruder was using to launch attacks and then e-mailed themselves evidence gathered from the remote host. Although this approach is convenient, it complicates the chain of custody, makes it more difficult to confirm the integrity of the digital evidence, and may not work at all if the e-mail is not delivered. Therefore, when collecting evidence from a remote machine, use multiple methods to obtain two or more copies of the evidence. For instance, display the contents of text files on screen so that they are recorded by whatever logging program the examiner is using and transfer files directly from the remote host to a collection system whenever possible.

Ultimately, the measures one takes to preserve digital evidence depend on the type of evidence, the severity of the crime, and the importance of the evidence to the investigation. In some situations, it is sufficient to take print screens and make a copy of information from the Internet. In other situations, like when there are too many files to copy individually, or when the charges are especially serious such as murder, it becomes necessary to seize the entire computer that contains the materials.

For instance, in certain cases, it is possible that someone else was using the suspect's home computer. While actively monitoring the suspect's Internet activities, investigators can simultaneously serve a search warrant on the suspect's house in an effort to catch him/her red-handed. However, it is likely that the suspect's system would contain enough evidence to implicate him/her and active monitoring might only provide corroborating evidence. While such corroborating evidence is useful, active monitoring is time consuming, invasive and costly and should only be used as a last resort when additional corroborating evidence is needed to build a solid case or when this information might reveal other victims or targets.

Most network analysis tools can interpret files in tcpdump format, making it the *de facto* standard. Collecting network traffic also involves special considerations. If the IP address of interest is already known, it is a simple matter to capture network traffic relating only to that computer. However, when a dial-up connection is involved, it is necessary to determine which IP address has been assigned to the account of interest.[4] Similarly, when IP addresses are assigned dynamically to hosts on a network, it may be necessary to monitor traffic from a specific MAC address. In other cases it may be necessary to monitor all traffic on a network. In any case, capturing network traffic can result in large files making it advantageous to start a new file regularly, naming each file uniquely, calculating hash values of each file, and storing files on secure media.

[4]Carnivore can determine which IP address is assigned to the account of interest by monitoring RADIUS authentications in network traffic (IITRI, 2000). Using other tools, it is also possible to monitor TACACS logs to determine which IP address is assigned to the account of interest.

When capturing network traffic, it may be desirable to limit the amount and types of information that is collected. For example, digital investigators may only be authorized to monitor Web traffic. Although network capture tools can be configured to only collect Web traffic, some of these tools assume that certain ports are involved while other tools actually recognize the protocols. Such filtering is made more difficult when protocols resemble each other – some peer-to-peer protocols are based on HTTP and some instant messaging programs try to resemble Web traffic to bypass firewall rules. Therefore, collect first and filter and analyze later whenever possible, and be sure that you know what assumptions the tools are making before narrowing the collection. When it is necessary to filter, take the approach of capturing everything and only excluding what is not required rather than beginning from an exclusionary position and selectively capturing certain traffic.

15.4 FILTERING AND DATA REDUCTION

Investigations involving computers often result in a large amount of data, much of it unrelated to the crime under investigation. Also, when dealing with files containing captured network traffic, there may be privileged or confidential information that forensic examiners are required to ignore or remove. Therefore, data filtering and reduction are an essential part of any investigation involving networks, enabling a more efficient and thorough forensic analysis of the digital evidence.

Filtering out irrelevant data from log files may be as simple as extracting entries that match certain criteria such as a certain time period, an IP address, or failed logon events. For instance, the following output shows only failed logon events relating to the user "eco" extracted from a Windows NT Event Log using ntlast utility.[5]

[5]http://www.foundstone.com

```
C:\>ntlast -f -u eco –file e:\case1\dc2\sec.evt

eco     WORKSTN13     MY-DOMAIN     Sun  Jan 19 11:00:11 am 2003
eco     WORKSTN10     MY-DOMAIN     Wed Jan 15 05:39:39 pm 2003
```

When examining established connections through a Cisco PIX firewall, it may be desirable to focus on one host rather than review every connection;

```
pix01# show conn foreign 192.168.0.232 255.255.255.255
7354 in use, 24529 most used
TCP out 192.168.0.232:3129 in 172.16.1.23:80 idle 0:12:04 Bytes 45235 flags UIO
TCP out 192.168.0.232:3130 in 172.16.1.23:22 idle 0:00:01 Bytes 4395 flags UIO
TCP out 192.168.0.232:3131 in 172.16.1.23:443 idle 0:00:54 Bytes 9935 flags UIO
```

However, this approach to collecting evidence from a firewall violates the recommendation provided in the previous sections – collect first and filter and analyze later. Therefore, it is advisable to display all connections, logging the results into a file, and then searching these results for the entries of interest. As another example of data reduction, the following output shows windump being used to extract data relating to one IP address from a file containing network traffic relating to many computers.

```
E:\case1\networktraffic>windump -r monitor1–01192003.dmp host 64.4.45.7
00:08:07.534671 64.4.45.7.80 > 192.168.1.102.1037: S 1378721726:1378721726(0)
ack 250897286 win 17316 <mss 1322,nop,nop,sackOK>
00:08:07.688663 64.4.45.7.80 > 192.168.1.102.1037: P 1:155(154) ack 338 win 16979
00:08:07.689768 64.4.45.7.80 > 192.168.1.102.1037: F 155:155(0) ack 338 win 16979
00:08:07.839232 64.4.45.7.80 > 192.168.1.102.1037: . ack 339 win 16979
00:08:07.942829 204.60.0.2.53 > 192.168.1.102.1038: 6 1/4/4 A 64.4.45.7 (208) (DF)
00:08:08.067639 64.4.45.7.80 > 192.168.1.102.1039: S 2707800119:2707800119(0)
ack 251070441 win 17316 <mss 1322,nop,nop,sackOK>
00:08:08.240567 64.4.45.7.80 > 192.168.1.102.1039: P 1:435(434) ack 410 win 16907
00:08:08.244832 64.4.45.7.80>192.168.1.102.1039: . 435:971(536) ack 410 win 16907
00:08:08.245727 64.4.45.7.80>192.168.1.102.1039:.971:1073(102) ack 410 win 16907
00:08:08.371354 64.4.45.7.80 > 192.168.1.102.1039: . 1073:1609(536) ack 410 win
<cut for brevity>
```

Most commercial sniffers have the ability to create filters, only displaying packets that match certain criteria. Alternatively, ranking hosts based on the amount of data that they are sending and receiving can reveal one host that is involved in a suspiciously large amount of data transfer as shown in Table 15.2.

SOURCE IP	DESTINATION IP	SOURCE BYTES	DESTINATION BYTES
192.168.0.5	207.68.162.250	49900	230869
192.168.0.5	207.68.162.24	47819	146996
192.168.0.5	65.54.228.250	12212	158032
192.168.0.5	207.68.172.245	12963	48012
192.168.0.5	65.54.208.222	11217	40002
192.168.0.5	208.185.54.22	2304	42975

Table 15.2

Connections between hosts, ordered by total number of application bytes transferred. Data extracted from tcpdump file (available on book Web site) using Argus "ramon -c -A -M Matrix"*. The same summary can be obtained using the NetIntercept "Traffic Load" report (available on the Web site).*

Similarly, viewing the number of connections between hosts may be useful for traffic analysis as shown in Table 15.3.

CONNECTIONS	SOURCE IP ADDRESS	DESTINATION IP ADDRESS
81	192.168.0.5	207.68.162.24
31	192.168.0.5	207.68.162.250
9	192.168.0.5	65.54.228.250
8	192.168.0.5	207.68.177.125
7	192.168.0.5	65.54.208.222

Table 15.3

Communication between hosts, ordered by number of connections. Data extracted from tcpdump file using the NetIntercept "Top N" report (available on book Web site).

15.5 CLASS/INDIVIDUAL CHARACTERISTICS AND EVALUATION OF SOURCE

As networks evolve, they contain an ever increasing number of different types of data, making it difficult for any one person to be familiar with all of them. Fortunately, as with other forms of digital evidence, class characteristics can be used to differentiate Web page from e-mail messages and Web server logs from e-mail server logs. Additionally, class characteristics can reveal which program was used to create a given piece of digital evidence and whether it was created on Windows, Mac OS, or UNIX. Furthermore, digital evidence on networks can contain characteristics, such as IP and MAC addresses, which are effectively individual characteristics in some situations. Together, these class and individual characteristics can be used to evaluate the source of digital evidence on a network.

Header lines in e-mail messages demonstrate how class characteristics, individual characteristics, and evaluation of source are useful when dealing with network related data. The following header indicates that the message was sent from a Mandrake (mdk) Linux machine with an Intel 586 processor running X11 and an e-mail client based on Mozilla version 4.75. If the computer that was assigned IP address 192.168.187.18 can be located, these class characteristics can be used to substantiate the connection to the computer.

```
Return-Path: <harasser@threat.net>
Received: from attack.threat.net (attack.threat.net [192.168.187.18])
      by lsh110.siteprotect.com (8.9.3/8.9.3) with SMTP id MAA21755
      for <eco@corpus-delicti.com>; Wed, 29 Jan 2003 12:38:30 -0600
To: eco@corpus-delicti.com
Date: Wed, 29 Jan 2003 13:32:19 -0500
Message-ID: <1043865139.9860@attack.threat.net>
X-Mailer: Mozilla 4.75 [en] (X11; U; Linux 2.2.17–21mdk i586)
From: harasser@threat.net
Subject: Your Worst Nightmare!
```

Even when this information is fabricated as detailed in Chapter 18, these characteristics can be used to search the Internet or a suspect's computer for messages with the same characteristics. Furthermore, when one employee targets another employee in their organization, computer systems on the organization's network may contain related digital evidence.

Entries in a Web server access log provide another illustrative example of class characteristics and evaluation of source in network related data. The following log entry indicates that the "project21.html" page was accessed from IP address 172.16.1.19 using a Web browser that is based on Mozilla version 4.75, configured to use English (en), running on a Windows 2000 computer.

```
2003-01-23 12:52:40 172.16.1.19 – 192.168.1.3 80 GET /documents/
project21.html – 200 Mozilla/4.75+[en]+(Windows+NT+5.0;+U)
```

Notably, class characteristics such as the Web browser and machine type can be falsified in the Web server request. The following log entries from the same Web server show an intrusion attempt via a well-known vulnerability in Microsoft Internet Information Server (IIS). The variations in Web browser version and computer type (e.g. DigiExt, Compaq) relating to a single source IP address (137.56.97.25) indicate that this information is being fabricated. Although these class characteristics conceal properties of the attacking system, they may reveal which program was used to launch the attack. Comparing these class characteristics with those in various exploit programs may result in a match. The match may be with a certain version of the Nimda worm or, if an individual launched the attack, this information could be used to search the offender's computer to find the tool he/she used.

```
2003-01-23 12:59:02 137.56.97.25 - 192.168.1.3 80 HEAD
/winnt/system32/cmd.exe /c+dir+c:/ 403
Mozilla/4.0+(+compatible;+[fr];+Windows+NT5.0;+athome020+)
2003-01-23 12:59:02 137.56.97.25 - 192.168.1.3 80 HEAD /cgi-
bin/..%5c../..%5c../..%5c../winnt/system32/cmd.exe /c+dir+c:/ 403
Mozilla/4.7+(+compatible;+MSIE+5.0;+AOL+5.0;+DigiExt+)
2003-01-23 12:59:02 137.56.97.25 - 192.168.1.3 80 HEAD
/msadc/..%2f..%2f..%2f..%2fwinnt/system32/cmd.exe /c+dir+c:/ 500
Mozilla/4.0+(+compatible;+[fr];+Windows+NT5.0;+DigiExt+)
2003-01-23 12:59:02 137.56.97.25 - 192.168.1.3 80 HEAD
/msadc/..à/€/à/€/à/€/¯../winnt/system32/cmd.exe /c/+dir+c:/ 404
Mozilla/4.7+(+compatible;+MSIE+5.0;+Windows+NT5.0;+Compaq+)
```

The impressions that buffer overflows leave on a system provide another illustrative example of class characteristics and evaluation of source in network related data. A buffer overflow is a common approach to breaking into computer systems. When a program fails to limit the length of an input value, it may be possible to give the program a larger than expected input value that causes it to write the extraneous information into the computer's memory. By carefully constructing the unexpectedly large input value, this weakness in the program can be exploited to cause the computer to execute commands and give an intruder access to the system. For instance, the following fragment of a log file recovered from a compromised host indicates that the attack was launched from IP address 192.168.1.231 and exploited a vulnerability in the FTP server.

```
Jan 24 17:07:22 target ftpd[567]: FTP session closed
Jan 25 00:21:54 target ftpd[576]: ANONYMOUS FTP LOGIN FROM
attacker.corpX.com [192.168.1.231],
□□□□□□□□□□□□□□□□□□□□□□□□□□□□□□□□□□□□□□□□□□□□□□□□□□□□□□□□□□□□□□□□□□□□□□
□□□□□□□□□□□□□□□□□□□□□□□□□□□□□□□□□□□□□□□□□□□□□□□□□□□□□□□□□□□□□□□□□□□□□□
□□□□□□□□□□□□□□□□□□□□□□□□□□□□□□□□□□□□□□□□□□□□□□□□□□□□□□□□□□□□□□□□□□□□□□
□□□□□□□□□□□□□□□□□□□□□□□□□□□□□□□□□□□□□□□□□□□□□□□□□□□□□□□□□□□□□□□□□□□□□□
□□□□□□□□□□□□□□□□□□□□□□□□□□□□□□□□□□□□□□□□□□□□□□□□□□□□1À1Û1É°FÍ□1À1ÛC‰ÙA
°?Í□ëk^1À1É□^^A¨F^DÍ¹ÿ^A°'Í□1À□^^A°=Í□1À1Û□^^H‰C^B1ÉþÉ1À□^^H°^LÍ□þÉuó1À^F
^I□^^H°=Í□þ^N°0þÈ^F^D1À^F^G‰v^H‰F^L‰ó□N^H□V^L°^KÍ□1À1Û°^AÍ□è□ÿÿÿ0bin0sh1.
.11
Jan 24 17:22:54 target inetd[448]: pid 576: exit status 1
```

Although intruders can use fake source IP addresses in packets when they do not require a response from the target system, the source IP address in this instance (192.168.1.231) could not be forged because this exploit uses TCP to return a command prompt to the intruder. Searching for this IP address in intrusion detection system logs and other network logs detailed in Chapter 17 may reveal other intrusion attempts. Examining other targeted systems for deleted log fragments similar to the one above may help identify other compromised systems. Additionally, if the intruder's personal computer can be obtained and a program for exploiting FTP servers is found, it can be compared to determine if it is consistent with the above log entry.

In addition to helping evaluate the source of an event, log files can contain class characteristics that are useful for determining which tools were used – similar to toolmark analysis in the physical world. When digital evidence examiners have difficulty determining what tool was used, they may find exemplars for comparison on the Internet, particularly on information security mailing lists. On mailing lists like Bugtraq,[6] information security professionals submit samples of log files associated with certain tools to help others detect attacks.

[6]*http://www.securityforcus.com*

Useful class characteristics can also be found in TCP/IP network traffic. In fact, signature-based intrusion detection systems rely on characteristics of network traffic to classify attacks. For instance, Snort[7] detects successful attacks against IIS Web servers by looking for packets from port 80 containing the term "Volume Serial Number," indicating a successful directory listing via the vulnerable Web server. The resulting intrusion detection system alert shown here contains the date, time, IP addresses, and other information about the packet discussed in Chapter 17.

[7]*http:// www.snort.org*

```
[**] [1:1292:1] ATTACK RESPONSES http dir listing [**]
01/23-12:59:02.865832 192.168.1.3:80 -> 137.56.97.25:25587
TCP TTL:127 TOS:0x0 ID:8817 IpLen:20 DgmLen:243 DF
***AP*** Seq: 0x5E3A36C3 Ack: 0x58C4137F Win: 0x4313 TcpLen: 32
TCP Options (3) => NOP NOP TS: 16339694 242252
```

Similarly, Snort detects network traffic that may be associated with the DeepThroat Trojan horse program by looking for packets from port 2140 containing the sentence "Ahhhh My Mouth Is Open." Signature-based intrusion detection systems are flexible enough to be useful in a wide variety of investigations, not just computer intrusions.

CASE EXAMPLE

Someone in the organization was apparently using a shared computer to view pornographic Web sites. The default page displayed by the Web browser on the shared machine was set to a pornographic site that another employee was directed to and found offensive. The offended employee filed a sexual harassment complaint with Human Resources and an investigation was opened. Although an examination of the machine confirmed that it was used to view pornographic Web sites regularly, it was not clear who was responsible. In an effort to catch the person responsible in the act of viewing pornography from that machine, the organization's main intrusion detection system was reconfigured to alert the investigator when specific sites were accessed from that machine. That afternoon, the intrusion detection system sent several alert messages to the investigator and he was able to walk over to the responsible individual and resolve the problem with the assistance of Human Resources and the individual's supervisor.

In addition to detecting specific words in a packet, intrusion detection systems can be configured to look for other kinds of class characteristics, including items in the TCP/IP header and sequences of bytes in the payload. For instance, Snort uses the following internal rule to detect possible buffer overflow attempts targeting UNIX printer daemons, examining all packets to port 515 for a pattern of bytes that is associated with a known exploitation of this vulnerability shown in bold.

```
alert tcp $EXTERNAL_NET any -> $HOME_NET 515 (msg:"EXPLOIT LPRng
overflow"; flow:to_server,established; content: "|43 07 89 5B 08 8D 4B 08 89 43
0C B0 0B CD 80 31 C0 FE C0 CD 80 E8 94 FF FF FF 2F 62 69 6E 2F 73 68 0A|";
reference:cve,CVE-2000-0917; reference:bugtraq,1712; classtype:
attempted-admin; sid:301; rev:4;)
```

Notably, this intrusion detection system alert only indicates an intrusion attempt via the LPRng printer daemon – the target system may have a newer version of the software that is not vulnerable to this attack. In fact, any of these intrusion detection system alerts may be a false alarm (a.k.a. false positive), triggered by an innocent packet that coincidentally contains the class characteristics that Snort is looking for. Therefore, further investigation is required to confirm that an attack actually occurred and that the attack was successful at gaining unauthorized access to the target host.

The popular port scanner called nmap also uses class characteristics in TCP/IP packets returned by a host to determine its operating system (Fyodor 1998).

```
C:\> nmap -sS -PT -PI -O -T 3 192.168.0.2
Starting nmap V. 3.00 ( www.insecure.org/nmap )
Interesting ports on HOST101 (192.168.0.2):
(The 1600 ports scanned but not shown below are in state: closed)
Port        State    Service
139/tcp     open     netbios-ssn
Remote operating system guess: Windows Millennium Edition (Me), Win 2000,
or WinXP
Nmap run completed -- 1 IP address (1 host up) scanned in 2 seconds
```

The class characteristics of network traffic for different TCP/IP stacks that are usually associated with particular operating systems (a.k.a. OS fingerprints) are contained in the nmap-os-fingerprints file that is installed with the nmap software. If the meaning or significance of a class characteristic is not clear, it may be necessary to experiment.

Investigators can also use class characteristics to better understand unusual packets that were specifically created to cause computers to crash. Determining how these packets differ from regular ones can help investigators to understand what is happening. The characteristics of these packets can also be used to determine which tool was used. If the same type of uniquely fabricated packet is used to crash several Web servers in an organization – the likelihood is that the same individual is responsible for all of the incidents. Knowing that a single individual is targeting certain Web servers may provide some insight into the motivation of the offender that would not have been possible without the linkage.

15.6 EVIDENCE RECOVERY

Recovering digital evidence such as deleted system or network log files from a server involves the techniques provided in Part 2 of this text. Deleted system log fragments can be found in unallocated space by searching for characteristics such as the date or message fields (e.g. "Mar 3," "LOGIN"). Also, it may be possible to repair corrupt UNIX "wtmp" log files or NT Event log files or at least extract some useful information from uncorrupted portions. Notably, it is possible for the "wtmp" file to become corrupted in a way that is not obvious and, when processed uncritically, can associate the wrong user account with the wrong connection. This emphasizes the importance of verifying important log entries before using them to form conclusions.

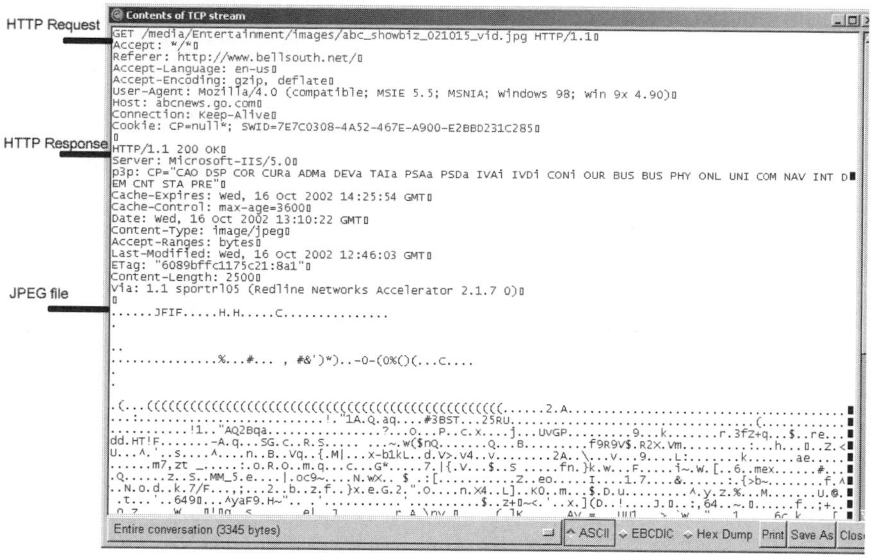

Figure 15.4

Ethereal (www.ethereal.com) used to reconstruct a TCP Stream relating to one component of a Web page being downloaded.

It is also be possible to recover digital evidence from network traffic. Network traffic relating to a single machine may contain e-mail communications, downloaded files, Web pages viewed, and much more. Interesting items can be recovered from network traffic by extracting individual packets and combining them. For instance, Figure 15.4 shows a network sniffer called Ethereal being used to reconstruct a TCP stream and display the contents of the communication. In this instance, the connection was a request to a Web server for a JPEG image. In this process of reconstruction, Ethereal takes data collected on the physical layer, extracts only the relevant packets from the transport and network layers, and displays the application layer protocol; a HTTP GET request for one image on a Web page.

Ethereal was not designed with evidence collection in mind but it is still useful for examining network traffic. The "Save As" option at the bottom right of the screen can be used to save the data to a file that can be opened with a Web browser, image viewer, or some other suitable software. However, the resulting exported file often contains data that prevent other programs from displaying the file correctly (such as the HTTP request data in Figure 15.4). Although this gives a sense of what communication was occurring, it does not show data as the user saw them.

Other tools for examining network traffic can reconstruct and display files from packets in network traffic more effectively. For instance, NetIntercept provides an images view that arranges all graphics files extracted from network traffic in a gallery or thumbnail arrangement, allowing digital evidence examiners to view them more efficiently. NetIntercept and similar tools can also reconstruct Web pages, enabling digital evidence examiners to view pages as the user saw them, as discussed in Chapter 16. Different network traffic analysis tools can reconstruct and display different types of data including e-mail, FTP, and Instant Messenger with varying degrees of success. So, when an individual downloads a compressed file from an FTP server or IRC, it may be desirable to recover this file from a network capture and examine its contents. Certain data formats are harder to reconstruct from network traffic, requiring special purpose tools. For instance, Review has a module for interpreting and displaying X sessions as detailed in Chapter 4 of the Handbook of Computer Crime Investigation (Romig 2001).

Some commercial tools (e.g. NetIntercept, NetDetector[8]) have many more analysis features and some are even marketed as digital evidence processing tools. The visualization capabilities of these tools help make examinations of digital evidence from networks more efficient. Regardless of the tool used, when collecting and analyzing network traffic using these systems, digital investigators must take some additional steps to document important details that are not recorded by these tools – such as the MD5 value of tcpdump files containing network traffic, the number of packets dropped, and actions taken by the examiner during analysis of data (i.e. no logs of examiners' actions are created by these tools).

[8] *http://www.niksun.com*

15.7 INVESTIGATIVE RECONSTRUCTION

The fundamentals of investigative reconstruction covered in Chapter 5 do not change when networks are involved. For instance, it may be necessary to perform a relational reconstruction to discern patterns in evidence obtained from a network. For instance, Figure 15.5 shows network traffic represented as host-to-host connections, highlighting one host that is generating the most activity and deserves further attention.

Creating this type of link diagram showing client–server connections can help identify important systems. For instance, in computer intrusion investigations, first focusing on the attacker's IP address can reveal which hosts were targeted and then examining traffic from each target can show which systems were compromised. Examining traffic from a compromised target can give investigators a general sense of what the attacker did on the system.

However, the reconstruction process can be more challenging when networks are involved. A criminal or victim can be several (virtual) places on a network at any given time, making the reconstruction process more complicated and arduous. For instance, a computer intruder may be sharing information with accomplices on IRC while they are breaking into computers around the world. Also, because it is difficult to obtain all relevant digital evidence on a network, there are often gaps in parts of the crime reconstruction.

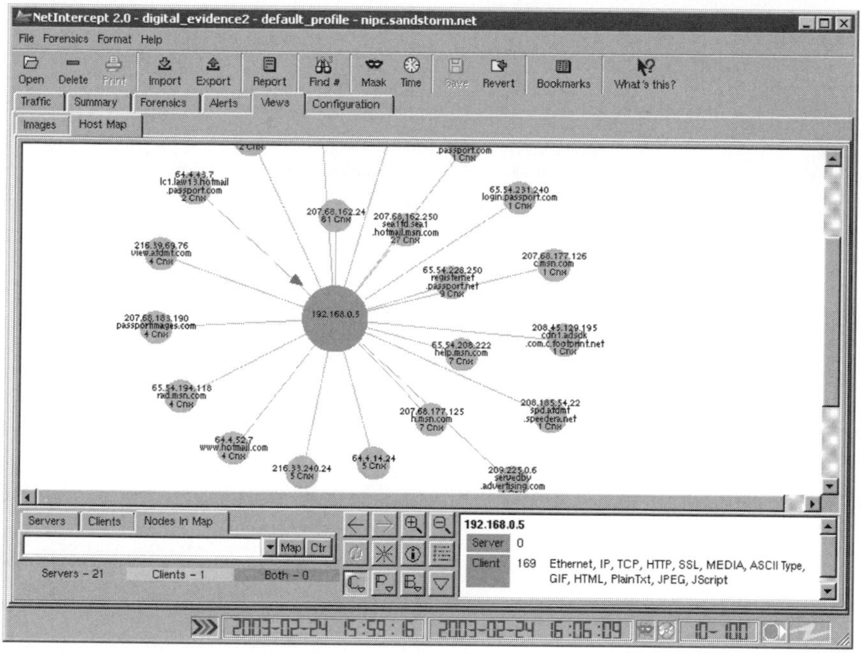

Figure 15.5

Network traffic depicted in IP address–IP address connections creating a circular mesh using NetIntercept.

CASE EXAMPLE

In an intellectual property theft case, one suspect has been identified but his contact within the organization is unknown. Most of the prime suspect's activities during the key time period are known except for details of his connections to Hushmail and Ziplip. Evidence on his hard drive indicates that he received stolen data at the time but it cannot be determined who sent them. Also, log files on the victim organization's network indicate that the prime suspect used a second dial-up account to access the Internet, connect to the organization's systems, and steal information but the Internet Service Provider for this second account does not have related log files. Without these intermediate log files, the continuity of offense cannot be established and the activities cannot be attributed to the offender.

An offender can also use the Internet to conceal his actual location by connecting through computers in other parts of the country or world. Computer intruders use this technique, launching their attack from a compromised computer in a distant location to hide their IP address and geographic location. Also, a Virtual Private Network (VPN) securely extends a local area network to anywhere in the world, providing an encrypted tunnel from the individual's computer at a remote location to the local network. In this way, people can connect their computers to a remote VPN server and obtain an IP address on that network, giving the impression that their computers are on the remote network (Figure 15.6).

Developing relational reconstructions is made more difficult by the mobility of hosts and changeability of networks. Computers can be moved, IP addresses reassigned, DNS entries changed, and individuals can connect to a computer

When AOL users access Web pages and some other Internet resources (AOL IM), their connections pass through proxies that AOL uses to manage network bandwidth but that conceals the individual's actual IP address. Other types of connections do not pass through these proxies (e.g. a Telnet connection to a server on the Internet) and so disclose users' IP addresses that can be tied to an AOL user account.

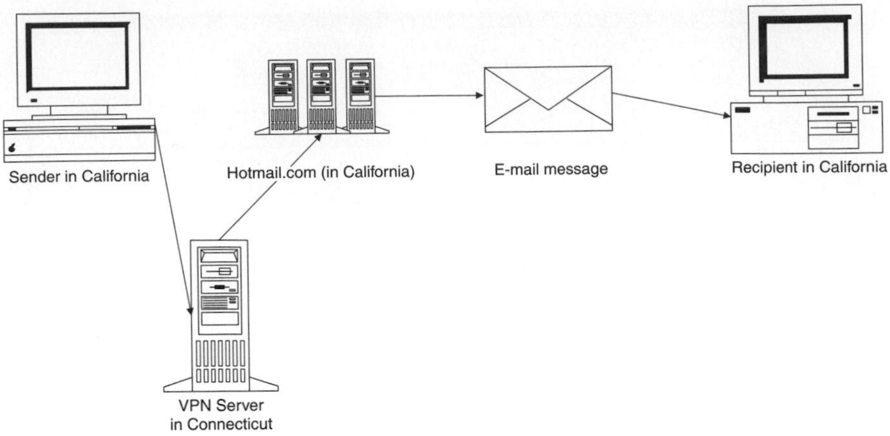

Figure 15.6

VPN connection makes an offender in California appear to be in Connecticut, throwing investigators off track and giving the victim a false sense of security.

remotely, or through a number of systems. Therefore, before assuming that an individual was in a particular location simply based on an IP address or the current location of the computer, examine the alternative possibilities closely. Furthermore, be careful not to assume too much from a log entry. A connection attempt recorded in network logs does not necessarily imply that an individual gained access to the system. Additional corroborating data is needed to determine if the individual successfully entered the system. Also, a functional analysis may reveal that the computer in question was configured to prevent such access.

Fortunately, networks often contain multiple sources of corroborating data that can be used to fill in any gaps, improve the fidelity of a reconstruction, and generally increase the certainty of what occurred. An intrusion investigation involving a Linux server compromised via FTP demonstrates the value of corroborating sources of evidence on a network.

CASE EXAMPLE

A computer intrusion was quickly detected by Tripwire when several system components were replaced using a rootkit (e.g. /bin/login, /usr/bin/du, /usr/bin/top, /usr/bin/killall, /usr/bin/find) The following entry in /var/log/secure showed a connection to the FTP server at the time:

```
Apr 24 22:50:34 ftpserver in.ftpd[2103]: connect from 62.30.247.138
```

There was a corresponding entry in /var/log/wtmp as shown here:

```
ftp ftp pc-62-30-247-138-do.blueyonder.co.uk [62.30.247.138] Tue Apr 24
22:50-22:50 (00:00)
```

This unauthorized connection was partially supported by the following entry in /var/log/messages, the only difference being the time stamp.[9]

```
Apr 25 02:50:40 ftpserver ftpd[2103]: ANONYMOUS FTP LOGIN FROM
pc-62-30-247-138-do.blueyonder.co.uk [62.30.247.138], guest@here.com
```

Knowing that the intruder could have altered logs on the compromised host, digital investigators checked the intrusion detection system logs for a corresponding entry but did not find in one. However, they did find an entry for a different time and source.

```
[**] FTP-site-exec [**]
04/25-02:48:44 04/25-02:49:37 63 62.122.10.221 -> 192.168.2.6S: 4158 D: 21
```

To get a more detailed picture of what occurred, the digital investigators searched the NetFlow logs for all connections to and from the compromised computer. They found that the original connection from blueyonder.co.uk at 22:50:34 was part of a broader scan for FTP servers, which was not logged by the intrusion detection system. The NetFlow logs also showed that the actual intrusion occurred at 02:47:12 from 62-122-10-221.flat.galactica.it and that the intruder downloaded a patch from RPMfind.net and fixed the vulnerability. Intruders often fix the vulnerability they exploit to prevent other intruders from gaining unauthorized access and to hide the fact that the system may be compromised (if computer security professionals scan the system for vulnerabilities it will not raise an alarm).

The intrusion detection system and NetFlow logs provided more reliable sources of digital evidence (C4 on the Certainty Scale discussed in Chapter 7) than the tampered logs on the compromised host (C0). Rather than the intrusion coming from the United Kingdom, the intrusion actually originated in Italy.

[9] *The particular FTP exploit used in this intrusion often inserts an incorrect time stamp, possibly because it is using the time on the computer used to launch the attack.*

Piecing together the large amounts of data that are common in network investigations can also be a challenge. One approach is to only extract portions that seem relevant to the investigation. Consider a harassment case in which the offender was reading the victim's e-mail via a Web proxy.

CASE EXAMPLE

Starting with the e-mail server logs shown below, digital investigators determined when the offender was accessing the victim's account and that he was connected through a Web proxy.

```
Apr 4 18:12:29 mailsrv imapd4[18788]: Login user= tsmith host=
www-proxy.domain.net [10.10.2.10]
Apr 4 18:16:03 mailsrv imapd4[18788]: Logout user= tsmith host=
www-proxy.domain.net [10.10.2.10]
Apr 5 17:52:47 mailsrv imapd4[19405]: Login user= tsmith host=
www-proxy.domain.net [10.10.2.10]
Apr 5 17:56:14 mailsrv imapd4[19405]: Logout user= tsmith host=
www-proxy.domain.net [10.10.2.10]
Apr 6 19:01:56 mailsrv imapd4[19956]: Login user= tsmith host=
www-proxy.domain.net [10.10.2.10]
Apr 6 19:04:42 mailsrv imapd4[19956]: Logout user= tsmith host=
www-proxy.domain.net [10.10.2.10]
```

Extracting the portions of the Web proxy logs that corresponded to the e-mail server logs, digital investigators found the offender's IP address. As an example, the following simplified log segment from April 6, 2003 shows the e-mail of a victim of harassment being accessed through the Web proxy from IP address 172.16.34.14.

```
172.16.34.14, anonymous, 4/6/02, 19:01:24, WWW-PROXY, mailsrv.ispX.com,
GET, http://mailsrv.ispX.com/login.html, 200
172.16.34.14, anonymous, 4/6/02, 19:02:02, WWW-PROXY, mailsrv.ispX.com,
GET, http://mailsrv.ispX.com/tsmith/inbox.html, 200
172.16.34.14, anonymous, 4/6/02, 19:03:27, WWW-PROXY, mailsrv.ispX.com,
GET, http://mailsrv.ispX.com/tsmith/message13.html, 200
172.16.34.14, anonymous, 4/6/02, 19:04:36, WWW-PROXY, mailsrv.ispX.com,
GET, http://mailsrv.ispX.com/tsmith/message14.html, 200
```

The offending IP address was a DSL account and the ISP provided investigators with the subscriber information, including his home address. This individual was the victim's ex-boyfriend who used a Web proxy to conceal his IP address while connecting to the victim's e-mail account. A search of his computer revealed incriminating Web browser history logs and portions of the victim's e-mail messages, confirming that the suspect's computer had been used to access the victim's e-mail account. In conclusion, the harasser's computer was located using e-mail server and Web proxy server logs (C-value C4) and implicating evidence was found on his computer (C-value C5), indicating that it was used to commit the offense.

The main problem with only extracting portions of logs is that important details might be missed. For instance, in the previous example, Web proxy logs from prior days might have shown the harasser accessing the victim's e-mail many times over an extended period, demonstrating persistent and intentional spying as opposed to a single, isolated event.

Another approach to dealing with large amounts of network related data is to reconstruct smaller, more manageable portions of the crime separately before combining them into complete crime reconstruction. For example, when criminal activity is spread out over an extended period of time, prioritizing and focusing on several critical periods and locations before combining them into a larger reconstruction will provide clues and leads more quickly than trying to reconstruct the entire crime all at once.

CASE EXAMPLE

A computer intruder broke into several servers over a period of months. It was not initially clear that the same individual had compromised all of these servers. The commonalities between these intrusions were only apparent after individual timelines were created using log files and file date–time stamps from each of the compromised systems. A rough timeline of the entire incident was constructed, providing an overview of events, but the individual timelines for each system were also useful to investigators in the long run because they contained more details.

It may not be possible to identify critical periods in a crime without performing some analysis on all available log files. Logs from routers, firewalls, intrusion detection systems, and other sources may only reveal important patterns when combined.[10] For instance, when an intruder is targeting systems on a network, firewall logs may only show a few denied connection attempts that do not cause alarm on their own. Similarly, when viewed independently, system logs on the targeted hosts may not cause alarm. However, when combined with router and intrusion detection system logs, it may become clear that the denied connections were part of a more widespread series of attacks against several systems on the network. When performing temporal analysis on multiple log files, it is generally more efficient to combine them before sorting them and analyzing them for patterns.

[10]Commercial software is available for combining and analyzing log files but they are often limited to a few log formats or require customization to accommodate new log formats. Using such tools may be justified if they help digital investigators analyze log files they regularly encounter in many investigations. However, few tools surpass Perl and UNIX for special purpose tasks such as analyzing log files that are only encountered occasionally.

However, before combining log files, it is crucial to correct for time zone differences and system clock discrepancies. Even log files from a single system can contain date–time stamps with different time zones. For instance, Microsoft's Internet Information Server logs are in GMT by default whereas the NT Event Logs generally use the local time. Internet service providers like AOL have been known to adjust date–time stamps in their logs into British Summer Time instead of GMT, resulting in a 1-hour discrepancy. Additionally, it may be necessary to rearrange certain log files before combining them with others. For instance, some logs are ordered by end time (e.g. pacct, NetFlow) and may provide a clearer picture of events when they are sorted by start time.

In some cases, it may be necessary to determine how a criminal was able to commit the crime. For example, when an intruder breaks into a computer that appears to be secure, digital investigators may need to conduct a detailed functional reconstruction or even a reenactment to determine if an unknown vulnerability was exploited or if the intruder had inside information such as a password to the system. Whenever possible, as part of the functional reconstruction of a crime, investigators should replicate the process that created the digital evidence. When asked to testify that a certain process created a given piece of digital evidence, investigators may be asked if they verified the process or even to provide a demonstration. Additionally, trying to replicate the process can improve digital investigators' understanding of evidence and the criminal or victim. In a missing persons investigation, there was a question regarding how much an individual deliberated over a goodbye e-mail message. Creating a test e-mail message and comparing the time stamps in the header may indicate how long it took the author to compose the message. For instance, the time in the Message-ID line of the following message indicates that it was started at 1019 hours on November 19 and the other times in the header indicate that it was sent at 1103 hours, a difference of 44 minutes.

```
Received: from mail.corpX.com (mail.corpX.com [192.168.5.18])
        by lsh110.siteprotect.com (8.9.3/8.9.3) with ESMTP id KAA09889
        for <eco@corpus-delicti.com>; Tue, 19 Nov 2002 10:03:36 -0600
Received: from localhost (sysadmin@localhost)
        by mail.corpX.com (8.11.6/8.11.6) with ESMTP id gAJG3W725027
        for <eco@corpus-delicti.com>; Tue, 19 Nov 2002 11:03:32 -0500
Date: Tue, 19 Nov 2002 11:03:32 -0500 (EST)
From: sysadmin <sysadmin@mail.corpX.com>
To: eco@corpus-delicti.com
Subject: Test time
Message-ID: <Pine.LNX.4.44.0211191019020.14986-100000@mail.corpX.com>
```

15.7.1 BEHAVIORAL EVIDENCE ANALYSIS

When examining digital evidence, particularly on networks, it is important to keep in mind that we are looking at effects of human activities and trying to reconstruct associated behavior and intent. People are creatures of habit to a certain degree – we seek the illusion of order, stability, and certainty in many areas of life. Our daily activities often revolve around things like our family, friends, meals, exercise, work, and entertainment. These activities can reflect our needs and, to some degree, our personalities and exposure to risk. For instance, bartenders and taxi drivers are at high risk of robbery and assault but also have access to a large number of potential victims. If someone becomes a victim, it is likely to occur through some aspect of his or her regular activities. If there is no clue how someone became a victim, some evidence may be missing or the targeting may have been opportunistic. Opportunistic is not to say random because the offender selected the victim with a purpose and for certain reasons, whether it was the time, place, or victim's appearance. Offenders have patterns in life and crime – again, these patterns as seen in evidence can reveal their needs.

Log files are a particularly rich source of behavioral evidence because they record so many actions. Using the information in these log files it is often possible to determine with a high degree of detail what an individual did or was trying to achieve. An appreciation for patterns of activity in log files can help digital investigators differentiate between an automated worm and a computer intruder gaining unauthorized access to a computer. In some cases it is possible to discern *modus operandi* behaviors from log file that can be used to determine if the same computer intruder was responsible for multiple intrusions. Patience, familiarity with data processing tools, and some understanding of the underlying technology are required to sift through large log files for the few pieces of relevant information but the effort will pay off in the long run as we become more reliant on technology.

It is often worthwhile to think about what the individual would have to do in order to achieve a given result, breaking activity into smaller segments and looking for signs of these segments. For instance, a computer intruder generally performs some level of surveillance of a target before attempting to break into the system. This approach can improve one's understanding of events, lead to additional sources or evidence, and give an indication of planning. Online sexual offenders often groom their victims to gain their trust – this can be a complex and prolonged process that can generate significant amounts of digital evidence.

CASE EXAMPLE

Individuals break into Web sites and vandalize the pages in retaliation for a perceived wrong and/or to assert their power over the owner(s) of the site. An obvious part of investigating this type of occurrence is to examine the log files of the Web server that was broken into for information about the intruders. Of course, this is obvious to intruders as well, so if they cannot delete the log files on the Web server they often break in from another computer that they have compromised. Typically, intruders will delete all of the digital evidence on the host they use to break into the Web server making it difficult for an investigator to track them down.

Fortunately, investigators can take advantage of a vandal's behavior and the Web server access log to narrow the pool of suspects. A vandal usually looks at the page after (and sometimes before) modifying it. The Web server access log contains IP addresses of computers that accessed the Web page. Therefore, by looking at entries in the log file around the time of the vandalism, investigators often find the IP address of the vandal. In many cases vandals use the browser on their personal computer to view the Web page so the IP address in the Web server access log is a direct link, bypassing any intermediate hosts that the vandal used to break into the Web server. Although it is not conclusive, this IP address can help investigators reconstruct the crime and find suspects.

Keep in mind that the same individual behavior can mean different things in different situations, so, rather than considering items of evidence in isolation, it is necessary to consider all activities together to gain insight into their overall meaning. Some individuals view Web pages via a Web proxy because the resources they are interested in are only accessible through the proxy. Some individuals use Web proxies to conceal their identities.

To understand how digital evidence on networks reflects behavior, it is instructive to consider some examples. When thieves target an organization's computer systems, their actions leave behind digital evidence that can reveal their intent, skill level, and knowledge of the target. Network logs may show a broad network scan prior to an intrusion, suggesting that the individual was exploring the network for vulnerable and/or valuable systems. This exploration implies that the individual does not have much prior knowledge of the network and may not even know what he/she is looking for but is simply prospecting. Conversely, thieves who have prior knowledge of their target will launch a more focused and intricate attack. For instance, if a thief only targets the financial systems on a network, this directness suggests that the intruder is interested in the organization's financial information and knows where it is located.

So, if the targeting is very narrow – the thief focuses on a single machine – this indicates that he/she is already familiar with the network and there is something about the machine that interests him/her. Similarly, time pattern analysis of the target's file system can show how long it took the intruder to locate desired information on a system. A short duration is a telltale sign that the intruder already knew where the data was located whereas protracted searches of files on a system indicates less knowledge.

The sophistication of the intrusion and subsequent precautionary acts help determine the perpetrator's skill level. The thief's knowledge of the target and his/her criminal skill can be very helpful in narrowing the suspect pool, particularly when only a few individuals possess the requisite knowledge and skills – suggesting insider involvement.

15.8 REPORTING RESULTS

Although the involvement of networks in a digital evidence examination does not necessarily change the structure of a final report, conveying results clearly becomes more complicated when networks are involved because more computers are involved, there are complex interactions, and all of the complexities must be simplified for decision makers. Diagrams can provide an overview of events and presenting digital evidence through the visualization tools used to perform the examination and analysis can help convey more technical aspects of a case in easy to understand terms.

When dealing with large cases involving hundreds of computers, it is useful to create a main report describing the overall examination and several more focused reports dealing with logical groupings of machines. For instance, if computers from three organizations were examined, it can be helpful to write separate reports relating to each organization. Alternatively, if a group of computer intruders gained unauthorized access to several hundred machines, it can be helpful to write separate reports relating to each type of machine (e.g. Solaris, Linux, Windows) to explain fully the different actions taken on each type of system.

15.9 SUMMARY

Connecting computers together is inherently risky. An individual can gain unauthorized access to a distant network. Anyone can intercept transmissions between networks. Additionally, connecting networks enables individuals, including criminals, to communicate in ways that were not possible before, resulting in a new set of problems. However, for every disadvantage there is an equal and opposite advantage. With the proper authority and precautions, digital investigators can gain access to and collect evidence from distant networks. Digital investigators can intercept digital evidence as it travels over a network, and computer networks enable digital investigators to communicate with each other and observe criminal activity and communication like never before.

The ultimate challenge for digital investigators is to follow cybertrails swiftly and thoroughly to find pockets of evidence before they are lost forever. This is challenging not only because evidence on a network is distributed and dynamic, but also because every network is different with unique combinations of hardware and software. Many networks have grown by a process of accretion, laying new technologies on top of old in a fairly haphazard manner. The result is almost organic: an entity that often seems to have a mind of its own. By learning how computer networks function and how forensic science can be applied to computer networks, we can take advantage of digital evidence and address the growing problem of cybercrime. Without an understanding of where information can be found on networks, digital investigators are guaranteed to waste a significant amount of time and are likely to lose valuable digital evidence. Additionally, without an understanding of how networks function, forensic network analysts will have a harder time making sense of any data they obtain from a network.

However, in some cases, even the people who are responsible for maintaining a network do not understand it completely. Therefore, it is unrealistic to expect an investigator to have full knowledge of a network before, or even after, an investigation. The most that can be expected of an investigator is to understand how computers and networks function in general and to have a familiarity with a variety of technologies and operating systems. Having a solid understanding of how networks function in general will enable an investigator to understand many different types of networks and will help determine when and what kind of expert is needed.

REFERENCES

Fyodor (1998) "Remote OS detection via TCP/IP Stack FingerPrinting", (Available online at http://www.insecure.org/nmap/nmap-fingerprinting-article.txt)

Romig S. (2001) "Incident Response Tools", Casey, E. (editor) *Handbook of Computer Crime Investigation*, London: Academic Press

Sommer P. (1997) "Downloads, Logs and Captures: Evidence from Cyberspace", Journal of Financial Crime, October, 1997, 5JFC2 138-152, (Available online at http://www1.bcs.org.uk/DocsRepository/03900/3968/logs.htm)

United States Department of Justice (2002) "Searching and Seizing Computers and Obtaining Electronic Evidence in Criminal Investigations" (Available online at http://www.cybercrime.gov/searchmanual.htm)

CASES

Bach v. Minnesotta (2002) Appeals Court, 8th Circuit, Case number 02-1238 (Available online at http://www.epic.org/privacy/bach/)

United States v Hilton (1997) District Court, Maine, Case Number 97-78-P-C (Available online at http://www.med.uscourts.gov/opinions/carter/2000/gc_06302000_2-97cr078_us_v_hilton.pdf)

CHAPTER 16

DIGITAL EVIDENCE ON PHYSICAL AND DATA-LINK LAYERS

The physical and data-link layers provide the foundation for everything else on a network. The physical layer is the medium that carries data – such as the cables, radio waves, microwaves, or lasers. The data-link layer joins a computer with the physical layer, and includes the transmission method (e.g. CSMA/CD) as mentioned in Chapter 14. Network Interface Cards (NICs) are part of the data-link layer – connecting computers to the network cables. Each NIC has an unique address (MAC address) that can be used to determine which host was used to commit a crime.

Network eavesdropping is the most common approach to gathering digital evidence on the data-link and physical layers. With the help of a network monitoring tool (a sniffer), investigators and criminals can capture large amounts of information as it travels through a network. This approach to collecting network traffic is comparable to making a bitstream copy of a hard drive – a sniffer can capture every byte transmitted on the network. As with any bitstream copy, files and other useful digital evidence can be extracted from network traffic using specialized tools. For example, digital investigators can use a sniffer to monitor a computer intruder or child pornographer on a network and recover toolkits, images, e-mail attachments, IRC communications with cohorts, and anything else the offenders transmitted on the network.

Equipment and programs for collecting digital evidence on the physical layer are discussed in this chapter. Although this network traffic resides at the physical layer, it contains data relating to the other network layers like TCP/IP and HTTP traffic (recall Figure 14.12). Therefore, to interpret captured network traffic it is necessary to have a solid understanding of the network, transport, and application layers. Tools for interpreting network traffic are presented in this chapter and the other network layers are discussed in more detail in Chapters 17 and 18.

Routers and other network devices also store data relating to the data-link layer such as MAC addresses. These addresses can indicate which computer was used to commit a crime. Although a MAC address is usually directly

ISBN: 0-12-163104-4

associated with the NIC in a computer, on many systems it can be changed to any value. This chapter describes where this information is stored and how it can be collected.

The most effective way to learn about the data-link layer as a source of evidence is to examine a specific example in detail. This chapter describes Ethernet in detail to provide a sense of how a network technology functions. Ethernet is a good example because it is one of the most widely used network technologies. Also, a familiarity with Ethernet makes it easier to understand how other network technologies operate – the 802.11 protocols are based on Ethernet. To highlight the similarities and differences between Ethernet and other network technologies, Ethernet is briefly compared to Asynchronous Transfer Mode (ATM). ATM is quickly becoming the standard for large-scale high-speed networking.

16.1 ETHERNET

As described in Chapter 14, specific combinations of NIC, cable, and transmission method are called *network technologies*. For instance, Ethernet cables, Ethernet cards, and the method that Ethernet cards use to transmit data (CSMA/CD) are jointly referred to as *Ethernet*. Ethernet is one of the most widely used network technologies and it has gone through several revisions. Some networks still use the original Ethernet technology that was created at Xerox PARC in the 1970s. However, most networks now use one of the newer versions of Ethernet (i.e. 10Base5,10BaseT, 100BaseT).

16.1.1 10Base5

The 10Base5 standard closely resembles the original Ethernet, relying on a continuous piece of thick ($\frac{1}{2}$inch) yellow coaxial cable – the ether. The technology is called 10Base5 because:

1 it can transmit data at 10 Mbits per second;

2 only one computer can transmit while the other listens (this is known as baseband);

3 the maximum recommended cable length is 500 meters (thus the 5 in 10Base5).

To connect a computer to a 10Base5 cable, a transceiver is poked into the cable's yellow plastic sheath at a particular point, indicated with a black mark, essentially tapping into the ether. The transceiver is then connected to the NIC inside the computer using a drop cable. The technical name for this drop cable is Attachment Unit Interface (AUI) (Figure 16.1).

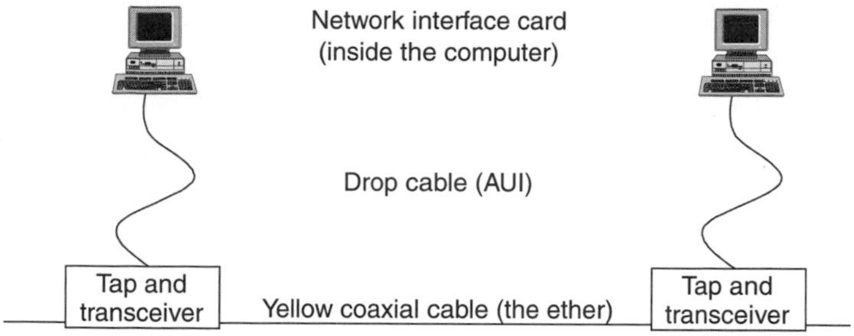

Figure 16.1

Old Ethernet configuration (modern configurations are conceptually the same).

16.1.2 10/100/1000BaseT

The most popular forms of Ethernet are 10BaseT and 100BaseT because they are cheaper and less cumbersome. These network technologies do not require a separate tap, transceiver, and drop cable, only a NIC and cable. 10BaseT and 100BaseT use unshielded twisted-pair (UTP) cables similar to regular telephone cords (two pairs of copper wires twisted together to reduce electrical interference). Unlike the thick yellow cables used by 10Base5, UTP cables are cheap and easy to bend around corners. However, UTP can only carry data about 100 meters whereas a 10Base5 cable can carry data for up to 500 meters. These cables are used to connect hosts to a central hub or switch that transmits data between hosts. A switch is analogous to a train system that enables trains to transfer from one track to another using a switching mechanism (Figure 16.2).

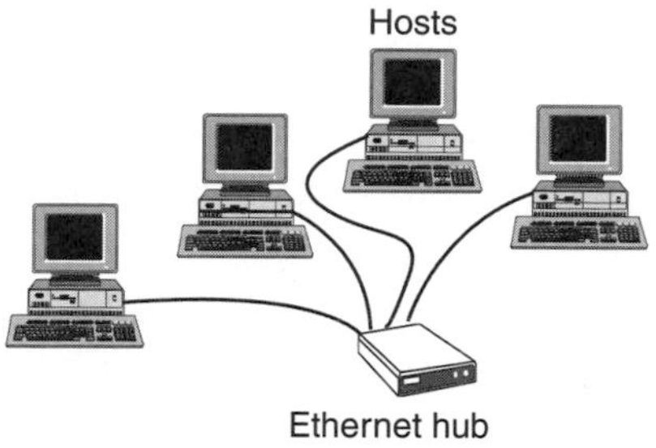

Figure 16.2

Computers on a 10BaseT network plugged into a hub.

The more recent version of Ethernet is 100BaseT, basically the same as 10BaseT except faster. Newer computers are using the latest advance in Ethernet technology; 1000BaseT. Table 16.1 summarizes the main distinguishing features of these standards.

Table 16.1

Different types of Ethernet.

IEEE 802.3 STANDARD	CABLE	MAX CABLE LENGTH (meters)	THROUGHPUT (Mbps)
10Base5 (thick Ethernet)	1/2" Yellow coaxial	500	10
10BaseT (twisted-pair Ethernet)	Twisted pair	100	10
100BaseT	Twisted pair	100	100
1000BaseT	Twisted pair	100	1000

16.1.3 CSMA/CD

Although Carrier Sense Multiple Access with Collision Detection (CSMA/CD) is a mouthful, the concept is straightforward: it is a "listen before acting" access method. Recall the analogy of the polite dinner conversation described in Chapter 14. At a polite dinner party, an individual who has something to say waits for a break in the conversation before speaking. If two people start to speak at the same time, they both stop for a moment before starting to speak again. Similarly, when two computers using Ethernet start to transmit data at the same time, they both sense that the other host is transmitting and they both stop for a random period of time before transmitting again. This method of communication works well as long as there are not too many hosts connected to the same wire. Having too many hosts on the network will result in many collisions and not enough successful communication.

16.2 LINKING THE DATA-LINK AND NETWORK LAYERS – ENCAPSULATION

In addition to connecting computers to the network, the data-link layer prepares data for their journey through the physical layer. For example, before sending an IP packets, Ethernet adds a header and checksum (a number used to verify the integrity of the data), encapsulating the packet in an *Ethernet frame.* Table 16.2 shows the segments of an IP packet encapsulated in an Ethernet frame.

Why are two types of addresses required – an IP address and a MAC address? Each address serves a different purpose. Put simply, Ethernet enables communication between hosts on the same network using MAC addresses while TCP/IP enables communication between hosts on different networks using IP addresses. Computer applications use TCP/IP to communicate, regardless of the network technology involved and computers themselves use the local network technology to exchange data. So, before an IP packet can be transmitted through the physical and data-link layers, it

Table 16.2

An IEEE 802.3 standard Ethernet frame (shaded) encapsulating an IP packet.

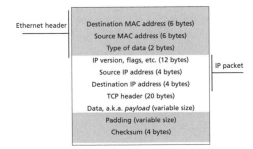

Ethernet header	Destination MAC address (6 bytes)	
	Source MAC address (6 bytes)	
	Type of data (2 bytes)	
	IP version, flags, etc. (12 bytes)	IP packet
	Source IP address (4 bytes)	
	Destination IP address (4 bytes)	
	TCP header (20 bytes)	
	Data, a.k.a. *payload* (variable size)	
	Padding (variable size)	
	Checksum (4 bytes)	

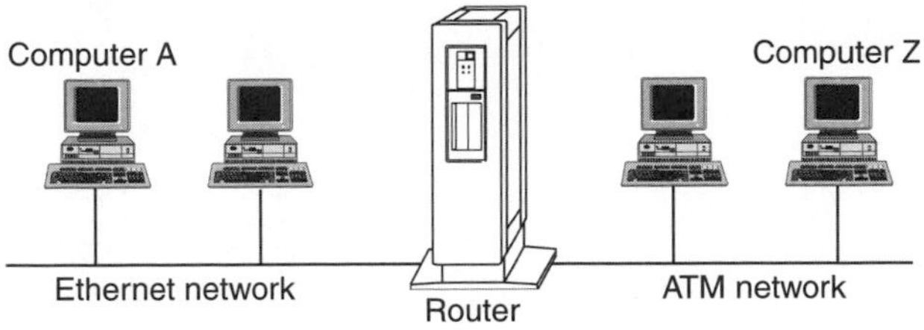

Figure 16.3

Computer A sending data to computer Z.

must be encapsulated in the local language (e.g. Ethernet, ATM, FDDI). For instance, at the data-link layer, Ethernet uses a particular kind of MAC addresses (e.g. 08-00-56-12-97-A8) to direct data, encapsulating IP packets into Ethernet frames as shown in Table 16.2.

Recall from Chapter 14, when a computer on one Ethernet network needs to send information to a computer on another network, it must send the information through a router.

In Figure 16.3, to deliver data to host Z, host A must first encapsulate data from the application layer, addressing packets and delivering them to the router. So, host A puts the data in an IP packet addressed to host Z and then encapsulates the IP packet in an Ethernet frame addressed to the router. When it receives the frame, the router peels off the Ethernet header and sees host Z's IP address. Once it sees that the IP packet is addressed to host Z on an ATM network, the router re-encapsulates the packet in an ATM cell (the ATM equivalent of an Ethernet frame) and sends it directly to host Z.

When host Z receives the ATM cell, it does the opposite of what host A did to send the data. The data-link layer on host Z peels off the ATM header and passes the IP packet to the TCP/IP software. Then, the TCP/IP software peels off the TCP and IP headers and passes the data to the appropriate application (e-mail, Web, Usenet, IRC, etc.).

One key point about MAC addresses is that they do not go beyond the router. Unlike IP addresses, MAC addresses are only used for communication between computers on the same network. Therefore, when a packet is sent through the Internet, it does not contain the MAC address of the computer that created it, only that of the local router that delivered it. If logs of network traffic are kept (e.g. Argus logs), investigators may be able to track data back to their source using MAC addresses.

CASE EXAMPLE

An organization noticed a large spike in their outbound network traffic, indicating that a denial of service attack was being launched from one of their hosts (192.168.0.7). However, when this host was examined, nothing unusual was found, suggesting that the attack had been launched from a different host, using the IP address 192.168.0.7 to misdirect investigators. Fortunately, the following Argus logs were available (only a small selection of the thousands of log entries are shown here).

In this example, Argus was installed on the same physical network segment. On larger networks, Argus can monitor multiple segments using proxy ARP and can record Virtual Local Area Network (VLAN) tags that identify which VLAN the data relate to.

```
% ra -m -t 01:00 – 08:00 -r /var/log/argus/argus.out – udp and host 192.168.0.1
01:03:17 udp 0:0:e2:7a:c3:5b 0:10:2f:1d:cd:ef 192.168.0.7.32769 <-> 172.16.102.45.80
03:03:19 udp 0:0:e2:7a:c3:5b 0:10:2f:1d:cd:ef 192.168.0.7.32769 <-> 172.16.102.45.80
03:21:16 udp 0:0:e2:7a:c3:5b 0:10:2f:1d:cd:ef 192.168.0.7.32769 <-> 172.16.102.45.80
05:03:24 udp 0:0:e2:7a:c3:5b 0:10:2f:1d:cd:ef 192.168.0.7.32769 <-> 172.16.102.45.80
07:03:25 udp 0:0:e2:7a:c3:5b 0:10:2f:1d:cd:ef 192.168.0.7.32769 <-> 172.16.102.45.80
07:51:58 udp 0:0:e2:7a:c3:5b 0:10:2f:1d:cd:ef 192.168.0.7.32769 <-> 172.16.102.45.80
```

These logs show a computer with MAC addresses (00:00:e2:7a:c3:5b) using the IP address in question. This system was located – an IBM Thinkpad running Linux that had been compromised and used as a launch pad for the denial of service attack. The other MAC address in these Argus logs belongs to the local switch, not the target of the attack.

MAC addresses can also sometimes be used to classify the type of machine. For instance, Ethernet MAC addresses comprise 12 hexadecimal digits (e.g. 00-10-4B-DE-FC-E9). The first six hexadecimal digits, called the Organizationally Unique Identifier (OUI), refer to the vendor of the NIC and the last six digits are the serial number for the particular NIC. Table 16.3 lists a small selection of vendors and their associated Ethernet MAC address prefix.[1] Note that large companies such as Cisco and 3Com use different identifiers for different product lines.

Ethereal uses this OUI information to classify network addresses. For instance, Figure 16.4 shows Ethereal being used to monitor traffic between a

[1] *A more complete list can be found at http://www.cavebear.com/CaveBear/Ethernet/vendor.html and a searchable database of these vendor codes can be found on the IEEE Web site at http://standards.ieee.org/regauth/oui/index.shtml. Keep in mind that vendors sometimes use other vendor's cards, such as a 3COM card in a Cisco device.*

Table 16.3

MAC addresses of different manufacturers.

PREFIX	MANUFACTURER	PRODUCT (WHEN APPLICABLE)
001007	Cisco Systems	Catalyst 1900
00100B	Cisco Systems	
00100D	Cisco Systems	Catalyst 2924-XL
001011	Cisco Systems	Cisco 75xx
00101F	Cisco Systems	Catalyst 2901
001029	Cisco Systems	Catalyst 5000
00102F	Cisco Systems	Cisco 5000
00104B	3Com	3C905-TX PCI
00105A	3Com	Fast Etherlink XL in a Gateway 2000
006097	3Com	
080020	Sun	
0001AF	Motorola	
080056	Stanford University	
08005A	IBM	
0001E6	Hewlett-Packard	
3C0000	3Com	Dual function (V.34 modem + Ethernet) card
444553	Microsoft	Windows95 internal "adapters"

Nokia Wireless Access Point and several hosts, including an Apple system (OUI 003065).

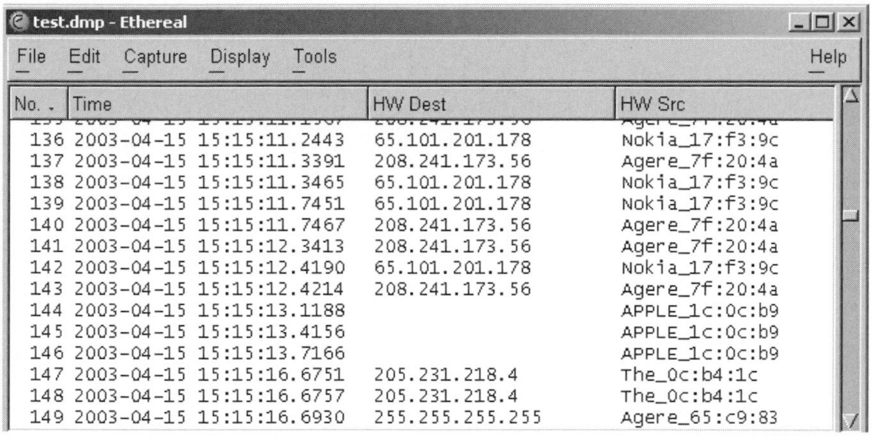

No.	Time	HW Dest	HW Src
136	2003-04-15 15:15:11.2443	65.101.201.178	Nokia_17:f3:9c
137	2003-04-15 15:15:11.3391	208.241.173.56	Agere_7f:20:4a
138	2003-04-15 15:15:11.3465	65.101.201.178	Nokia_17:f3:9c
139	2003-04-15 15:15:11.7451	65.101.201.178	Nokia_17:f3:9c
140	2003-04-15 15:15:11.7467	208.241.173.56	Agere_7f:20:4a
141	2003-04-15 15:15:12.3413	208.241.173.56	Agere_7f:20:4a
142	2003-04-15 15:15:12.4190	65.101.201.178	Nokia_17:f3:9c
143	2003-04-15 15:15:12.4214	208.241.173.56	Agere_7f:20:4a
144	2003-04-15 15:15:13.1188		APPLE_1c:0c:b9
145	2003-04-15 15:15:13.4156		APPLE_1c:0c:b9
146	2003-04-15 15:15:13.7166		APPLE_1c:0c:b9
147	2003-04-15 15:15:16.6751	205.231.218.4	The_0c:b4:1c
148	2003-04-15 15:15:16.6757	205.231.218.4	The_0c:b4:1c
149	2003-04-15 15:15:16.6930	255.255.255.255	Agere_65:c9:83

Figure 16.4

Ethereal classification of NIC addresses.

This type of class characteristic can be useful for narrowing a search on a network – knowing that the suspect used an Apple system can make it easier to locate the computer in question.

16.2.1 ADDRESS RESOLUTION PROTOCOL (ARP)

Computers on a network do not necessarily know each other's MAC addresses. For example, when a computer wants to send an IP packet, it only knows the IP address of the destination host. To discover the MAC address of the destination host, a computer simply asks every other host on the network: is this your IP address? The host with that IP address responds with its MAC address. This simple exchange is called the Address Resolution Protocol (ARP).

Although ARP is part of TCP/IP, it is generally considered a part of the data-link layer. The easiest way to think about ARP is to imagine it straddling the network and data-link layers (Figure 16.5).

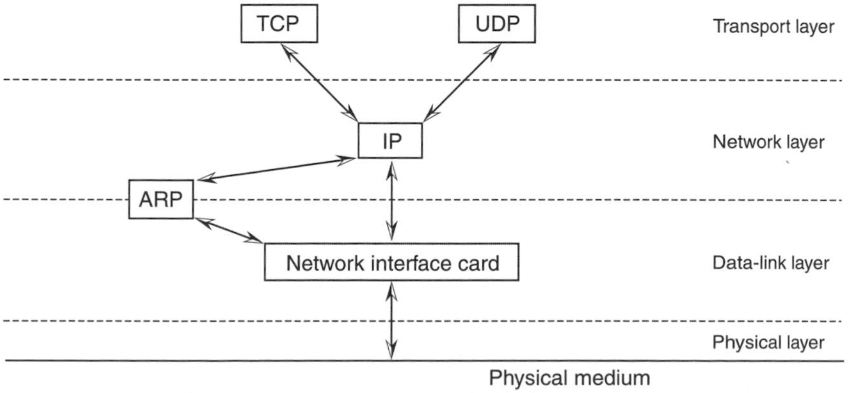

Figure 16.5

Summary diagram of TCP/IP separated by OSI layer.

This address discovery process might seem like a lot of effort that could be replaced by a list of IP → MAC address associations. However, every computer would have to have such a list and whenever a computer was added to the network, the list on each computer would have to be updated. As a compromise, computers keep a temporary list of IP → MAC address associations. So, two computers that communicate frequently will not have constantly to remind each other of their respective IP addresses. This temporary list is called an ARP table (a.k.a. ARP cache) and can be viewed on Unix and Windows NT/2000/XP machines using the **arp -a** command as shown here:

```
:~% arp -a
Net to Media Table
Device          IP Address              MAC Addr
-------         ---------------         
e0              192.168.1.1             08:00:20:75:d3:fb
e0              192.168.1.3             08:00:20:1c:1f:67
e0              192.168.1.4             08:00:20:1c:6a:ff
e0              192.168.1.9             00:60:83:24:1f:4d
e0              192.168.1.23            08:00:20:7d:40:9c
e0              192.168.1.33            08:00:20:80:fe:34
e0              192.168.1.39            08:00:20:7f:17:3c
e0              192.168.1.45            08:00:20:7d:e3:94
e0              192.168.1.53            00:04:ac:44:3f:4e
e0              192.168.1.75            08:00:20:1c:5b:df
e0              192.168.1.103           08:00:20:87:2c:73
e0              192.168.1.144           08:00:20:86:4a:cf
e0              192.168.1.134           08:00:20:87:a5:bb
e0              192.168.1.232           08:00:20:86:e2:5c
e0              192.168.1.234           08:00:20:7e:2d:ef
```

So, if a criminal reconfigures his computer with someone else's IP address to conceal his identity, the local router would have an entry in its ARP table showing the criminal's actual MAC address associated with someone else's IP address. If the record in the ARP table is not used for a while (usually between 20 minutes and 2 hours), it is deleted. Notably, IPv6 addresses contain the MAC address of the network interface they are associated with.

16.2.2 POINT TO POINT PROTOCOL AND SERIAL LINE INTERNET PROTOCOL

The use of modems to connect computers to the Internet deserves a quick mention here. Many people dial into an ISP to connect to the Internet – transmitting data over a copper telephone line instead of an Ethernet or fiber optic cable. This type of connection is much less sophisticated than network technologies like Ethernet, FDDI, and ATM. An addressing scheme is not required since the modem in a person's home is connected directly to one of their ISP's modems through telephone wires. All that is required is a simple method of encapsulating IP packets and sending them over the telephone wires. Several protocols do just this, including Point to Point Protocol (PPP) and Serial Line Internet Protocol (SLIP). Although it is open to debate, think of PPP and SLIP as on the data-link layer and the serial line that they use as on the physical layer in a dial-up connection. Notably, many broadband Internet providers are using PPP over Ethernet (PPPoE) to establish a PPP connection using a variation of the Ethernet protocol.

16.3 ETHERNET VERSUS ATM NETWORKS

Recall from Chapter 14 that ATM uses fiber optic cables and specialized equipment (ATM switches) to enable computers to communicate at very high rates (Gbits per second). ATM networks were originally developed by the telecommunications industry to handle multimedia communications (combined video, voice, and data). Therefore, it is no coincidence that ATM works like voice telephone systems. Switches establish circuits between computers on a network (like a telephone call) and ATM network addresses use the same standard as telephone numbers – they have a local network number and then a prefix (like an area or country code) for communication between distant networks.[2]

[2]*ATM addresses contain information that is used for routing so there is some network layer functionality in ATM. However, for the purposes of this text it is sufficient to think of ATM as the physical and data-link layers.*

Notice that this circuit establishment is different from Ethernet. Like Ethernet, ATM encapsulates data into what are called *ATM cells*. However, ATM cells are not addressed in the same way as Ethernet frames. Instead of addressing a cell using the MAC address of the destination computer, ATM uses a number that identifies the circuit that the ATM network has established between two computers. Two computers will use the same circuit for the duration of their communication.

Although ATM uses a form of ARP (called ATMARP) to discover Machine Access Control (MAC) addresses, the approach that ATM takes is slightly different. Instead of allowing individual computers to respond to ARP requests, ATMARP uses a central server to keep track of IP → MAC address associations. This central server responds to all ARP requests on a given ATM network.

Although there are some differences between Ethernet and ATM, the digital evidence on each is similar. There are log files, MAC addresses, ARP tables, and encapsulated data travelling through the network cables – all of which can be a source of digital evidence.

16.4 DOCUMENTATION, COLLECTION, AND PRESERVATION

A common approach to collecting digital evidence from the physical layer is using a sniffer. Sniffers put NICs into "promiscuous mode" forcing them to listen in on all of the communications that are occurring on the network.

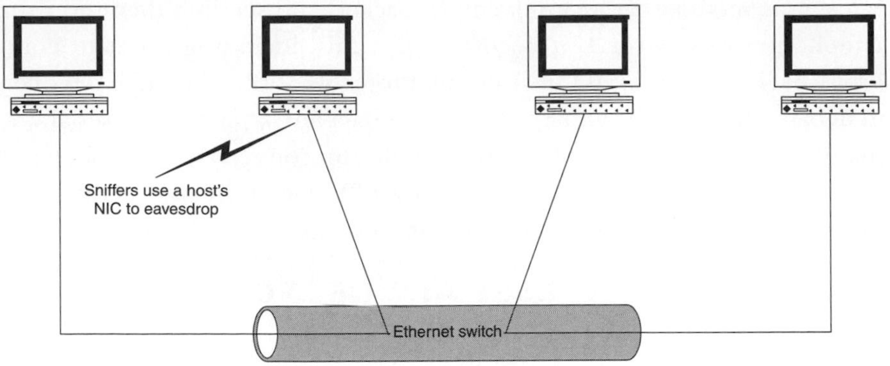

Figure 16.6

Computers connected at the physical level are vulnerable to eavesdropping.

Because switches prevent one host on the network from monitoring other hosts' traffic, computer intruders often simply monitor traffic to and from the computer they have broken into. Some computer intruders have been known to record themselves unwittingly with their own sniffer when they return to examine the captured traffic. This is analogous to someone setting up a video camera to tape an area, returning to check that the camera is working (recording themselves in the process) and leaving the camera to tape more activities. Obtaining such a recording makes it easier to track an intruder (Figure 16.6).

Other criminals take steps to protect themselves against eavesdropping using encryption. It is virtually impossible to break strong encryption. For example, computer intruders who are aware that investigators might try to monitor sessions will encrypt them using software like Secure Shell (SSH). However, even if data are encrypted, collecting and analyzing the network traffic can be informative. For instance, if hundreds of packets containing encrypted data were traveling between two individuals while one of them committed a crime, the second person may well be an accomplice and there may be probable cause to search the second person's computer or property.

Collecting network traffic using a sniffer can be invasive and resource consuming, very much like wire-tapping and there are strict laws that must be adhered to when intercepting communications as described in Chapter 3. It is possible to limit the invasiveness of this evidence collection method by only recording packet header information, not the contents (a.k.a. payload). Some operating systems come with sniffers (e.g. tcpdump on Linux, snoop on Solaris) but these are not necessarily the best platforms to use. Operating systems like Windows and Linux are not particularly efficient at capturing network traffic on high-speed networks and become overloaded, failing to collect important data. Windows systems may be suitable for 10BaseT segments and Linux may be suitable for 100BaseT networks. The most reliable operating systems for collecting Gigabit network traffic are OpenBSD and FreeBSD (Garfinkel 2002).

16.4.1 SNIFFER PLACEMENT

Sniffers can be used on a network in a variety of ways – to appreciate the limitations of each approach consider a computer intrusion investigation. After an intruder gains unauthorized access to a Linux host, investigators could use tcpdump on the compromised system to collect network traffic to and from the compromised host. However, using the compromised system to collect evidence may destroy other evidence on the system. Furthermore, the intruder could have modified the tcpdump program to conceal or destroy evidence. Instead, investigators could use a nearby host on the same network segment to monitor traffic to and from the compromised host. However, this approach to collecting network traffic as evidence is only effective when computers are connected with a hub. Recall that a switch prevents one host on the network from monitoring traffic to other hosts.

When a switch is involved, one approach is to utilize a feature in switches called Switched Port Analyzer (SPAN). A *SPANned* port (a.k.a. mirrored port) enables eavesdropping by copying network traffic from one port on the switch to another. However, a SPANned port only copies valid Ethernet packets, does not duplicate all error information, and the copying process receives lower priority than routine data transmission that may increase dropped Ethernet frames. These shortcomings are a concern when collecting evidence because they can interfere with a complete and accurate copy of the network traffic. To avoid these shortcomings, a hardware tap such as those made by Finisar[3] or NetOptics[4] can be used to connect more than one device to the switch port of interest. In this way, a sniffer can collect an exact copy of network traffic and any error information relating to the switch port can also be collected. Error information is important from a documentation standpoint because it shows if any frames were dropped. The main limitation of using a SPANned port or a hardware tap is that the sniffer cannot see local traffic between computers on the same subnet, only traffic entering and leaving the subnet through the switch. Special switches are available that can be configured to give a sniffer access to all traffic passing through the switch, including local traffic.

[3]*http://www.finisar.com*
[4]*http://www.netoptics.com*

In the previous discussion, a sniffer was being installed on the same physical network segment as the compromised host. However, a sniffer can be installed at different locations on a network to capture specific information. For instance, if investigators are interested in traffic to and from an individual's home computer, they can install a sniffer on the suspect's Internet Service Provider network. The DCS1000 (a.k.a. Carnivore) used by the FBI can detect which IP address is assigned to a given dial-up user and monitor only traffic to and from that IP address. In other situations, when all traffic entering a large network might contain digital evidence, a sniffer can be placed near the main point of entry to the network such as the Internet border. Some organizations install Argus probes and intrusion detection systems (essentially special purpose sniffers) at such points on their network to detect attempted intrusions and other anomalies. Logs from these systems can be very useful in an investigation and if more organizations maintained such logs it would be much easier to track down offenders. Although an organization may have the legal right to monitor network traffic it may have policies against such monitoring given the potential privacy violation.

Be aware that it is not possible to use a sniffer when connected to a network via a modem. Unlike NICs, modems cannot be put into promiscuous mode. Furthermore, for a sniffer to work, the computer must be on the same network as the computers being sniffed. Since there are only two modems connected to a dial-up connection (one at each end) there are no other computers to sniff.

16.4.2 SNIFFER CONFIGURATION

[5] *http://www.tcpdump.org*

As noted at the beginning of this chapter, sniffers can capture entire frames, so this form of eavesdropping also collects evidence from the transport and network layers. However, by default some sniffers (e.g. **tcpdump**[5]) only capture 68 bytes of each Ethernet frame, resulting in an incomplete copy of network traffic. Therefore, when collecting evidence, it is important to configure whichever sniffer is being used to collect complete frames. Most modern Ethernet networks use maximum frame size of 1514 bytes but higher speed networks such as ATM have larger Maximum Transfer Units (MTU). To ensure that the entire frame is collected, it is generally advisable to configure sniffers with a large maximum value such as 65535 bytes (Ethereal uses 65535 as a default).

When collecting network traffic, the *de facto* standard is to store the data in a tcpdump file with a ".dmp" extension. For instance, the following command stores all network traffic in a tcpdump file named case 001-04032003-01.dmp and also specifies a maximum size of 65535 bytes:

```
examiner1% tcpdump -w case001-04032003-01.dmp -s 65535
tcpdump: listening on eth0
^C
5465763 packets received by filter
0 packets dropped by kernel
examiner1% md5sum case001-04032003-01.dmp
3bd1154c4f3cb6813c074e404cf9ca10 case001-04032003-01.dmp
```

Once the collection process is complete, the MD5 value of the tcpdump file can be calculated to document its integrity and the data can be preserved on CD-ROM or some other write-only medium.

16.4.3 OTHER SOURCES OF MAC ADDRESSES

As noted earlier, ARP tables contain MAC addresses that can be useful in an investigation. Some organizations keep log ARP information on their network using tools like ARPwatch[6] to detect suspicious activities such as an individual reconfiguring a host with another IP address to misdirect investigators or ARP table poisoning – a technique for sniffing on switched networks. If there are no such ARP logs, investigators might be able to obtain relevant IP → MAC address associations from the ARP table on a router using a command like `show ip arp`. Although every host on a network has an ARP cache, the ARP table on a router is the most useful because it contains the IP → MAC address associations for all of the hosts it communicated with recently. As discussed in the previous chapter, the collection of volatile data such as the ARP table can be documented by taking photographs or print screens, cutting and pasting the contents into a file, or using the logging capabilities of a program like Hyperterminal when connecting to routers and other network devices.

[6] *ftp://ftp.ee.lbl.gov*

Some organizations maintain a list of authorized MAC addresses along with information about the system owners. This information is used for security purposes, making it more difficult for malicious individuals to connect a computer to the network. For instance, MAC addresses are used by the Dynamic Host Configuration Protocol (DHCP is discussed in the next chapter) to assign IP addresses to authorized computers on a network. If the MAC address is not registered with the DHCP server, it will not be automatically assigned an address. This is not foolproof from a security standpoint since the malicious individual could simply configure their computer with an IP address on the network. Therefore, some organizations take the added precaution of configuring their switches and 802.11 Access Points to only accept certain MAC addresses. Again, this is not foolproof since the malicious individual could reconfigure his/her computer with a recognized MAC address but each layer of security makes unauthorized activities more difficult.

These security measures can be useful from an investigative standpoint. If only a limited number of MAC addresses were permitted to connect to a given device, this can limit the suspect pool in an investigation to those authorized computers. Also, even if a DHCP server does not keep a permanent log of each request that it received, it does maintain a database of the most recent requests along with the associated MAC addresses and IP addresses. This DHCP database can be queried to determine the MAC address of the computer that was assigned a given IP address during a given period. For instance, the following DHCP lease shows that the computer with hardware address 00:e0:98:82:4c:6b was assigned IP address 192.168.43.12 starting at 20:44 on April 1, 2001 (the date format is "weekday yyy/mm/dd hh:mm:ss" where 0 is Sunday):

```
lease 192.168.43.12 {
starts 0 2001/04/01 20:44:03;
ends 1 2001/04/02 00:44:03;
hardware ethernet 00:e0:98:82:4c:6b;
uid 01:00:e0:98:82:4c:6b;
client-hostname "oisin";
}
```

The OUI "00e098" in this MAC address indicates the NIC is made by AboCom Systems, Inc., Taiwan, Republic of China, providing a useful class characteristic.

CASE EXAMPLE

An employee received a harassing e-mail message that was sent from a host on the employer's network with IP address 192.168.1.65. The DHCP server database indicated that this IP address was assigned to a computer with MAC address 00:00:48:5c:3a:6c at the time the message was sent. This MAC address was on the organization's list of MAC addresses but was associated with a printer that had been disconnected to the network. However, examining the router's ARP table revealed that the IP address 192.168.1.65 was being used by another computer with MAC address 00:30:65:4b:2a:5c. Although this MAC address was not on the organization's list, there were only a few Apple computers on the network and the culprit was soon found.

16.5 ANALYSIS TOOLS AND TECHNIQUES

It is useful to understand what the network traffic looks like in its most basic form. An actual Ethernet frame (encapsulating an IP packet) looks like this in hexadecimal:

```
08 00 5a 47 43 58 08 00 20 21 fb 7d 08 00 45 00 00 1d c0 fa 00 00 3c 11
00 a2 0a 17 2d 43 0a 17 2d 4414 0e 0f d4 00 0d 3c bc 72 6f 6f 74 00 00 00
00 00 00 00 00 00 00 00 00 00
```

As noted in Table 16.2 showing the general Ethernet frame structure, the bytes represent the following.

Table 16.4

Break down of an Ethernet frame in hexadecimal.

08 00 5a 47 43 58	Source Ethernet address (OUI IBM Corporation)
08 00 20 21 fb 7d	Destination Ethernet address (OUI Sun Microsystems)
08 00	denotes the fact that this frame contains an IP packet
45 00 00 1d c0 fa 00 00 3c	part of the IP header (version, length, etc.)
11	indicates that the packet contains UDP data (11; 17 decimal) not TCP data (06), etc.
00 a2	checksum used to verify that the packet was not damaged in transit
0a 17 2d 43	source IP address (10.23.45.67)
0a 17 2d 44	destination IP address (10.23.45.68)
14 0e 0f d4 00 0d 3c bc	UDP source port (5134), destination port (4052), header length and checksum
72 6f 6f 74	The word "root" in hexadecimal
00 00 00 00 00 00 00 ...	The rest is padding

When analyzing network traffic, it is generally desirable to know what time events occurred. The tcpdump format includes date–time stamps for each frame that was captured but some tools, including **tcpdump** itself, only display the time and not the date.[7] For instance, using **tcpdump** to view the file named *hotmail-02242003.dmp* – available on the Web site associated with this book – does not display the date, only the time.

[7] The date–time stamps in tcpdump files are stored in UNIX epoch time – a 32-bit hex value representing the number of seconds since January 1, 1970.

```
examiner1% tcpdump -r hotmail-02242003.dmp
15:59:15.501154 192.168.0.5.32769 > 192.168.0.1.53: 6342+ A?
www.hotmail.com. (33) (DF)
```

Looking at the beginning of the same tcpdump file shows a date–time value of A3875A3E, which equates to Monday, February 24, 2003 15:59:15 GMT-0500:

```
D4C3B2A1 02000400 00000000 00000000  |  ╘├█í   ....   ....   ....   |  16
DC050000 01000000 A3875A3E A2A50700  |  ■...   ....   úçZ>   óÑ..   |  32
4B000000 4B000000 0030AB1D CDEF0000  |  K...   K...   .0½.   =∩..   |  48
E28AC46B 08004500 003D750D 40004011  |  Γè─k   ..E.   .=u.   @.@.   |  64
444CC0A8 0005C0A8 00018001 00350029  |  DL└¿   ..└¿   ..Ç.   .5.)   |  80
A6DE18C6 01000001 00000000 00000377  |  ª▐.╞   ....   ....   ...w   |  96
77770768 6F746D61 696C0363 6F6D0000  |  ww.h   otma   il.c   om..   | 112
010001A3 875A3E54 AE07003C 0000003C  |  ...ú   çZ>T   «..<   ...<   | 128
000000FF FFFFFFFF FF0030AB 1DCDEF88  |  ...           .0½   .=∩ê   | 144
63110900 00000C01 01000001 03000431  |  c...   ....   ....   ...1   | 160
```

Because this file was created on an Intel system, the date–time values are in little-endian format (e.g. A3875A3E) whereas a tcpdump file created on a Solaris machine has date–time values in big-endian format (e.g. 3E5A87A3).

16.5.1 KEYWORD SEARCHES

In some cases, it may be sufficient during an examination to search a tcpdump file for a specific keyword. For instance, usernames and passwords for file transfer, e-mail, and other services can be found by searching the keywords "USER," "PASS," and "login" as shown here using a simple UNIX utility called **ngrep**:[8]

[8] http://ngrep.sourceforge.net

```
examiner1% ngrep -w 'USER|PASS|login' -t -x -s 65535 -I case02-04032003.dmp
input: case02-04032003.dmp
match: ((^USER|PASS\W) | (\WUSER|PASS$) | (\WUSER|PASS\W))
#####################################
T 2003/04/03 10:07:39.066816 192.168.0.5:32788 → 172.16.1.10:21 [AP]
  55 53 45 52 20 61 72 67     6f 6e 69 6d 6f 6e 0d  0a   USER argonimon..
##########
```

```
T 2003/04/03 10:08:01.956350 192.168.0.5:32788  →  172.16.1.10:21 [AP]
  50 41 53 53 20 70 61 73   73 77 6f 72 64 2d 72   65  PASS password-re
  76 65 61 6c 65 64 0d 0a                               vealed..
##########
T 2003/04/03 10:24:59.182353 192.168.0.5:32869  →  172.16.1.23:143 [AP]
  32 20 6c 6f 67 69 6e 20   22 6e 61 6d 65 22 20 22   2 login "name" "
  70 61 73 73 77 6f 72 64   2d 72 65 76 65 61 00 00   password-revea..
  09 01 00 00                                           ....
###################################################exit
```

Similarly, when looking for connections to IRC, searching for nicknames and channel names may provide all of the information that a digital investigator requires. In the aforementioned "hotmail-02242003.dmp" file, searching for packets containing the keyword "POST" can reveal the act of the suspect sending a message (Figure 16.7). The "HTTP POST" command corresponds to the act of sending a Hotmail message.

Although **tcpdump** and Argus do not have a keyword search feature, they can be used in combination with **grep** to find items of interest.

16.5.2 FILTERING AND CLASSIFICATION

When dealing with large amounts of data involving many hosts, it is often necessary to focus the examination on certain protocols or traffic to and from specific hosts. The **tcpdump** program enables filtering based on certain criteria but uses the **libpcap** filter syntax, which is complex. For instance, the following **tcpdump** arguments can be used to examine traffic from a single host (192.168.0.5) to a given network (any IP address starting with

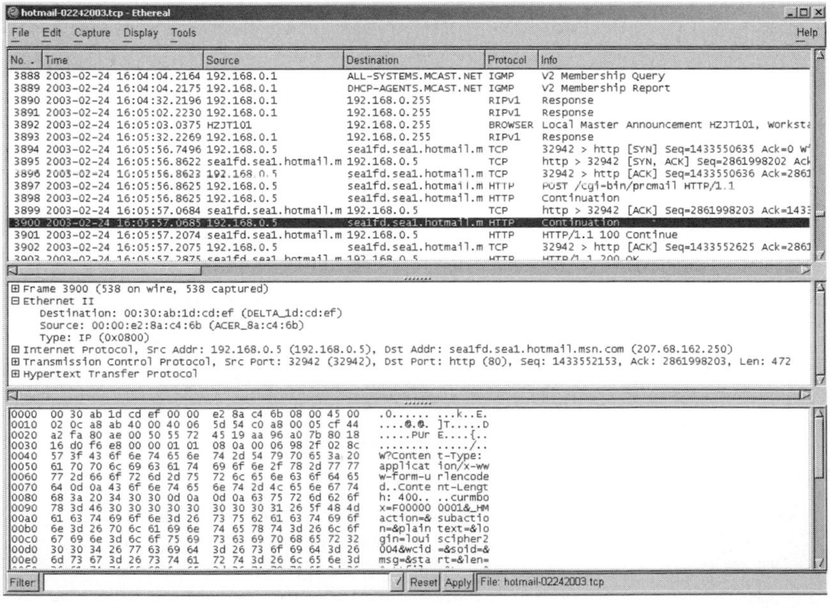

Figure 16.7

Ethereal showing packet in "hotmail-02242003.dmp" file containing the keyword "POST," corresponding to the act of sending the message through Hotmail.

172.16.), excluding traffic to ports 21, 53, and 80:

```
# /usr/sbin/tcpdump -nex -s 65535 -r case001-04032003-02.dmp src host
192.168.0.5 and dst net 172.16.0.0/16 and dst port not (21 or 53 or 80)
```

Additionally, tcpdump can only recognize and extract a limited number of protocols, including TCP and UDP. To extract only Web traffic, for instance, one might look for traffic to port 80 but this would miss relevant Web traffic if the server was using a different port, such as 8080. Argus can be used to examine tcpdump files and uses a similar filter syntax as tcpdump but has more options and keeps track of session state information. Ethereal provides more filtering functionality using a slightly less complex syntax and supports more protocols. For instance, the above filter can be implemented in Ethereal using the following syntax:

```
ip.src == 192.168.0.5 and ip.dst == 131.243.0.0/16 and not (ftp or dns or http)
```

Although Ethereal supports more protocols than tcpdump, it makes some assumptions about the expected behavior of protocols that prevent it from automatically classifying traffic that does not meet these basic assumptions. For instance, Ethereal does not automatically recognize and classify FTP traffic when a port other than the default port (21) is used. However, once the digital evidence examiner correctly classifies the FTP traffic, Ethereal can be instructed to interpret the data using the "Decode As" feature on the Tools menu.

Some commercial products have more features than these free tools that facilitate traffic filtering and classification. For instance, Figure 16.8(a) and (b) shows NetIntercept being used to locate and view the same information shown in Figure 16.7.

NetIntercept's graphical user interface allows the examiner select criteria for filtering such as source and destination IP addresses within a certain time period. Also, NetIntercept interprets protocols rather than simply making assumptions based on default ports. By interpreting protocols, this tool can extract noteworthy elements (e.g. usernames, passwords, files, credit card numbers) and store them in a database to facilitate examination and analysis. This protocol analysis feature is also useful for finding traffic that violates expected behavior such as an FTP server running at a non-standard port. NetIntercept lists all such anomalies in the Alerts section and can generate a printable report of this information. This protocol anomaly detection feature is conceptually similar to the file signature mismatch detection provided by

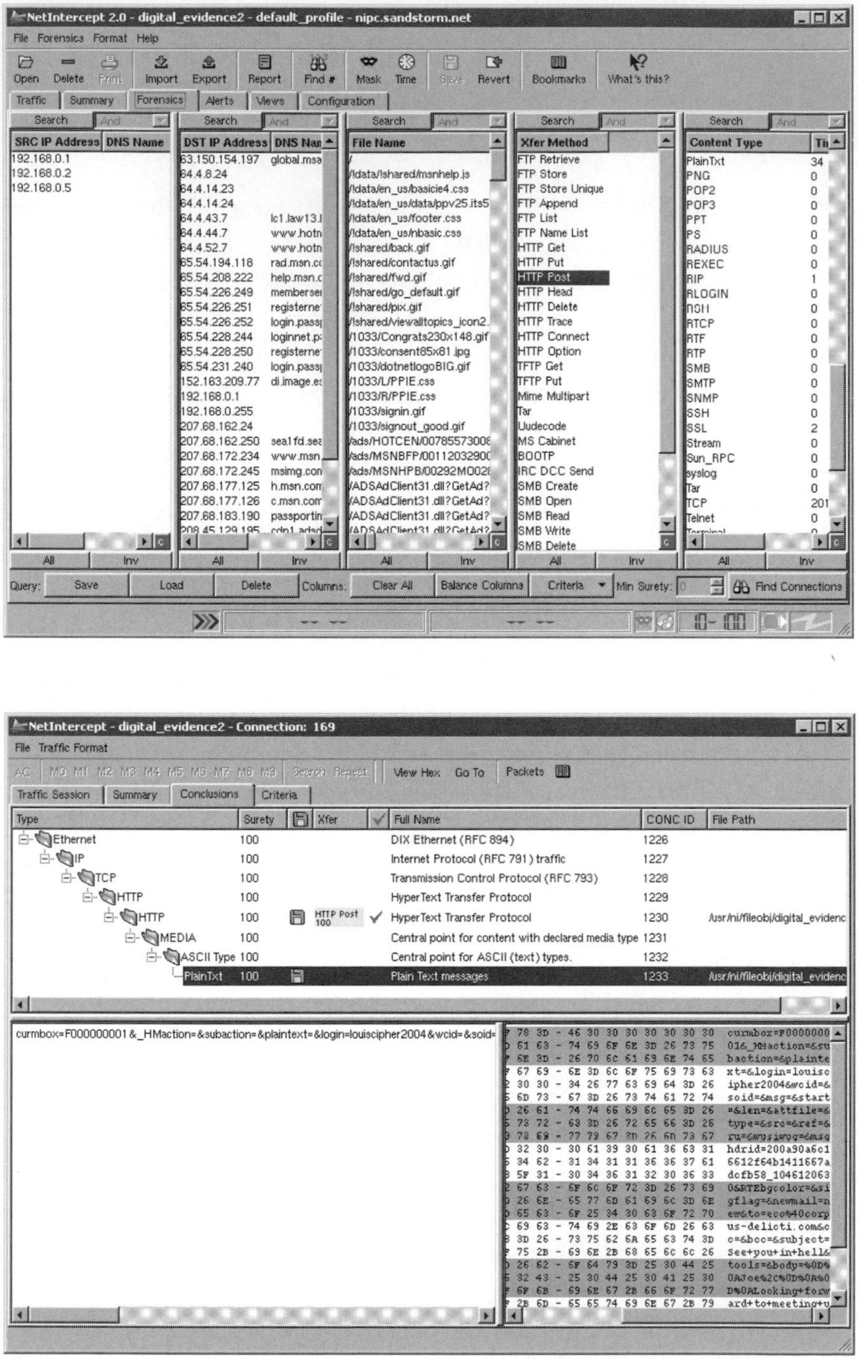

Figure 16.8

(a) Using the NetIntercept forensics view to examine network traffic and locate important items such as an "HTTP POST." (b) Using NetIntercept to view the same packet as Figure 16.7 containing the "POST" keyword.

most media examination tools like FTK and EnCase. NetIntercept can generate other useful reports from network traffic, including traffic statistics and an inventory of components in Web traffic that is conceptually similar to an inventory of files on a disk.

16.5.3 RECONSTRUCTION

It is often desirable to reconstruct related packets into complete messages or sessions. For example, data contained in captured frames might be reassembled to form an e-mail message or Web page. Ethereal can be used to reconstruct streams in a rudimentary way (recall Figure 15.4), but can be cumbersome for large amounts of data and has some limitations from a digital evidence examination standpoint. For instance, Figure 16.9 shows the

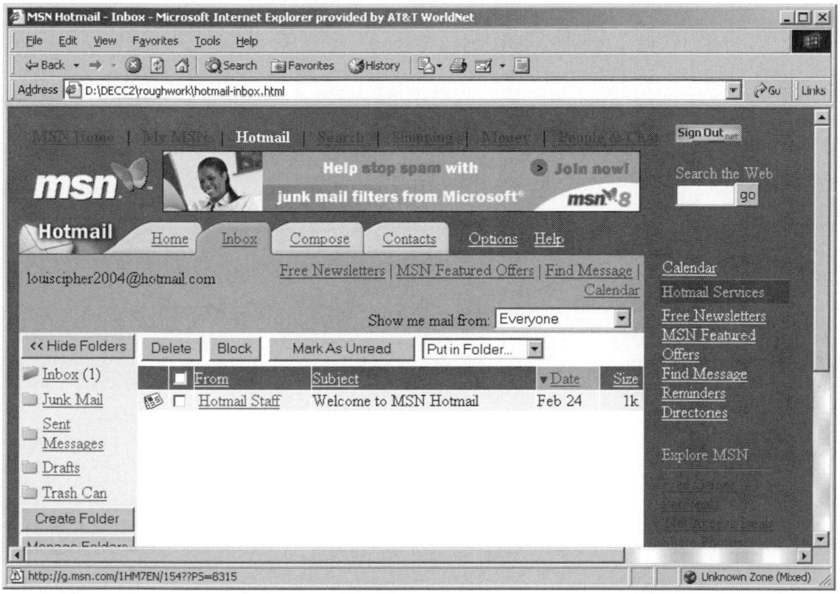

Figure 16.9

Hotmail Inbox recovered using Ethereal.

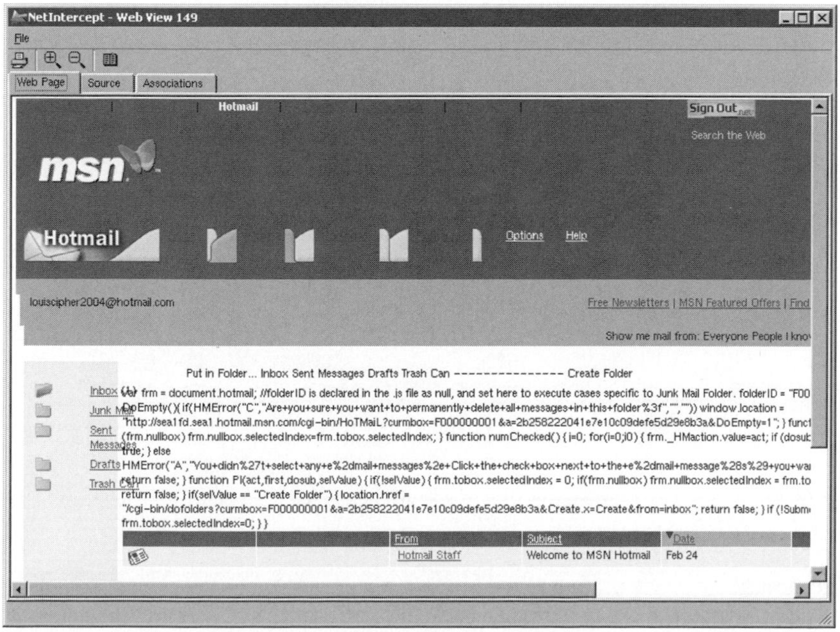

Figure 16.10

Hotmail Inbox extracted from a tcpdump file and displayed using NetIntercept.

Hotmail Inbox recovered from the "hotmail-02242003.dmp" file using Ethereal. The banner advertisement at the top of the Web page was not present in the original traffic – it was automatically updated from the Internet when the reconstructed page was opened in a Web browser. At the very least, this spoliation of the evidence should be avoided by performing the examination on a computer that is not connected to the Internet. This also demonstrates the importance of understanding the limitations and quirles of tools being used to examine digital evidence.

Some commercial tools are specifically designed for digital evidence examination and provide more visualization features that make it more efficient to examine large amounts of network traffic. For example, NetIntercept can also reconstruct and extract content from network traffic, such as Web pages, files transferred using FTP, and Word documents contained in MIME encoded e-mail attachments. Figure 16.10 shows the

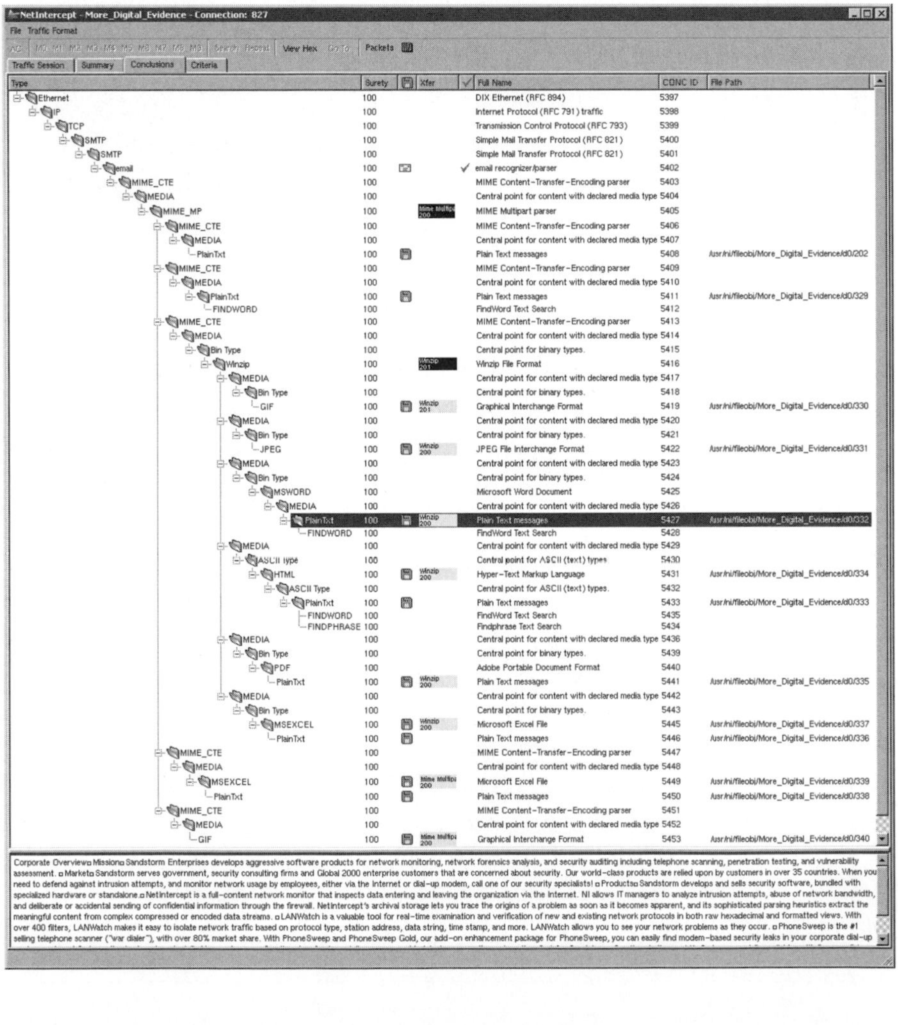

Figure 16.11

MIME encoded e-mail attachments containing data in a ZIP file extracted from a tcpdump file and displayed using NetIntercept.

Hotmail Inbox shown in Figure 16.9 but reconstructed and displayed using NetIntercept. Notably, the banner advertisement at the top of the Web page is the original one from the "hotmail-02242003.dmp" file. Also, to protect the examiner's machine from malicious code, NetIntercept displays reconstructed Web pages in a protective viewer that does not execute scripts but does display them in raw form to facilitate analysis. Figure 16.11 shows NetIntercept displaying the content of several word documents and other files stored in a Zip file that was attached to an e-mail message. By decoding attachments and compressed archives in the way, NetIntercept can perform keyword searches on their content.

16.6 SUMMARY

The physical and data-link layers are one of the richest sources of digital evidences on a network. Data-link layer addresses (MAC addresses) are more identifying than network layer addresses (e.g. IP addresses) because a MAC address is usually directly associated with the Network Interface Card in a computer whereas an IP address can be easily reassigned to different computers. Eavesdropping can provide a large amount of evidence that can give investigators a detailed view of what a criminal is doing. Also, data captured using a sniffer can be very useful for reconstructing a crime or verifying that other sources of digital evidence contain accurate information. For example, if the accuracy of log files that summarize events is in doubt, data captured using a sniffer can be used to corroborate entries in the logs.

Until recently, logs of activities at the physical and data-link layers were rarely kept. Logging every piece of information that passes through a network, including all of the ARP requests and replies, can result in very large log files. However, as disk space becomes cheaper and monitoring tools like Argus developed, more organizations are retaining such logs. Without these kinds of logs, it is more difficult to obtain digital evidence from the physical and data-link layers because the majority of the data are transient. The ARP table on most computers only keeps entries for 20 minutes, DHCP database entries are regularly overwritten, and data traveling through the network is only available for capture for a fraction of a second.

REFERENCES

Garfinkel S. (2002) "Network Forensics: Tapping the Internet", O'Reilly Network (Available online at http://www.oreillynet.com/pub/a/network/2002/04/26/nettap.html)

Graham R. (2000) "Sniffing (network wiretap, sniffer) FAQ", (Available online at http://www.robertgraham.com/pubs/sniffing-faq.html)

CHAPTER 17

DIGITAL EVIDENCE AT THE NETWORK AND TRANSPORT LAYERS

For a communication system to work it must have an addressing mechanism. Often, there is also a need for some form of verification that a message has reached its destination. Take a postal service as an example. Addresses are used to direct letters and, when necessary, the postal service will inform the sender when a letter has been delivered. Similarly, computer networks require an addressing scheme and sometimes a method for confirming that information has been delivered. The network and transport layers are responsible for these important aspects of computer networks.

Activities on the network and transport layers generate information that is often critical in an investigation. Log files contain information about activities on the network, when they occurred, and the addresses of the machines involved. State tables contain information, including IP addresses, about current or very recent connections between hosts. The IP addresses in log files and state tables can be used to determine the point of origin of a crime, thus leading investigators to likely suspects. Additionally, these sources of digital evidence are useful for investigative reconstruction and are crucial for establishing the continuity of offense.

Processing and analyzing evidence on the network and transport layers is like digging into the glue that holds a network together. This digging can turn up a lot of information but you have to be willing to roll up your sleeves and get your hands dirty. In other words, you have to become familiar with the technical details of these layers to take advantage of them as a source of digital evidence.

To understand how the networks and transport layers work it is helpful to examine a specific example. TCP/IP is a good example because it is the most commonly used implementation of the network and transport layers – it is a fundamental part of the Internet. This chapter provides an overview of how TCP/IP and related systems, such as the Domain Name System, work. This chapter also describes how TCP/IP can be involved in crimes and discusses how forensic science can be applied to digital evidence on the network and transport layers. Analogies are used to clarify technical concepts and many

Digital Evidence and Computer Crime Second Edition
ISBN: 0-12-163104-4

minute details are omitted for the sake of simplicity. References are provided at the end of the chapter for investigators wishing to learn more about TCP/IP.

In addition to describing TCP/IP in detail, this chapter provides a brief overview of cellular data networks. Cellular phones and other hand-held devices can be used to access the Internet and they depend on computer networks that are similar to the Internet in many respects. These similarities are emphasized to enable investigators to generalize their knowledge of the network and transport layers and use that knowledge to understand other internetworks.

17.1 TCP/IP

TCP/IP is a combination of protocols that includes the Internet Protocol (IP), Transport Control Protocol (TCP), and User Datagram Protocol (UDP). IP functions at the network layer, addressing and routing data. TCP operates on the transport layer – acknowledging receipt of information and resending information when necessary. UDP is a very simple protocol that some applications use instead of TCP when an acknowledgement of receipt is not desired or when acknowledgements are handled by the application. These transport layer protocols are designed to ameliorate the common problems that arise on a network, including hardware failure, network congestion, and data delay, loss, and corruption and sequencing errors (Figure 17.1).

When a large number of hosts are competing to use the same wires and hardware on a network, some fair method of sharing these resources is necessary. To enable equal sharing of the network, TCP and UDP break data into small packets (a.k.a. datagrams) before they are transmitted.

Figure 17.1

TCP/IP diagram with OSI layers superimposed.

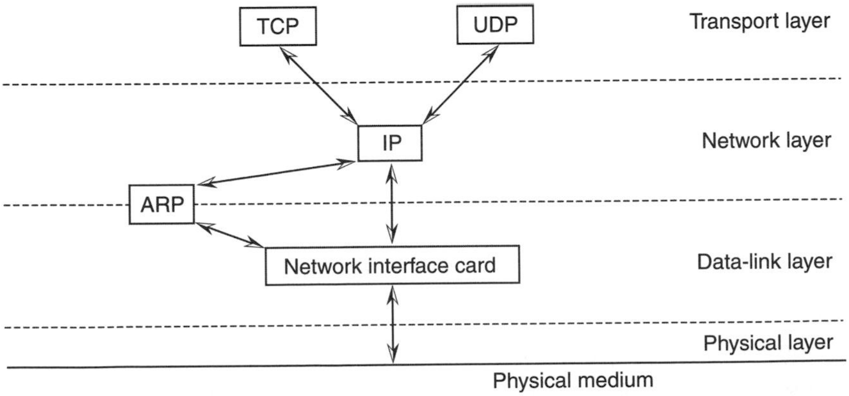

Breaking data into packets prevents large messages from monopolizing the network and enables two hosts to open multiple lines of communication on a single physical wire. For example, two hosts can exchange e-mail, Web pages, and Usenet messages simultaneously by breaking the information into packets and putting the packets on the network, entrusting routers to direct packets to their destination where they are reconstituted. This type of network is called a *packet-switched* network to differentiate it from the more expensive and reliable circuit-switched networks.

> Circuit-switched networks operate by forming a dedicated connection (circuit) between two points. The US telephone system uses circuit switching technology – a telephone call establishes a circuit from the originating phone through the local switching office, across trunk lines, to a remote switching office, and finally to the destination telephone... The advantage of circuit switching lies in its guaranteed capacity: once a circuit is established, no other network activity will decrease the capacity of the circuit. One disadvantage of circuit switching is cost: circuit costs are fixed, independent of traffic. For example, one pays a fixed rate for a phone call, even when the two parties do not talk.
>
> Packet-switched networks, the type used to connect computers, take an entirely different approach... The network hardware delivers the packets to the specified destination, where software reassembles them into a single file again. The chief advantage of packet-switching is that multiple communications among computers can proceed concurrently, with intermachine connections shared by all pairs of machines that are communicating. The disadvantage, of course, is that as activity increases, a given pair of communicating computers receives less of the network capacity. That is, whenever a packet-switched network becomes overloaded, computers using the network must wait before they can send additional packets. (Comer 1995)

17.1.1 INTERNET PROTOCOL AND CELLULAR DATA NETWORKS

On the network layer, the Internet Protocol (IP) is primarily responsible for addressing and routing information. After TCP breaks data into packets, IP addresses each packet and adds some other information (recall Table 16.2). Cellular digital packet networks use network layer protocols like IP to address packets. Although GPRS does not quite follow the OSI model, it supports TCP/IP using a tunneling protocol. The following scenario describes the potential of wireless packet-switched networking if you were traveling between Los Angeles and Las Vegas:

> You boot up your notebook computer with its CDPD wireless modem enroute to your office in Los Angeles. The ride from Las Vegas to Los Angeles will take several hours, but you can't wait. You've got to check your e-mail for an important message regarding your biggest client. Let's look at the concepts that allow you to do this.
>
> When your wireless modem initiates a connection, a registration process is started that provides your remote device with access to your home carrier's wireless network. Your

> wireless modem is homed to a specific router that will keep track of your location and all messages intended for you will be forwarded to that router.
>
> When you move out of your home [region], this home router will forward your packets to another router, which in turn directs traffic within the group of [neighboring regions] you are in at that particular time. This method keeps routing updates to a minimum and allows you to roam freely, from [region] to [region] or city to city. (Henry and DeLibero 1996)

17.1.2 IP ADDRESSES

Each computer attached to the Internet has a unique address, called an IP address. Each IP address comprises of two parts, the network number and the host number. The network number is a unique number that identifies a computer network attached to the Internet and the host number is a unique number that identifies a computer on that network. This is conceptually the same as a telephone number that has an area code and a local number (Figure 17.2).

Figure 17.2

IP addresses are conceptually the same as telephone numbers.

To accommodate networks of different sizes, three classes of addresses were agreed upon (Table 17.1).

Table 17.1

IP address classes.[a]

	IP ADDRESS RANGE	EXAMPLE NETWORK # (N) AND HOST # (H)
Class A	1.0.0.0–126.0.0.0	124.11.12.13 is network 124, host 11.12.13
Class B	128.0.0.0–191.0.0.0	156.134.15.16 is network 156.134, host 15.16
Class C	192.0.0.0–223.0.0.0	192.132.12.13 is network 192.132.12, host 13

[a]Several IP address ranges (10.0.0.0–10.255.255.255, 172.16.0.0–172.31.0.0, and 192.168.1.0–192.168.1.255) are set aside for private use and are not used in the same way as other IP addresses.

These classes of IP addresses are like real estate on the Internet. Class A is prime Internet real estate because it can accommodate up to 16,777,214 hosts, whereas a Class C network can only fit 254 hosts. The larger Class A and Class B networks are usually divided into *subnets* to make them more manageable. The most common subnet size is 254 hosts but *subnet masks* permit few hosts per subnet.

Although each computer on the Internet has a unique IP address, computers can be reconfigured with a different IP address quite easily, enabling criminals to misdirect investigators. What prevents an offender from changing the IP address of his computer prior to committing a crime, making it appear to come from another host on the network? The answer depends on the circumstances. For instance, when a dial-up connection is used (e.g. PPP), the ISP assigns an IP address to the connection. Under these circumstances, it is not possible for the offender to reconfigure his computer with another IP address. When a computer is connected to an Ethernet network, it can be configured with any IP address. However, routers segregate networks into subnets, and the offender can only reconfigure his computer with another IP address on the same subnet.[1]

[1] *Some routers are configured insecurely to permit outgoing packets from a masquerading host that is configured with an IP address that is not on the same subnet. However, TCP responses to these packets would be sent to the actual network that contains this IP address and not to the masquerading host. Although a bi-directional TCP connection cannot be established, this flaw can be used to launch a denial of service attack, making it appear to originate from a different network.*

17.1.3 DOMAIN NAME SYSTEM

Although computers work well with numbers, people are more comfortable with names. For convenience, the Domain Name System (DNS) was created to assign names to IP addresses. For example, the canonical name for 64.39.2.185 is "cirrus.rackspace.com" as shown here using nslookup – a command that comes with Windows and UNIX for querying the DNS:

```
C:\> nslookup 64.39.2.185
Name:      cirrus.rackspace.com
Address:   64.39.2.185
Aliases:   www.rackspace.com
```

Notably, this IP address also has a secondary "alias" entry in DNS (www.rackspace.com). Whenever a name is used to refer to a computer (e.g. typing the name of a Web site into a browser), the DNS works behind the scenes to determine the associated numerical IP address.

Another useful tool for querying DNS is called dig (Domain Information Groper), available on UNIX systems and in the NetScanTools Pro for Windows.[2] The following dig results for the above IP address show its name and authoritative DNS servers. Authoritative DNS servers are the servers that all other servers in DNS rely on for the correct information relating to a given host:

[2] *http://www.nwpsw.com*

```
% dig -x 64.39.2.185

; <<>> DiG 9.2.1 <<>> -x 64.39.2.185
;; global options: printcmd
;; Got answer:
;; ->>HEADER<<- opcode: QUERY, status: NOERROR, id: 64879
;; flags: qr rd ra; QUERY: 1, ANSWER: 1, AUTHORITY: 2, ADDITIONAL: 0
```

```
;; QUESTION SECTION:
;185.2.39.64.in-addr.arpa.          IN      PTR

;; ANSWER SECTION:
185.2.39.64.in-addr.arpa. 86400     IN      PTR     cirrus.rackspace.com.

;; AUTHORITY SECTION:
2.39.64.in-addr.arpa.    86400      IN      NS      ns2.rackspace.com.
2.39.64.in-addr.arpa.    86400      IN      NS      ns.rackspace.com.

;; Query time: 89 msec
;; SERVER: 192.168.0.1#53(192.168.0.1)
;; WHEN: Mon Apr 7 18:21:38 2003
;; MSG SIZE rcvd: 111
```

It is sometimes possible to obtain a list of all machines in the DNS belonging to a specific organization (a.k.a. domain or zone) by performing a *zone transfer* as shown in Figure 17.3 using NetScanTools Pro.

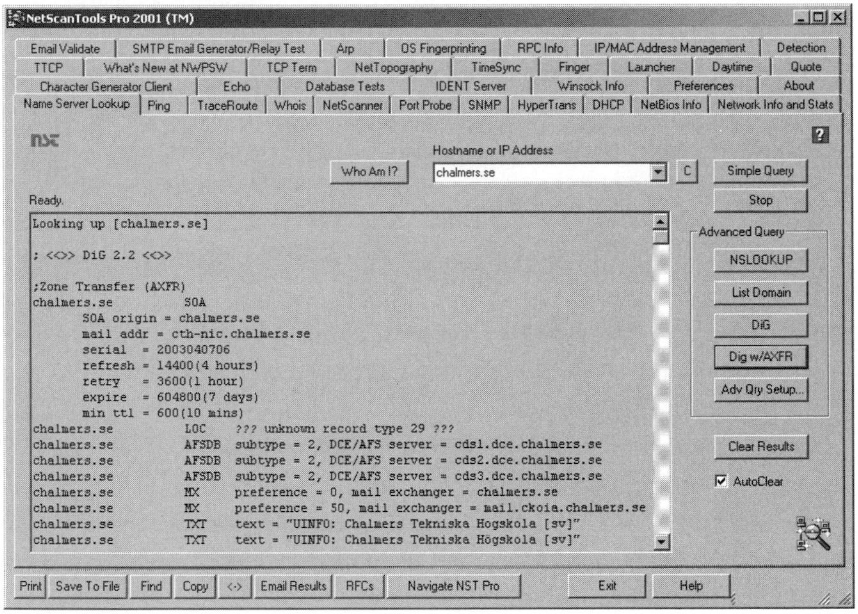

Figure 17.3

A zone transfer using NetScanTools Pro requires the DNS server to be set to one of the target system's DNS servers under Advanced Query Options (accesses using the "Adv Qry Setup" button).

A zone transfer can be obtained on UNIX using the command **dig @ns.domain.com domain.com AXFR**. However, because computer intruders can use information in a zone transfer to plan an attack on a network, some DNS servers do not permit this type of query.

17.1.4 IP ROUTING

Once addressed, a packet is ready to venture out onto the Internet where it will be directed to the destination specified in the IP header. For example, when a computer in Baltimore sends information to yale.edu in New Haven, the information must pass through several intermediate routers. The IP software on each router contains a routing table that it uses to determine where to send information (Figure 17.4).

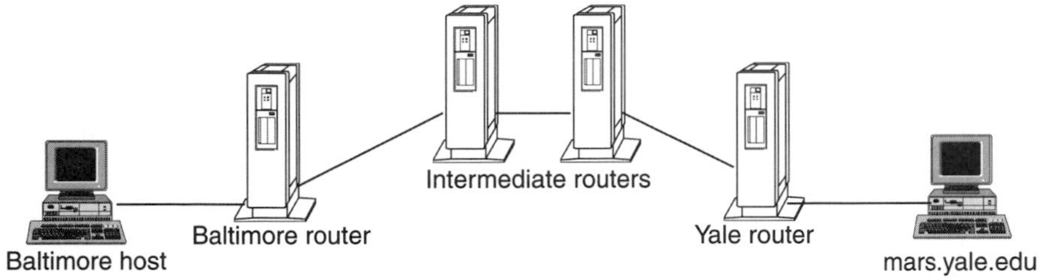

Figure 17.4
IP Routing.

An analogy might clarify how routing tables work. Imagine someone driving a car from Baltimore to New Haven and reaching a junction with three signs. One sign indicates that Philadelphia is straight ahead, another sign indicates that Atlantic City is to the right, and a third sign indicates that all other locations are to the left. Therefore, the driver goes right and continues until reaching another junction. The driver repeatedly follows the road signs until finding one that says "New Haven," indicating that the destination city has been reached. All that remains is for the driver to find the specific building that he/she is looking for. Routing tables are the road signs on the information superhighway. When a packet is traveling from Baltimore to New Haven, the routers that it passes through are like junctions and the routing tables are used to determine where the packet should go next to reach its destination. When the packet finally reaches the network that it is destined for, all that remains is for a router to direct the packet to the correct host. To extend the analogy, networks use different protocols for short and long distance routing just as people use different road signs when travelling short and long distances.

A program called **traceroute** provides a list of routers that information passes through to reach a specific host. For instance, the route that a packet takes between a host in Baltimore and yale.edu is shown here:[3]

```
% traceroute yale.edu
traceroute to yale.edu (130.132.59.127), 30 hops max, 40 byte packets
 1 a6-0-0-1710.q-esr1.balt.verizon-gni.net (151.196.4.194) 126.933 ms 17.403 ms
   18.702 ms
 2 dca-edge-04.inet.qwest.net (63.238.58.233) 18.934 ms 39.274 ms 24.343 ms
 3 dca-core-02.inet.qwest.net (205.171.9.65) 20.827 ms 85.062 ms 19.051 ms
 4 ewr-core-03.inet.qwest.net (205.171.8.182) 24.504 ms 95.07 ms 25.54 ms
 5 ewr-core-02.inet.qwest.net (205.171.17.33) 24.121 ms 23.582 ms 22.059 ms
 6 bos-core-01.inet.qwest.net (205.171.8.28) 31.766 ms 27.12 ms 27.171 ms
 7 bos-edge-02.inet.qwest.net (205.171.28.14) 28.826 ms 28.482 ms 29.089 ms
 8 63.145.0.14 (63.145.0.14) 32.776 ms 32.485 ms 31.323 ms
 9 greed.net.yale.edu (130.132.1.39) 109.16 ms 37.569 ms 36.242 ms
10 yale.edu (130.132.59.127) 112.104 ms 32.962 ms 53.772 ms
```

[3] *Basically,* **traceroute** *obtains this information by sending ICMP echo requests (a.k.a. ping) to each intermediate router and displaying the details of the corresponding ICMP echo replies.*

The traceroute program is useful for getting a rough idea of which routers were involved in the transport of information on the Internet. Intermediate routers may have relevant digital evidence in log files as discussed in later chapters. Also, the path that the data took can clarify which borders and boundaries were crossed during the perpetration of a crime. Special purpose programs like Visual Route attempt to superimpose traceroute results on a map to provide related geographical information. However, this geographical information is usually quite general and can be incorrect. Therefore, when seeking digital evidence from a specific router, use Whois databases, described in Chapter 18, to obtain contact information for the people responsible for that router and contact them directly to determine exactly where the desired data are located.

It is a common misconception that routers are more intelligent, finding the "best" route between hosts. Although this is technically possible it is rarely practised at present. Similarly, many people make the mistake of thinking that two packets will take different routes traveling between the same two hosts on the Internet. As can be seen when using traceroute, the route between two hosts remains the same even though the Internet was designed to be flexible. Packets can be forced to take a different path by changing the routing table on one of the intermediate routers, effectively creating a detour. This type of detour can be created manually (e.g. by a network administrator or computer intruder) or using protocols such as BGP and OSPF. However, network administrators only make such changes once in a while and once such a change is made, all packets will follow the same detoured path. Therefore, it is safe to assume that all packets traveling between the host in Baltimore and Yale University take the same route, making it much easier to establish the continuity of offense and locate digital evidence relating to a limited number of intermediate routers. Over longer periods of time, routes change as network administrators make improvements.

17.1.5 SERVERS AND PORTS

When a computer receives packets of an e-mail message, a Web page, and a Usenet message at the same time, how does it distinguish between the different types of data? How does the host know which packets contain pieces of the e-mail and which packets contain pieces of the Web page? Computers use numbers, called *ports*, to distinguish between different types of data.

To clarify, imagine a single computer running an e-mail server and a Web server, each listening for network connections on their default ports (25 and 80, respectively). When the computer receives packets with the number 25 in the port field (Figure 17.5), it assumes that they are e-mail related. If the packets are not e-mail related, the e-mail server will not know what to do with the data and will return an error, crash, or do nothing at all.

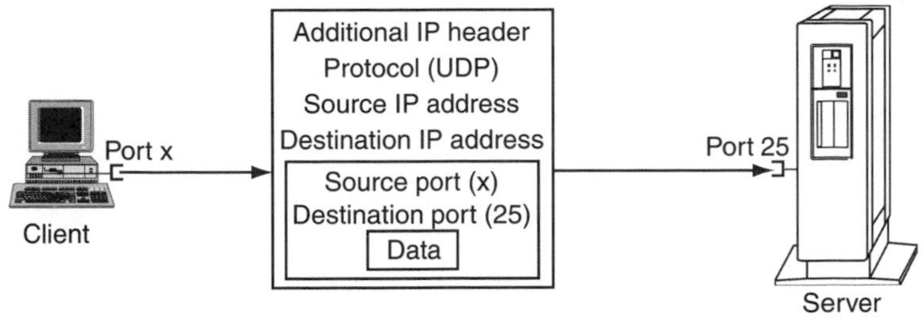

Figure 17.5

UDP packet with port number in the heading being transmitted to a server.

Similarly, the computer receives packets with the number 80 in the port field, it assumes that the packets are intended for the Web server. However, any server can be configured to listen at any port so these port associations are not definitive.[4]

Any host, even a personal computer in someone's home, can function as a server on the Internet. In fact, Windows desktops come with a server that listens for network connections on port 139 and enables resource sharing over networks using NetBIOS. For instance, using a program like nmap to scan a Windows XP machine remotely for listening ports gives the following results:

[4] *A more complete list of port associations is available at: Internet Assigned Numbers Authority (http://www.iana.org/assignments/port-numbers) and in the "services" file that comes with* nmap.

```
remote-scanning-machine% nmap 192.168.0.4

Starting nmap V. 3.00 (www.insecure.org/nmap)
Interesting ports on      (192.168.0.4):
Port          State     Service
135/tcp       open      loc-srv
139/tcp       open      netbios-ssn
445/tcp       open      microsoft-ds
5000/tcp      open      UpnP
31337/tcp     open      unknown
5800/tcp      open      vnc-http
5900/tcp      open      vncs

Nmap run completed — 1 IP address (1 host up) scanned in 34 seconds
```

The above port scan results indicate that another server, called Virtual Network Computer (VNC),[5] is listening for connections on port 5800 and 5900. The VNC program permits full remote control of a computer and has legitimate uses such as remote system administration. However, computer intruders also use VNC and similar programs (e.g. SubSeven, Back Orifice) to gain full remote control over hosts they have broken into.

[5] *http://www.realvnc.com/*

Information about listening ports and any associated connections can be obtained using the netstat command. For instance, executing netstat on the same Windows XP host (192.168.0.4) that was just scanned with nmap produces the following output:

```
C:\>netstat -ano -p tcp
Active Connections
Proto   Local Addresses      Foreign Address      State         PID
TCP     0.0.0.0:135          0.0.0.0:0            LISTENING     912
TCP     0.0.0.0:445          0.0.0.0:0            LISTENING     4
TCP     0.0.0.0:1028         0.0.0.0:0            LISTENING     4
TCP     0.0.0.0:5000         0.0.0.0:0            LISTENING     1124
TCP     0.0.0.0:5800         0.0.0.0:0            LISTENING     2760
TCP     0.0.0.0:5900         0.0.0.0:0            LISTENING     2760
TCP     192.168.0.4:139      0.0.0.0:0            LISTENING     4
TCP     192.168.0.4:1540     0.0.0.0:0            LISTENING     4
TCP     192.168.0.4:1540     192.168.0.2:139      ESTABLISHED   4
TCP     192.168.0.4:5900     172.16.0.15:2512     ESTABLISHED   2760
```

The last connection (in bold) shows that a remote computer (172.16.0.15) is connected to the Windows XP system via VNC on port 5900. Additionally, the second to last line indicates that the Windows XP host is accessing a shared resource on another Windows host (192.168.0.2) using NetBIOS (port 139). Although it is not evident from this information alone whether these connections are legitimate or suspicious, it is clear that someone has full remote control of this Windows XP system via VNC and can access some information on a neighboring host (192.168.0.2) via the NetBIOS connection. This example also demonstrates the importance of correlating data from multiple sources to obtain a more complete picture of what is going on.

17.1.6 CONNECTION MANAGEMENT

Remember that on a packet-switched network, computers are not connected using dedicated circuits. Instead, to make large-scale internetworking more reliable, TCP creates what are called *virtual circuits* (a.k.a. *TCP streams*), establishing, maintaining, and terminating connections between hosts. To establish a virtual circuit, TCP performs a three-way handshake (Figure 17.6). First, host A asks host B for a connection by sending what is commonly known as a SYN packet.[6]

[6] *A SYN packet contains the special SYN bit that indicates that host A wants to synchronize sequence numbers with host B. TCP uses sequence numbers to keep packets in order.*

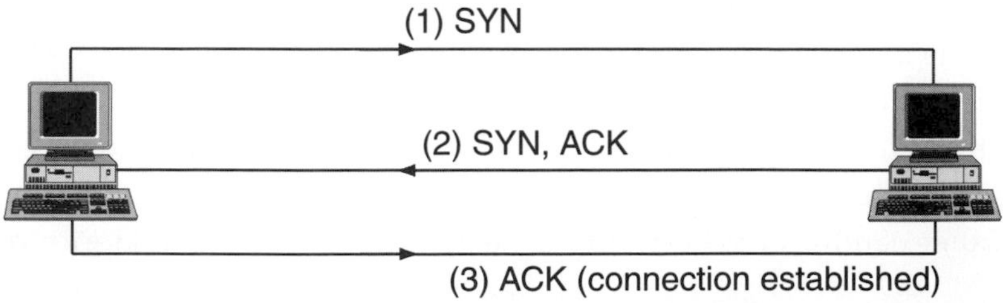

Figure 17.6

TCP establishing a connection using a three-way handshake.

Second, host B acknowledges host A's request by returning a packet containing the special acknowledgment (ACK) bit (this acknowledgment packet also contains a SYN bit to enable the host to synchronize). Third, host A sends a packet containing data (with the ACK bit) to host B thus establishing a connection.

Once a connection is established, TCP has the very important responsibilities of verifying that a packet reaches its destination, reassembling packets into their original form, and controlling the rate at which data are transmitted – making sure that data are not sent faster than the receiver can process it.

The concept behind TCP's connection management is simple – it keeps a record of everything that it sends until it receives an acknowledgment that the information reached its destination. If TCP does not receive an acknowledgment after a set amount of time, it assumes that the information was lost and resends it. So, if one packet is lost or damaged in transit, TCP will resend just that packet, not the entire message.

As simple as this may seem, it is actually quite ingenious. If a major portion of a network is destroyed, TCP assumes that the network will be repaired quickly and continues to retransmit data – patiently waiting for an acknowledgment. If the network is not repaired quickly, TCP will eventually stop trying to resend information. However, if the network is repaired quickly, TCP will resume communication between two hosts despite the interruption. This differs from a telephone call, which is terminated when the connection is broken. When two hosts have finished communicating, TCP terminates the connection by sending a packet containing the FIN or RST bits.[7]

Keep in mind that TCP streams are bi-directional, enabling a host to both send and receive data. Each TCP stream comprises two flows, one for receiving data and the other for sending data. This aspect of TCP can be clearly seen in router NetFlow logs showing a connection to a Hotmail account from the client (192.168.1.105):

```
examiner1% flow-cat /netflow/2002/2002-08/2002-08-28/ft-v05.2002-08-28.213000-0400 | flow-filter -Skiosk -f./
kiosk.acl | flow-print -f5
Start             End               Sif  SrcIPaddress   SrcP  DIf  DstIPaddress    DstP  P  Fl  Pkts  Octets
0828.21:38:19.94  0828.21:38:19.94  2    192.168.1.105  0     19   66.113.201.11   2048  1  0   1     60
0828.21:38:57.715 0828.21:39:01.339 2    192.168.1.105  1925  13   64.4.53.7       80    6  3   6     609
0828.21:39:01.539 0828.21:39:02.495 2    192.168.1.105  1927  13   64.4.53.7       80    6  3   18    1172
0828.21:39:02.299 0828.21:39:05.439 2    192.168.1.105  1928  13   64.4.53.7       80    6  3   15    1081
0828.21:39:02.323 0828.21:39:05.723 2    192.168.1.105  1929  13   216.33.150.251  80    6  3   8     652
<cut for brevity>
```

Corresponding flows to the client are listed here, using the -D (destination) option of the **flow-filter**[8] command instead of -S (source).

[7]There are some nuances to the way that TCP uses sequence numbers and controls the rate at which data are sent that are beyond the scope of this text. Additional information about TCP can be found in Comer's Internetworking with TCP/IP Vol I (Comer 1995) and Steven's TCP/IP Illustrated (Stevens 1994).

[8]http://www.splintered.net/sw/flow-tools/

```
examiner1% flow-cat /netflow/2002/2002-08/2002-08-28/ft-v05.2002-08-28.213000-0400 | flow-filter -Dkiosk -f ./kiosk.acl
| flow-print -f5
Start              End                Sif  SrcIPaddress    SrcP  DIf  DstIPaddress   DstP  P  Fl  Pkts  Octets
0828.21:38:11.597  0828.21:38:11.597  11   66.113.201.11   0     4    192.168.1.105  0     1  0   1     60
0828.21:38:50.245  0828.21:38:53.869  11   64.4.53.7       80    4    192.168.1.105  1925  6  3   5     514
0828.21:38:54.69   0828.21:38:55.25   11   64.4.53.7       80    4    192.168.1.105  1927  6  3   26    12085
0828.21:38:54.833  0828.21:38:57.969  11   64.4.53.7       80    4    192.168.1.105  1928  6  3   17    6795
0828.21:38:54.853  0828.21:38:58.257  11   216.33.150.251  80    4    192.168.1.105  1929  6  3   8     3041
<cut for brevity>
```

Each NetFlow entry in the above output contains the start and end times of the flow, source and destination, IP addresses, and port numbers, followed by the number of packets in each flow, a number representing the protocol (e.g. 1 for ICMP, 6 for TCP, 17 for UDP), a number representing the combination of TCP flags in each flow, the number of packets, and the number of bytes (a.k.a. octets) transmitted, respectively.

17.1.7 ABUSES OF TCP/IP

Computer intruders have used their knowledge of TCP to gain unauthorized access to systems. One approach, called IP spoofing, was first described by Robert Morris (Morris 1985), father of Richard Morris Jr. – the creator of the first Internet worm and one of the first individuals to be prosecuted under the Computer Fraud and Abuse Act. IP spoofing takes advantage of the fact that many organizations configure certain hosts on their network to trust other hosts simply based on an IP address. With this kind of host-based authentication, in a computer that receives instructions that appear to come from a trusted IP address the instruction will be accepted without question. This trust arrangement is efficacious when two or more hosts on their network communicate so frequently that it is infeasible to require a password to be entered by a person every time the computers need to exchange data. However, a clever computer intruder can take advantage of this intercomputer trust in the following way to execute a command on the trusting computers without being prompted for a password:

1 The intruder disables the trusted computer using a denial of service attack.

2 The intruder sends a SYN packet to the trusting computer but forges the source IP address so that it appears to come from the trusted computer.

3 The trusting computer will send an ACK packet to the trusted computer and will be expecting an ACK packet in return to finalize the TCP connection. However, the trusted computer is unable to respond because it was disabled in step 1. Instead, the intruder sends an ACK packet with a forged source IP address, making it appear to come from the trusted computer.

4 The trusting computer thinks that it has established a legitimate connection with the trusted computer. The intruder can then send forged packets that appear to be coming from the trusted computer containing commands that the trusting computer will execute.

There is one nuance to IP spoofing that is important to be aware of – the intruder must be able to predict the TCP sequence numbers that the trusting computer is expecting in packets it receives. Newer operating systems use less predictable sequence numbers to make it more difficult to carry out this type of attack.

One of the most highly publicized IP spoofing attacks occurred in December 1994 when Kevin Mitnick broke into Tsutomu Shimomura's computers. Shimomura's description of the subsequent investigation and digital evidence he found hints at how challenging such investigations can be. Shimomura's computers were named "Osiris" and "Ariel." After gaining access to the computers, the intruder bundled the cellular telephone software that he wanted, a compressed file called oki.tar.Z. The intruder deleted the compressed file after transferring a copy to another machine that he had broken into.

> One of [the pieces of evidence] was a mysterious program, Tap, that I had seen when I peered into Osiris's memory the day before. It was a transient program that someone had created and placed in my computer's memory for a specific task. When the computer was turned off or rebooted it would vanish forever. And what about the ghost of the file oki.tar.Z, whose creation suggested that someone was after cellular telephone software ... There was another crucial discovery from looking at Ariel's data; the intruder had tried to overwrite our packet logs, the detailed records we keep of various packets of data that had been sent to or from our machines over the Internet. The erased log files revealed that in trying to overwrite them the intruder hadn't completely covered over the original file. It was as if he had tried to hide his footprints in the sand by throwing buckets of more sand on top of them. But here and there, heels and toes and even a whole foot were still visible. (Shimomura and Markoff 1996)

A more active abuse of TCP is session hijacking (a.k.a. man-in-the-middle attack), enabling an individual to take control of someone else's connection to a server. Basically, by monitoring traffic using a sniffer and then manipulating the TCP stream, it is possible to insert commands that will be executed on the server or even take the session over entirely. This attack has been automated by tools like Hunt[9] and Ettercap[10] but is made more difficult by using encryption.

[9] *http://lin.fsid.cvut.cz/~kra/index.html*

[10] *http://ettercap.sourceforge.net/*

17.2 SETTING UP A NETWORK

To better understand how all of this fits together, imagine that Henrietta the Hacker wants to set up an Internet café. Henrietta purchases several computers, a wireless (802.11) access point, and a switch to connect them together using some networking technology (e.g. Ethernet). She also purchases a firewall to filter traffic between the café network and the Internet. However, she still has to connect her network to the global Internet.

The first step to getting on the map, as it were, is to obtain an IP address on the Internet. Henrietta could apply to a registry like the American Registry for Internet Numbers[11] for a Class C block of IP addresses but it is more cost effective to select an Internet Service Provider that already has a block of IP addresses and will assign her one of them for a fee. One public IP address is sufficient because Henrietta can configure her café network using one of the private blocks of IP address mentioned earlier (e.g. 10.0.0.0–10.255.255.255, 172.16.0.0–172.31.0.0, and 192.168.1.0–192.168.1.255). Most firewalls can perform Network Address Translation (NAT), enabling the network administrator to connect multiple hosts to the Internet via one public IP address. Henrietta's network is depicted in Figure 17.7.

[11] *http://www.arin.net/registration/index.html*

Now, suppose that a customer, Keith the Thief, comes into the café with his laptop and connects to the Internet through Henrietta's network. When Keith requests any information from the Internet (e.g. a Web page) this

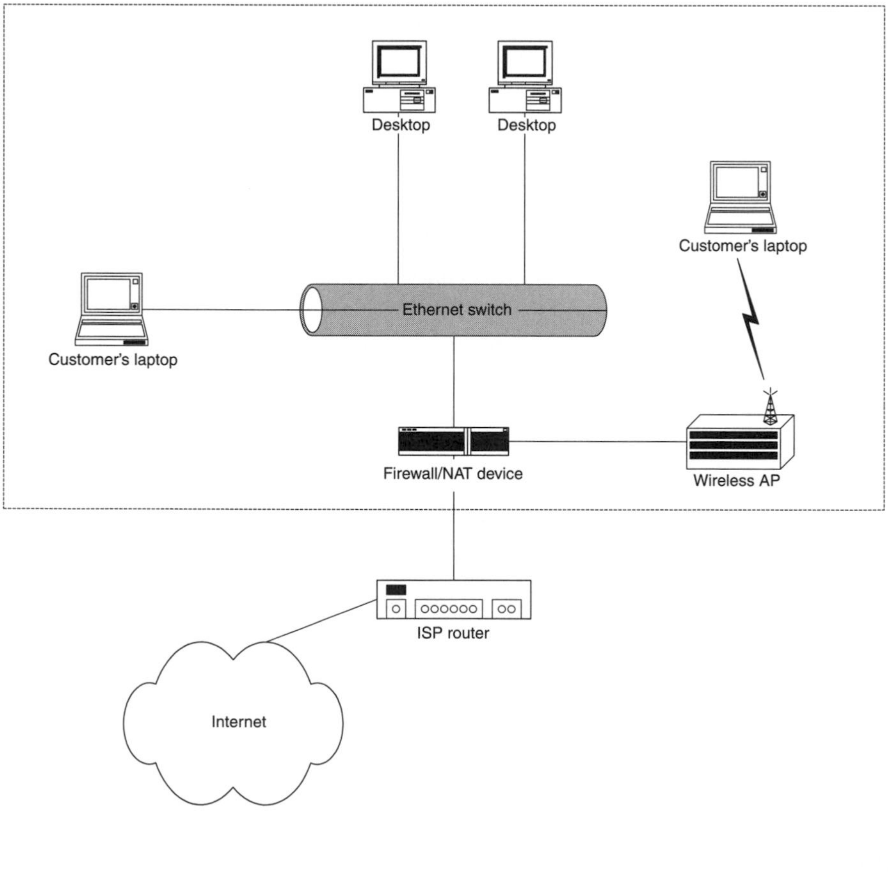

Figure 17.7

Internet café with several kiosks, Ethernet ports for customer laptops, and a wireless access point connected together with an Ethernet switch and connected to an ISP's router by a firewall performing NAT.

information will first pass through Henrietta's ISP and firewall before going to his laptop. Similarly, any information that Keith sends out (e.g. e-mail) will pass through Henrietta's firewall and her ISP's router before reaching the Internet. There are two obvious implications of this arrangement.

First, Henrietta the Hacker could observe and keep a log of all of Keith the Thief's activities. Second, most things that Keith sends through the Internet will indicate that they originated from Henrietta's café so someone could contact her in relation to his activities on the Internet.

Unfortunately, many NAT devices do not maintain logs of traffic that pass through them, making it more difficult to determine which computer was involved in a crime originating from this type of network. This is why more organizations are using Argus to maintain logs of network activities. Even when it is possible to determine which computer was used in an Internet café or public library, it can be difficult to associate an individual with the computer. However, it is not impossible as the following cases demonstrate:

CASE EXAMPLE

In 2000, Jeff Vijay, a man who was convicted in 1994 for stalking his ex-girlfriend and her new husband in Michigan, was accused of sending the same couple threatening e-mail messages from a public-access computer at a San Jose library where Vijay's mother worked. The threatening messages had a return e-mail address "death4u@alumni.com" and contained language similar to notes and voice mail messages attributed to the man in 1994, including the same threats and misspellings. During a preliminary hearing, a judge ruled that there was not enough evidence in the new case to prove that the suspect had been using the library computer at the time the threatening messages were sent. However, when the case went to trial, the jury quickly concluded that Vijay had sent the threatening e-mails and found Vijay guilty. (Romano, B. "Internet stalking charges dropped" Published Sunday, April 9, 2000, in the San Jose Mercury News

Also in 2000, a University of Iowa student admitted to sending a bomb threat via e-mail as well as several racist e-mail threats. The messages were tracked back to a computer in a campus building and a hidden camera was installed to determine who was sending the messages (Tribune 2000).

17.2.1 STATIC VERSUS DYNAMIC IP ADDRESS ASSIGNMENT

One decision that Henrietta had to make when requesting an IP address for her Internet café was whether to ask the ISP for a static or dynamic IP address. With a static IP address her network would always have the same IP address. One advantage of a static IP address is that it can be assigned a name of her choosing, such as "www.cafe-henrietta.com," enabling her to create a Web site for her Internet café.[12] If Henrietta did not need a static IP address, a less expensive alternative is to have her ISP assign her with a different IP address periodically. This approach enables an ISP to reassign

[12]This type of domain name can be obtained through registrars like Network Solutions (http://www.networksolutions.com). Once a domain name has been registered, any ISP can enter it into their DNS servers to associate the name with an IP address on their network.

IP addresses to their customers whenever necessary to make more efficient use of them. This type of dynamic IP assignment has become the norm for many ISPs that provide Internet access to a large number of people. Additionally, within her own small network, Henrietta could use dynamic IP addresses to make it easier for customers to connect their laptops to her network.

Notably, this dynamic assignment can make it more difficult to determine who was using an IP address at a given time. Fortunately for investigators, ISPs often maintain a log of dynamic IP address assignments, listing who was assigned a particular IP address during a specific period.

CASE EXAMPLE

In an extortion case, the offender sent messages through Hotmail from an Internet café to ensure that the e-mail headers did not contain an IP address that could be connected to him. However, when investigators obtained logs from Hotmail they found that the blackmailer had established and accessed his Hotmail account through a dial-up account. They were able to trace the identity of the offender using information relating to the dial-up account obtained from the ISP.

[13] http://www.dyndns.org
[14] http://www.no-ip.com

Services like DynDNS[13] and No-IP[14] provide DNS service for dynamic IP address, enabling Henrietta to select a name like "cafe-henrietta.dyndns.org" and update the dynamic DNS record whenever her dynamic IP address changes. Criminals use dynamic DNS service to run illicit servers using dynamic IP addresses, enabling cohorts who know the name (e.g. "illicit.dyndns.org") to access the server while making it difficult for investigators who do not know the name to locate the server each time the dynamic IP address changes.

Notably, these dynamic DNS records are different from the names that an ISP gives their dynamic IP addresses in their DNS servers. For instance, the following DNS query shows the IP address 151.196.245.139 is assigned one name by DynDNS and another by the ISP (Verizon):

```
C:\>nslookup cases.dyndns.org
Name: cases.dyndns.org
Address: 151.196.245.139
C:\>nslookup 151.196.245.139
Name: pool-151-196-245-139.balt.east.verizon.net
Address: 151.196.245.139
```

This example also demonstrates that some dynamic IP addresses have the abbreviations of cities and/or geographic regions that can be helpful in determining a rough location for an IP address.

17.2.2 PROTOCOLS FOR ASSIGNING IP ADDRESSES

Some networks use the Bootstrap Protocol (BOOTP) and others use the Dynamic Host Configuration Protocol (DHCP) for assigning IP addresses to all hosts, even ones with static IP addresses. These protocols are used to prevent computers from being configured with incorrect IP addresses. Sometimes computers are misconfigured accidentally, causing two computers to interfere with each other. Also, sometimes individuals purposefully assign their computers with someone else's IP address to hide their identity. Using BOOTP or DHCP prevents these situations from occurring by centrally administering IP addresses.

BOOTP and DHCP are quite similar – both require hosts to identify themselves (using its MAC address) before obtaining IP addresses. When a computer is booting up, it sends its MAC address to the BOOTP or DHCP server. If the server recognizes the MAC address it sends back an IP address and makes a note of the transaction in its log file. The server can be configured to assign a specific IP address to a specific MAC address thus giving the effect of static IP addresses.

All of these acronyms can be confusing but the idea is simple. A central computer keeps track of which hosts are using which IP addresses. Under certain circumstances, the log files on these central BOOTP and DHCP servers will show the times a specific computer is connected to and disconnected from the network. This could be used to determine when a computer dialed into a network or when a host that is usually part of the network was turned on and turned off.

17.3 TCP/IP RELATED DIGITAL EVIDENCE

Given the central role that TCP/IP plays in networks, it should come as no surprise that IP addresses, port numbers, TCP flags, and other TCP/IP related data accumulate in many places. Understanding how to find and exploit these sources of digital evidence is central to investigating crime on networks. As noted in the previous chapter, sniffer logs contain TCP/IP related information.

CASE EXAMPLE

While investigating a UNIX computer intrusion, investigators found a program called **router** that they did not recognize. Examining the contents of this binary file revealed that it was a Portuguese sniffer, specially designed to capture usernames and passwords, that saved captured data in a file named "/etc/.X0" as shown here:

```
Erro abrindo socket Erro setando flags da placa Erro setando modo promiscuo
-------+ [%d bytes]+ ---------+[%d segs]+ -------- [RST]
[Fim de coneccao]
%s %s => %s [%d]
%c eth0 w+ /etc/.X0
Erro abrindo %s macunaim@hotmail.com joao@localhost localhost
------- [Sniffer Terminado]
```

In addition to usernames and passwords to other systems on the network, the "/etc/.X0" file contained evidence of several unauthorized Telnet connections from Brazil using a stolen account. Ironically, the intruder had recorded his crime and IP address with his own sniffer:

```
Tue Mar 18 18:54:52 2003
mx1.corpZ.com.br → server1.corpX.com [23]
#'vt100!stolenaccount
password
w
dnsmail 43876537
id
cd /
------+ [60 segs]+

Fri Mar 21 05:18:45 2003
dialup34.corpX.com → server1.corpX.com [23]
!#' 38400,38400username
password
pine
term=vt100
pine
-------+ [60 segs] +
------- [Fim de coneccao]
```

Searching unallocated space for class characteristics of this sniffer log, the digital evidence examiner was able to find similar incriminating fragments of an older sniffer log that the intruder had deleted.

Although TCP/IP data can be captured using a sniffer, it is not feasible to capture all network traffic in all situations, making it necessary to rely on other sources of evidence such as log files that show past connections and in state tables that show recent and current connections between hosts. Several examples of log files and state tables containing this type of information have been mentioned in passing. The following sections discuss these and other useful sources of TCP/IP related information in more detail, demonstrating how they can be useful in an investigation.

17.3.1 AUTHENTICATION LOGS

Authentication logs are very useful because they show which account was associated with an activity and often contain an associated IP address or telephone number, substantially narrowing the suspect pool.

CASE EXAMPLE (SHINKLE 2002):

An unusual lead developed during a serial homicide investigation in St Louis when a reporter received a letter from the killer. The letter contained a map of a specific area with a handwritten X to indicate where another body could be found. After investigators found a skeleton in that area, they inspected the letter more closely for ways to link it to the killer. The FBI determined that the map in the letter was from Expedia.com and immediately contacted the site to determine if there was any useful digital evidence.

The Web server logs on Expedia.com showed only one IP address (65.227.106.78) had accessed the map around May 21, the date the letter was postmarked. The ISP responsible for this IP address was able to provide the account information and telephone number that had been used to make the connection in question similar to the information shown here:

```
Username: MSN/maurytravis
UUNET Resllerer: MSN
IP address assigned: 65.227.106.78
Time of connection: 19:53:34 May 20
Time of disconnect: 22:24:19 May 20
ANI information: (212) 555–1234
```

Both the dial-up account and telephone number belonged to Maury Travis. Investigators arrested Travis and found incriminating evidence in his home, including a torture chamber and a videotape of himself torturing and raping a number of women, and apparently strangling one victim. Travis committed suicide while in custody and the full extent of his crimes may never be known.

Internet dial-up logs such as those used in the Travis case are generally created by RADIUS or TACACS authentication servers. Other network devices such as Virtual Private Network (VPN) concentrators also use RADIUS or TACACS to authenticate users. Organizations use these centralized authentication servers to make account administration easier rather than having different user accounts on each system. Network administrators can search the associated authentication logs to obtain the type of information mentioned in the Travis case; that is, which user account was assigned an IP address at a given time. For instance, the following RADIUS logs were generated by Microsoft Internet Authentication Server (IAS) running on a machine named IAS-SERVER (172.16.1.45) when the "ianjones" account in the CORPX domain was used to connect through a VPN concentrator

(172.16.1.219) from 64.252.248.133:

```
172.16.1.219,CORPX\ianjones,03/08/2003,17:46:04,IAS,IAS-SERVER,
5,7029,6,2,7,1,66,64.252.248.133,61,5,4108,172.16.1.219 4116,0,4128,CORPX
VPN,4129,CORPX\ianjones,25,311 1 172.16.1.4510/08/2002 14:38:34
22348,4127,3,4130,corpx.com/Users/ianjones,4136,1,4142,0

172.16.1.219,CORPX\ianjones,03/08/2003,17:46:04,IAS,IAS-SERVER,25,311 1
172.16.1.4510/08/2002 14:38:34
22348,4130,corpx.com/Users/ianjones,6,2,7,1,4108,172.16.1.219,
4116,0,4128,CORPX
VPN,4129,CORPX\ianjones,4120,0 × 0259414C45,4127,3,4149,Allow access if
dial-in permission is enabled,4136,2,4142,0

172.16.1.219,CORPX\ianjones,03/08/2003,17:46:07,IAS,IAS-SERVER,
5,7029,6,2,7,1,8,
172.16.19.53,25,311 1 172.16.1.45 10/08/2002 14:38:34
22348,40,1,44,E0D03B6B,66,64.252.248.133,45,1,41,0,61,5,4108,172.16.1.219,
4116,0,4128,CORPX VPN,4136,4,4142,0
```

These log entries contain the IP address assigned to the connecting host by VPN concentrator (172.16.19.53) along with other connection details (Microsoft 2000). The corresponding logout was recorded as shown here:

```
172.16.1.219,CORPX\ianjones,03/08/2003,17:55:12,IAS,IAS-SERVER,
5,7029,6,2,7,1,8,
172.16.19.53,25,311 1 172.16.1.45 10/08/2002 14:38:34 22348,40, 2,42,
36793575,43,
6837793,44,E0D03B6B,46,35619,47,417258,48,59388,49,1,66,64.252.248.133,
45,1,41,0,61,5,4108,172.16.1.219,4116,0,4128,CORPX VPN,4136,4,4142,0
```

This VPN connection and IP address assignment is depicted in Figure 17.8.

Some organizations use a centrally administrated mechanism like Kerberos to handle authentication for all of their hosts and applications, logging all authentication requests in a log file on the Kerberos server. These logs include the date and time of the authentication request as well as the IP address and user name making the request:

```
May 12 10:23:52 kerberos1 krb5kdc[2324](info):
AS_REQ 192.168.19.4(88): ISSUE: authtime 1052829558,
user/ianjones@CORPX.COM for krbtgt/CORPX.COM@CORPX.COM
```

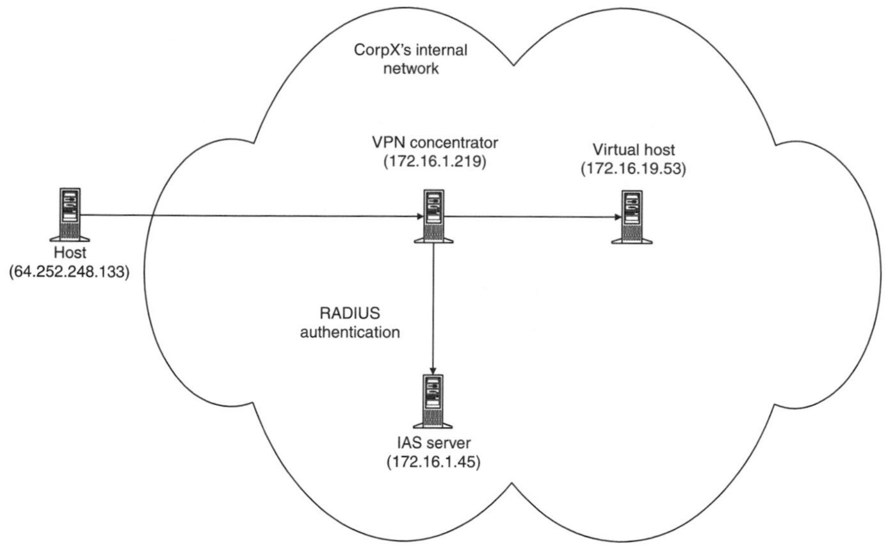

Figure 17.8

VPN concentrator (172.16.1.219), IAS server (172.16.1.45), and connecting host (64.252.248.133; 172.16.19.53).

These types of centralized authentication systems can be a very useful and reliable source of digital evidence because they correlate events from multiple sources on the network and store the log files on a system that is generally more secure than other hosts on the network. Windows Security Event Logs can also be configured to record which accounts logged in when, and Windows 2000 Active Directory facilitates centralized authentication mechanisms like Kerberos.[15]

E-mail, Web, and other Internet servers may also have authentication logs useful for connecting online activities with an individual. For instance, the following logs from an e-mail server show the account "eco" being used to check e-mail from IP address 10.10.2.10, once at 11:01 on February 6 and a second time at 15:02:[16]

```
Feb 6 11:01:26 mailsrv ipop3d[26535]: Login user=eco
host=dialup.domain.net [10.10.2.10]
Feb 6 11:01:28 mailsrv ipop3d[26535]: Logout user=eco
host=dialup.domain.net [10.10.2.10]
Feb 6 15:02:48 mailsrv ipop3d[244]: Login user=eco
host=dialup.domain.net [10.10.2.10]
Feb 6 15:02:49 mailsrv ipop3d[244]: Logout user=eco
host=dialup.domain.net [10.10.2.10]
```

Multiuser systems often have records of which accounts logged in when. The following segment shows that an intruder used an account named "toor"

[15]Depending on the configuration, the Windows Security Event Log may not contain IP addresses of remote systems. Unless Kerberos related logging is enabled, the Event log only records the NetBIOS name of remote systems. Notably, Kerberos authentication does not have to be in use for the advanced logging feature to work.

[16]Post Office Protocol (POP) and Internet Message Access Protocol (IMAP) servers both enable clients to read their e-mail remotely and both have similar authentication logs.

to log into a UNIX system from a Pacbell dial-up account:

```
% last toor
toor  pts/0  Wed Mar 31   18:27   ppp-90.scrm01.pacbell.net - 18:30   (00:12)
toor  ftp    Wed Mar 31   18:28   ppp-90.scrm01.pacbell.net - 18:27   (00:11)
```

Windows NT/2000/XP systems maintain similar authentication logs but they usually only contain the NetBIOS name of the connecting system, not the IP address.

Many other servers have their own authentication mechanisms and associated logs. In some instances, particularly when dealing with customized applications, it is necessary to obtain the assistance of someone familiar with the system to locate and comprehend these logs.

17.3.2 APPLICATION LOGS

Many applications have log files, other than authentication logs, containing information about peoples' activities on a network. For instance, the following FTP transfer logs ("xferlog") show the user account and IP address used to delete files on the server:

```
Nov 14 00:17:23 fileserver1 ftpd[2536]: user32 of 202.180.75.79 [202.180.75.79]
deleted /d2/project13/data1.xls
Nov 14 00:17:24 fileserver1 ftpd[2536]: user32 of 202.180.75.79 [202.180.75.79]
deleted /d2/project13/data2.xls
Nov 14 00:17:24 fileserver1 ftpd[2536]: user32 of 202.180.75.79 [202.180.75.79]
deleted /d2/project13/report1.doc
Nov 14 00:17:25 fileserver1 ftpd[2536]: user32 of 202.180.75.79 [202.180.75.79]
deleted /d2/project13/report2-final.doc
Nov 14 00:17:25 fileserver1 ftpd[2536]: user32 of 202.180.75.79 [202.180.75.79]
deleted /d2/project13/report2-rev2.doc
Nov 14 00:17:26 fileserver1 ftpd[2536]: user32 of 202.180.75.79 [202.180.75.79]
deleted /d2/project13/report2-rev1.doc
```

Similarly, each time a Web server receives a request from a client, it records the client's IP address in its access log along with the date, time, and what the client requested. In addition to showing the request from an IP address used by a suspect, Web access logs can be used to determine which IP address accessed a specific page during a certain time, as in the Maury Travis case. A few other common examples are provided here to demonstrate how they can be used in an investigation.

CASE EXAMPLE

When an individual defaces a Web page, he/she usually views it shortly before and after the defacement to check his/her work as can be seen in the following

Web server log entries:

```
04:17:33 216.67.71.92 HEAD /msadc/msadcs.dll 200
04:19:20 216.67.71.92 GET /default.html 404
04:19:32 216.67.71.92 GET /default.htm 200
04:19:36 216.67.71.92 GET /images/spacer.gif 200
04:19:40 216.67.71.92 GET /line.gif 200
04:19:50 216.67.71.92 GET /images/image1.gif 200
04:19:59 216.67.71.92 GET /msadc/msadcs.dll 200
04:20:33 216.67.71.92 POST /msadc/msadcs.dll 200
04:20:37 216.67.71.92 GET /default.htm 200
04:20:39 216.67.71.92 GET /Default.htm 200
```

The first entry shows a scan for known vulnerabilities locating a vulnerable DLL (msadcs.dll) on a Web server. Two minutes later the intruder attempts to view the default page, misspelling it the first time. The intruder breaks in and replaces the default page, the actual page replacement is not logged by the Web server because it is not uploaded through the Web server. The last two entries show the intruder checking the default page again to view the defacement. In cases where an intruder launches an attack through another compromised machine, he/she may still view the page using the Web browser on his own machine.

Many marketing companies make their money by examining the Web pages that a particular individual views and using this information to learn about his/her interests. This same approach can be useful in an investigation for determining who was using a specific computer at a certain time. Web server logs, like their corresponding Web browser history and cached files on a personal computer, can provide strong circumstantial evidence that a particular individual was responsible for the activity in question.

To better understand how to extract behavioral information in log files it is useful to compare routine behavior with more anomalous behavior. When an individual sends an e-mail message, this action is recorded in the e-mail server log file as shown here:

```
Feb 7 15:05:30 mailsrv sendmail[1257]: PAA01257: from=(eco@
corpus-delicti.com), size=793, class=0, pri=30793, nrcpts=1,
msgid=(4.2.0.58.19991013150621.0099fa90@mailsrv.corpus-delicti.com),
proto=ESMTP, relay=dialup.domain.net [10.10.2.101]

Feb 7 15:05:31 mailsrv sendmail[1259]: PAA01257: to=bturvey@
corpus-delicti.com, delay=00:00:03, xdelay=00:00:00, mailer=relay,
relay=mail.domain.net. [10.10.2.11], stat=Sent (PAA00253 Message accepted
for delivery)
```

Note that a single message creates two entries in an e-mail server log, containing source and destination details, and both containing the same message ID (e.g. PAA01257). In this instance, the IP address of the sender was 10.10.2.101. Compare this normal activity with the following log entries that show someone forging an e-mail message using the SMTP forgery method detailed in Chapter 18:

```
Oct 15 01:20:09 mailserver sendmail[27941]: BAA27941:
from=forged.from@home.net, size=114, class=0, pri=30114, nrcpts=1,
msgid=<199910150518.BAA27941@mailserver>, proto=SMTP,
relay=host1.domain.net [20.134.161.6]

Oct 15 01:20:10 mailserver sendmail[28214]: BAA27941:
to=target@ayyahoo.com, delay=00:01:14, xdelay=00:00:01, mailer=esmtp,
relay=192.168.1.50, stat=Sent (BAA08487 Message accepted for delivery)
```

The forger evidently made a typo in the target e-mail address and the resulting e-mail header will contain associated backspaces and other characters (e.g. target\bl@ayyahoo.com) when viewed in hexadecimal format.

There are many different commercial network applications and some organizations make their own in-house applications with unique logging mechanisms. Therefore, it is sometimes necessary to perform research and even a functional reconstruction to understand what actions relate to specific log entries.

CASE EXAMPLE

An organization's primary server was targeted by a denial of service attack that lasted for several days. The log files indicated that dozens of machines had been involved in the attack. However, when investigators examined some of the attacking machines, it became clear that some of the machines seized had not been involved in the attack and the date–time stamps in the application server logs were misleading. Using a similar server to perform a functional reconstruction, it was determined that log entries were not made when a request was initially received. Instead, each request was held in a queue that was processed sequentially and a log entry was only made when the request was processed. Because the denial of service attack had created a large queue on the server, it had taken several hours for requests to be processed and associated log entries to be generated. Therefore, the log entries did not accurately reflect when each portion of the attack had occurred.

17.3.3 OPERATING SYSTEM LOGS

Most operating systems can maintain logs of noteworthy events such as system reboots, errors, modem usage, and network interface cards being put into promiscuous mode by a sniffer. Because they were initially designed with networks in mind, log files on UNIX systems generally retain more TCP/IP

Table 17.2

Log files on various types of UNIX.

FILE	DESCRIPTION
Aculog	If modems are attached to the computer, this log contains a record of when the modems were used to dial out
Authlog or secure	On some systems these files contain security related logs including information relating to authentication on the system such as logon attempts
Lastlog	This log file contains a record of each user's most recent login (or failed login)
loginlog	Records failed logins
Syslog	The syslog file (sometimes called "messages" or "system" depending on the type of UNIX and its configuration) is the main system log file. Some servers, such as Sendmail and SSH on UNIX, can be configured to log into the syslog file and these main log files often contain information that is also found in other log files, e.g. failed logins. Additionally, routers and firewalls are usually configured to add their logs to the syslog file on a remote logging server
utmp and utmpx	These files contain a record of all users currently logged into a computer. The "who" command accesses this file
wtmp and wtmpx	These files contain a record of all of the past and current logins and records system startups and shutdowns. The "last" command accesses this file
xferlog	This file contains a record of all files that were transferred from a computer using the File Transfer Protocol (FTP)

related information than Windows NT Event Logs. Table 17.2 describes the most common system logs on UNIX machines. Newer versions of UNIX usually store their log files in "/var/adm" or "/var/log" whereas older versions store them in "/usr/adm." However, the location of these logs is configurable in "/etc/syslog.conf" and can be on a remote syslog server.

The following entries in the syslog file relate to the Brazilian intruder encountered earlier, showing several unauthorized connections, one corresponding to the entry in his sniffer log on March 18. The intruder attempted to login again on March 25 and entered the stolen password twice before realizing it had been changed and that his crime had been discovered:

```
% grep "corpZ\ |promiscuous" syslog
Mar 1 23:50:29 server1 login: LOGIN ON 0 BY stolenaccount FROM
mx1.corpZ.com.br
Mar 7 19:08:49 server1 login: LOGIN ON 1 BY stolenaccount FROM
mx1.corpZ.com.br
Mar 7 19:13:37 server1 kernel: device eth0 left promiscuous mode
Mar 7 19:14:21 server1 kernel: device eth0 entered promiscuous mode
Mar 18 18:55:27 server1 login: LOGIN ON 1 BY stolenaccount FROM
mx1.corpZ.com.br
Mar 25 21:09:53 server1 login[29708]: FAILED LOGIN 1 FROM mx1.corpZ.com.br
FOR stolenaccount, Authentication failure
Mar 25 21:10:11 server1 login[29708]: FAILED LOGIN 2 FROM mx1.corpZ.com.br
FOR stolenaccount, Authentication failure
```

Most UNIX system log files contain information about incoming traffic, but not outgoing traffic. This makes it relatively easy to determine what an individual was doing to a computer but makes it difficult to determine what an individual was doing from the computer. To overcome this limitation some system administrators install host-based firewalls (e.g. IPFilter, IPChains, ZoneAlarm) on their computers that logs details about noteworthy incoming and outgoing network connections.

17.3.4 NETWORK DEVICE LOGS

Because of their central role, network devices often generate logs that provide an overview of activities on a network. Such an overview can help investigators gain an initial understanding of what occurred and which hosts were involved. The overview of network activity that these logs provide can be very detailed, showing activities that were not recorded in other logs. Even when the activities were recorded by other systems, logs from network devices can be used for corroboration, providing independent sources of digital evidence relating to the same events.

CASE EXAMPLE

An organization found a host on their network was apparently compromised using a new exploit that was not detected by the intrusion detection system. NetFlow logs were examined to gain a clearer understanding of how the host had been compromised. The NetFlow logs showed that, at approximately 12:25 A.M. on October 21, adsl-61-105-217.msy.bellsouth.net (208.61.105.217) targeted the SSH daemon on the compromised machine. This reconnaissance activity corresponded with the following system log entries from one of the computers:

```
Oct 21 00:29:25 hostA sshd[18967]: connect from 208.61.105.217
Oct 21 00:29:25 hostA sshd[18967]: log: Connection from 208.61.105.217
port 4584
Oct 21 00:29:34 hostA sshd[18967]: fatal: Did not receive ident string.
```

At about 02:15 A.M. on October 21, NetFlow logs showed that 66.28.12.53 accessed the SSH server. This corresponded with the following buffer overflow recorded in the syslog file of the compromised host:

```
Oct 21 02:16:24 hostA sshd[18997]: connect from 66.28.12.53
Oct 21 02:16:24 hostA sshd[18997]: log: Connection from 66.28.12.53 port 2974
Oct 21 02:16:24 hostA sshd[18997]: log: Could not reverse map address
66.28.12.53.
Oct 21 02:16:25 hostA sshd[18998]: connect from 66.28.12.53
Oct 21 02:16:25 hostA sshd[18998]: log: Connection from 66.28.12.53 port 2975
Oct 21 02:16:25 hostA sshd[18998]: log: Could not reverse map address
66.28.12.53.
<cut or brevity>
Oct 21 02:18:29 hostA sshd[19119]: fatal: Local: crc32 compensation attack:
network attack detected
```

At this stage, the intruder installed an IRC bot and French ident daemon to reply to IRC servers with a name other than root. Many IRC servers will not accept connections from the root account on a machine because, recognizing it as a sign of compromise:

```
Oct 21 02:46:37 hostA in.ident2[28529]: error: setuid(-2): Paramètre invalide
Oct 21 02:46:37 hostA in.ident2[28529]: error: cannot reduce self's rights
```

The intruder also replaced SSH with a Trojaned version that captured passwords in a file named "/usr/lib/libfl.so.3." The Trojaned SSH daemon also had a backdoor associated with the user name "smiley":

```
# strings sshd
Rhosts with RSA authentication disabled.
RSA_new failed
BN_new failed
Warning: keysize mismatch for client_host_key: actual %d, announced %d
RSA authentication disabled.
Password authentication disabled.
smiley
/usr/lib/libfl.so.3
user: %s
password: %s
rcvd SSH_CMSG_AUTH_TIS
```

Additionally, the intruder replaced "/bin/login" with a Trojaned version that appeared to allow access to a machine if the client's DISPLAY variable is set to "smiley." Scanning the network for other systems with the same backdoors uncovered two more compromised machines. The intruder had attacked these systems from a different IP address, which is why they did not show up in the original examination of the NetFlow logs. Unfortunately, the original NetFlow logs had not been preserved, only the results from the initial examination. By the time the extent of the attacker's penetration was realized, the original NetFlow logs had been overwritten. Without the original NetFlow log files it was not possible to obtain an overview of what the attacker had done with the other compromised systems or if the intruder had gained access to other systems on the network. Also, log files from the IRC bot were encrypted, preventing investigators from obtaining additional information about the intruder.

Because, network devices like routers and firewalls have a limited amount of memory to store logs, they are usually configured to send a copy of their logs to a remote log server for permanent storage. For instance, a router might send system logs to one remote log server and NetFlow logs to a collector on a different host. In most situations, router logs contain limited information about the operation of the router, whereas NetFlow logs generally contain information about every flow through the router.

Consider a situation in which Corporations X's primary router suddenly stops routing all traffic and the "enable" password used to configure the system has been changed, suggesting a serious system failure or sabotage. The following logging details from the router indicate that logs are stored permanently on a remote log server with IP address 172.16.3.2 and show some recent logs are still stored temporarily in memory:

```
oisin% date
19:48:05 UTC Fri Apr 11 2003
oisin% telnet route-server.backbone.net
...
route-server>show clock
*00:16:05.378 UTC Sat Apr 12 2003
route-server>show logging
Syslog logging: enabled (0 messages dropped, 5 messages rate-limited,
0 flushes, 0 overruns)
   Console logging: level debugging, 1577 messages logged
   Monitor logging: level debugging, 22 messages logged
   Buffer logging: level debugging, 175 messages logged
   Logging Exception size (8192 bytes)
   Trap logging: level informational, 1586 message lines logged
      Logging to 172.16.3.2, 429 message lines logged
Log Buffer (50000 bytes):
*Apr 7 18:47:02: %SYS-5-CONFIG_I: Configured from console by vty0
(172.16.21.4)
*Apr 7 18:51:01: %SEC-6-IPACCESSLOGP: list telnet-log permitted tcp
172.16.21.4(64628) -> 172.16.24.66(23), 300 packets
*Apr 8 00:13:18: %SEC-6-IPACCESSLOGP: list telnet-log permitted tcp
172.16.19.53 (36182) -> 172.16.24.66 (23), 126 packets
*Apr 9 02:18:41: %SEC-6-IPACCESSLOGP: list telnet-log permitted tcp
172.16.19.53 (64805) -> 172.16.24.66 (23), 118 packets
*Apr 9 02:19:01: %SYS-5-CONFIG_I: Configured from console by vty0
172.16.19.53
```

Each log entry begins with the date and time, followed by classification codes detailed in Cisco IOS (2000). The first entry in this router log indicates that someone (the network administrator in this case) connected to the router from 172.16.21.4 using Telnet and reconfigured it on April 7 at 18:47 hours. The next two log entries show the administrator connecting using Telnet from the same IP address to check the router. The last connection and reconfiguration on April 9 from 172.16.19.53 was not authorized and was cause for concern. This IP address was associated with the organization's

VPN server mentioned in the Authentication Logs section. Recall that the RADIUS logs (Section 17.3.1) indicated that the "ianjones" account was used to commit this offense.

Note that the router clock is inaccurate and the date–time stamps must be adjusted to correct the error.[17] Fortunately, when these logs are sent to the remote server for permanent storage, the server adds a date–time stamp using its own clock as a reference. This can result in unusual looking log entries on the server such as this one, where the server time zone is GMT:

[17]Router clocks are notoriously unreliable and it is not uncommon to find that a router clock is off by several days. To address this problem, many routers are configured to automatically correct the clock on a regular basis using the Network Time Protocol (NTP).

```
Apr 8 21:51:41 [route-server] 1435: *Apr 9 02:19:01: %SYS-5-CONFIG_I:
Configured from console by vty0 172.16.19.53
```

Some organizations also configure their routers to block certain traffic and maintain a log of denied connections, essentially functioning as a firewall. A sample log entry generated by a Cisco Private Internet eXchange (PIX) firewall when it blocks an unauthorized connection is shown here:

```
Jun 14 10:00:07 firewall.secure.net %PIX-2-106001: Inbound TCP connection
denied from 10.14.21.57/41371 to 10.14.42.6/22 flags SYN
Jun 14 10:00:47 firewall.secure.net %PIX-2-106001: Outbound TCP connection
denied from 10.14.42.5/41371 to 10.10.4.16/22 flags SYN
```

The format of these log entries is similar to those of a router, starting with the date and time, followed by the name of the firewall, the PIX alert information (Cisco PIX 2000), the action, source, and destination. Different firewalls have slightly different formats that are described in the product documentation.

17.3.5 STATE TABLES

State tables contain information about the current or very recent state of connections between computers. Data in state tables are quite transient – inactive entries are usually cleared in less than an hour. As noted in the previous chapter, the ARP table on every host contains IP addresses relating to recent communications. Also, firewalls, routers, and many other pieces of network equipment maintain a state table of active and recent connections. For instance, on a Cisco PIX firewall these connections can be listed using the `show conn detail` command. This information can be used to corroborate other evidence and establish the continuity of offense. As mentioned earlier in this chapter, current and recently terminated TCP/IP connections on a server of personal computers can be viewed using the `netstat` command.

CASE EXAMPLE

A man who was using ICQ to harass a woman believed that he could not be caught because he had configured his ICQ client to hide his IP address. However, the woman consulted with a computer expert and learned that if she could initiate a TCP/IP connection with the man's computer, she could view his IP address using the netstat command. So, the next time the woman was harassed by this man, she sent an ICQ instant message to him, and used netstat to obtain his IP address. The woman contacted his Internet Service Provider and the harassment stopped. This method of finding an individual's IP address is not limited to ICQ. If the harasser had used IRC, AOL IM, or any other application that uses TCP/IP to transfer data the same method could have been used to track him down.

Another example of state tables; for recent outgoing NetBIOS connections, Windows maintains a list of NetBIOS names and their resolved IP addresses in the NetBIOS name table. For instance, in the earlier example involving VNC (Section 17.1.5), the name table on the Windows XP machine running the VNC server (192.168.0.4) had one NetBIOS connection to 192.168.0.2 as shown here:

```
C:\> nbtstat –c

              NetBIOS Remote Cache Name Table

Name               Type          Host Address      Life [sec]
-------------------------------------------------------------
WORKSTN2      <20> UNIQUE        192.168.0.2       567
```

Incoming NetBIOS connections can be viewed using the **net session** command but the associated IP address is not displayed. For instance, executing this command on WORKSTN2 (192.168.0.2) in the aforementioned VNC example shows which user account was used to establish the NetBIOS session but only provides the NetBIOS name of the Windows XP machine (WORKSTN1). Recall that the associated IP address may be obtainable using **netstat**:

```
C:\>net session

Computer        User name        Client Type              Opens Idle time
--------------------------------------------------------------------------
\\WORKSTN1      USER1            Windows 2002 2600        0 00:00:23

The command completed successfully.
```

Similarly, UNIX maintains a list of remote machines that are connected to Network File System shares that can be displayed using the **showmount – a** command as shown here:

```
[nfs-server]# showmount -a
All mount points on case:
192.168.0.101:/shared-drive
```

On the client, the **mount** command shows all remote shares that are being accessed, which generally corresponds to the information in /etc/fstab mentioned in Chapter 11:

```
[nfs-client]# mount
<entries relating to local drives cut for brevity>
/mnt on 192.168.0.7:/ remote/read/write/nosetuid/dev=2f80002 on Thu Apr
10 08:31:19 2003
```

These commands are used in computer intrusion investigations to determine which machines have made connections to a given system. These commands are also useful for locating potential sources of digital evidence on networks as discussed in Chapters 10 and 11.

CASE EXAMPLE

In the process of executing a search warrant to seize a suspect's home computer in a child pornography investigation, the digital evidence examiner notices that the system has an Ethernet connection to a small router. The router had several other Ethernet cables, suggesting that there are other computers in the vicinity. Before shutting the suspect's system down, the examiner uses the netstat -an, nbtstat -c and net session commands to document NetBIOS connections to and from the suspect's system. In addition to listing several connections to other systems on the suspect's home network, these commands showed a computer on the Internet connecting to the suspect's system using an account called "GERY." The digital evidence examiner used the net file command and found that a file containing child pornography was being accessed by the user "GERY":

```
C:\>net file

ID         Path                   User name         #Locks
-------------------------------------------------------------------------------

2          D:\pictures\joey01.zip   GERY              0
The command completed successfully.
```

This information provided probable cause to obtain a warrant for the remote computer that belonged to one of the suspect's online cohort who manufactured and traded child pornography.

17.3.6 RANDOM ACCESS MEMORY CONTENTS

TCP/IP related data may be found in RAM on any host, including servers, routers, firewalls, and dial-up terminal servers. By extracting the contents of RAM it may be possible to obtain IP addresses and other useful data relating to network activity. For instance, in one case a computer intruder used a stolen account to install an IRC bounce (BNC) bots that enables individuals to connect to IRC via a compromised host, thus concealing their actual IP address from other people on IRC. Although the traffic between the clients and IRC bot was encrypted, it was possible to obtain some information by examining the contents of memory:

```
% /mnt/cdrom/static-binaries/solaris/last stolenaccount
stolenaccount pts/18 Apr 18 12:34 mail.almustaqbal.com.lb – 12:58 (00:24)
% /mnt/cdrom/static-binaries/solaris/ps -ef | grep stolenaccount
root  3485  2432  0  18:05:03  pts/17  0:00  grep    stolenaccount
root  3430  2387  0  18:04:37  pts/10  0:00  script  stolenaccount.04182003
stolenaccount  9455   1    0    Apr 17 ?   0:01   ./tcsh unf
stolenaccount  13961  1    0    Apr 17 ?   0:01   ./bnc

% /mnt/cdrom/static-binaries/solaris/gcore –o /mnt/evidence/core 9455
gcore: /mnt/evidence/core.9455 dumped
% /mnt/cdrom/static-binaries/solaris/gcore –o /mnt/evidence/core 13961
gcore: /mnt/evidence/core.13961 dumped
% cd /mnt/evidence
% strings – core.9455 | more
<cut for brevity>
PART #cavite
a QUIT :sTiLL dA oNe i wAnT…sTiLL dA oNe i LoVeCAVITE's WebSite
(www.cavitechannel.com)
8.244 PRIVMSG #cavite :
0, 0**********
4, 4***
8, 8***
1, 12********* GoodByE all *********
8, 8***
4, 4***
:CuCuMbEr-!v2000@210.23.248.244 PRIVMSG #cavite :
<cut for brevity>

DjCuRe 210.23.248.165 graz.at.Eu.UnderNet.org djcure H :4G
:McLean.VA.us.undernet.org 352 boseman #cavite SMuRF 210.23.248.163
Amsterdam.NL

.Eu.UnderNet.org explorer2 H :4
2, 15
:McLean.VA.us.undernet.org 352 boseman #cavite ofm_cap nova4117.
i-next.net Manha
```

```
ttan.KS.US.Undernet.Org Jhayr H :3 FERNANDO JOSE
:McLean.VA.us.undernet.org 352 boseman #cavite ~clarice web.cyworld.net
Arlingto

n.VA.US.Undernet.Org Clarimace H :3 Pls join #Li
lpid.bnc
fuckj00
<cut for brevity>
```

Network devices may also contain some TCP/IP related information in RAM that is not available from the command line. It may be possible to recover such data but the process of dumping the contents of memory varies with each device. For instance, the procedure for obtaining a memory dump of a Cisco router is detailed in (Cisco 2002). It is also possible to extract the contents of RAM by physically connecting special equipment to it but this is expensive and rarely feasible for network devices.

17.4 SUMMARY

Watching information move around the Internet is like watching ants work. Tiny entities move around quickly, bumping into each other and occasionally getting lost or damaged, but an overall order is maintained by TCP. These activities generate entries in log files and state tables of servers and personal computers, intermediate routers and firewalls, and other hosts on the network. These and other sources of digital evidence can be located and collected using the methodologies and techniques provided in Chapter 15. The resulting digital evidence can be used to corroborate Web browser history, e-mail messages, and other activities on related hosts.

There are several challenges that investigators encounter when dealing with TCP/IP as evidence. For instance, IP headers only contain information about computers, not people, so it is difficult to prove that a specific individual created a given packet. However, an investigator can use the source IP address to get closer to the point of origin of the crime. Knowing the point of origin of TCP/IP traffic can also help identify suspects. For example, only a small group of individuals might have access to a given computer or the ability to use a specific IP address (e.g. in a home or college dormitory).

Another challenge arises when criminals change their IP address frequently (using dynamic IP addresses). Individuals who exchange illegal information and materials by turning their personal computers into file servers can avoid detection by regularly changing the IP address of the server. For instance, by dialing into a large ISP, such a criminal will be assigned an IP address that others then use to connect to the computer being used as a file server. After a few hours, the criminal might decide that it is time to move. Disconnecting and redialing will often result in the criminal being assigned a different IP address. The only difficulty on the criminal's end is notifying a select group of people using the criminal's computer as a file server about the new IP address. Investigators find it difficult to find and monitor these roaming servers. However, once found, the IP address of a server can lead investigators to the culprit.

Another significant challenge arises when information in the IP header is falsified. It is possible to create a packet with a false source IP address making it appear that data are coming from one computer when it is actually coming from another. For example a malicious program will purposefully insert a false source IP address into packets, before interrupting service on a network (e.g. by flooding a network with data or crashing a central machine on the network). When the administrators of the flooded network try to track down the culprit, they find that the information in the packets is false – making it difficult to trace information back to the sender. When a source IP address has been falsified, tracking becomes a lengthy and tedious process of examining log files on all of the routers that the information passed through. When multiple ISPs are involved, the time and effort that it takes to get everyone's cooperation is rarely justified and there is a high probability that the trail will be too cold to follow. Additionally, if one ISP does not maintain logs, it may not be possible to establish the continuity of offense and track down the source of the attack.

Yet another challenge is that few networks are designed to make evidence collection simple. Evidence is scattered and there is rarely one person in an organization who has access to, or even knows about, all of the possible sources of digital evidence on their network. Also, every network is unique, comprising many different components that are sometimes held together by little more than the digital equivalent of duct tape. Therefore, it is impractical to create a general checklist of all potential sources of evidence with an associated method of collection. As was mentioned before, as digital evidence becomes utilized more, some organizations will develop digital evidence maps of their networks to save time and protect themselves against liability. In the absence of such a map, looking for digital evidence on a network is a matter of exploration and interviewing knowledgeable people.

REFERENCES

Cisco (2002) "Creating Core Dumps", Internetworking Troubleshooting Handbook, 2nd Edition (Available online at www.cisco.com/univercd/cc/td/doc/cisintwk/itg_v1/tr19aa.htm)

Comer, D. E. (1995) Internetworking with TCP/IP Volume I: *Principles, Protocols, and Architecture*, 3rd edn. Upper Saddle River, NJ: Prentice Hall.

Henry P., De Libero G. (1996) "Strategic Networking: From LAN and WAN to Information Superhighways", Massachusetts: International Thomson Publishing Company.

Microsoft (2000) "Interpreting IAS-formatted log files", Microsoft Windows 2000 Server Documentation (Available online at http://www.microsoft.com/windows2000/en/server/help/sag_ias_log1a.htm)

Morris R. T. (1995) "A Weakness in the 4.2BSD UNIX TCP/IP Software," Bell Labs Computer Science Technical Report 117 (February 25, 1985) (Available online at http://www.eecs.harvard.edu/~rtm/papers.html)

Route, Daemon (1997) "Juggernaut," *Phrack 50* (Available online at http://www.phrack.com/show.php?p=50&a=6).

Shinkle P. (2002) "Serial Killer Caught By His Own Internet Footprint", St. Louis Post-Dispatch, 6-17-2002.

Shimomura T. and Markoff J. (1996) "Takedown: The Pursuit of Kevin Mitnick, America's Most Wanted Computer Outlaw – by the man who did it". New York: Hyperion.

Stevens W. R. (1994) TCP/IP Illustrated, Volume 1 Boston: Addison Wesley.

Tribune News Services (2000) "Black student charged with racist e-mail threats at college", April 21, 2000.

CHAPTER 18

DIGITAL EVIDENCE ON THE INTERNET

The growth of the Internet has greatly increased the ways that computers can be involved in a crime and creates many potential sources of digital evidence. Feeling protected by some level of anonymity, individuals often do things on the Internet that they would only imagine in the physical world and express thoughts that they would otherwise keep to themselves. What many people do not realize is that eavesdropping on a network is elementary and servers on the Internet retain a significant amount of information about individuals' activities, creating a cybertrail similar to a paper trail in the physical world.

Some of these data are transient, only remaining on servers for a few seconds, minutes, or days while other forms of digital data can be retrieved years later. These digital data can tell us about an individual's private thoughts and interests, patterns of behavior, whereabouts at a specific time – information that can be very useful in an investigation. As such, it is important for anyone who is involved with criminal investigation, prosecution, or defense work to be comfortable with the Internet as a source of evidence.

This chapter focuses on investigating criminal activity on the application layer of the Internet. Case examples are used to give a practical understanding of how the main services on the Internet can be involved in criminal activity and how they can be a source of digital evidence. The discussions of the Internet's application layer in this chapter can be generalized to any network, such as a company's internal network. Collecting digital evidence at the application layer is like taking a surface scraping of a network. For every piece of digital evidence found at the application layer, there is more related data in other layers of the network that can be obtained as discussed in previous chapters.

18.1 ROLE OF THE INTERNET IN CRIMINAL INVESTIGATIONS

When the Internet is involved in a crime, it generally fits in the categories of Instrumentality or Information as Evidence. For example, online sex

Digital Evidence and Computer Crime Second Edition
ISBN: 0-12-163104-4

offenders, cyberstalkers, computer intruders, and fraudsters use the Internet as an instrument to commit their crimes. Also recall the Cassidy case mentioned in Chapter 7 in which she was convicted of using the Internet to persuade a man to kill her husband. When the Internet is used in such an active way, treating the Internet as an instrumentality of an offense appropriately elevates the importance of digital evidence in the case, potentially increasing the attention it receives and the care with which it is processes.

CASE EXAMPLE (KANSAS v. ROBINSON 2001):
Robinson first used newspaper personal ads to acquire victims and then used the Internet proactively to extend his reach (Fatal Bondage 2001). Robinson also used the Internet reactively to conceal his identity online, often hiding behind the alias "Slavemaster." John E. Robinson used the Internet to con some of his victims into meeting him, at which time he allegedly sexually assaulted some and killed others (Judge 2001). Investigators found five computers in Robinson's home and information on the Internet relating to Robinson's online nickname "Slavemaster." Robinson was found guilty on several counts and sentenced to death in Kansas but still faces murder charges in Missouri.

Interestingly, Robinson's use of the Internet reflects the *modus operandi* he used to acquire victims in the physical world, posing as a respectable businessman interested in a relationship.

When the Internet plays a less active role in a crime, it is more useful to categorize it as "information as evidence." For example, digital evidence on the Internet can simply indicate that a crime has occurred and provide investigative leads.

CASE EXAMPLE (NORTH DAKOTA v FROISTAD 1998):
In one homicide case, involving arson, the Internet played several roles in the investigation. On March 22, 1998 in his e-mail based support group, Larry Froistad made the following confession about killing his
5-year-old daughter, Amanda, 3 years before:

> My God, there's something I haven't mentioned, but it's a very important part of the equation. The people I'm mourning the loss of, I've ejected from my life. Kitty had to endure my going to jail twice and being embarrassed in front of her parents. Amanda I murdered because her mother stood between us. I let her watch the videos she loved all evening, and when she was asleep I got wickedly drunk, set our house on fire, went to bed, listened to her scream twice, climbed out the window and set about putting on a show of shock, surprise and grief to remove culpability from myself. Dammit, part of that show was climbing in her window and grabbing her pajamas, then hearing her breathe and dropping her where she was so she could die and rid me of her mother's interferences.

Froistad, a 29-year-old computer programmer, was arrested and extradited from California to North Dakota where the crime occurred. He apparently confessed again while in police custody. However, upon mature reflection, Froistad pleaded innocent to the charge of murder, a charge that can lead to life imprisonment but not execution, since North Dakota does not have a death penalty. His lawyers initially argued that someone else could have sent the e-mail messages and that Froistad was mentally ill. However, during a forensic examination of Froistad's computer, numerous child pornography references were discovered along with three short AVIs (computer videos) depicting children involved in sexual acts with adults. Also discovered were references by Froistad to a sexual relationship with his daughter and admissions to sexual contact with her. This additional evidence provided a motive for the murder and raised the charges to child exploitation resulted in the death of a minor, potentially subjecting Froistad to more severe Federal penalties, including death. In response to this prospect, the defendant pled guilty to the Federal charges and received a ten year sentence, and also pled guilty to murder in state court which resulted in a 40 year sentence.

Internet-related data has also been used to locate offenders and missing persons even when the Internet did not play a role in the crime. A simple letter can have associated digital evidence on the Internet that can be used to identify an offender, as in the Maury Travis case example in the previous chapter. Also, the Internet can simply provide a meeting place for individuals who commit a crime in the physical world. For instance, Ruth Stabler and Frank Dobson met online and developed a relationship that culminated in Dobson killing Stabler's husband.

18.2 INTERNET SERVICES: LEGITIMATE VERSUS CRIMINAL USES

The Internet provides the infrastructure for many different services. Most people are familiar with services such as e-mail and the World Wide Web. Although many of us use these Internet services, we rarely access them directly. Instead we use applications (computer programs) that make it easier to use the services on a network. For example, many people use the Netscape Navigator application to access Web pages stored on distant Web servers. Similarly, Eudora is an application used to access e-mail on distant e-mail servers. The underlying services are comprised of application layer protocols, many of which are defined in Request For Comment (RFC) documents.[1] Although there are thousands of Internet services and applications, the process of understanding the Internet can be simplified by considering its five main services:

[1] *http://www.ietf.org/rfc.html*

- World Wide Web (WWW or Web)
- E-mail
- Newsgroups (a.k.a. Asynchronous Discussion Groups)
- Synchronous (Live) Chat Networks
- Peer-to-Peer (P2P)

The last two categories are growing rapidly, with more people communicating using live chat applications such as Microsoft Netmeeting, AOL IM, and Yahoo IM, and sharing music, video, and other media using applications like KazaA.[2]

[2]*http://www.kazaa.com*

Internet services like the Web, Usenet, and IRC retain information about people, organizations, and geographical areas. People use the Internet to communicate, explore new ideas, and make purchases from the comfort of their homes. Many organizations use the application layer of their private networks to facilitate communication between employees and to make sales, payroll, and other routine financial transactions more efficient. This combination of social and financial activity makes the application layer an attractive place for criminals. Con artists find a large number of marks through e-mail, Usenet, and the Web. Sexual offenders have a wide selection of hunting grounds (e.g. chat networks) and victims to choose from on the Internet. Stalkers use Internet services to obtain information about their victims and sometime harass their victims using the Internet. Thieves break into private networks of organizations and steal credit card numbers and trade secrets. Hate groups use the Internet to communicate, publish, and threaten.

Only a limited amount of research has been performed to quantify and analyze criminal activity on the Internet. Some of the resulting assertions about crime on the Internet have been based on limited data and are unverifiable.

CASE EXAMPLE (CARNEGIE MELLON UNIVERSITY 1995):
The Georgetown University Law Review published a research paper by Martin Rimm, a student at Carnegie Mellon University (CMU). The paper described and classified the sexually oriented materials circulating on the Internet and quantified the relative amounts of obscene and illegal materials versus other kinds of materials. Rimm's study generated a great deal of interest, reaffirming many people's view that the Internet was primarily used to exchange pornographic materials. Time magazine was so taken with the results that they published a special issue entitled Cyberporn featuring Rimm's study. The CMU administration was so concerned that their computer systems were being used to distribute illegal materials, they temporarily removed all sexually explicit images from the newsgroups on their servers. Ultimately, the study did not fare well under academic scrutiny – the research methodology and data analysis was flawed.

To gain a better understanding of how the Internet facilitates criminal activity, researchers conducted an exploratory study of two Usenet groups, one relating to lock picking and safe cracking and the other dedicated to undermining satellite television encryption mechanisms (Mann and Sutton 1998). Other studies have focused on child pornography and child exploitation on the Internet (Durkin and Bryant 1999). In fact, entire research groups, such as COPINE[3], have been established to address the growing concern of online child exploitation.

[3] *http://copine.ucc.ie*

There are some general assertions that can be made about crime on the Internet. The Web does not contain much direct evidence of criminal activity because there is such a high risk of detection. Much of the illegal activity on the Web is carefully hidden (e.g. password protected), and only available to trusted individuals. Criminals utilize Usenet to collaborate and to distribute pornography of all kinds including child pornography. Criminals feel relatively safe on Usenet because they can conceal their identities and can prevent their messages from being archived, thereby reducing the risk of detection. Criminals that are determined to avoid detection while using the Internet use more private services like e-mail, realtime chat, and peer-to-peer networks. One informal study found that 6% of the requests on a peer-to-peer network appeared to be for child pornography (Palisade Systems 2003). However, this study was based on file names rather than content and probably does not reflect the actual amount of child pornography on these systems.

18.2.1 THE WORLD WIDE WEB

The Web first became publicly available in 1991 and has become so popular that it is often mistakenly referred to as the Internet. Other Internet services including e-mail, Usenet, and synchronous chat networks are now accessible through Web pages. Web pages make it easier for individuals to interact with other Internet services – hiding the complexity with a user-friendly facade.

The popularity and rapid growth of the Web is mainly due to its commercial potential. Using the Web, organizations and individuals alike can make information and commodities available to anyone in the world. Before 1990, some of this information was only available through less user-friendly programs like WAIS, FTP, Archie, Veronica, and Gopher. The Web incorporated these older services and continues to grow, producing the largest information repository in human history. As the Web becomes more widely used to make monetary transactions, associated criminal activities grow. In addition to using the Web to steal from individuals and even steal their identities for profit, some criminals have established Web sites to sell prescription drugs in violation of international customs law. Additionally, some criminals use the Web to provide information to and communicate with fellow criminals. For example, there are an increasing number of recipes for illegal substances on the Web.

CASE EXAMPLE (UNITED STATES v. REEDY 2000):
In 1999, US Postal Inspectors found the Landslide Web site advertising and conspiring to distribute child pornography. The Texas company associated with the site, Landslide Productions, Inc., was owned and operated by Thomas and Janice

Reedy. The US Department of Justice estimates that the Reedys made more than $1.4 million from subscription sales of child pornography in the one month that the Landslide operation was in business. Customers could subscribe to child pornography Web sites through a Ft. Worth post office box, or via the Internet. Landslide also offered a classified ads section on its site, allowing customers to place or respond to personal ads for child pornography (USPS 2001). Although the Web sites and related digital evidence were located in Indonesia and Russia, when digital evidence examiners obtained Thomas Reedy's computer, they found more than 70 images of child pornography and a list containing the identities of thousands of Landslide customers around the world. The resulting investigation was called Operation Avalanche. Thomas Reedy was sentenced to life in prison, and Janice Reedy was sentenced to 14 years in prison.

Some Web sites that have an illegal purpose attempt to obfuscate their actual location by using Web redirection services (e.g. www.kickme.to). This type of redirection simply embeds the page within a frame and can be seen clearly by viewing the source HTML through a Web browser or from the server directly as shown here:

```
% telnet illicit.kickme.to 80
Trying 64.235.234.138 ...
Connected to ns2.dynamicname.com.
Escape character is '^]'.
GET /index.html HTTP/1.1
Host: illicit.kickme.to
HTTP/1.1 200 OK
Date: Sun, 25 May 2003 13:16:50 GMT
Server: Apache/1.3.27 (Unix) PHP/4.1.2
Vary: Host
X-Powered-By: PHP/4.1.2
Transfer-Encoding: chunked
Content-Type: text/html
2e9
<!DOCTYPE HTML PUBLIC "-//W3C//DTD HTML 4.0 Transitional//EN">
<HTML>
<HEAD>
        <TITLE>Illicit Site</TITLE>
        <SCRIPT>
        <!--
        if(top!=self)
        top.location.href=self.location.href;
        //--->
        </SCRIPT>
```

```
</HEAD>
    <!-- frames -->
    <FRAMESET ROWS="100%,*" FRAMEBORDER="no" FRAMESPACING="0">
        <FRAME NAME="REDIRECTION_MAIN"
SRC="http://server1.somewhereelse.com/illicit" MARGINWIDTH="0"
MARGINHEIGHT="0" SCROLLING="auto" FRAMEBORDER="0">
        <FRAME NAME="AD_BOTTOM" SRC="/ad.html" MARGINWIDTH="0"
MARGINHEIGHT="0" SCROLLING="auto" FRAMEBORDER="0">
    </FRAMESET>
</HTML>
0
Connection closed by foreign host.
```

Other Web sites use redirection to forward the individual to a completely different server so investigators must remain alert and verify which server they are connected to when collecting digital evidence. Another common obfuscation approach used by fraudsters to obtain credit card information is to send e-mail posing as a legitimate business (e.g. Paypal, eBay) instructing individuals to submit their account information and credit card number to a URL like "http://www.paypal.com@bylink.net," giving the impression that data is being sent to Paypal when, in fact, it is being sent to "bylink.net."[4] By using this type of URL fraudsters are taking advantage of a feature in the HTTP protocol, described in RFC1738, that supports a username and password in the format "http://username:password@www.website.com."

[4]To obfuscate the actual site, some fraudsters do not put the name of the fraudulent server in the misleading link. Instead they use the IP address or decimal equivalent such as http://www.paypal.com@209.15.160.99 or http://www.paypal.com@3507462243

18.2.2 E-MAIL

E-mail, as the name suggests, is a service that enables people to send electronic messages to each other. Provided a message is correctly addressed, it will be delivered through cables and computers to the addressee's personal electronic mailbox. Every e-mail message has a header that contains information about its origin and receipt. It is often possible to track e-mail back to its source and identify the sender using the information in e-mail headers. Even if some information in an e-mail header is forged it can contain information that identifies the sender. For example, although the following header was forged to misdirect prying individuals, it still contains information about the sender, ec30@is4.nyu.edu.

```
Received: from NYU.EDU by is4.nyu.edu; (5.65v3.2/1.1.8.2/26Mar96-0600PM) id
AA08502; Sun, 6 Jul 1997 21:22:35 -0400
Received: from comet.connix.com by cmcl2.NYU.EDU (5.61/1.34) id AA14047;
Sun, 6 Jul 97 21:22:33 -0400
```

```
Received: from tara.eire.gov (ec30@IS4.NYU.EDU [128.122.253.137]) by
comet.connix.com (8.8.5/8.8.5) with SMTP id VAA01050 for
<eoghan.casey@nyu.edu; Sun, 6 Jul 1997 21:21:05 -0400 (EDT)
Date: Sun, 6 Jul 1997 21:21:05 -0400 (EDT)
Message-Id: ,199707070121.VAA01050@comet.connix.com
From: fionn@eire.gov
To: achilles@thessaly.gov
Subject: Arrangements for Thursday's battle: spears or swords
```

E-mail is one of the most widely used services on the Internet and is one of the most important vehicles for criminal activity, offering a high level of privacy, especially when encryption or anonymous services are used, making it difficult to determine if e-mail is being used to commit or facilitate a crime. Although an e-mail message can be intercepted at many points along its journey or collected from an individual's computer, personal e-mail is usually protected by strict privacy laws, making it more difficult to obtain than many other forms of digital evidence. Even if investigators can obtain incriminating e-mail, it can be difficult to prove that a specific individual sent a specific message. For instance, an individual can easily claim that he/she did not send the message.

CASE EXAMPLE (CBS 2001):
When Fahad Naseem was initially arrested in connection with the kidnapping and killing of journalist Daniel Pearl, he admitted to sending ransom e-mails using his laptop. The laptop and handwritten versions of the e-mails were found in his possession. However, Naseem later retracted his confession and his defense attorney claimed that logs from Naseem's ISP indicated that his account was not connected to the Internet at the time the e-mails were sent. To shed further doubt on Naseem's involvement the defense claimed that the laptop produced in court had a different serial number than the one recorded in police records and that other documentation relating to the computer was inconsistent. For instance, documentation indicated that FBI agent, Ronald Joseph, was examining the laptop between February 4–7, whereas documents indicated that the laptop was not seized until February 11. However, the court denied the appeal, including the following explanation.

> The leading of Shaikh Naeem to the recovery of the laptop being used through connection No. 66 from his system as the house of accused Fahad Naseem on 11/02/2002 was provided to [Ronald Joseph] who had examined the same and conducted the forensic examination and formulated his report which was conveyed to the investigation from the Consulate General of the United States of America vide Ex.49/3, on examining the report, he has categorically stated that the Black Soft Computer came with "Proworld" written on the exterior and upon opening the case a Dell Latitude Cpi laptop was found on it. The laptop was identified in the report produced by this witness to be of model PPL with Serial No. of ZH942 and located inside the laptop was an IBM travel star hard driver [sic] which was stated to have been removed from the laptop and viewing the label on the hard drive model, the drive was identified as 4.3 GB of storage capacity and the Model No. was determined by this witness to be OKLA24302 with a serial number of 4/1000N81834 . On examining articles 1 and 2 of Ex 49 compared with the

Mushernama recovery of the laptop in juxtaposition with the computer Forensic Examination report and identifying the numbers of the same, there is no doubt whatsoever that this Laptop is the same equipment which was recovered from the possession of accused Fahad Naseem on 11/02/2002. The Forensic Examination report is also ex.49/B. It would be seen that the said report reflects the laptop to have been made available to this witness on 4/02/2002 as suggested by the defense. Availability of the laptop at the American Consulate on 4/02/2002 is not only unnatural but impossible because of the fact that complainant Marianne Pearl had filed the complaint with the police on 4/02/2002 (ex-53/A) at 2345 hours which had in fact set the ball rolling at the hands of the Investigating Agency. (DAWN Group 2002)

18.2.3 NEWSGROUPS

Newsgroups are the online equivalent of public bulletin boards, enabling asynchronous communication that often resembles a discussion. Anyone with Internet access can post a message on these bulletin boards and come back later to see if anyone has replied. Most newsgroups are part of a free, global system called the User's Network (Usenet) that began in 1979.

Because Usenet messages are broadcast to millions of people around the world, it is the perfect medium for individuals to communicate with a huge audience. Criminals use this global forum to exchange information and commit crimes, including defamation, copyright infringement, harassment, stalking, fraud, and solicitation of minors. Also, child pornography and pirated software is advertised and exchanged through Usenet to a limited degree. Offenders subscribe to newsgroups that attract potential victims (e.g. alt.abuse-recovery, alt.teens).

CASE EXAMPLE

Sharon Lopatka was killed by a man she met on the Internet first through Usenet and then in a BDSM channel on IRC. Interestingly, nobody who knew Sharon in person, including her husband, suspected that she was involved in this type of activity or even had such an interest.

Subject:	>>>> Wanna Buy My Worn...Pantyhose...and Panties????
From:	nancyc544@aol.com (NancyC544)
Date:	1996/05/15
Message-ID:	<4nduca$2j4@newsbf02.news.aol.com
Newsgroups:	alt.pantyhose
organization:	America Online, Inc. (1-800-827-6364)
reply-to:	nancyc544@aol.com (NancyC544)
sender:	root@newsbf02.news.aol.com

Hi! My name is Nancy. I am 25, have Blonde hair, green eyes am 5'6 and weigh 121. Is anyone out there interested in buying my worn...pantyhose...or....panties? This is not a joke or a wacky internet scam. I am very serious about this. I live in the U.S. but I can ship them anywhere in the world. If you are serious you can e-mail me at: nancyc544@aol.com

Like e-mail, Usenet messages have headers containing information about the sender and the journey that the message took. However, the format of the headers in Usenet is slightly different from e-mail. As with e-mail, the header can be modified to make it more difficult to identify the sender. With training and practice, investigators can learn to extract a great deal of information from Usenet.

18.2.4 SYNCHRONOUS CHAT NETWORKS

Live conversations between users on the Internet exist in many formats (e.g. text, audio, video), a huge variety of topics, and take place 24 hours a day. There are many organizations such as AOL and Yahoo that provide large chat areas as well as Instant Messaging programs, and some ISPs have small chat areas for their customers. Additionally, there are more obscure chat areas on the Internet that can be accessed using Telnet (e.g. Multiuser Domains, Telnet Talkers).

One of the largest chat networks is Internet Relay Chat (IRC), started in 1988. IRC can be accessed by anyone on the Internet using free or low-cost software.[5] Because it is not necessary to pay or even register, IRC is effectively anonymous and, therefore, attractive to criminals. IRC is made up of separate networks such as Undernet, DALnet, Efnet, and IRCnet and no single organization controls all of them. Each subnet is simply a server, or combination of servers, run by a different group of people. Although they are all part of IRC, the subnets are physically separate. So, connecting to the Undernet subnet does not give access to chat rooms (a.k.a. channels) on DALnet. IRC allows individuals to create their own, self-titled rooms as shown in Figure 18.1 and some people choose not to have their channels listed, making them more difficult to locate.

[5] http://www.irchelp.org

There are thousands of chat rooms in operation worldwide on IRC at any given time. Many IRC chat rooms exist to facilitate the discussion of unlawful activities and the exchange of illegal materials. Computer intruders gather in IRC chat channels to share information, ranging from general intrusion techniques to passwords of compromised systems. Child pornographers meet to exchange materials and IRC has even been used to broadcast live sessions of children being sexually abused. Some channels are plainly visible and some can even be found through search engines on the Web.[6] However, many channels are difficult to find because they are dealing with illegal activity, and may be acessed by invitation only, or protected by a password.

[6] http://searchirc.com/

There are chat channels with names like "#carderz" and "#cardz" dedicated to selling stolen credit cards or trading them for equipment, compromised computers, and other items that are considered valuable. For example, Carlos Salgado was convicted of hacking into computer systems, stealing tens of thousands of credit cards, and selling them on IRC using the

CHANNEL NAME	PARTICIPANTS	DESCRIPTION
#0!!!!!!!ltlgirlsexchat	12	Sexy and Friendly FANTASY CHAT Channel for YOUNG GIRLS and those that love them!!! No snuff, torture, rape, force, extreme, mom/son channels. No trading, invites, on-joins or spam. 15 minutes between trolling messages. Girls under 20 can type !girl for a plussy.
#0!!!!bifem-dogsex	13	Welcome to #0!!!!bifem-dogsex LadyMary's friendly channel! 18+ Only ! We do not approve of rape and pedophile/underage channels - please leave immediately. DO NOT message anyone unless you ask!!!
#cracks	19	#cracks is now open. Serial Search !serial program name .New channel format. Absolutely NO files in the channel. This channel is for chat/search only, so it does NOT break Dalnets new AUP. :D
#masterccs	35	Welcome In The Official #CC Channel I Trading , Pasting Illegal Informations is NOT Permited ! I We are not responsible of normal users activities ! I EnJoY !!
#mp3cablez	80	-=M=P=3=C=A=B=L=E =Z=- Best High Speed Servers On Phazenet New/Pre_Release Movies Classic Rock Box Sets Zipped Albums Karaoke Christian Roms And More Always Open Slots
#192+mp3albums	127	www.mp3albums.ca FUCK THE RIAA. To share type !serv <MrStatic> novus, you like sniffing the exercise bike seat?

Figure 18.1

A list of a few IRC chat channels.

nickname SMAK. Other channels are dedicated to trading pirated music, videos, and software (a.k.a. *warez*).

IRC has a direct client connection (DCC) feature that allows two individuals to have a private conversation and exchange files without being seen by anybody. As the name suggests, DCC establishes a direct connection between personal computers, bypassing the IRC network, leaving little or no digital evidence on the IRC servers. Fortunately for digital evidence examiners, remnants of IRC sessions can sometimes be salvaged from unallocated or swap space as discussed in Part 2 of this text. Also, some offenders keep personal logs of the direct, private communications that they have on IRC. This ability to chat privately and transfer files over a more secure connection is very powerful and can lead to a level of criminal activity that gives meaning to the name that inspired the subnet name; Undernet. DCC could be thought of as an underworld of the Internet because it is the least visible part of IRC.

Another feature of IRC, called "fserve" (short for fileserver), enables people to make files on their personal computers available to many other IRC users. Many of the people trading files on IRC (e.g. pornography and pirated software) use this feature. One of the most sophisticated and popular fserves is Panzer.[7]

[7] *http://www.filetrading.net/irc/fileservers/panzer.htm*

ICQ ("I seek you") is another large, free chat network that anyone on the Internet can use but, unlike IRC, it has a registration process. After completing a registration form with details like name, e-mail address, and personal interests, each individual is assigned a user identification number (UIN) for the ICQ network. Some people provide identifying information when they register, but many do not, making it more difficult to connect an individual with an ICQ number.

Instead of gathering in chat rooms, most ICQ users seek each other out and jointly agree to have a conversation. While this limits contact with others on the ICQ network, it enables more private conversations than on other chat networks. In this respect, misconduct facilitated by ICQ is more difficult to detect because a third party cannot participate in ICQ conversations unless invited. However, unlike direct chat on IRC, ICQ directs messages through a central system where they can be monitored. Notably, ICQ network also has asynchronous discussion boards and some chat rooms that can be accessed using a Web browser.[8]

[8] *http://www.icq.com*

The privacy, immediacy, and impermanence of synchronous chat networks make them particularly conducive to criminal activity. Also, the potential for direct contact with potential victims is appealing to some criminals. For instance, sex offenders can obtain victims immediately, leaving very little digital evidence. Even though chat sessions are not automatically archived or searchable by the public, a surprising amount can be learned from the activities in the millions of online chat rooms. Although it can be a challenge to locate and identify criminal on chat networks, criminals let their guard down, feeling protected by the perceived anonymity making these chat networks useful resources for investigators.

18.2.5 PEER-TO-PEER NETWORKS

A host on a peer-to-peer network can simultaneously function as server and client (a.k.a. servent), downloading files from peers while allowing peers to download files from it. The two most popular peer-to-peer networks, KaZaA and Gnutella, use protocols based on HTTP to exchange data. By design, many of these applications have a limited amount of information that can be useful to investigators. When individuals first connect to a peer-to-peer network, they are only required to select a unique username. Although the choice of username may be sufficiently unique to search for related information on the Internet, there is very little to go on other than the IP address.

When a file is being downloaded from a peer, the associated IP address can be viewed using `netstat`. However, some peer-to-peer clients can be configured to connect through a SOCK proxy to conceal the peer's actual IP address. While most peer-to-peer systems transfer files using a single connection, a KaZaA peer can download fragments from multiple peers and reassemble them into a complete file. Figure 18.2 shows search results in the KaZaA Media Desktop – the "+" beside an item indicates that it is available from multiple locations and can be downloaded in fragments. Newer peer-to-peer networks like eDonkey are implementing this capability to download

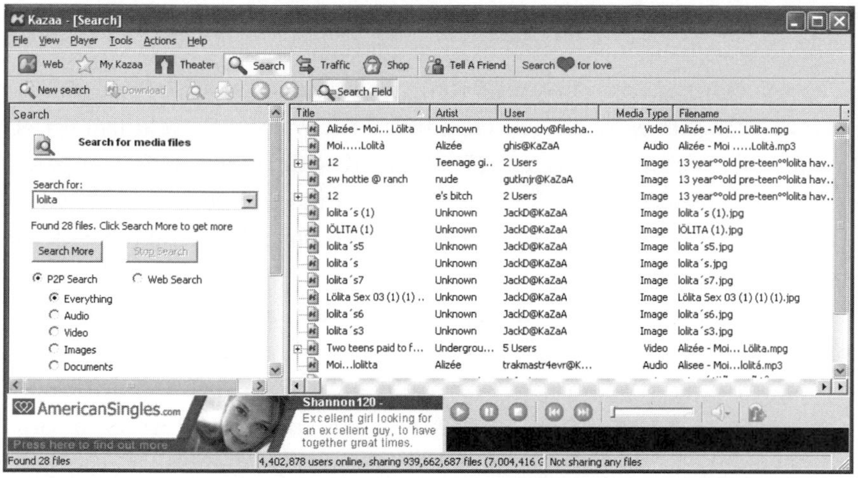

Figure 18.2
KaZaA Media Desktop (KMD).

pieces of a file from multiple sources. This fragmentation feature does not conceal the sources of the file fragments but does make it more difficult for digital evidence examiners to recover complete files from network traffic. The KaZAlyser[9] utility is useful for extracting information from computers that were used to exchnge files via KaZaA, such as file names, times and IP addresses.

[9]http://www.sandersonforensics.co.uk

KazaA has one feature that can be beneficial from an investigative standpoint – whenever possible it obtains files from peers in the same geographic region. Therefore, if investigators find a system with illegal materials, there is a good chance that it is nearby.

18.3 USING THE INTERNET AS AN INVESTIGATIVE TOOL

An important aspect of following the cybertrail in an investigation is to search for related information on the Internet such as a victim's Web pages or Usenet messages, an offender's e-mail address or telephone number, and personal data in various online databases. Because the Internet contains so much loosely ordered information, searching for something in particular can be like looking for a needle in a haystack. This is why it is crucial to learn how to search the Internet effectively. In addition to becoming familiar with various search tools, it is necessary to develop search strategies.

One method of searching for digital evidence on the Internet is to look for online resources in a particular geographical area. For instance, if a victim or unknown offender lives in San Francisco, there is likely to be a higher concentration of related information in that area. Searching online telephone directories, newspaper archives, bulletin boards, chat rooms, and other resources dedicated to San Francisco can uncover unknown aspects of a known victim's online activities and can lead to the identity of a previously unknown offender. Search engines that focus on a particular country (e.g. www.google.it, ie.altavista.com) can also be useful for a geographically focused search.

Another strategy is to search within a particular organization. For instance, if a victim or offender is affiliated with a particular company or school, there is likely to be a higher concentration of personal information in associated online resources. As with a geographically focused search, looking through an organization's online telephone directory, internal bulletins or newsletters, discussion boards or mailing lists, and other publicly accessible online resources can lead to useful information. Additionally, it may be possible to query systems on an organization's network for information about users. Although it is permissible to access information on an organization's computer systems in non-invasive ways, care should be taken not to cross the line into unauthorized access.

Besides searching for real names, nicknames, full e-mail addresses, and segments of e-mail addresses, it can be productive to focus searches around unusual interests, searching areas on the Internet that the victim or suspect frequented. Given the difficulty in making informed guesses of where a victim or offender might go on the Internet, this type of search usually develops from a lead. For instance, interviews with family and friends, or an examination of a victim's computer may reveal that she subscribed to a particular newsgroup and frequented a particular IRC chat room to arrange sexual encounters. An offender or victim may have left traces of their activities in these online areas. Searching these areas can be particularly productive if the offender and victim communicated with each other in a public area on the Internet, revealing connections between them.

In addition to the traces of activities that remain on the Internet, online witnesses who used the same areas may have logs of the activities on their computers. For instance, in the Sharon Lopatka case, participants in the AOL and IRC channels that the victim and offender frequented recalled that both of them did not employ "safe-words" to prevent injury during rough sex (Cairns 1996). As another example, after apprehending an offender, some digital evidence examiners will contact people who the offender was in contact with on the Internet (e.g. sent e-mail, AOL Buddy list). By sending a letter to these individuals informing them of the situation and asking them for any related information, it is possible to locate witnesses and other victims. In some cases, victims of a common offender seek each other out to form online support networks. These associations can be helpful to the victims. They can also be useful to investigators because the networks make identifying and contacting victims easier. However, sharing information about the criminal activity and the offender among victims who are also potential witnesses may complicate matters when the time comes for them to testify.

Notably, these search strategies are not mutually exclusive and can be effectively combined to locate the majority of available information on the Internet regarding the search subject. Whichever combination of search strategies is used, investigators should document important searches, indicating when, where, and how specific items were found. Handwritten notes combined with the investigator's Web browser history are generally sufficient to show when, where, and how information was located. Also, because information on the Internet can change at any moment, screenshots and copies of Web pages are useful for documenting what investigators saw at the time. Some tools for capturing a Web site efficiently and fairly completely are:

- Web Whacker: www.webwhacker.com
- Adobe Acrobat: www.adobe.com
- Teleport: www.tenmax.com/teleport/pro/home.htm
- Httrack: www.httrack.com
- Web Copier: www.maximumsoft.com
- Snagit: www.techsmith.com
- Anawave's WebSnake: http://www.websnake.com/
- Htdig: http://www.htdig.org
- Surfsaver: www.surfsaver.com/download
- Wget: http://www.gnu.org/software/wget/wget.html
- Black Widow: www.softbytelabs.com/BlackWidow

Some of these tools will not copy subpages of a Web site if links to these subpages are encoded in a scripting language that the tool does not understand. Therefore, it is advisable to test a tool to ensure that it is adequate for the task and inspect the resulting files to verify that they are satisfactory. Any files that are generated during the search process should be inventoried, documenting file names, MD5 values, and date–time stamps.

18.3.1 SEARCH ENGINES

Search engines are among the most useful tools for finding information on the Internet. Although search engines are not particularly difficult to use, there is some skill involved in using them effectively. Each search engine has different contents, archiving methods, search features, and limitations. Therefore, if is important to understand how each search engine works and which ones are best suited for particular tasks.

Many search engines, like Altavista, actively update themselves by running programs that search the Web incessantly for new data. As a result, they can turn up recent information but lack older, outdated data.[10] Google compensates for this shortcoming by retaining a copy of Web pages it has found – this "cached" information is useful when the original is gone.

[10] *An archive of many Web pages can be found at http://web.archive.org/*

Google is also capable of searching Word documents and PDF files that other search engines overlook. Additionally, Google has a searchable archive of Usenet messages stretching back to 1981. Another unique feature of Google is its search algorithm (PageRank), which estimates the relevance and quality of data based on the number of links to the data from other sources on the Web. It is important to be aware of how each search engine attempts to "help" with a search so that this "help" can be utilized when it is useful and avoided when it is not.

Investigators can employ the language of the search engines they are using to create more narrowly focused searches. For example, some search engines understand words like AND, OR, NOT, and NEAR. Some search engines also allow symbols such as "−" to exclude terms for the search and "+" to include terms. For instance, in Altavista, the following commands can be used to find documents containing the words "unsolved" and "homicide" but not the words "mystery" or "mysteries:"

```
+homicide +unsolved −mystery −mysteries
homicide AND unsolved AND NOT myster*
```

Some offenders protect themselves by using computer-smart nicknames such as En0ch|an instead of Enochian. The zero instead of an "o" and the pipe (|) instead of an "i" confound search algorithms. In such cases, clever use of search engine syntax (e.g. AND, OR, NEAR) is required. Search engines can also be useful for finding connections on the Web. For instance, pages containing links to a suspect's Web site can be found by searching Google or Altavista using the syntax "link:www.suspectswebpage.com." For additional discussion about utilizing advanced features of search engines see SearchEngineWatch.[11]

[11] *http://www.searchenginewatch.com/facts/index.php*

Keep in mind that searching for obviously illegal terms will rarely turn up anything illegal. Many Web sites use illegal terms to attract interest, but actual criminals make some effort to hide their activities using euphemisms. For instance, some offenders use the terms "lolita" or "nature shots" to refer to images of children, or "family fun" to refer to incest. These euphemisms may turn up during the initial searches, in which case it will be necessary to expand the search using this new knowledge and gradually narrow the search again. Also, individuals who want their Web pages to be excluded by search engines can simply place "robots.txt" files on their Web sites.

Metasearch engines such as Copernic and Metacrawler enable individuals to search multiple search engines simultaneously from a single site. Because they utilize many other search engines, metasearch engines can be useful for brainstorming or finding very specific details. However, since metasearch engines tend to usurp control of the search, their results can be incomplete or can contain unrelated entries. As a result, metasearch engines make it more difficult to determine why certain pages were included in the results, making it difficult to explain to others how the page was found. Search results may contain pages that are unrelated to the subject in question but that contain some of the keywords. Failing to explain exactly how a particular piece of evidence was found can weaken a case. Furthermore, the large number of hits that are common in metasearch engines can be overwhelming and can hinder an investigation.

Although metasearch engines can be useful when searching for very specific details (e.g. occurrences of a telephone number on a Web page), it is important to also search specialized search engines or databases (e.g. telephone directories) when looking for fine details.

18.3.2 ONLINE DATABASES (THE INVISIBLE WEB)

There are many databases on the Web containing data within specific subject areas. For example, online databases contain information about sex offenders, missing children, individuals' assets and credit history, and medical information. Many of these databases can be located using search engines but the information they contain can only be queried directly. For instance, using Google or Altavista for "sex AND offender AND database" leads to various Sex Offender Registries around the United States. Some databases are organized on the following Web sites, making them easier to find.

- InvisibleWeb: http://invisibleweb.com
- Internets: http://www.internets.com
- JournalismNet: http://www.journalismnet.com
- PowerReporting: http://www.powerreporting.com

There are also online databases, such as AutoTrack and KnowX, containing a wide variety of information about individuals but these databases charge fees for use.

Whois databases are particularly useful for investigations involving the Internet. Whois databases are maintained by Internet registrars and contain the names and contact information of people who are responsible for the many computer systems that make up the Internet. These databases can reveal who is responsible for a particular Web site, including their name, telephone number and address. There are separate Whois databases for different countries – some of the main databases are listed here and others can be found at Allwhois.[12]

[12] *http://www.allwhois.com*

- United States (NetSol): http://www.netsol.com/cgi-bin/whois/whois
- United States (ARIN): http://whois.arin.net/whois/index.html
- Europe: http://www.ripe.net/db/whois.html
- Asia: http://whois.apnic.net/

Some registrar databases only have information on high-level domains while others have information on IP addresses. For instance, to find the contact information for "www.wsex.com," search Netsol whereas to find contact information for the associated IP address (207.42.132.101), search ARIN. Note that these databases have slightly different contact information for the World Sports Exchange.

Domain name: www.wsex.com	IP Address: 207.42.132.101
Registrant: Big Green (WSEX-DOM)	ISP: Cable & Wireless Antigua
	SPRINT-CF2A87
Woods Center #11	
St. Johns Antigua	OrgName: World Sports Exchange
AG	OrgID: WSE-9
	Address: Friar's Hill Road
Domain Name: WSEX.COM	Address: Woods Center, St John's
	City:
Administrative Contact:	StateProv:
holowchak, jason (NZHOWTMQZI)	PostalCode:
jasonholowchak@hotmail.com	Country: AG
hodges bay	
st. johns, na na	NetRange: 207.42.132.96-207.42.132.127
AG	CIDR: 207.42.132.96/27
268-480-3861 123 123 1234	NetName: CWAG-207-42-132-96
Technical Contact:	NetHandle: NET-207-42-132-96-1
Hanson, Spencer (SH2534)	Parent: NET-207-42-132-0-1
spencer@WWW.WSEX.COM	NetType: Reassigned
World Sports Exchange Ltd	Comment:
Ryan's Place, High Street	RegDate: 2001-04-20
St. John's	Updated: 2001-04-20
AG	
268 480-3888	TechHandle: MH1271-ARIN
	TechName: Hayden, Matthew
Record expires on 19-Sep-2009.	TechPhone: (268)-480-3888
Record created on 18-Sep-1996.	TechEmail: jay@wsex.com
Domain servers in listed order:	
NS.WSEX.COM 207.42.132.101	
NS2.JASONHOLOWCHAK.COM 207.42.132.119	
NS.JASONHOLOWCHAK.COM 66.216.122.143	

Sites such as Geektools[13] facilitate searches by providing a single interface to many **Whois** databases. It is also possible to search some **Whois** databases for other fields such as names and e-mail addresses. Some individuals use services like Domain by Proxy[14] to prevent their contact information from being placed in the Whois database system.

[13] *http://www.geektools.com*

[14] *http://www.domainsbyproxy.com*

18.3.3 USENET ARCHIVE VERSUS ACTUAL NEWGROUPS

Archives such as Google Groups contain millions of messages from tens of thousands of newsgroups. These archives are invaluable tools for investigators because they contain a vast amount of detailed information about individuals and their interactions. By searching this archive, it may be possible to learn about a person's interests, personality, and much more. However, these archives are not comprehensive and should not be depended on completely when dealing with Usenet. Few archives include message attachments and anyone can specify that they do not want their postings to be archived. Any newsgroup posting with "x-no-archive: yes" as its first line will be ignored by archiving software. Also, there are private newsgroups that are not archived.

Therefore, it is important for investigators to become familiar with and involved in the actual newsgroups related to an investigation rather than rely entirely on the archives. As well as seeing information that is not archived by Google Groups (e.g. images and other file attachments), it is useful to see discussions develop and progress, get to know the characters of the participants, and observe patterns of a particular group's behavior. Additionally, investigators may be able to observe offenders of their local community in newsgroups dedicated to a specific geographic region.

18.4 ONLINE ANONYMITY AND SELF-PROTECTION

It is important for investigators to become familiar with online anonymity to protect themselves, and to understand how criminals use anonymity to avoid detection. In addition to concealing obvious personal information like name, address, and telephone number, some offenders use IP addresses that cannot be linked to them. Such IP addresses can be obtained by using free ISPs that allow individuals to dial into the Internet without requiring them to identify themselves. Other ISPs unintentionally provide this type of free, anonymous service when one of their customer's dial-up accounts is stolen and used by the thief to conceal his identity while he commits crimes online. Public library terminals and Internet cafes are other popular methods of connecting to the Internet anonymously.

Investigators should use anonymity to protect themselves while searching for criminals on the Internet, particularly when conducting an undercover investigation. Online undercover investigations can be used in many types of criminal activity including online gambling. When investigating online gambling it is necessary to create several undercover identities to make transactions and gather intelligence into the supporting organizations and networks. Undercover identities are also used to purchase drugs on the Internet and stolen hardware through online auction sites. In child exploitation cases, undercover investigators may pose as children or as pedophiles to gather evidence in a case as described in Chapter 21. Computer intruders can be tracked on IRC, counterfeiters can be ferreted out, and fraudsters can be apprehended all with the assistance of online undercover identities.

18.4.1 OVERVIEW OF EXPOSURE

In their book *Investigating Computer Crime,* Clark and Diliberto demonstrate the dangers of online investigations by outlining the problems they encountered during one online child exploitation investigation.

> 1 Telephone death threats.
> 2 Computer (BBS) threats.
> 3 Harassing phone call (hundreds).
> 4 Five Internal affairs complaints.
> 5 Complaints to district attorney, state attorney general, and FBI.
> 6 Surveillance of officer.
> 7 Videotaping of officer off duty (of officer giving presentation in church on subject of "dangers of unsupervised use of computers by juveniles").
> 8 Video copied and sent to militant groups.
> 9 Multimillion dollar civil suits filed.
> 10 Tremendous media exposure initiated by suspects.
> 11 Hate mail posted on Internet resulting in many phone calls.
> 12 Investigator's plane tickets canceled by computer.
> 13 Extensive files made on investigators and witnesses, including the above computerized information: name, address, spouse, date of birth, physical, civil suits, vehicle description, and license number.
> 14 Above information posted on BBS.
> 15 Witnesses' houses put up for sale and the bill for advertising sent to witnesses' home addresses by suspects.
> 16 Witnesses received deliveries of products not ordered, with threatening notes inside.
> 17 Hundreds of people receiving personal invitation to witness's home for a barbeque (Put out by computer).
>
> And much more! After 18 months of this, when all was said and done, the suspect was sentenced to 6 years, 4 months in state prison. All the complaints against the investigator were found to be unfounded, and the investigator was exonerated of any wrongdoing. (Clark and Diliberto 1996)

Simply conducting research to gather intelligence online most likely will not open an investigator to these types of attacks. However, the above testimonial highlights the imperative that when conducting an investigation involving Internet usage and technically savvy targets, proper, predetermined protocol must be followed. Chapter 19 discusses undercover best practices in more detail and, in addition to following applicable jurisdictional policies, attorneys should be consulted prior to conducting online undercover investigations.

18.4.2 PROXIES

One approach to concealing one's IP address while surfing the Web is to direct all page requests through a proxy. Web servers that are accessed via a proxy record the IP address of the proxy rather than that of one's computer. Commercial Web proxies like Anonymizer.com are available and there are many machines on the Internet that act as proxies either accidentally or by design. Additional information about Web proxies are available at

- http://www.all-nettools.com/privacy/anon.htm
- http://inetprivacy.com/a4proxy/
- http://anon.inf.tu-dresden.de/

When offenders use Web proxies to conceal their identities, it makes tracking more difficult because investigators must obtain information from the server running the proxy to determine the actual IP address of the offender. These logs may even be available on systems that are specifically designed to protect the identity of users. For instance, a now defunct anonymous proxy service called "SafeWeb" debunked the commonly held belief that their anonymizing service did not retain log files.

> ...what do we do with the logs? Every night we tar them up, ship them to a central machine, compile stats on how many clients we served and how many ads we served, gpg the logs, and store them for 7 days. After that they get deleted, unless someone manages to supena (sic) them. In which case we pull out only the entrys associated with the supena (sic), and keep them around until we're actually served with said supena (sic).

It is also possible to connect to IRC or ICQ through a proxy that does not just handle Web traffic, such as a Wingate or SOCKS proxy. Increasingly, individuals who want to hide their IP address on chat networks are finding misconfigured hosts with open proxies and are using them without authorization. It can be difficult to obtain log files from these misconfigured proxies when they are located in another country. To address this growing problem, many IRC networks will not allow connections from hosts that are running a proxy server.

18.4.3 IRC "bots"

Individuals can make it more difficult to locate them on IRC by using the invisibility feature.[15] However, the invisibility feature does not conceal the individual from others in the same channel, so this offers limited protection. One advanced aspect of IRC that some offenders use to conceal their actual IP address are "bots." These programs can function like proxies and can be used to perform various tasks from administering a channel to launching denial of service attacks. "Eggdrop" is one of the more commonly used IRC bots and can be configured to use strong encryption (blowfish) that conceals the contents of

[15] *http://www.mirc.co.uk/faq6.html#section6–26*

its logs and configuration files making it necessary to examine network traffic to observe nicknames, passwords, etc. The IRCOffer bot is also widely used to share pirated software, movies and other illegal materials. Another popular type of bot is a "bouncer" (BNC for short) that allows an individual to connect to IRC via the machine that is running the BNC bot. When an individual is connected to IRC via a BNC bot, only the IP address of the computer running the BNC bot is visible – the individual's actual IP address is not visible on IRC.

18.4.5 ENCRYPTION

To protect their Internet communications, some individuals encrypt data using PGP or specialized e-mail services such as Hushmail[16] and Zixmail.[17] Others use the secure e-mail standard (S/MIME) that is integrated into many e-mail clients. The encryption keys used in S/MIME are usually stored on an individual's system, protected by a password. For instance, by default, Netscape stores these keys in a file called "key3.db". However, these keys can also be generated and stored on a hardware device such as an iButton[18] or iKey.[19] These devices are portable and will destroy the encryption keys they contain if they are tampered with.

[16] http://www.hushmail.com
[17] http://www.zixmail.com/
[18] http://www.ibutton.com/
[19] http://www.rainbow.com

Some IRC clients support encryption, making it more difficult for investigators to monitor communications and recover digital evidence.

CASE EXAMPLE (ORCHID CLUB/OPERATION CATHEDRAL):
A major investigation into an online child pornography ring that started with the online chat room called Orchid Club and expanded to a chat room called Wonderland Club has involved hundreds of offenders around the globe. Interestingly, when the Wonderland Club members learned that they were under investigation, they did not disperse but began using more sophisticated concealment techniques such as encryption and moving to different IRC servers frequently. The use of encryption significantly hindered investigators. In one instance, a suspect's computer was sent from the UK to the FBI in an effort to decrypt the contents but to no avail. Overall, the level of prosecution in this case was low relative to the number of individuals involved.

Additionally, Trojan horse programs can be configured to encode traffic between the client and server. For instance, by default, each packet sent between a Back Orifice client and server is XOR-ed with a known pattern (XOR is a simple binary operation). However, these packets begin with same pattern of bytes and intrusion detection systems can be configured to determine the key and decrypt the traffic. Therefore, more technically proficient intruders will configure Back Orifice to use a plugin with stronger encryption.

In general, it is not feasible to decrypt network traffic and it is more effective to seek and recover digital evidence from the end points of the communication. Computer intruders have realized this – rather than attempting to obtain credit cards as they are transmitted between the client and server through an encrypted Secure Socket Layer (SSL) connection, intruders target the end points. Computer intruders usually steal credit cards by installing a Trojan program on individuals' systems and monitoring their keystrokes, or by breaking into the server and stealing the file or database that contains credit card information. Similarly, when intruders cannot obtain passwords using a sniffer because traffic is being encrypted using SSH, they target the end point, replacing the SSH server software with a version that records passwords in a file. Alternatively, intruders target the original SSH server software before it is distributed (CERT 2002).

18.4.5 ANONYMOUS AND PSEUDONYMOUS E-MAIL AND USENET

Individuals who are more technically savvy and are especially interested in concealing their identity, send messages through anonymous or pseudonymous services. For instance, when e-mail is sent through an anonymous remailer, identifying information is removed from the e-mail header before sending the message to its destination. The most effective anonymous remailers (e.g. Mixmaster and Cypherpunk) are quite sophisticated and make it very difficult to determine who sent a particular message. For instance, the following message was sent through the "anon.efga.org" remailer.

```
Received: from server1.efga.org by is4.nyu.edu;
(5.65v3.2/1.1.8.2/26Mar96-0600PM) id AA09406; Sat, 9 Aug 1997 00:43:54 -0400
Received: (from anon@localhost) by server1.efga.org (8.8.5/8.8.5) id AAA08333;
Sat, 9 Aug 1997 00:44:06 -0400
Date: Sat, 9 Aug 1997 00:44:06 -0400
Message-Id: <BEDPZMcwd925FWA/mG0Tyg==@JawJaCrakR>
To: ec30@is4.nyu.edu
Subject: Test
From: Anonymous <anon@anon.efga.org>
Comments: This message was remailed by a FREE automated remailing service.
For additional information on this service, send a message with the subject
"remailer-help" to remailer@anon.efga.org. The body of the message will be
discarded. To report abuse, contact the operator at admin@anon.efga.org.
Headers below this point were inserted by the original sender.
```

However, even when these types of remailers are used, evidence transfer occurs – the sender transfers something in the message, the message leaves something behind with the sender, and intermediate machines that handle the message may have useful information. The sender may disclose something personal or the message may contain class characteristics that give a clue about its origin. The sender's computer may retain fragments of the message, the encryption key used to sign the message, or a clear connection to the remailer used.

CASE EXAMPLE (USDOJ 1999):
Carl Johnson used anonymous e-mail to threaten notable figures, including federal judges by posting to an e-mail list entitled Cyberpunks. Johnson used a system called "Assassination Politics" – a computerized gambling operation where participants "predicted" the date of death of the Government employee, with the assassination payoff being funneled to the assassin as proceeds from the bet as described in one of his messages.

> Leading eCa$h candidate for dying at an opportune time to make some perennial loser "Dead Lucky" are: e$ 2,610.02 J. Kelley Arnold, United States Magistrate Judge, Union Station Courthouse, 1717 Pacific Avenue, Tacoma, Washington ... I feel it is necessary to make a stand and declare that I stand ready and willing to fight to the death against anyone who takes it upon themselves to try to imprison me behind an ElectroMagnetic Curtain. This includes the Ninth District Court judges ... I will share the same "DEATH THREAT!!!" with Judges Fletcher, Nelson and Bright that I have shared with the President and a host of Congressional and Senatorial representatives.

Johnson used several aliases and anonymous remailers when posting to the mailing list and in one message he sent his private PGP key to the list. Johnson's use of remailers and encryption ultimately implicated him – authorities matched the PGP digital signature on e-mail messages to an encryption key discovered on his computer. Interestingly, because he sent his key to the mailing list, many people had access to the private PGP key that was used to implicate him. So, the connection between Johnson and the digital signature that what was used to implicate him was not a one-to-one match. Nonetheless, the court held that the Government's technical evidence was sufficient to prove that Johnson wrote the messages and found him guilty.

Intermediate servers may contain timestamped logs that show where data was received from and where it was forwarded. Using these fragments of information it may be possible to narrow the suspect pool and then focus an investigation on a few individuals. Some remailers make efforts to minimize information transfer that could be used to link a message with its sender but none are perfect.

Truly anonymous remailers do not enable the sender to receive a response to their messages because there is no way to connect the message back to the individual who sent it. For this reason, true anonymous services are only useful when an individual does not want to maintain two-way communication.

> Anonymity means you have no reputation or persistence – in essence, you have no identity and people can't establish long-term relationships with you.

> Pseudonymity – creating persistent alter-egos that cannot be associated with your true identity – lets you access the full power and resources of the Internet, and establish long-term relationships, without sacrificing your privacy. (http://www.freedom.net/faq/pseudo.html)

Because most people using e-mail want a response, they use pseudonymous servers such as Asarian-host to conceal their actual identities as shown in the following Usenet message.

```
Path: news.ycc.yale.edu!pln-
e!extra.newsguy.com!lotsanews.com!newsfeed1.earthlink.net!
uunet!uunet!in1.uu.net!rutgers!usenet.logical.net!news.dal.ca!torn!howland.
erols.net! newsfeed.berkeley.edu!su-news-
hub1.bbnplanet.com!news.bbnplanet.com!news.alt.net!
anon.lcs.mit.edu!nym.alias.net!mail2news
Comments: To protect the identity of the sender, certain header fields are not
shown. Anonymous email addresses for asarians can be requested by filling in
the appropriate form at: http://asarian-host.org/emailform.html
Message-ID: <199809212245.QAA16547@asarian-host.org>
Posted-Date: Mon, 21 Sep 1998 16:45:21 -0600 (MDT)
Date: Mon, 21 Sep 1998 18:40:36 -0400
From: "lisa"
Reply-To: lisa@REMOVE_THIS.asarian-host.org
Organization: Asarian-host.org
Subject: cutting
Newsgroups: alt.abuse.recovery
Comments: Anonymous USENET posting by Asarian-host, using Email Gateway:
mail2news@anon.lcs.mit.edu Mail-To-News-Contact: postmaster@nym.alias.net
```

Some remailers keep logs of the actual e-mail addresses of individuals, but many remailers will perish rather than make such concessions, even when illegal activity is involved. There is a possibility that investigators can compel a pseudonymous remailer to disclose the identity of the sender but it requires significant effort since their business is to protect the identity of their users.

CASE EXAMPLE

A pseudonymous remailer in Finland named anon.penet.fi was compelled to disclose the identities of subscribers as a result of actions of the Church of Scientology (COS). During the investigation, anon.penet.fi operator Johan Helsingius was heard as a witness. He was asked to reveal the pseudonymous accounts used to disseminate private COS documents, but refused. A legal battle followed, Julf was required by the courts to reveal identifies, and he ultimately discontinued the remailing service.

18.4.6 FREENET

An anonymous information sharing system that is accessed via a Web browser, called Freenet,[20] is becoming increasingly popular among child pornographers and other criminals. Figure 18.3 shows the Java Freenet client that can access information via Web links or using "keys" similar to URLs that are associated with each file on the network.

[20] *http://freenet.sourceforge.net*

Each computer that joins Freenet becomes a node on the network, storing files that others can download. Freenet uses strong encryption and regularly moves data from one computer to another, making it difficult to determine where the information originated. This concealment activity makes it difficult to establish the continuity of offense, making it necessary to evaluate their source based on characteristics of the files and their contents as described in Chapter 9.

In addition to concealing data, encryption is used to protect users legally as explained on the Freenet FAQ:

> to keep operators from having to know what information is in their nodes if they don't want to. This distinction is more a legal one than a technical one. It is not realistic to expect a node operator to try to continually collect and/or guess possible keys and then check them against the information in his node (even if such an attack is viable from a security perspective), so a sane society is less likely to hold an operator liable for such information on the network.

Freenet also supports Near Instant Messaging (NIM) as well as online discussions via a program called Frost. Other applications are being developed to make Freenet more usable.

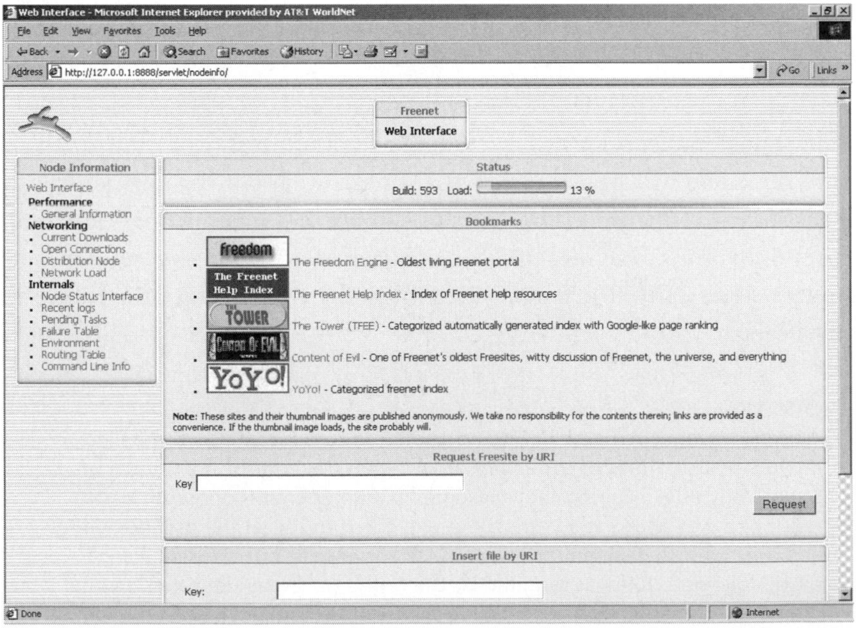

Figure 18.3

Java client providing links to Freenet.

18.4.7 ANONYMOUS CASH

Anonymous cash services like V-Cash and InternetCash implement a simple concept that can be useful to individuals who want to protect their privacy. Individuals can purchase anonymous cash through one of these services and then use it to purchase products from vendors that accept this form of currency. Another form of online currency are e-metals (e.g. e-gold, e-silver) that are backed by precious metals and are accepted by various online vendors and in some eBay auctions. In fraud cases that involve anonymous cash, it is quite difficult to identify the offender because of the added layer of protection.

18.5 E-MAIL FORGERY AND TRACKING

It is often possible to track e-mail back to its source and potentially identify the sender using the information in e-mail headers. In addition to learning how to extract information from e-mail headers, it is important to understand how e-mails can be forged. The main use of forged e-mail is to give the receiver a false impression. For example, the sender might pose as the recipient's boss or friend. Some offenders forge e-mail in an effort to conceal their identity. However, this approach to anonymity is ineffective because forgeries usually contain the sender's IP address.

Before delving into e-mail forging and tracking, it is necessary to understand how a message is created and transmitted. Electronic mail is similar to regular mail in many ways. There are computers on the Internet, called Message Transfer Agents (MTA) (Figure 18.4), which are the equivalent of post offices for electronic mail. When an e-mail message is sent, it first goes to a local MTA. Just as a post office stamps letters with a postmark, the local MTA puts the current time and the name of the MTA along with some technical information, at the top of the e-mail message. This e-mail equivalent of a postmark is called a "Received" header. The message is then passed from one MTA to another until it reaches its destination.

Every MTA that receives the message puts a received header at the top of the message. A simple analogy to this is a stack of pancakes; newer ones are on top.

One approach to preserving a complete copy of an e-mail message, including headers, is to save it to file and calculate the MD5 value of the file. Notably, printing an e-mail message will not usually show the header information unless it is displayed. Most e-mail applications can display e-mail headers. For example, while viewing an e-mail message in Netscape Mail select the View – Headers – All menu item or Options – Show Headers on older vesions, in Outlook Express select the File – Properties menu item and click on the Details tab, in Outlook select the View – Options menu item, in Eudora click on the "Blah, Blah, Blah" button at the top of the message, and in Pine type H.

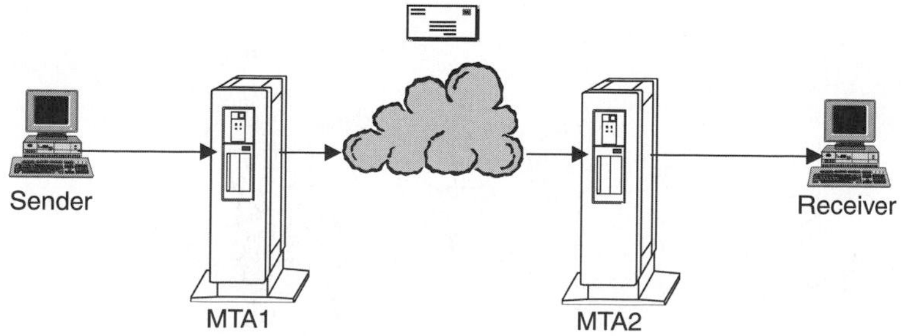

Figure 18.4
Message Transfer Agent.

This means that the last computer to handle the message is listed at the top of the header, and the first computer is listed near the bottom. Therefore, to track an e-mail message back to the sender, simply retrace the route that the e-mail traveled by reading through the e-mail's Received headers.

E-mail forgery takes advantage of how MTAs exchange messages using Simple Mail Transport Protocol (SMTP). Remember that a protocol is just an agreed upon way of "speaking" and, as the name suggests, the Simple Mail Transfer Protocol is quite simple. In four broken English sentences (helo, mail from, rcpt to, data) one MTA (mta.sending.com) can instruct another MTA (mta.receiving.com) to pass an e-mail message on to its destination. Using the same broken sentences, an individual can command an MTA directly using Telnet on a Windows machine by clicking on the Start button, selecting Run, typing "telnet mta.sending.com 25". This instructs Telnet to connect to port 25 on the MTA and permit SMTP commands to be typed and sent to the MTA as shown here:

```
% telnet 192.168.201.11 25
Trying 192.168.201.11...
Connected to 192.168.201.11.
Escape character is '^]'.
220 mta.sending.com ESMTP Sendmail 8.11.6/8.11.6; Sat, 10 May 2003
14:58:57 -0500
helo forgery.com
250 mta.sending.com Hello forgery.com, pleased to meet you
mail from: forger@forgery.com
250 2.1.0 forger@forgery.com... Sender ok
rcpt to: louiscipher2004@hotmail.com
250 2.1.5 louiscipher2004@hotmail.com... Recipient ok
data
354 Enter mail, end with "." on a line by itself
Received: from fake.com ([10.12.227.15]) by mta.nonexistent.com (MSMTP
4.04) with SMTP id g5BK2m642810 for jane.doe@corpX.com; Sat, 10 May 2003
16:00:00 -0500
From: Joe Smith <joe.smith@corpX.com>
To: Jane Doe <jane.doe@corpX.com>
Date: Sat, 10 May 2003 15:12:16 -0400
Message-ID: <069601c31728$122ee620$9eef7222@jxsdqfofq>
I am coming to get you.
Joe
.
```

In the SMTP session shown above, the **helo** command introduces the sending host. The "**mail from:**" command specifies where bounces and receipts will be delivered, regardless of what the "From" line contains. The "**rcpt to:**" command specifies where the e-mail message will be delivered, regardless of what the "To" line contains. The **data** command begins the message and fake headers can be entered here. The body of the message should be separated from any headers by at least one blank line. The body of the message is terminated by a single "." on a line by itself, resulting in the following message.

```
Received: from mta.sending.com ([192.168.201.11])
    by mc1-f7.law16.hotmail.com with Microsoft SMTPSVC(5.0.2195.5600);
    Sat, 10 May 2003 13:02:26 -0700
Received: from forgery.com ([172.16.237.235])
    by mta.sending.com (8.11.6/8.11.6) with SMTP id h4AK1l531700 for
    louiscipher2004@hotmail.com; Sat, 10 May 2003 15:01:53 -0500
Received: from fake.com ([10.12.227.15])
    by mta.nonexistent.com (MSMTP 4.04) with SMTP id g5BK2m642810 for
    jane.doe@corpX.com; Sat, 10 May 2003 16:00:00 -0500
From: Joe Smith <joe.smith@corpX.com>
To: Jane Doe <jane.doe@corpX.com>
Date: Sat, 10 May 2003 15:12:16 -0400
Message-ID: <069601c31728$122ee620$9eef7222@jxsdqfofq>
Return-Path: forger@forgery.com
I am coming to get you.
Joe
```

The alert examiner will see that the forged Received header is not consistent with the other headers. First, the date–time stamp in the forged header is 1 hour later than the other date–time stamps in the message. Second, the forged header indicates that the message was accepted by "mta.nonexistent.com" in which case the next Received header should show the message being passed from "mta.nonexistent.com" to "mta.sending.com." However the next header contains no reference to "mta.nonexistent.com" and instead reveals the sender's actual IP address (172.16.237.235). The ISP responsible for the sender's IP address could use this information to determine which user account was used to send the message. To hide their IP address, some e-mail forgers send messages by connecting to an SMTP relay via a proxy as shown here:

```
% telnet proxy.isp.com 3128
    Trying proxy.isp.com...
    Connected to proxy.isp.com.
    Escape character is '^]'.
    CONNECT smtp.relay.com:25 HTTP/1.0
[hit return twice]
    Host: smtp.relay.com:25
    HTTP/1.0 200 Connection established
    HELO [YOUR DOMAIN]
    MAIL FROM: [YOUR EMAIL ADDRESS]
    RCPT TO: [YOUR EMAIL ADDRESS]
    DATA
    Testing for an open squid proxy
    .
```

This approach makes it even more difficult to determine the originating IP address because the Web proxy effectively conceals this information. Although some Web proxies add a "X-Forwarded-For" header containing the sender's IP address, this information is not retained in an e-mail header.

18.5.1 INTERPRETING E-MAIL HEADERS

Unless a re-mailer or advanced forging technique has been used, a key piece of information that can lead to the sender's identity will be stored somewhere in the message. The trick is to find that key piece of information among the mass of misleading information. For e-mail tracking purposes, the two most useful e-mail headers are the "Message ID" and "Received" headers. A Message-ID is required to be globally unique – no two different messages will ever have the same Message-ID. Some MTAs construct the Message-ID using the current date and time, the MTA's domain name, and the sender's account name. For instance, a message sent on December 4, 1999 from mail.corpX.com by user13 might have the following Message-ID header:

```
Message-Id: <user13120499152415–00000153@mail.corpX.com>
```

The Message-ID cannot always be relied on since it can be forged as shown in the previous section. Although forged Received headers can be inserted into a message to confuse investigators, some of the headers at the top of the message must be valid because they were added by MTAs that delivered the message.

In some cases, a Received header will contain the sender's e-mail address. In other cases, a Received header will contain the IP address of the originating computer and it may be necessary to contact someone at the ISP responsible for the IP address to find out who was using the computer in question at the time the message was sent. For instance, many individuals attain "pseodonymity" by using non-identifying e-mail addresses (e.g. Hotmail, Netaddress) but they are unaware that the e-mail headers of these messages contain the IP address of the originating computer. For instance, the following Hotmail message contains the originating IP address in two places.

```
Return-Path: <louiscipher2004@hotmail.com>
Received: from hotmail.com (f14.pav1.hotmail.com [64.4.31.14])
     by mta.receiving.com (8.9.3/8.9.3) with ESMTP id UAA06245
     for <john.doe@receiving.com>; Wed, 28 Aug 2002 20:42:17 -0500
Received: from mail pickup service by hotmail.com with Microsoft SMTPSVC;
     Wed, 28 Aug 2002 18:42:08 -0700
Received: from 192.168.12.48 by pv1fd.pav1.hotmail.msn.com with HTTP;
     Thu, 29 Aug 2002 01:42:08 GMT
X-Originating-IP: [192.168.12.48]
From: "Louis Cipher" <louiscipher2002@hotmail.com>
To: john.doe@receiving.com
Subject: Look behind you
Date: Wed, 28 Aug 2002 21:42:08 -0400
Message-ID: <F148Bi89QtpfTYSl1q400015c21@hotmail.com>
X-OriginalArrivalTime: 29 Aug 2002 01:42:08.0339 (UTC)
FILETIME=[494ED230:01C24EFD]
I'm watching you
Louis Cipher
____________________________________________________________
Send and receive Hotmail on your mobile device: http://mobile.msn.com
```

Hotmail and many other similar services keep logs that can be useful for identifying the sender. In one case, by tracing a Hotmail message to a library computer in Berkeley, investigators located a fugitive named Troy A. Mayo who was wanted for questioning in the death of a pregnant teenager. Keep in mind that a Web proxy can be used to hide the IP address of the originating computer making it much more difficult to determine the actual source of the message. When a proxy is used, the message header will contain the IP address of the proxy server and it would be necessary to obtain access logs from the proxy server to determine the actual origin of the message.

18.6 USENET FORGERY AND TRACKING

Usenet is made up of news servers all over the world that communicate using the Network News Transport Protocol (NNTP). Each server subscribes to a selection of newsgroups and stores a copy of each Usenet newsgroup it subscribes to. There is no centralized server that coordinates Usenet – it is a cooperative network.

More, specifically, when a message is posted to a newsgroup, it is initially stored on only one news server. At a prearranged time, this news server automatically sends the message – along with all of the other new messages that it has – to a prearranged set of neighboring servers. These servers add their names to the message header and pass the messages on to other servers, and so on. In this way, messages are eventually passed along to all of the other people who participate, to create the global Usenet network. Like e-mail, the path a Usenet message takes can often be traced back to the computer used to send it. To better understand Usenet messages, it is helpful to have a basic understanding of NNTP.

The NNTP command commands that news server servers use to exchange messages are defined in RFC 977. For instance, the **group** command tells the server which newsgroup the message is intended for. The **post** command indicates the beginning of the actual message. Take a moment to read the description of the post command in this RFC:

> If posting is allowed, response code 340 is returned to indicate that the article to be posted should be sent. Response code 440 indicates that posting is prohibited for somc installation-dependent reason.
>
> If posting is permitted, the article should be presented in the format specified by RFC850, and should include all required header lines. After the article's header and body have been completely sent by the client to the server, a further response code will be returned to indicate success or failure of the posting attempt.

Note that the server allows any header lines to be entered, allowing individuals to forge Usenet messages. However, the message header will often contain the originating IP address. For example, the following shows a forged Usenet message being created by connecting to port 119 on a news server and entering NNTP commands.

```
% telnet news.sending.com:119
200 news.sending.com NNRP server INN 1.4unoff4 05-Mar-96 ready
(posting ok).
group alt.test
211 1280 633804 635463 alt.test
post
340 Ok
Subject: Usenet forgery
Path: none!nada
From: forger@forgery.com
Newsgroups: alt.test
This is a forged Usenet message.
.
240 Article posted
quit
205
```

This resulted in the following message – the header contains the IP address of the originating computer (192.168.10.4).

```
Path: news.ycc.corpX.com!pln-
e!extra.newsguy.com!lotsanews.com! howland.erols.net!
newsfeed.concentric.net!news.sending.com!none!nada
From: forger@forgery.com
Newsgroups: alt.test
Subject: Usenet forgery
Date: 27 Sep 1998 17:37:13 GMT
Message-ID: <6ult49$fha@news.sending.com>
NNTP-Posting-Host: 192.168.10.4
This is a forged Usenet message.
```

The following section describes how to interpret the header information in a Usenet message and determine the origin.

18.6.1 INTERPRETING USENET HEADERS

A standard Usenet message consists of several header lines, each consisting of a keyword followed by a colon and some additional information. The required header lines in a Usenet message are "From," "Date," "Newsgroups,"

"Subject," "Message-ID," and "Path." Other optional header lines such as "NNTP-Posting-Host" and "X-Trace" are often added to help determine the origin of the message. One of the most useful lines for tracking messages is the Path line. As stated in RFC 1036:

> 2.1.6. Path
>
> This line shows the path the message took to reach the current system. When a system forwards the message, it should add its own name to the list of systems in the "Path" line. The names may be separated by any punctuation character or characters (except "." which is considered part of the hostname). Thus, the following are valid entries:
>
> cbosgd!mhuxj!mhuxt
>
> cbosgd, mhuxj, mhuxt
>
> @cbosgd.ATT.COM,@mhuxj.ATT.COM,@mhuxt.ATT.COM
>
> teklabs, zehntel, sri-unix@cca!decvax
>
> (The latter path indicates a message that passed through decvax, cca, sri-unix, zehntel, and teklabs, in that order.) Additional names should be added from the left. For example, the most recently added name in the fourth example was teklabs. Letters, digits, periods and hyphens are considered part of host names; other punctuation, including blanks, are considered separators.
>
> Normally, the rightmost name will be the name of the originating system. However, it is also permissible to include an extra entry on the right, which is the name of the sender. This is for upward compatibility with older systems.
>
> The "Path" line is not used for replies, and should not be taken as a mailing address. It is intended to show the route the message traveled to reach the local host.

However, some part of the Path header may be a forgery. Copies of the message from multiple sources will show which portions are forged – the forged portion of the path will remain constant while the true path will vary depending on which servers the message passed through. Another useful header for tracking is the Message-ID. As with e-mail, the Message-ID is usually added by the first news server that receives the message but can be forged. The NNTP-Posting-Host and X-Trace headers often show the actual source, but this can be forged as well. NNTP-Posting-Host is an extension not mentioned in the original RFC but described in RFC 2980 as follows:

> 3.4.1 NNTP-Posting-Host
>
> This line is added to the header of a posted article by the server. The contents of the header is either the IP address or the fully qualified domain name of the client host posting the article. The fully qualified domain name should be

> determined by doing a reverse lookup in the DNS on the IP address of the client. If the client article contains this line, it is removed by the server before acceptance of the article by the Usenet transport system.
>
> This header provides some idea of the actual host posting the article as opposed to information in the Sender or From lines that may be present in the article. This is not a fool-proof methodology since reverse lookups in the DNS are vulnerable to certain types of spoofing, but such discussions are outside the scope of this document.

Not all servers include the optional "NNTP-Posting-Host" or "X-Trace" lines, making it more difficult to determine the source of a message. In such cases, it may be necessary to look for "rough edges" in the message that can be used to search for related information on the Internet. A rough edge is any aspect of a message that may be repeated in other messages from the same individual. A rough edge might be an unusual misspelling of a word, a choice of online nickname, or the way an individual signs a message. In one case, each message that an individual posted to Usenet contained the following line at the bottom of the text.

```
Get paid to read email: http://www.sendmoreinfo.com/SubMakeCookie.cfm?
Extract-69381
```

The Extract-ID is a unique number assigned to each individual who uses the Sendmoreinfo.com service. Searching for other messages containing this Extract-ID led to the identity of the sender.

18.7 SEARCHING AND TRACKING ON IRC

There are two general reasons for wanting to track an individual on IRC: (1) investigators become aware of the person through IRC and want to learn more about him/her and (2) investigators learn about the person and suspect that he/she uses IRC. Before tracking anyone on IRC, it is necessary to configure some form of logging to document the search. For instance, in mIRC logging can be configuration as shown in Figure 18.5.

Including the date in the file name is a good practice from an evidence gathering standpoint and the "Timestamp logs" feature records the date and time of all lines in a log file, making it easier to keep track of when events recorded in the logs occurred.

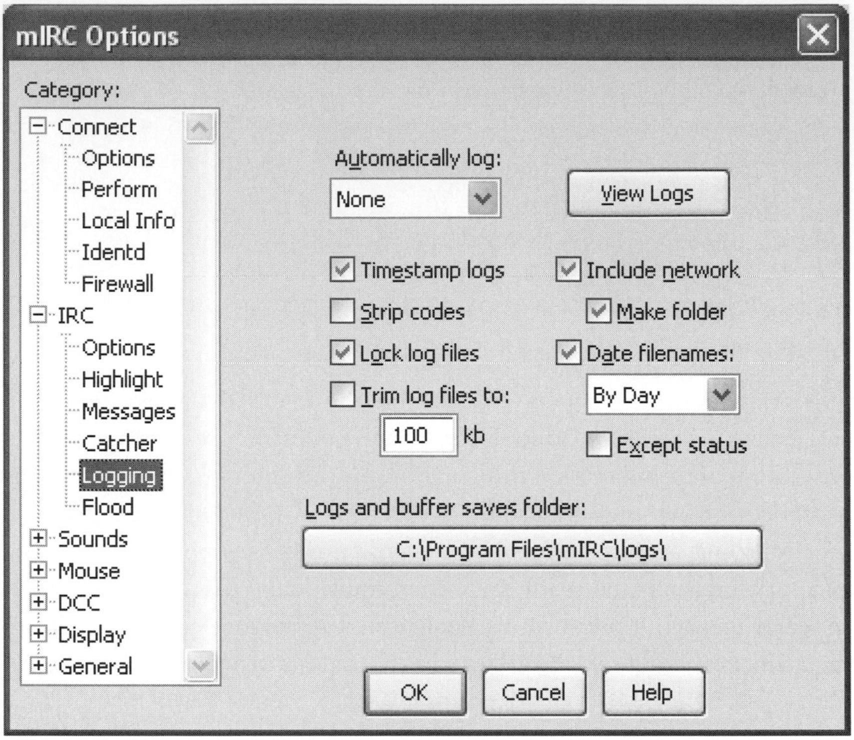

Figure 18.5

Logging configuration, accessed via the File – Options menu item.

When a broad search of a particular IRC subnet is required, the who command is most useful. The who command can search for any word that might occur in a person's hostname or nickname, or can be used to search for people in a particular region. For instance, Figure 18.6 shows the who command being used to find all Verizon users from Baltimore (*east.balt. verizon.net).

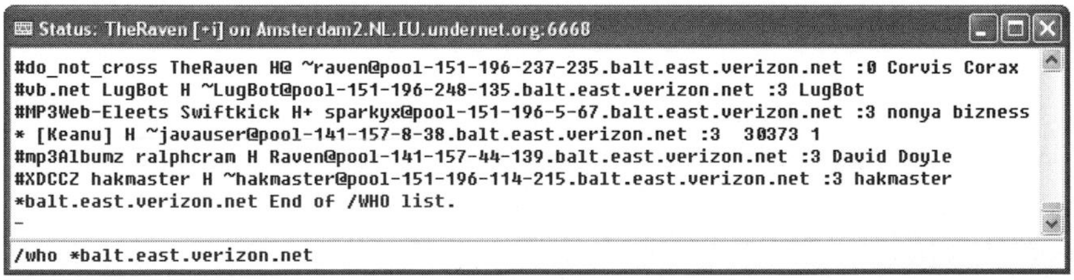

Figure 18.6

Results of the who command on IRC.

Similarly, it is possible to search for individuals in a specific country using commands "/who *.se" or "/who *.ie" for all individuals in Sweden and Ireland, respectively. As another example, the command "/who *raven*," finds all users with the word "raven" in their nickname or hostname.

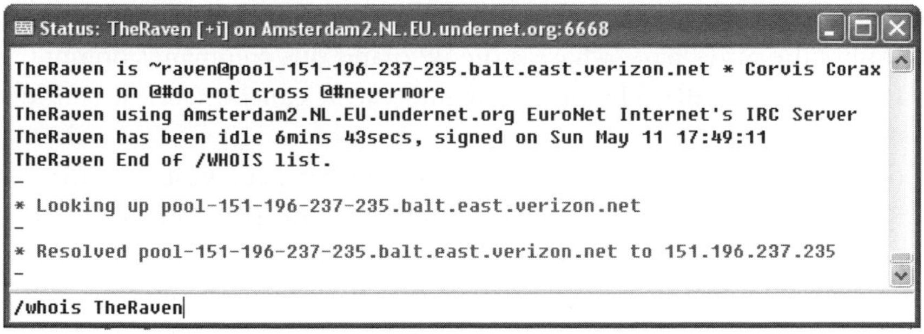

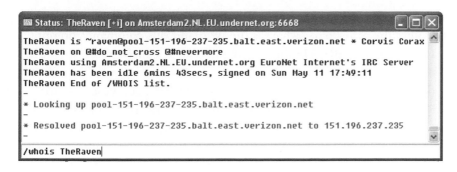

Figure 18.7

Results of the whois and dns commands on IRC.

When a particular individual of interest has been found on IRC, the **whois** command can provide additional details. The **whois** command on IRC is not the same as the Whois databases mentioned earlier. The **whois** command uses a person's IRC nickname to get information like the person's IP address and, if he/she provides it, e-mail address. Figure 18.7 shows information obtained about an IRC user named "TheRaven" using **whois**, listing channels TheRaven is in (#nevermore, #do_not_cross) and, more importantly, the computer he/she is connecting from (pool-151-196-237-235.balt.east.verizon.net). The IP address associated with this host name was obtained using the command "**/dns TheRaven**."

Additional information about these and other IRC commands are detailed at the The IRC Command Cosmos.[21] Note that it is not advisable to use the

[21] *http://www.irchelp.org/irchelp/misc/ccosmos.html*

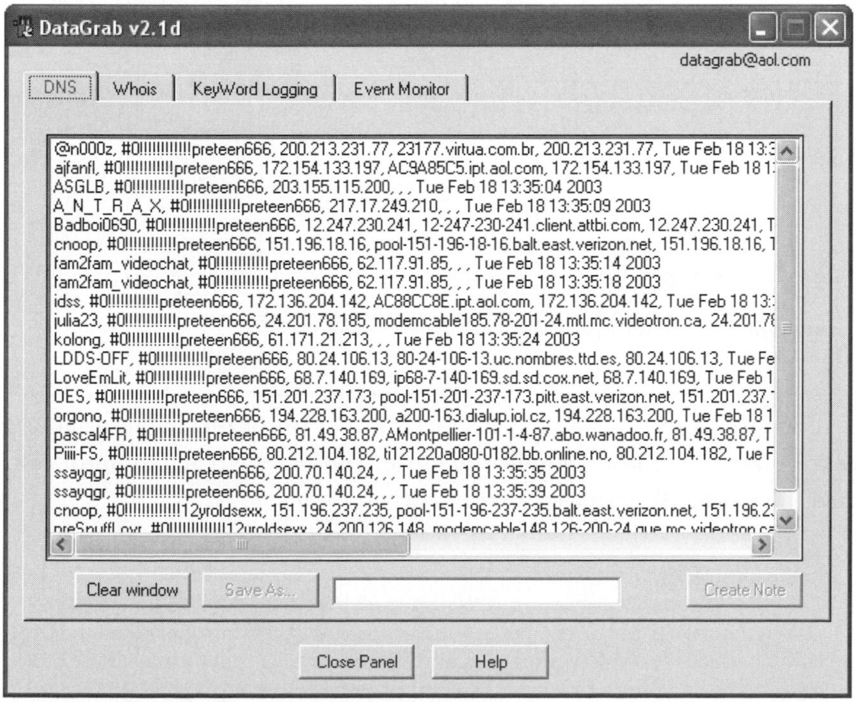

Figure 18.8

DataGrab.

finger command on IRC to gather information about an individual because it notifies the other party whereas the **who** and **dns** commands do not.

If a particular IRC channel is of interest, it can be fruitful to use an automated program that continuously monitors activity in that channel. A utility called DataGrab[22] facilitates monitoring activities on IRC and gathering whois and DNS information. Figure 18.8 shows DataGrab being used to gather DNS information about all participants in a channel called "#0!!!!!!!!!!!!preteen666," saving the date–time stamped results into text file. The "KeyWord Logging" feature can be configured to record information whenever a particular word occurs in the chat room that is being monitored.

[22] *http://members.aol.com/datagrab/*

Chat Monitor[23] is another useful tool for automatically monitoring specific IRC channels and looking for anyone connecting from particular countries. Figure 18.9 shows Chat Monitor logging individuals who are participating in the IRC channel called "#0!!!!!!!!!!!!preteen666."

[23] *http://www.surfcontrol.com*

Figure 18.9
Chat Monitor.

Chat Monitor can also be configured with a list of nicknames that are of interest using its "Buddy Monitor" feature. Additionally, Chat Monitor can be used to analyze IRC logs for a particular user's activities.

CASE EXAMPLE

During a routine security audit, a Windows 98 host was found running BO2K. When the owner of the computer was informed that the intruder could monitor all of her activites, she was shocked and noted that this could explain how her credit card had been stolen and used to subscribe to pornographic Web sites.

A preliminary digital evidence examination uncovered an ".exe" entry in Registry in the RunServices key. Additionally, an unknown service named "ae.exe" was running. The executable was located in "C:\Windows\System\ae" along with IRC chat and DCC logs, indicating that it was an IRC bot. One file named "finger.txt," included the following details about the bot that would be provided to anyone who fingered the host.

```
[default]
::::::::::::::::::::::::::general:info:::::::::::::::::::::::::::::::::::::::::::::::::
::: hi! my ip is 135.223.23.5 and right now i'm on irc.concentric.net as nautilus
::: i have 0 chats. i have 0 queries.
::: i have 0 sends. i'm on 7 channels.
::: use /finger help@135.223.23.5 for more information type shit.
::::::::::::::::::::::::::::::::::::::::::::::::::::::::::::::::::::::::::::::::::::::::
[help]
```

A log file revealed the following activities of one of the intruders, nicknamed "epitaph:"

- Sep 12 07:25:09: epitaph logged into the compromised machine from 1Cust226.tnt1.sierra-vista.az.da.uu.net with the username root and password puritycontrol
- Sep 13 11:13:33: epitaph connected from 1Cust226.tnt1.sierra-vista.az.da.uu.net, replaced some files (e.g. autoexe.bat) and deleted files in the McAfee folder to disable the antivirus software, preventing it from detecting the Trojan program

Another log file showed what appeared to be the same intruder connecting to the IRC bot using the nickname "aeon." The intruder's cohorts who connected to the IRC bot called her Julz or Julie and one log entry in the IRC bot contained the e-mail "jgraham@usr07.primenet.com." The intruder called the IRC bot as "julian v1.5" and described it as "a small project made in boredom." Using an undercover account, investigators connected to the IRC server that the bot was connected to (irc.concentric.net) and started observing the intruder and her cohorts. Additionally, the investigators searched the Internet for rough edges in the log files like "ae.exe," "epitaph," "aeon," "jgraham@usr07.primenet.com," and "julian v1.5." They also performed a geographically focused search in the Sierra Vista region of Arizona. Their search uncovered a Web page "http://www.primenet.com/~jgraham/" that contained a link to a Web page associated with "aeon." Using finger on the Sam Spade page to query the Primenet server about the jgraham account returned the following:

```
09/15/02 16:55:26:
finger jgraham@usr07.primenet.com (206.165.6.207)
Login: jgraham Name: John Graham
Directory: /user/j/jgraham Shell: /bin/bash
Mailbox last read: Sept 15 12:31:24 2002
Currently logged in via na02.fhu-130 IPnet: 208-50-51-49.nas2.fhu.primenet.com
```

The last line indicated that someone was logged into the Primenet server using this account from "208-50-51-49.nas2.fhu.primenet.com." Using finger on Sam Spade to query the host directly returned the following:

```
09/15/02 17:17:04:
finger @208-50-51-49.nas2.fhu.primenet.com (208.50.51.49)
if your name is joshua gabbard, you're a dungpunching faggot.
also: www.subweb.net
www.subweb.net/index.htm
subweb I: the eye of the nephilim
```

Notably, the nickname "nephilim" occurred in IRC logs on the compromised host. Whois "www.subweb.net" did not reveal anything useful.

Repeating these steps again the following day, Whois "www.subweb.net" had been updated and contained the intruder's name, home address and telephone number and finger revealed the following:

```
09/16/02 23:00:26:
finger @208-50-51-162.nas2.fhu.primenet.com
(208.50.51.162)
::::::::::::::::::::::::::::::::julian:info::::::::::::::::::::::::::::::::::::::::::::::::::::::::::::
what is "julian"? a small project done in countless hours of boredom. "julian"
itself is an acronym for, jag's universal liberally inclined artifical nerd.
originally, julian had moods and "intelligent" reactions as per those moods.
however, due to a conflict of productive interest, julian was completely rebuilt,
less the moods. a better interface was designed and more controls were
implemented. the moods may be back in the summer of 2002, provided julian's
author is still unemployed.
use /finger help@208.50.51.162 for more information type shit.
...
::: current channels for julian1 on irc.east.gblx.net:6667 as of 19:59:46
:::::::::::::::::::::::::::::::::::::::::::::::::::::::::::::::::::::::::::::::::::::::::::::::::::
::: 1) #terrorism +tn (no topic set) 2 ops, 2 nonops, 4 total.
::: 2) #julian +tn (no topic set) 1 ops, 0 nonops, 1 total.
```

Although these IRC channels were not plainly visible on IRC, searching for the known nicknames of the intruder and her cohorts (e.g. "/whois epitaph," "/whois aeon") revealed that they were connected to these channels from several compromised hosts. All of the information gathered indicated that the intruder was a high school student in the Sierra Vista region of Arizona. Because she was a minor and the cost of the damages was lower than the legal threshold, the intruder was not arrested but received a warning.

18.8 SUMMARY

Criminal activity on the Internet can generate a significant amount of information at the application layer, including Web pages, Usenet posts, e-mail messages, and IRC logs. In addition to extracting information from these sources of digital evidence, it is important to apply the lessons from previous chapters, seeking related server logs and possibly monitoring network traffic, to establish continuity of offense and locate the offender. Also keep Locard's exchange principle in mind, looking for transfer of digital evidence between the offender's computer and other systems on the Internet to help attribute online activities to the offender. It can be more difficult to establish continuity of offense when offenders attempt to conceal their activities or identity on the Internet. This is particularly true when Freenet is involved, making it necessary to rely on class and individual characteristics, searching image databases for similar characteristics.

When following the cybertrail, remember that one of the main limitations of the Internet as a source of evidence is that it generally only has the latest version of information. If a Web page is modified or someone retracts a Usenet post, the old information is usually lost. Because it cannot be assumed that evidence will remain on the Internet for any duration, it should be collected as quickly as possible. It is also important to remember that not all activities on the Internet are automatically archived (e.g. IRC). If you are fortunate to be in the right place at the right time, witnessing live interactions can greatly benefit an investigation. Otherwise, you might be lucky enough to find Internet chat logs when you search a suspect's computer. Either way, these live interactions contain a wealth of behavioral information about the individuals who are involved.

REFERENCES

CBS News (2002), "Daniel Pearl Killers Appeal", July 17, 2002 (Available online at http://www.cbsnews.com/stories/2002/05/31/attack/main510651.shtml).

CERT (2002), "Trojaned OpenSSH Distribution" (Available online at http://www.cert.org/advisories/CA-2002-24.html)

Clark F. and Diliberto K. (1996) *Investigating Computer Crime*, CRC Press.

Cairns G. (1996) "Snuffsex," Australian Broadcasting Corporation's NewsRadio Network.

DAWN group (2002), "Defence disputes video's validity: Daniel Pearl case", 17 May 2002, (Available online at http://www.dawn.com/2002/05/17/top2.htm).

Durkin K. F. and Bryant C. D. (1999) "Propagandizing pederasty: a thematic analysis of the on-line exculpatory accounts of unrepentant pedophiles". *Deviant Behavior: An Interdisciplinary Journal*, 20, 103–127.

Mann D. and Sutton M. (1998), "Netcrime: More Change in the Organization of Thieving", *British Journal of Criminology*, 38(2).

National News (2002), "Text of Daniel Pearl case verdict", July 17, 2002 (Available online at http://www.jang.com.pk/thenews/jul2002-daily/17-07-2002/national/n6.htm).

Palisade Systems (2003) "Porn Tops File Sharing Usage" (Available online at http://www.palisadesys.com/news&events/p2pstudy.pdf).

United States Department of Justice (1999), "Man convicted of threatening Federal judges by Internet e-mail" (Available online at http://www.usdoj.gov/criminal/cybercrime/johnson.htm)

United States Postal Service (2001), "Multimillion-Dollar Child Pornography Enterprise Dismantled".

CASES

Froistad v. United State (2001) Supreme Court, North Dakota, Case Number 20010111 (Available online at http://www.court.state.nd.us/court/briefs/20010111.aeb.htm).

United States v. Reedy (2000) District Court, Northwest District of Texas, Fort Worth Division, Case number 400-CR-0540Y (Available online at http://news.findlaw.com/hdocs/docs/reedy/usreedy51700indct.pdf).

PART 4

INVESTIGATING COMPUTER CRIME

## CHAPTER 19

INVESTIGATING COMPUTER INTRUSIONS

...the safecracker has been portrayed as a masked, bewhiskered, burly individual whose daring was matched only by his ruthlessness in disposing of interference. This legend undoubtedly had its origin in the facility with which the safecracker could be caricatured by cartoonists. His safe, mask, blackjack, and flashlight have come to be the picturesque symbols of the professional criminal. By this intimate association, the safe burglar has acquired in fiction the attributes of character corresponding to the physical properties of the safe itself – steely toughness of fiber and impregnability to moral suasion. Historically, this picture may have been true, but modern criminal society is far more democratic. The safecracker category, for example, includes all races, colors, and creeds: the skilled craftsman and the burglar; the timid and the bold; the lone wolf and the pack member; the professional criminal and the young amateur trying his wings; the local thug and the strong boy from a distant city. The occupation of safecracker has proved so remunerative to some practitioners, that its membership has swollen beyond the limits imposed by any of the restrictions of qualifications in the form of skill.

(O'Hara, 1970)

New ways to interfere with and break into computers seem to be developed every day. Although it takes a certain degree of skill to find new ways to implement these attacks, once a new method of attack is developed, it is often made available on the Internet. Programs that automatically exploit a vulnerability are commonly called *exploits*, and many of them are freely available at sites like SecurityFocus.[1] With a little knowledge of computer networks, almost anyone can obtain and use the necessary tools to be a nuisance – or even dangerous (e.g. breaking into a computer and erasing its contents). It takes skill and experience, however, to break into a computer system, commit a crime, and cover one's tracks.

[1] *http://www.securityfocus.com*

Individuals break into computers for a wide range of purposes, including stealing valuable information, eavesdropping on users' communication's, harassing administrators or users, launching attacks against other systems,

Digital Evidence and Computer Crime Second Edition
ISBN: 0-12-163104-4

storing toolkits and stolen data, and defacing Web sites. Some individuals view computer intrusions as victimless crimes. However, whether a computer intruder purloins proprietary information from an organization, misuses a computer system, or deletes the contents of an individual's hard drive, people are affected in a very real way. If, for example, a computer intruder changes prescription information in a pharmacy database, tampers with critical systems at an airport, disables an emergency telephone service, or damages other critical systems, the ramifications can be fatal.

In many cases, only people who are intimately familiar with a specific computer system possess the skills required to break into or tamper with it. As a result, individuals inside an organization commit a significant percentage of computer crimes (CSI 2003). However, the number of attacks from the Internet is increasing. Computer intrusions have become such a problem that it is considered to be a national security risk by many developing countries. Despite the seriousness of this problem, many organizations are reluctant to report intrusions to law enforcement for a variety of reasons.

Given the growing threat, it is important to track down the perpetrators of these crimes, bring them to justice, and discourage others from following in their footsteps. Even if an organization decides not to prosecute an individual who targets their systems, a thorough investigation can help determine the extent of the damage, prevent future attacks, and mitigate any associated liability to shareholders, customers, or other organizations that were attacked. This chapter discusses how to investigate computer intruders and presents ways to determine an intruder's intent, motivations, and skill level.

19.1 HOW COMPUTER INTRUDERS OPERATE

The most straightforward way to break into a computer is to steal or guess a password. However, if this is not a viable option, an intruder can usually gather enough information about a system to gain access to it. The most basic way to gather information about a system is to use a port scanner as shown here:

```
% probe_tcp_ports 192.168.52.2
Host 192.168.52.2, Port 7 ("echo" service) connection ... open.
Host 192.168.52.2, Port 9 ("discard" service) connection ... open.
Host 192.168.52.2, Port 13 ("daytime" service) connection ... open.
Host 192.168.52.2, Port 19 ("chargen" service) connection ... open.
Host 192.168.52.2, Port 21 ("ftp" service) connection ... open.
Host 192.168.52.2, Port 23 ("telnet" service) connection ... open.
Host 192.168.52.2, Port 25 ("smtp" service) connection ... open.
Host 192.168.52.2, Port 53 ("domain" service) connection ... open.
Host 192.168.52.2, Port 69 connection ... open.
```

```
Host 192.168.52.2, Port 79 ("finger" service) connection ... open.
Host 199.168.52.2, Port 110 ("pop" service) connection ... open.
```

This basic TCP port scanner shows that, in addition to running an e-mail server on port 25, this computer has a number of other servers, including an FTP server on port 21 for people to transfer files to and from the computer, a finger server on port 79 that can give out information about individuals with accounts on the machine, and a POP server for users to check their e-mail remotely. The operating system and server version can often be inferred from this type of port scan, or using a more advanced port scanner like nmap. Knowing the operating system and services that are running on a computer is often all that is required – because certain services on certain operating systems are known to be vulnerable. For instance, the following shows an exploit that is freely available on the Internet being used to gain unauthorized access to an FTP server:

```
% wuftpd-exploit -t 192.168.7.25 -s 0
Target: 192.168.7.25 (ftp/<shellcode>): RedHat 6.2 with wuftpd 2.6.0(1)
Return Address: 0x08075844, AddrRetAddr: 0xbfffb028, Shellcode: 152

loggin into system..
USER ftp
331 Guest login ok, send your complete e-mail address as password.
PASS <shellcode>
230-Next time please use your e-mail address as your password
230 Guest login ok, access restrictions apply.
STEP 2 : Skipping, magic number already exists: [87,01:03,02:01,01:02,04]
STEP 3 : Checking if we can reach our return address by format string
STEP 4 : Ptr address test: 0xbfffb028 (if it is not 0xbfffb028 ^C me now)
STEP 5 : Sending code.. this will take about 10 seconds.
Press ^\ to leave shell
Linux ftp-server.corpX.com 2.2.14-5.0 #1 Tue Mar 7 20:53:41 EST 200 0 i586
unknown
uid=0(root) gid=0(root) egid=50(ftp) groups=50(ftp)
w
  8:54am  up 3 days, 12:21,   0 users,  load average:   0.12,    0.09,  0.03
USER     TTY        FROM              LOGIN@     IDLE   JCPU   PCPU   WHAT
last
ftp      ftpd7718      intruder.isp.com   Wed   Sep 20 08:52 still     logged in
ftp      ftpd7291      helpsrv.smut.com   Tue   Sep 19 15:13 still     logged in
reboot   system boot   2.2.14-5.0         Sat   Sep 16 20:33           (3+12:21)
ftp      ftpd1120      203.235.121.105    Sun   Sep 10 04:08 - down    (1+21:32)
ftp      ftpd833       mail2.txinc.com    Sat   Sep 9  21:39 - down    (2+04:02)
reboot   system boot   2.2.14-5.0         Sat   Sep 9  12:21           (2+13:20)

wtmp begins Thu Sep  7 17:59:03 2000
```

When intruders cannot access a system through known security holes, they use less technical methods to gain access. Intruders sometimes even dig through garbage for useful information. Intruders also try to get information using *social engineering* and *reverse social engineering*. Social engineering refers to any attempt to contact legitimate users of the target system and trick them into giving out information that can be used by the intruder to break into the system. For example, calling someone and pretending to be a new employee who is having trouble getting started can result in useful information like computer names, operating systems, and even some information about employee accounts. Alternatively, pretending to be a computer technician who is trying to fix a problem can also lead to useful information. There are many different ways to do this, including calling people claiming to be looking into a problem or going into the organization to look around. Some people will even make the mistake of giving out their passwords.

Reverse social engineering is any attempt to have someone in the target organization contact you for assistance. Instead of contacting them, they contact you. For example, sending a memo with a "new" technical support e-mail can result in a flood of information. The advantage of reverse social engineering is that the user is less likely to be suspicious and report the incident. When people seek help from an intruder who resolves their problems, they are less likely to be suspicious and are unlikely to have any reason to report the incident to anyone.

Table 19.1 summarizes the various methods of approach/attack. The categories are not mutually exclusive – intruders may employ several of these attack vectors to achieve their goals.

Many of the attack methods in Table 19.1 were discussed in earlier chapters such as buffer overflows in Chapter 15, session hijacking in Chapter 16, and IP spoofing in Chapter 17.

Table 19.1

Different attack methods. (Dunne, Long, Casey 2000)

ATTACK VECTOR NAME	DESCRIPTION
Authentication bypass	Gaining access while avoiding standard authentication
Authentication failure	Taking advantage of authentication systems which "fail open"
Buffer overflows	Exploiting stack memory overwriting in networked server programs
Password cracking	Brute-force, reverse-engineering, and "dictionary" based methods used to discover account passwords
Password sniffing	Capturing account passwords via a network "tap"
Session hijacking	Piggybacking on authorized user connections from the Internet into internal hosts and networks
Social engineering	Impersonation of authorized personnel to gain access or network passwords
Spoofing	Having a computer masquerade as a different "trusted" computer to gain access
Trojan horses	Malicious programs such as BackOrifice can provide "back doors" (unauthorized avenues for access) into hosts from the Internet

After intruders gain access to a computer, they may be able to compromise the administrator account (known as "root" on UNIX systems) thus getting unrestricted access to the entire system. In fact, certain security holes allow computer intruders to break into a computer and get root access in one step. With unlimited access to the system, it is possible for an offender to modify any information on the computer, thus removing traces of an intrusion. Intruders may change the system clock, delete log files, and replace system components. There are specific computer programs, called *rootkits*, which automate the process of hiding a break-in enabling a low skilled offender to exhibit higher skilled behavior. For example, the Rootkit[2] project is developing such a program for Windows machines. More sophisticated rootkits such as Knark, Sebek, and Suckit on Linux are emerging, making intrusion investigations even more challenging. Increasingly, criminals are using strong encryption on UNIX systems such as encrypted RAM disks, Blowfish encryption in IRC eggdrop bots, encrypted executables using Teso Burneye (Phrack 58, 2001), and using other "anti-forensic" tools to make digital evidence examinations more difficult (Phrack 59, 2002).

[2]*http://www.rootkit.com*

Once an intruder has gained access to one computer on a network, it may be possible to gather additional information about a network and obtain passwords to other systems using a sniffer.

19.2 INVESTIGATING INTRUSIONS

As detailed in Chapter 4, the first step when investigating an incident is to determine if there actually was one – there must be a corpus delicti. Computers and networks are complex systems that can be misunderstood or that can malfunction, resulting in false incident reports. To determine what occurred, investigators should interview the individuals who witnessed the incident, those who reported it, and anyone else who was involved. Whenever possible, interviews should be conducted in person or by telephone in a discrete manner. A lack of caution in the initial stage of an investigation can alert an offender and can result in workplace rumors or media leaks that cause more damage than the incident.

CASE EXAMPLE

In one incident, an organization detected employees from a competitor's network gaining unauthorized access to a server. Sufficient evidence was gathered to prove the illegal activity and to identify the competitor's employees who had committed the crime. To avoid publicity and preserve a good relationship with the competitor, the victim organization decided to resolve the problem through private communication rather than through legal action. However, an employee in the victim organization leaked the story to the press, creating a national scandal that caused more damage than the incident itself.

Once the nature and severity of an incident has been determined, it is advisable to inform legal counsel, human resources, managers, public relations, and possibly law enforcement as outlined in the organization's incident/emergency response plan. Keep in mind that it may not be possible to trust the network that the offender has targeted, so encryption should be used for all incident-related communications and activities on the network. Also, be aware that it can take years to resolve some incidents so it is crucial to document all actions taken in response to an incident, including all communications. Detailed notes are useful for recalling, explaining the incident years later, and it may be necessary to re-interview certain people or call on them to testify to clarify certain details. Additionally, noting the dates and times of events, including the time it took to recover systems helps calculate cost of damage.

Investigating computer intrusions usually involves a large amount of digital evidence. Investigators must search large log files for relevant entries, examine programs to determine their purpose closely, and explore the network for additional clues. In addition to being technically challenging, there is often pressure on an investigator to resolve the problem quickly. Relevant log files and state tables might be erased at any moment and the system owners/users want to gain access to the information on the system. It is often necessary to interpret digital evidence instantaneously to determine where additional evidence might be found.

Under such conditions, especially when several computers are involved, it is easy to overlook important digital evidence, neglect to collect digital evidence properly, document the investigation inadequately and jump to incorrect conclusions. The most effective approach to managing this kind of complex, high-pressure, error-prone investigation is to use standard operating procedures (SOPs) with associated forms to collect the most common sources of digital evidence. Having a routine method for quickly preserving digital evidence for future examination leaves investigators with more time to deal with the nuances and peculiarities of individual incidents. These procedures should employ the concepts covered in Parts 1, 2 and 3 of this book.

A common challenge that arises during intrusion investigations is the need to protect the target systems against further attack. Investigators may even be asked to remove a backdoor and repair the target system before they have collected evidence from the system. Whenever possible, evidence should be preserved prior to repairing the target system or altering its state in any other way. It is usually feasible to protect the target system by isolating it on the network while it is being processed as a source of evidence. In some cases, it may be viable to isolate a system simply by unplugging its network cable. However, when the system is a critical component of a network, it may be necessary to involve network administrators to reconfigure a router or firewall, partially isolating the system but permitting vital connections to enable an organization to remain in operation.

CASE EXAMPLE

A routine vulnerability scan of a network detected Back Orifice running on a Windows 2000 server. Because of the critical role that this server played in the organization, a rapid response and recovery was required. The organization was unwilling to take the server offline because this would disrupt business operations. They wanted the server to be fixed quickly and were not concerned with apprehending the culprit. Investigators determined that the server had been compromised via IIS and found Web server access logs that corresponded with the initial intrusion containing the intruder's IP address. Additionally, they found that the Back Orifice executable was named "wlogin.exe" and was installed as a service named "WinLogin" as shown in the following Registry key:

```
D:\>regdmp
\Registry
<cut for brevity>
(HKLM\System\CurrentControlSent\Services)
    WinLogin
      Type = REG_DWORD 0x00000110
      Start = REG_DWORD 0x00000004
      ErrorControl = REG_DWORD 0x00000000
      ImagePath = REG_EXPAND_SZ ' "C:\WINNT\System32\wlogin.exe" '
      DisplayName = WinLogin
```

Furthermore, NT Application Event logs showed that Norton AntiVirus had detected Back Orifice but had not been able to remove it:

```
D:\>dumpel -c -l application
<cut for brevity>
1/19/2002,12:32:48 AM,4,0,20,Norton AntiVirus,N/A,CONTROL, Unable to restore
C:\WINNT\system32\wlogin.exe from backup file after clean failed.
1/19/2002,1:09:11 AM,1,0,5,Norton AntiVirus,N/A, CONTROL, Virus Found!Virus
name: BO2K.Trojan Variant in File: C:\WINNT\Java\w.exe by: Scheduled scan.
Action: Clean failed : Quarantine succeeded : Virus Found!Virus name:
BO2K.Trojan Variant in File: C:\WINNT\system32\wlogin.exe by: Scheduled
scan. Action: Clean failed : Quarantine failed :
1/19/2002,1:09:11 AM,4,0,2,Norton AntiVirus,N/A, CONTROL, Scan Complete:
Viruses:2  Infected:2  Scanned:62093  Files/Folders/Drives Omitted:89
```

The intruder had also installed an IRC bot in C:\WINNT\Java that contained several possible leads including IP addresses, nicknames, and IRC channel passwords. However, because the priority was to recover the system, this evidence was collected hastily and the Trojan horse program was removed. After removing the rogue service from the Registry, the server was rebooted to ensure that all remnants of the process were eliminated. Unfortunately, the domain controller did not reboot successfully. Attempting to fix the problem had effectively done more damage than the intruder, interrupting business operations while attempting to restore the server. After some pandemonium, the system was restored from backup, a lengthy process resulting in a prolonged interruption in business that the organization had hoped to avoid.

By the time the domain controller had been recovered, the organization was more interested in apprehending the culprit. Their concerns were exacerbated when they realized that the intruder could have obtained passwords from the server and used them to compromise other system on the network. Unfortunately, much of the evidence had been destroyed when the system was restored from backup and the Back Orifice executable had been erased by Norton AntiVirus. It was determined that there was too little evidence to apprehend and prosecute the intruder. Using the little information that they had preserved, the organization did their best to determine if the intruder had targeted any other systems on their network.

One of the more difficult decisions is whether to shut down a compromised system or collect some data from it beforehand. When investigating a computer intrusion, it is often desirable to capture and record system information that is not collected by a bitstream copy of the hard disk. For instance, it is useful to document current network connections, which user accounts are currently logged on, what programs are running in memory (a.k.a. processes), and which files these processes have open. Processes in memory, network state tables, and encrypted disks may contain valuable data that are lost when a system is shut down. However, examining a live system is prone to error and may change data on the system. One approach to minimizing these risks is to use automation – running a standard script that gathers basic information and saves it to external media. However, this does not address the possibility that the operating system is untrustworthy. Even when trusted tools are used to examine a computer, system calls can be intercepted and manipulated by a rootkit. Ultimately, investigators must weigh the importance of volatile data against the risk of operating the computer.

Notably, shutting a system down does not necessarily destroy all process-related data. Virtual memory, in the form of swap files, enables more processes to run than can fit within a computer's physical memory (RAM). Therefore, digital evidence from processes can be recovered even after the system is shut down. For instance, the following information was recovered from the Windows 2000 swap file "pagefile.sys" on a compromised Web server, showing the intruder (208.61.131.188) executing commands on the system via a vulnerability in the Web server:

```
COMPUTERNAME=WWW..............ComSpec=C:\WINNT\system32\cmd.exe
...............CONTENT_LENGTH=0................GA
TEWAY_INTERFACE=CGI/1.1.......HTTP_ACCEPT=image/gif, image/x-xbitmap, ima
ge/jpeg, image/pjpeg, */*...........HTTP_HOST=192.168.16.133...
.....HTTP_USER_AGENT=Microsoft URL Control - 6.00.8862.........
.....HTTP_CACHE_CONTROL=no-cache.....HTTPS=off.......INCLUDE=C:
\Program Files\Mts\Include...........INSTANCE_ID=1...LIB=C:\Prog
```

```
ram Files\Mts\Lib....LOCAL_ADDR=192.168.16.133.......NUMBER_OF_PROCES
SORS=1..........Os2LibPath=C:\WINNT\System32\os2\dll;........
...OS=Windows_NT...Path=C:\Perl\bin;C:\WINNT\system32;C:\WINNT;C:\Program
Files\Mts................PATH_TRANSLATED=c:\Inetpub\wwwroot..
............PATHEXT=.COM;.EXE;.BAT;.CMD;.VBS;.JS;.VBE;.JSE;.WSF;.WSH
........PROCESSOR_ARCHITECTURE=x86......PROCESSOR_IDENTIFIER=x86 F
amily 6 Model 5 Stepping 2, GenuineIntel..............PROCESSOR_LE
VEL=6...............PROCESSOR_REVISION=0502.........QUER
Y_STRING=/c+ping+172.16.81.74+-n+56000+-w+0+-l+56000........REMOTE_ADDR=
208.61.131.188......REMOTE_HOST5208.61.131.188......REQUEST_METHOD=G
ET..............SCRIPT_NAME=/msadc/../../../../../../winnt/system3
2/cmd.exe.....SERVER_NAME=192.168.16.133......SERVER_PORT=80..SERVE
R_PORT_SECURE=0............SERVER_PROTOCOL=HTTP/1.1........S
ERVER_SOFTWARE=Microsoft-IIS/4.0..............SystemDrive=C:..
<cut for brevity>
"c:\Inetpub\wwwroot\msadc\..\..\..\..\..\..\winnt\system32\cmd.exe" /c
ping 172.16.81.74 -n 56000 -w 0 -l 56000.
```

Other similar fragments, some in Unicode format, were also recovered showing the intruder launching denial of service attacks against many hosts on the Internet.

When dealing with a computer intrusion, do not assume that the incident is isolated – there may be other systems on the network that are involved.

CASE EXAMPLE

A system administrator found unusual files on a Windows NT server that he was responsible for. The host had been compromised via the IIS Web server and was running Serv-U FTP Server v3.0 ("c:\winnt\system32\setup\x2x\rundll16.exe") on ports 666 and 9669. The FTP server was being used to share pornography, feature length films, and other media stored in "d:\recycler\<sid>\COM1\database." The term "Pubstro" is sometimes used to refer to a Windows NT server that has been compromised and is being used to distribute files. Windows has difficulty with directories named "COM1" and "LTP1" because it associates these names with DOS drivers. This trick makes directory traversal during a live examination difficult. The FTP server configuration file referenced several other directories that did not appear to be present on the system including "d:\recycler\<sid>\ COM1\databaselRWAMLCDP" and "c:\IRWAMELCDP." Logs from the FTP server showing many connections from many hosts downloading files were located in "c:\winnt\system32\os2\dll\backup\." The intruder placed the pulist and kill utilities from the NT resource kit on the system with the FTP server along with a DLL called psapi.dll was in the directory along with the FTP server. The intruder also placed two executables, named nc.exe and bot.exe, in "c:\winnt\system32." A related configuration file contained the following lines:

#0dayvcd with password psA4C70E33CF55B74D5F1C21B8EE46DD8F
Vcd with password pt0BED4C47C1826BE160D6FA8E4F85A28F
admin with password qgEDA3C477AF1702713437C873A460F230

Another file named "msgtoadmin.txt" contained the following text:

Note To admin
Well what can I say. I broke in yes. But I'm not here to attack. I'm sorry for any inconvience this may have caused you. No viruses or worms have been installed. That is not my intension. I just love you bandwidth:) If you have read this you must have cought me. And once you have don't worry, I'm gone and won't bother you again. again sorry for any inconvience this may have caused. Have a good day,
X

Not taking the intruder's assurances to heart, the system administrator port scanned all of the systems on the network looking for ports 666 or 9669 and found several other similarly compromised systems. The administrator found other compromised systems by monitoring network traffic for distinctive terms like "#0dayvcd" and "RWAMLCDP."

When responding to an incident on a large network, port scanning may produce too many false positives in which case simple Perl scripts can be created to scan machines on a network and inspect their responses for specific class characteristics to determine if they are compromised. For instance, in the preceding case example, scanning for systems that displayed the distinctive Serv-U logon banner would be more accurate. Also, configuring an intrusion detection system to look for specific class characteristics in network traffic is an effective approach to finding other compromised systems.

19.2.1 PROCESSES AS A SOURCE OF EVIDENCE (WINDOWS)

When a computer program is executed it is instantiated in RAM, taking up system resources such as memory and file descriptors. Such an instantiation of a program is called a process because it is performing a task or process as it runs on the system. Examples of processes are winword.exe (Microsoft Word for Windows) and lsass.exe (Windows NT's local security administration subsystem). Default processes on Windows 2000 are listed on the Microsoft Web site.[3] However, when investigating a computer intrusion, it is not safe to assume that a program is legitimate simply based on its name or the port it is bound to. Computer intruders take advantage of an investigator's preconceived theories by making malicious programs resemble legitimate ones. It is trivial to name an executable inetinfo.exe and have it listen for network connections on port 80 to mislead investigators into assuming

[3] *http://support.microsoft.com/support/kb/articles/q263/2/01.asp*

that it is a Microsoft Internet Information Server (IIS). Only a closer inspection of the process, such as which files it has open, will reveal its true nature.

There are a number of utilities that enable investigators to gather information about processes that are currently running on a Windows NT/2000/XP computer. Commands such as **netstat** and **nbtstat** are installed with the operating system and other specialized tools that are freely available on the Web such as **fport**[4] and **handle**.[5] Although many of the details provided by utilities like **handle** may not be relevant to the investigation, small segments can reveal useful details about programs and files created by an intruder. The usefulness of these tools is best demonstrated through a detailed case example.

[4]*http://www.foundstone.com*
[5]*http://www.sysinternals.com*

CASE EXAMPLE

The following intrusion detection system logs show an attack against a critical UNIX machine (192.168.128.14) from another important Windows 2000 server (192.168.164.163) on the network:

```
[**] [1:1326:1] EXPLOIT ssh CRC32 overflow NOOP [**]
   04/24-03:28:43 192.168.164.163→192.168.128.14 S: 2445 D: 22

[**] [1:1326:1] EXPLOIT ssh CRC32 overflow NOOP [**]
   04/24-07:18:21   3 192.168.164.163→192.168.128.14 S: 2888 D: 22
```

A port scan of the Windows 2000 server, named "server1," showed many open ports, including one that gave a command prompt to anyone who connected using Telnet:

```
% bin/probe_tcp_ports 192.168.164.163
Port 80 (possibly http)
Port 135 (possibly rpc service)
Port 139 (possibly rpc service)
Port 443 (possibly https)
Port 445 (possibly netbios)
Port 1025
Port 1046
Port 1048
Port 1051
Port 1061
Port 1433   (possibly ms-sql)
Port 2025
Port 3372
Port 3389
Port 3497
Port 4362
Port 7904
Port 12323
Port 43958
```

```
% telnet server1 12323
Microsoft Windows 2000 [Version 5.00.2195]
(C) Copyright 1985–2000 Microsoft Corp.
C:\WINNT\system32>
```

The network cable was disconnected from server1 immediately to prevent further unauthorized remote access. A rapid response and recovery was desired to minimize the impact on business continuity. Management wanted to determine what the intruder changed on the system and what actions were necessary to remove all backdoors.

The output of the netstat command confirmed the ports that were seen with the remote port scan, but did not show the remote addresses of machines that were connected to this system because the network cable had been unplugged. The processes listed using Alt-Ctrl-Del included two unrecognized processes named sqldiagmsrv and sqldiagncv as shown in Figure 19.1. More details about these processes, like how long they had been running, could be obtained using pslist.[6]

[6]*http://www.sysinternals.com*

Figure 19.1

Unusual process viewed using Alt-Ctrl-Del.

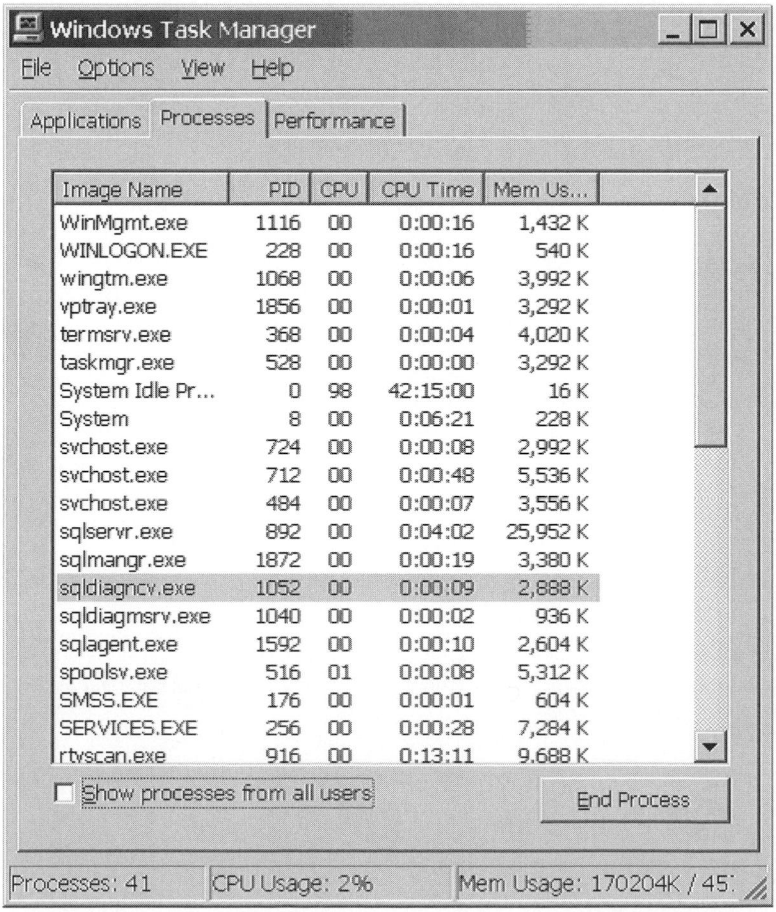

These unrecognized processes were examined more closely to determine what they were doing on the system. The fport command showed that c:\winnt\system32\sqldiagncv.exe was bound to port 12323.

```
D:\>fport
FPort v1.33 - TCP/IP Process to Port Mapper
Copyright 2000 by Foundstone, Inc.
http://www.foundstone.com
Pid   Process        Port   Proto Path
1152  inetinfo    -> 80     TCP   C:\WINNT\System32\inetsrv\inetinfo.exe
484   svchost     -> 135    TCP   C:\WINNT\system32\svchost.exe
1152  inetinfo    -> 443    TCP   C:\WINNT\System32\inetsrv\inetinfo.exe
8     System      -> 445    TCP
556   msdtc       -> 1025   TCP   C:\WINNT\System32\msdtc.exe
960   MSTask      -> 1027   TCP   C:\WINNT\system32\MSTask.exe
1152  inetinfo    -> 1028   TCP   C:\WINNT\System32\inetsrv\inetinfo.exe
892   sqlservr    -> 1029   TCP   C:\MSSQL7\binn\sqlservr.exe
8     System      -> 1031   TCP
892   sqlservr    -> 1433   TCP   C:\MSSQL7\binn\sqlservr.exe
1152  inetinfo    -> 2025   TCP   C:\WINNT\System32\inetsrv\inetinfo.exe
556   msdtc       -> 3372   TCP   C:\WINNT\System32\msdtc.exe
368   termsrv     -> 3389   TCP   C:\WINNT\System32\termsrv.exe
1152  inetinfo    -> 4362   TCP   C:\WINNT\System32\inetsrv\inetinfo.exe
1152  inetinfo    -> 7904   TCP   C:\WINNT\System32\inetsrv\inetinfo.exe
1052  sqldiagncv  -> 12323  TCP   c:\winnt\system32\sqldiagncv.exe
1068  wingtm      -> 43958  TCP   C:\WINNT\system32\wingtm.exe
484   svchost     -> 135    UDP   C:\WINNT\system32\svchost.exe
8     System      -> 445    UDP
256   services    -> 1026   UDP   C:\WINNT\system32\services.exe
516   spoolsv     -> 1030   UDP   C:\WINNT\system32\spoolsv.exe
916   rtvscan     -> 2967   UDP   C:\Program Files\NavNT\rtvscan.exe
1152  inetinfo    -> 3456   UDP   C:\WINNT\System32\inetsrv\inetinfo.exe
```

The handle command, which lists which system resources each process is using, showed that the sqldiagncv executable was running with SYSTEM level authority, allowing significant access to the system:

```
sqldiagncv.exe pid:   1052 NT AUTHORITY\SYSTEM
    18: File          C:\WINNT\system32
    e0: Section       \BaseNamedObjects\__R_0000000000f2_SMem__
```

The listdlls command showed the command line parameters that sqldiagncv was executed with as well as its associated dynamic link libraries:

```
sqldiagncv.exe pid: 1052
Command line: c:\winnt\system32\sqldiagncv.exe -l -d -p 12323 -t -e cmd.exe
Base        Size     Version          Path
0x00400000  0x13000                   c:\winnt\system32\sqldiagncv.exe
0x77f80000  0x7b000  5.00.2195.2779   C:\WINNT\System32\ntdll.dll
0x77e80000  0xb5000  5.00.2195.4272   C:\WINNT\system32\KERNEL32.dll
0x75050000  0x8000   5.00.2195.2871   c:\winnt\system32\WSOCK32.dll
0x75030000  0x13000  5.00.2195.2780   c:\winnt\system32\WS2_32.DLL
0x78000000  0x46000  6.01.9359.0000   C:\WINNT\system32\MSVCRT.DLL
0x77db0000  0x5c000  5.00.2195.4453   C:\WINNT\system32\ADVAPI32.DLL
0x77d40000  0x70000  5.00.2195.4266   C:\WINNT\system32\RPCRT4.DLL
0x75020000  0x8000   5.00.2134.0001   c:\winnt\system32\WS2HELP.DLL
0x785c0000  0xc000   5.00.2195.2871   C:\WINNT\System32\rnr20.dll
0x77e10000  0x64000  5.00.2195.4314   C:\WINNT\system32\USER32.DLL
0x77f40000  0x3c000  5.00.2195.3914   C:\WINNT\system32\GDI32.DLL
0x77980000  0x24000  5.00.2195.4141   c:\winnt\system32\DNSAPI.DLL
0x77340000  0x13000  5.00.2173.0002   c:\winnt\system32\iphlpapi.dll
```

Searching the Registry revealed that the **sqldiagncv** process was being started as a service named **sqldiagmsrv**:

```
sqldiagmsrv
    Type = REG_DWORD 0x00000010
    Start = REG_DWORD 0x00000002
    ErrorControl = REG_DWORD 0x00000001
    ImagePath = REG_EXPAND_SZ c:\winnt\system32\sqldiagmsrv.exe
    DisplayName = sqldiagmsrv
    ObjectName = LocalSystem
    Parameters
    Application = c:\winnt\system32\sqldiagncv.exe -l -d -p 12323 -t -e cmd.exe
```

The last write time of this Registry key was consistent with the intruder's other activities on the system:

[7] *Executable version of keytime.pl from http://patriot.net/~carvdawg/perl.html*

```
D:\> keytime7 system/currentcontrolset/services/sqldiagmsrv
HKEY_LOCAL_MACHINE\system/currentcontrolset/services/sqldiagmsrv, Wed
4/3/2002 14:21:09:971
```

A copy of the **sqldiagncv** executable was placed on an analysis system for further inspection and it quickly became apparent that it was **netcat**:

```
C:\WINNT\system32>sqldiagncv -h
[v1.10 NT]
connect to somewhere:   nc [-options] hostname port[s] [ports] ...
listen for inbound:     nc -l -p port [options] [hostname] [port]
options:
        -d              detach from console, stealth mode
        -e prog         inbound program to exec [dangerous!!]
        -g gateway      source-routing hop point[s], up to 8
        -G num          source-routing pointer: 4, 8, 12, ...
        -h              this cruft
        -i secs         delay interval for lines sent, ports scanned
        -l              listen mode, for inbound connects
        -L              listen harder, re-listen on socket close
        -n              numeric-only IP addresses, no DNS
        -o file         hex dump of traffic
        -p port         local port number
        -r              randomize local and remote ports
        -s addr         local source address
        -t              answer TELNET negotiation
        -u              UDP mode
        -v              verbose [use twice to be more verbose]
        -w secs         timeout for connects and final net reads
        -z              zero-I/O mode [used for scanning]
port numbers can be individual or ranges: m-n [inclusive]
```

In summary, the intruder used the Windows 2000 system to launch an attack against the SSH server on an internal UNIX machine, thus bypassing the firewall which did not allow connections to the SSH server from the Internet.

In some instances, it may be desirable to capture data in memory relating to a particular process using the **pmdump**[8] utility. For instance, the following commands show **pmdump** being used to copy the contents of memory relating to a Pretty Good Privacy (PGP) process:

[8] *http://ntsecurity.nu*

```
D:\>pslist pgptray
Name      Pid       Pri   Thd   Hnd   Mem     Elapsed Time
PGPtray   1332   8  7           150   1264    2:20:33.466
D:\>pmdump 1332 d:\evidence\pgptray.mem
```

The resulting memory dump file may contain a PGP passphrase or data in unencrypted form.

19.2.2 PROCESSES AS A SOURCE OF EVIDENCE (UNIX)

Although the full contents memory can be dumped into a file using dd (e.g. dd </dev/mem > host.mainmemory), this lumps everything together and does not provide information about separate processes. One approach to examining processes on UNIX systems is to use the ps command as shown here, specifying that all processes should be listed using command options like "ps -aux" for most versions of UNIX and "ps -ef" for others:

```
% ps -aux | more
USER  PID  %CPU  %MEM    SZ   RSS  TT    S  START     TIME   COMMAND
root    3   0.4   0.0     0     0  ?     S  Apr 25    64:39  fsflush
root  199   0.3   0.2  4800  1488  ?     S  Apr 25     2:14  /usr/sbin/syslogd
root 3085   0.2   0.2  2592  1544  ?     S  14:07:12   0:00  /usr/lib/sendmail
root    1   0.1   0.1  1328   288  ?     S  Apr 25     4:03  /etc/init -
root 3168   0.1   0.1  1208   816  pts/5 O  14:07:27   0:00  ps -aux
root 2704   0.1   0.2  2096  1464  ?     S  14:05:37   0:00  /usr/local/etc/sniffer
root  163   0.0   0.1  1776   824  ?     S  Apr 25     0:19  /usr/sbin/inetd -s
root  132   0.0   0.1  2008   584  ?     S  Apr 25     0:00  /usr/sbin/keyserv
root  213   0.0   0.1  1624   776  ?     S  Apr 25     0:16  /usr/sbin/cron
root  239   0.0   0.1   904   384  ?     S  Apr 25     0:07  /usr/lib/utmpd
```

Additional information about each process, including a list of files and sockets that they are using, can be obtained using the lsof utility. Much of the detail provided by lsof may not be useful in most cases, such as which libraries are being accessed by each process. However, lsof can be useful for finding programs and files created by an intruder and can be compared with the output from ps to find discrepancies caused by rootkits. If a particularly interesting process appears in this list like "sniffer" or "destroyer," an investigator might want to take a closer look. Some types of UNIX allow one to save and view the contents of RAM that is associated with a particular program using the "gcore" command.

Another approach to examining processes on a UNIX system is through the /proc virtual file system. For instance, the following files on a Linux system are linked with the command line parameters, memory contents, and other details associated with a running **netcat** process:

```
$ ls -l /proc/1104
total 0
-r--r--r--    1   eco   eco   0   May 17   12:36   cmdline
lrwxrwxrwx    1   eco   eco   0   May 17   12:36   cwd ->
/usr/local/bin
-r--------    1   eco   eco   0   May 17   12:36   environ
lrwxrwxrwx    1   eco   eco   0   May 17   12:36   exe ->
/usr/sbin/nc
dr-x------    2   eco   eco   0   May 17   12:36   fd
-r--r--r--    1   eco   eco   0   May 17   12:36   maps
-rw-------    1   eco   eco   0   May 17   12:36   mem
-r--r--r--    1   eco   eco   0   May 17   12:36   mounts
lrwxrwxrwx    1   eco   eco   0   May 17   12:36   root -> /
-r--r--r--    1   eco   eco   0   May 17   12:36   stat
-r--r--r--    1   eco   eco   0   May 17   12:36   statm
-r--r--r--    1   eco   eco   0   May 17   12:36   status
$ more /proc/1104/cmdline
/usr/sbin/nc-l-p31337-t
```

The **grave-robber** program in The Coroner's Toolkit can be used to collect process information and other system details, including the following:

```
-rw-r--r--    1   root   root   1558129   May 30   18:50   coroner.log
-rw-r--r--    1   root   root    154596   May 30   18:50   MD5_all
-rw-r--r--    1   root   root      5618   May 30   18:50   error.log
drwx------    2   root   root      4096   May 30   18:50   trust
drwx------    2   root   root      4096   May 30   18:50   user_vault
drwx------   10   root   root      4096   May 30   18:49   conf_vault
-rw-r--r--    1   root   root   2939919   May30    18:48   body
drwx------    2   root   root      4096   May 30   18:48   command_out
drwx------    2   root   root      8192   May 30   18:48   icat
drwx------    2   root   root      8192   May 30   18:47   proc
drwx------    2   root   root      4096   May 30   18:47   removed_but_running
drwx------    2   root   root     16384   May 30   18:47   pcat
-rw-r--r--    1   root   root     10470   May 30   18:45   body.S
```

The "coroner.log" documents each action taken by grave-robber along with the date and time. Extracted data, such as recovered files, and process memory obtained using pcat and from the /proc virtual file system are organized into directories. The output of certain commands like lsof and ps are saved in the "command_out" directory and a mactime database (a.k.a. body file) of all files on the system is created. System configuration files and other files of interest are also preserved. Additionally, grave-robber calculates the MD5 values of all files, including the file containing the MD5 values. Even though a log file is created when grave-robber is run, it is advisable to document the process by taking notes and using the script command as discussed in Chapter 15.

19.2.3 WINDOWS REGISTRY

As demonstrated in previous case examples, intruders use the Registry to ensure that programs they have installed stay running, even after the system is rebooted. For instance, Trojan horse programs often have associated entries in the Registry. The most common locations in the Registry for Trojans are:

```
HKEY_LOCAL_MACHINE\Software\Microsoft\Windows\Current Version\Run
HKEY_LOCAL_MACHINE\Software\Microsoft\Windows\Current Version\RunOnce
HKEY_LOCAL_MACHINE\Software\Microsoft\Windows\Current Version\RunOnceEx
HKEY_LOCAL_MACHINE\Software\Microsoft\Windows\CurrentVersion\RunServices
HKEY_CURRENT_USER\Software\Microsoft\Windows\Current Version\Run
HKEY_CURRENT_USER\Software\Microsoft\Windows\Current Version\RunOnce
HKEY_CURRENT_USER\Software\Microsoft\Windows\Current Version\RunOnceEx
HKEY_CURRENT_USER\Software\Microsoft\Windows\CurrentVersion\RunServices
```

This list is not exhaustive since intruders regularly think of new ways to utilize the Registry such as making entries in the following keys:

```
HKEY_LOCAL_MACHINE\System\CurrentControlSet\Control\Session
Manager\KnownDLLs
HKEY_CLASSES_ROOT\exefile\shell\open\command
HKEY_LOCAL_MACHINE\Software\Classes\exefile\shell\open\command
```

The default Registry value for version 7.2.1 of SubSeven is "WinLoader" and for Back Orifice 2000 it is "UMGR32," but these can be modified to make them harder to detect. Also, some Trojan horse programs do not just use the Registry. For instance, Subseven can be configured to start using entries in WIN.INI (e.g. run = subseven.exe) and SYSTEM.INI (e.g. shell = explorer.exe subseven.exe).

19.2.4 ACQUISITION OVER NETWORK

When it is necessary to make a bitstream copy of the hard drive on a compromised system, it is possible to do so over the network. For example, the following shows one partition on "host13" being copied using dd to a remote system called "examiner" via netcat:

```
host13# df -k
Filesystem          kbytes    used     avail    capacity   Mounted on
/proc                    0       0         0          0%   /proc
/dev/dsk/c0t3d0s0   134335   30698     90204         26%   /
/dev/dsk/c0t3d0s6   737894  461662    217201         69%   /usr
fd                       0       0         0          0%   /dev/fd
swap                139944   29044    110900         21%   /tmp
host13# dd bs = 4096 if = /dev/dsk/c0t3d0s0 | nc examiner 3416

examiner# nc -l -p 3416 > host13-c0t3d0s0.dd
examiner# mount -o ro,loop,ufstype = sun -t ufs host13-c0t3d0s0.dd /mnt/host13
```

There are other methods of acquiring a bitstream copy of a disk over a network, including using Open Data Duplicator (ODD)*.

*http://odessa.sourceforge.net

19.2.5 CLASSIFICATION, COMPARISON, AND EVALUATION OF SOURCE

When investigating computer intrusions, it is often necessary to inspect files closely to determine what they are and how to interpret them. One approach to classifying files placed on a system by an intruder is to search the Internet for files with similar characteristics. For instance, the denial of service attack tools that were used to attack Yahoo and other large Internet sites contain information that can be useful for locating the source of the attacks. For instance, the following lines can be extracted from a denial of service tool called "trin00." The IP addresses at the end indicate where the "trin00 master" programs are located on the Internet. The computers running the master programs may have useful digital evidence on them:

```
socket
bind
recvfrom
%s %s %s
alf3YWfOhw.V.
PONG
*HELLO*
10.154.101.4
192.153.76.84
```

In addition to classifying a certain piece of digital evidence, it is often desirable to find unique characteristics that differentiate a given piece of digital evidence from other, similar pieces of digital data. In particular, it is very desirable to be able to determine the source of a piece of digital evidence. For instance, being able to show that a given sample of digital evidence originated on a suspect's computer could be enough to connect the suspect with the crime.

CASE EXAMPLE (LONDON 2002):
21-year-old Samir Rana, nicknamed "t0rner," was arrested following a year-long investigation into the creation of the Linux rootkit called "t0rnkit" and on suspicion of being a leading member of the infamous hacker group "Fluffi Bunni." Investigators had copies of the rootkit, IRC chat logs, and other evidence indicating that the suspect was the creator of t0rnkit. It was also reported that the suspect owned the pink stuffed toy depicted in website defacements by Fluffy Bunny.

19.3 INVESTIGATIVE RECONSTRUCTION

...the safecracker has been portrayed as a masked, bewhiskered, burly individual whose daring was matched only by his ruthlessness in disposing of interference. This legend undoubtedly had its origin in the facility with which the safecracker could be caricatured by cartoonists. His safe, mask, blackjack, and flashlight have come to be the picturesque symbols of the professional criminal. By this intimate association, the safe burglar has acquired in fiction the attributes of character corresponding to the physical properties of the safe itself – steely toughness of fiber and impregnability to moral suasion. Historically, this picture may have been true, but modern criminal society is far more democratic. The safecracker category, for example, includes all races, colors, and creeds: the skilled craftsman and the burglar; the timid and the bold; the lone wolf and the pack member;

> the professional criminal and the young amateur trying his wings; the local thug and the strong boy from a distant city. The occupation of safecracker has proved so remunerative to some practitioners, that its membership has swollen beyond the limits imposed by any of the restrictions of qualifications in the form of skill. (O'Hara 1970)

Like their predecessors (safe crackers), individuals who break into computers for profit have been stereotyped to the extreme. Despite overwhelming evidence to the contrary, computer intruders have been stereotyped as white, middle class, obsessive antisocial males between 12 and 28 years old with an inferiority complex, and a possible history of physical and sexual abuse (Casey 2002). Several other attempts have been made to create statistical profiles of computer intruders using information from media reports, offender interviews, and anecdotal observations. Although these profiles may give a general overview of past offenders and might be useful for diagnosing and treating associated psychological disorders, they have little investigative usefulness. In fact, such inductive criminal profiles can mislead investigators, causing them to jump to incorrect conclusions about an offender.

A more effective approach to learning about an offender in a given crime is to perform an investigative reconstruction as detailed in Chapter 5. By objectively analyzing available evidence, learning about the victims, and recognizing significant aspects of the crime scenes, an investigator can discern patterns of behavior and can gain a better understanding of the relationships between the victim, offender, and crime scenes, ultimately leading to a clearer understanding of the offender.

19.3.1 PARALLELS BETWEEN ARSON AND INTRUSION INVESTIGATIONS[9]

It is useful to examine well-established disciplines, such as arson investigation, to gain insight into the problems we face today in computer crime investigations. Although computer crime is a new development, there are many similarities between a computer that contains evidence and an arson crime scene. Most essentially, in all cases, people are responsible for the actions that leave behind clues. Additionally, as noted in the opening quotation of this chapter, we are dealing with evidence that has deteriorated significantly. An arson investigator's task is to recover fragmentary evidence and use it to determine what occurred.

When computer intruders make no effort to conceal their activities, investigators can obtain information about the offender's behaviors from log files

[9]This section is adapted from Casey, E. (2003), Arson, Archaeology, and Computer Crime Investigation, Computer Fraud and Security.

and other available digital evidence. However, if significant evidence has been destroyed, it is more difficult to determine what the intruder intended and investigators must rely more heavily on crime scene characteristics and victimology to understand the incident. Arson investigators are familiar with this type of situation – similarities between arson and computer intrusions are shown in Table 19.2.

Despite a paucity of evidence and a chaotic crime scene, arson investigators have learned to examine a scene methodically for the kinds of clues that have been most useful for solving crimes in the past. Arson investigators look for several key crime scene characteristics, related to those discussed in Chapter 5, that are applicable to computer intrusions: point of origin, method of initiation, requisite skill level, nature, and intent (Table 19.3).

Table 19.2

Comparison of features in arson and computer crime.

FEATURE	ARSON	COMPUTER CRIME
Dimensional expansion	Evidence may be found far from the blast or may have been projected vertically onto roofs, into trees, etc.	Evidence may be located in distant hosts. Network monitoring systems may have relevant log files
Layering	Burnt, collapsed structures create layers of evidence	Deleted data on a computer disk are layered under active data
Tools	Accelerants, explosive materials, bomb fragments, and other items found at the crime scene may have class characteristics that help connect the crime to the perpetrator	Toolkits and other items found at a computer crime scene may have class characteristics that help connect the crime to the perpetrator
Secondary scenes	An arsonist's home or bomb maker's workshop generally have evidence that can be linked with the scene	Computers used by the offender to compile programs or launch an attack usually have evidence that can be linked to the scene
MO, signature, skill	The composition of an incendiary device can be unique to the offender, such as detonator or explosive mixture used, revealing the offender's skill level	Tools used by computer criminals can have unique characteristics introduced by the offender, revealing the offender's skill level

Table 19.3

Comparison of crime scene characteristics in arson and computer intrusions where "cwd" refers to the current working directory of a process (where it was started).

	POINT OF ORIGIN	METHOD OF INITIATION	REQUISITE SKILL LEVEL	NATURE AND INTENT
Arson	Warehouse window	Matches and crude fuse (cotton rag soaked in gasoline)	Low (simple Molotov cocktail – readily available materials)	Broad targeting (destroy warehouse)
Arson	Engine (front of car)	Electric arc (triggered by car ignition)	High (car bomb made with military grade explosives)	Narrow targeting (kill car driver)
Intrusion	SSH server (port 22)	Buffer overflow (CRC-32 compensation attack detector vulnerability)	Low (exploit freely available on Internet)	Targeting and intent unclear (need more details)
Intrusion	/tmp/.tmp (cwd of process)	Rootkit (t0rnkit) script	Medium (rootkit available on Internet)	Concealment (precautionary act)
Intrusion	/home/janedoe (cwd of process)	"sudo rm -rf ../johndoe/*"	Low (simple UNIX command)	Narrow targeting (delete user files)

Let us first consider the nature and intent of the crime. Computer criminals and arsonists alike may destroy evidence to cover their tracks, to retaliate against some perceived wrong, and/or to demonstrate their power. To determine whether destruction was intended to inflict damage or simply as a precautionary act, it is helpful to consider whether the targeting was broad or narrow. *Narrow targeting* refers to any destruction that is designed to inflict specific, focused, and calculated amounts of damage on a specific target such as targeting "/home/janedoe" in Table 19.3. *Broad targeting* refers to destruction that is designed to inflict damage in a wide reaching fashion. Rather than targeting a single individual by deleting their files, an intruder might delete information that is important to the entire organization, targeting the entire organization or what it represents as in the following case example:

CASE EXAMPLE (NEW JERSEY 1996):
Tim Lloyd, the primary system administrator for Omega Engineering Corporation, was originally fired for stealing expensive equipment. In retaliation, Lloyd executed time-delayed commands on Omega's primary server that deleted all of the company's important data and programs on a specific date. Specifically the method of initiation was a modified version of the DELTREE command ("FIX /Y F:*.*") to delete everything on the drive combined with the "PURGE F:\ /ALL" command to obliterate the deleted data. A high degree of skill was required to implement this narrowly targeted attack and the intent was to destroy all of Omega's important data and programs. Lloyd also erased all related backup tapes. Experts spent years recovering pieces of information from the servers, desktops, and even computers of ex-employees. Although the damage was extensive, this attack is considered narrowly targeted because it was designed to inflict a specific damage on a specific target. (Gaudin 2000)

In this case, the nature of the crime was malicious and Lloyd's intent was to punish his former employer for perceived wrongs.

To determine if the targeting was narrow or broad, it is helpful to determine intentional versus actual damage. This means learning as much about the configuration of the target computer as possible and the amount of damage incurred by the target. For example, programs like chroot limit the damage that can be done on a system if one application (e.g. a Web server) is compromised. An intruder who was hoping to damage a wide area of the computer would be thwarted by such restrictions. If the intruder destroys everything in the restricted area, this is likely evidence of broad targeting and the intruder might not have achieved his/her goal of destroying everything on the computer. On the other hand, if the intruder deletes a few files in the restricted area, this is evidence of narrow targeting and the intruder probably achieved his/her goal.

The Lloyd case example also demonstrated that, in addition to knowing the perpetrator's intent, determining who had access to the point of origin or method of initiation can lead to prime suspects. For instance, in Table 19.3 only a few people had access to the point of origin "/home/janedoe" and the method of initiation "sudo," reducing the suspect pool to Jane Doe and others with administrative privileges on the system. In the previous case example, digital evidence recovered from the damaged system immediately implicated Lloyd because he was the only individual with the requisite access to the point of origin and ability to create the destructive program.

Determining skill level can also lead to suspects. The skill level and experience of a computer criminal is usually evident in the methods and programs used to break into and damage a system. For instance, an offender who uses readily available software and chooses weak targets for little gain is generally less skilled and experienced than an offender who writes customized programs to target strong installations. A skilled computer criminal might create a time bomb specifically designed to destroy important data at a particular time or when a certain triggering event occurs as in the previous case example. Having said this, a skilled offender can successfully achieve specific goals using programs that exist on the system. Therefore, what is known about point of origin, method of initiation, and nature and intent of the destructive act should all be taken into account when assessing the offender's skill level.

Notably, precautionary acts – destroying data to conceal, damage, or destroy any items of evidentiary value – are not always very thorough. Items that an intruder intended to destroy can be examined by digital evidence examiners to exploit them for their full evidentiary potential, no matter how little debris is left behind. For example, if a small portion of a deleted file remains on a disk, this remaining digital evidence should be carefully reconstructed and examined to determine why the offender tried to destroy it.

19.3.2 CRIME SCENE CHARACTERISTICS

In addition to being a primary crime scene, computers can be secondary scenes in the form of launch pads, listening posts, or storage sites. Intruders use launch pads to hide their identities while committing other crimes (e.g. breaking into other computers, distribute illegal materials, cyberstalking). Also, intruders often use a launch pad when the target computer is difficult to compromise from outside a network but can be compromised from another computer on the same network. Intruders use listening posts to look for other likely targets on a network, and use storage sites to keep toolkits, stolen data, and other incriminating evidence. These secondary scenes can be a rich source of digital evidence that can be associated with a particular individual.

A computer intruder's method of approach and attack can reveal a significant amount about the offender's skill level, knowledge of the target, and intent. The concept of broad versus narrow targeting can also be useful when examining the method of approach and attack. For instance, network logs may show a broad network scan prior to an intrusion, suggesting that the individual was exploring the network for vulnerable and/or valuable systems. This exploration implies that the individual does not have much prior knowledge of the network and may not even know what he/she is looking for but is simply prospecting. Conversely, intruders who have prior knowledge of their target will launch a more focused and intricate attack. For instance, if an intruder only targets the financial systems on a network, this directness suggests that the intruder is interested in the organization's financial information and knows where it is located.

So, if the targeting is very narrow – the intruder focuses on a single machine – this indicates that he/she is already familiar with the network and there is something about the machine that interests him/her. Similarly, time pattern analysis of the target's file system can show how long it took the intruder to locate desired information on a system. A short duration is a telltale sign that the intruder already knew where the data were located, whereas protracted searches of files on a system indicates less knowledge. The intruder's knowledge of the target and criminal skill can be very helpful in narrowing the suspect pool, particularly when only a few individuals possess the requisite knowledge and skills suggesting insider involvement.

The sophistication of the intrusion and subsequent precautionary acts help determine the perpetrator's skill level.

CASE EXAMPLE

An organization received a complaint that one of their Solaris workstations was being used to launch attacks against others on the Internet. The organization was not particularly concerned by the complaint since the workstation did not contain valuable information and believe that the problem could be resolved with relative ease.

Examining the server revealed obvious signs of intrusion. The intruder had gained access through a vulnerability that had been widely publicized that week, added a new account, deleted log files but failed to cover tracks completely. In short, this intruder was noisy, lacked finesse, and was not interested in information on the system. These factors are consistent with a low skill intruder. However, a closer examination of the system revealed an oddity one month earlier:

```
# ls -altc /usr/ucb/ps | head
-rwsr-xr-x   1  root   sys    24356  Jun 6 17:20 ps
# ls -altc /usr/sbin/inetd | head
-r-xr-xr-x   1  root   root   39544  Jun 6 17:20 inetd
```

An analysis of the ps command showed that it had been compiled using a non-Sun compiler, indicating that the vendor had not created it. There were no unusual entries in log files from that time period but searching for other files created on that date led to a sniffer that was cleverly concealed within the system:

```
# ls -altc /kernel
-rw-r--r--   1    root   root   60      Jun 6   17:20 pssys
# more /kernel/pssys
1 "./update.hme -s -o output.hme"
# cd /usr/share/man/tmp
# ls -altc
total 156
-rw-r--r--   1    root   root   23787   Jun 12 07:52 output.hme
drwxr-xr-x   2    root   root   512     Jun 6   17:20 .
drwxr-xr-x   40   bin    bin    1024    Jun 6   17:20 ..
-rwx------   1    root   root   25996   Jun 6   17:20 update.hme
```

The sniffer output file "output.hme" contained the following entry, indicating that the intruder could have observed legitimate users on the network accessing valuable research data on another system:

```
-- TCP/IP LOG -- TM: Fri Jun 11 10:28:52 --
PATH: host01.corpY.com(64376) => server.corpY.com(ftp)
STAT: Fri Jun 11 10:30:45, 20 pkts, 135 bytes [DATA LIMIT]
DATA: USER james
    :
    : PASS smiley:)-99
    :
    : CWD researchdata
    :
    : GET research0302.dat
```

This intruder left almost no trace of the intrusion and used relatively sophisticated concealment techniques, suggesting a high skill level. Without additional evidence, it was not possible to determine how the intruder had gained access to the system. The most likely hypothesis was that this intruder used the same vulnerability exploited by the second intruder but knew about it several weeks before it became widely publicized. The cautious, focused nature of the attack suggested that the intruder had a particular goal and was monitoring network traffic to achieve this goal. However, without additional evidence it was not possible to determine if the intruder was interested in the research data or something else on the organization's network.

This case example demonstrates how the choice of secondary crime scene can be significant, leading to additional insights. The intruder deliberately selected the Solaris workstation as a listening post, revealing a high skill level and a specific interest in monitoring network traffic. In other cases, an intruder may select a computer to launch an attack because the computer itself is fast, it is connected to a fast network, it is easy to break into, it is located in a different country, or it is located near the target. Alternatively, an intruder may use a particular network to launch attacks because he/she has broken into computers on the network before and is confident that he will not be caught. If the intruder has broken into other systems on the network in the past, the organization may have archived digital evidence from those systems that can help apprehend the offender.

Seemingly minor details regarding the offender can be important. Therefore, investigators should get in the habit of contemplating what the offender brought to, took from, changed, or left at the crime scene. For instance, investigators might determine that an offender took valuables from a crime scene, indicating a profit motive. Alternatively, investigators might determine that an offender took a trophy or souvenir to satisfy a psychological need. In both cases, investigators would have to be perceptive enough to recognize that something was taken from the crime scene.

Although it can be difficult to determine if someone took a copy of a digital file (e.g. a picture of a victim or valuable data from a computer), it is possible. Investigators can use log files to glean that the offender took something from a computer and might even be able to ascertain what was taken. Of course, if the offender did not delete the log files investigators should attempt to determine why the offender left such a valuable source of digital evidence. Was the offender unaware of the logs? Was the offender unable to delete the logs? Did the offender believe that there was nothing of concern in the logs? Small questions like these are key to analyzing an offender's behavior.

CASE EXAMPLE

An organization believed that an ex-employee stole information prior to quitting on September 16, 2002. Investigators were asked to determine if the ex-employee had taken documents from his Windows 2000 workstation, a copy of the client contact database (clients.mdb), or anything related to a sensitive project called "ProjectX" stored on a UNIX file server (192.168.2.10). Investigators preserved digital evidence on the Windows 2000 and UNIX systems by making a bitstream copy of the hard drives. Logon/logoff records from the ex-employee's workstation indicate that he used the computer on September 16, 2002, between 08:50 A.M. and 09:10 A.M.:

```
C:\>ntlast /ad 16/9/2002 /v
Record Number: 18298
ComputerName: WKSTN11
EventID: 528 - Successful Logon
Logon: Tue Sep 16 08:50:58am 2002
Logoff: Tue Sep 16 09:10:00am 2002
Details –
        ClientName: user11
        ClientID: (0x0,0xDCF9)
        ClientMachine: WKSTN11
        ClientDomain: CORPX
        LogonType: Interactive
```

Investigators check the building security (card swipe) records to confirm that the ex-employee was in the vicinity of the computer at the time. These records show that the suspect entered the building at 08:45 A.M.

Further examination of the ex-employee's workstation shows that the "clients.mdb" file was accessed at 08:58:30 A.M. and that a related file named "clients.xls" was created shortly after in a temporary directory. The ex-employee's e-mail outbox shows the "clients.xls" was sent to a Hotmail address. Performing a functional reconstruction of the "Send To" feature in Microsoft Access suggests that the ex-employee used this method to e-mail the database. Another file named "private.doc" was accessed at around the same time as a shortcut file (with a ".lnk" extension) associated with the floppy drive (A:), suggesting that the file was copied to a floppy disk using the "Send To" feature of Windows. The last accessed date–time stamp of another shortcut file indicated that the SSH client on the machine was launched. Additionally, the following SSH key file associated with the UNIX file server had been accessed at the same time, suggesting that a connection was made to the server at that time:

```
C:\Documents and Settings\user11\Application Data\SSH\ HostKeys\
key_22_192.168.2.10
```

Logon records on the UNIX server show a corresponding logon session from the ex-employee's computer. A sensitive file named "projectX" was found on the server and had a last access date–time consistent with the logon session:

```
% last user11
user11 pts/77 wkstn11.corpx.com Sep 16 09:05–09:06 (00:01)
% ls -altu
-rwxr-xr-x 1 admin staff 8529583 Sep 16 09:05 projectX
```

A deleted copy of the "projectX" file was recovered from the ex-employee's workstation. Comparing the date–time stamps of this file with the copy on the server indicates that the file was copied from the server at 09:05 A.M. Specifically, the date–time stamps of deleted "projectX" file recovered from the ex-employee's workstation were:

Created: 09:05 A.M:

Accessed: 09:07 A.M:

Modified: 09/12/2002 10:07:07 A.M:

Also of note was an entry in the Registry (HKEY_USERS\<user11-sid>\Software\Windows\Explorer\RecentDocs\NetHood) indicating that a NetBIOS connection had been established between the ex-employee's workstation and a computer on a competitor's network. This Registry key had a Last Write Time of 09/13/2002 at 11:04 A.M. and network logs confirmed a connection at this time. Network logs also showed a NetBIOS connection from the ex-employee's computer to the competitor's network at 09:07 A.M. on September 16, 2002:

```
[**] Netbios Access [**]
09/16-09:07:03.313894 192.168.16.88:1576 → 172.16.14.3:139
TCP TTL:127 TOS:0x0 ID:61055 IpLen:20 DgmLen:231 DF
***AP*** Seq: 0x4A8908DB Ack: 0x5C6EFB75 Win: 0x431B TcpLen: 20
```

This connection was also recorded in the following NetFlow logs:

Start	End	SrcIPaddress	SrcP	DstIPaddress	DstP	P	Fl	Pkts	Octets
0916.09:07:04	0916.09:07:56	192.168.16.88	1576	172.16.14.3	139	6	3	9711	8693271

The fact that the number of bytes transferred is roughly equivalent to the size of the "projectX" file indicates that this file was transferred to the competitor's system.

This example demonstrates the usefulness of network level logs to corroborate important events. These types of corroborating evidence are especially important when investigating computer intrusions because automated toolkits enable even low skilled offenders to employ sophisticated concealment techniques on a compromised host.

19.3.3 AUTOMATED AND DYNAMIC MODUS OPERANDI

Toolkits that automate the actions required to break into a computer and destroy or conceal evidence of the intrusion provide an *automated modus operandi* that make multiple offenders almost indistinguishable. When every crime scene looks almost identical, it becomes more difficult to link cases committed by a single offender and to understand the unique motivations of different offenders. Although these toolkits reduce the amount of behavioral information that is available to investigators, it is possible to differentiate between automated actions and the offender's behavior.

CASE EXAMPLE

An organization became concerned when they detected an attack against a server that contained valuable intellectual property:

```
[**] FTP-site-exec [**]
09/14-12:27: 208.181.151.231 -> 192.168.12.54
09/14-12:28: 24.11.120.215   -> 192.168.12.54
09/14-12:33: 64.28.102.2     -> 192.168.12.54
```

The digital evidence examiner noted that the server's clock was 4 hours, 40 minutes fast but did not find any signs of compromise initially. There were no entries in the wtmp or syslog files at the time of the attack, no unusual processes were visible using ps, and the ls and find commands did not reveal anything alarming. However, comparing the output of ps and lsof uncovered several discrepancies, suggesting that the system was compromised.

The digital evidence examiner made a bitstream copy of the hard drive and observed two directories that had not been visible during the initial examination: "/usr/info/.t0rn" and "/usr/src/.puta/t0rnsniff." The examiner also found a modified copy of a rootkit named "Tornkit" that the intruder had used (Figure 19.2). Searching the Internet revealed that this rootkit was being used by intruders around the world and had become common enough to warrant an advisory from CERT.[10]

[10] *http://www.cert.org/incident_notes/IN-2000-10.html*

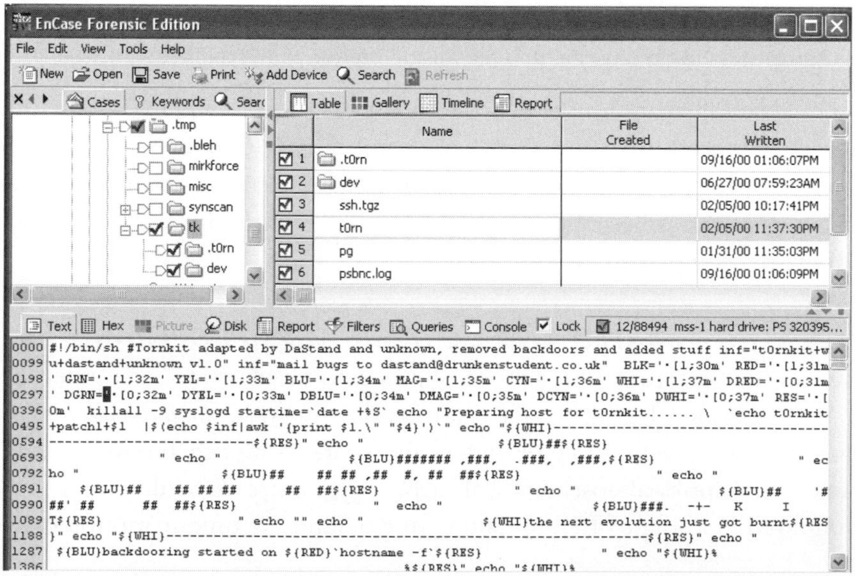

Figure 19.2

EnCase used to analyze Linux system showing rootkit installations script.

Searching unallocated space for deleted syslogs (taking into account the clock offset) uncovered the following entry showing a buffer overflow of the FTP server:

```
Sep 14 17:07:22 host1 ftpd[617]: FTP session closed
Sep 15 00:21:54 host1 ftpd[622]: ANONYMOUS FTP LOGIN FROM 231.efinityonline.com
[208.181.151.231],
□□□□□□□□□□□□□□□□□□□□□□□□□□□□□□□□□□□□□□□□□□□□□□□□□□□□□□□□□□□□□□□□□□□□□□□□□□□□□□□□□□
□□□□□□□□□□□□□□□□□□□□□□□□□□□□□□□□□□□□□□□□□□□□□□□□□□□□□□□□□□□□□□□□□□□□□□□□□□□□□□□□□□
□□□□□□□□□□□□□□□□□□□□□□□□1À1Û1É°Fí□1À1ÛC%ÙA°?í□ëk^1À1É□^^A^F^Df¹ÿ^A°'í□1À□^^A°=í□1À
1Û□^^H%C^B1ÉþÉ1À□^^H°^Lí□þÉu61À^F^I□^^H°=í□þ^N°0þÈ^F^D1À^F^G%v^H%F^L%ó□N^H□V^L°^Kí
□1À1Û°^Aí□è□ÿÿÿ0bin0sh1..11
Sep 14 17:22:54 host1 inetd[448]: pid 622: exit status 1
```

This and other recovered logs entries confirmed the source of initial intrusion. Other recovered log segments indicated that the intruder had been monitoring network traffic:

```
Sep 15 23:05:41 host1 kernel: device eth0 entered promiscuous mode
Sep 15 23:09:37 host1 kernel: device eth0 left promiscuous mode
Sep 15 23:09:39 host1 kernel: device eth0 entered promiscuous mode
Sep 15 23:10:22 host1 kernel: device eth0 left promiscuous mode
Sep 15 23:10:27 host1 kernel: device eth0 entered promiscuous mode
```

After performing an investigative reconstruction, it was concluded that the target was at high risk of intrusion and that the intruder was not aware of the valuable information on the server. The server was at high risk of intrusion because it was not protected by a firewall and was running an FTP server with a well-known vulnerability that was trivial to exploit. The intrusion was preceded by a broad scan of the network for systems with vulnerable FTP servers, suggesting that the intruder was not specifically targeting one particular server. The intruder's ignorance of the valuable contents of the server was further evident from date–time stamps on the file system – the sensitive data had not been accessed during the intrusion. Also, the intruder's primary intent was to use the compromised host to launch attacks against other systems, monitor network traffic for passwords, and connect to IRC. These activities were not consistent with a sophisticated attacker who was interested in stealing the information on the server.

More experienced intruders often have a preferred toolkit that they have pieced together over time and that have distinctive features. For instance, a compressed TAR file containing the following tools were found on several compromised machines, indicating that a single offender was responsible for all of the intrusions:

```
% tar tvf aniv.tar
-rw-r--r--   1   358400   Mar   8   17:02   BeroFTPD-1.3.3.tar.gz
-rw-r--r--   1   326      Mar   8   17:02   readmeformountd
-rw-r--r--   1   757760   Mar   8   17:02   root.tar.gz
-rwxr-xr-x   1   8524     Mar   8   17:02   slice2
-rw-r--r--   1   6141     Mar   8   17:02   mountd.tgz
-rw-r--r--   1   849920   Mar   8   17:02   rkb.tar.gzb
```

Also, some intruders personalize their toolkits with nicknames and comments. For instance, the following rootkit script recovered from several compromised Solaris systems contains the intruder's nickname and had been modified in later intrusions to use "/var/yp/..." instead of "/var/tmp/..." as a working directory:

```
#[*] – hacker nickname
#[*] – SunOS rootkit v1
echo "creating directories"
mkdir /var/yp/.../
mkdir /var/yp/.../old/
echo "switching directory..."
cd stuff
echo "moving files..."
mv * /var/yp/.../
echo "cleaning up..."
cd ..
rm -rf stuff
rm -rf s1.tar
```

So, in addition to being helpful for linking intrusions committed by the same individual, distinctive features of a toolkit can be viewed as both an MO- and signature-related behavior. However, keep in mind that the intruder may have been given a customized toolkit and may not have personalized it himself/herself.

In addition to the contents of a toolkit, the way a particular intruder uses a toolkit can be unique. For instance, it is sometimes possible to recover a list of the commands an intruder typed, revealing MO-related behavior as shown here:

```
% more .bash_history
w
pico /etc/passwd
mkdir /lib/.loginrc
cd /lib/.loginrc
/usr/sbin/named
ls
w
ls
/usr/sbin/named
ls
cd ~
```

```
ls
mv aniv.tar.gz /lib/.loginrc
cd /lib/.loginrc
tar zxf aniv.tar.gz
ls
cd aniv
ls
tar zxf rkb.tar.gz
ls
cd rkb
./install
```

These commands refer to a directory named "/lib/.loginrc" that was useful for linking several intrusions to the same offender.

To make case linkage even more difficult, offenders who use the Internet can change their *modus operandi* with relative ease. As offenders become more familiar with the Internet, they usually find new ways to make use of it to achieve their goals more effectively. An offender who uses the Internet creatively can change his *modus operandi* so frequently and completely that it is best described as dynamic. For instance, individuals who break into well-secured computer systems may have to develop a novel intrusion plan for each unique target. A *dynamic modus operandi* has also been seen when an offender is consciously trying to foil investigators.

19.3.4 EXAMINING THE INTRUDER'S COMPUTER

If all goes well in an investigation, the intruder's computers can be examined for evidence relating to the crime. Recalling Locard's Exchange Principle, during the commission of a crime, evidence is transferred between the offender's computer and the target. For instance, in one case the intruder's Windows NT computer contained the following digital evidence linking him with the compromised systems:

- lists of dial-up accounts and passwords, including the one used to commit crimes;
- nmap scans of target networks;
- lists of compromised hosts (trophy list and memory aid);
- list of UNIX commands executed on compromised hosts (memory aid);
- sniffer logs from compromised hosts (digital evidence transfer);
- directory listings from compromised UNIX hosts (digital evidence transfer);
- stolen data from compromised hosts, including credit cards and private e-mail;
- TAR file with class characteristics linking it to compromised UNIX host;
- toolkits found on compromised hosts;
- FTP and terminal emulator configuration files relating to compromised hosts;

- IRC logs showing suspect connecting to IRC from compromised hosts;
- IRC logs of suspect boasting about breaking into specific hosts;
- IRC logs of suspect communicating with accomplices.

When examining an intruder's computer, begin by searching for what is known such as the time periods of the intrusions, host names, IP addresses, and stolen user accounts. Searching for online nicknames may uncover remnants of online communications with accomplices and mention of other targets. The MD5 values of files found on the compromised hosts can be used to search for identical files on the intruder's hard drive. Any files that are found can be further analyzed for class characteristics that link them to a particular host. It may also be fruitful to look for evidence transfer such as directory listings from the compromised systems (e.g. in unallocated space or a swap file).

19.4 DETAILED CASE EXAMPLE[11]

[11]This case example is based on abstracted lessons from various investigations. Any resemblance to actual incidents is coincidental.

One Friday morning, a system administrator at Corporation X in New York noticed an unusual process named monitor on an important database server and found a hidden directory ("/usr/share/man/...") containing monitor and what appeared to be an associated sniffer log. One of Corporation X's digital evidence examiners responded, following standard operating procedures to confirm that the host had been compromised and to gather related evidence. A quick analysis of the digital evidence from the system revealed the point of origin, method of initiation, and intent. The intruder had broken in through a recently publicized vulnerability in the Oracle database software running on the server. The intruder had fixed the vulnerability to prevent others from exploiting it, installed a rootkit with a backdoor for regaining entry to the system, and started a sniffer to monitor network traffic. There was no evidence on the system that revealed the source of the attack or the intruder's IP address. Furthermore, Corporation X's firewall, intrusion detection system, and NetFlow logs did not appear to contain any entries that were obviously related to the intrusion.

The examiner informed Corporation X's management and attorneys of the developing situation and obtained approval to proceed. He then discovered several other compromised servers using an automated scanning tool configured to detect the intruder's backdoor. He did not have administrative access to many of these systems and it would have taken several days to gain physical access to some of them. Because there was an imminent danger that the intruder would return to delete the sniffer logs from these systems, the organization authorized the examiner to connect to the systems through the intruder's backdoor and collected digital evidence remotely as discussed in

Chapter 15. Connecting through the intruder's backdoor had the added advantage of concealing the fact that the examiner was connected to the system, reducing the risk of alerting the intruder to his presence:

```
examiner1% script host32-062202-case14524
Script started on Sat Jun 22 13:58:15 2002
examiner1% ssh -l backdoor_account host32.corpX.com
Last login: Thu Jun 20 07:15:55 on pts/2
# w
  1:58pm up 83 day(s), 8:56, 0 users, load average: 0.02, 0.02, 0.07
User  tty  login@  idle  JCPU  PCPU  what
# ps -ef
     UID   PID  PPID  C  STIME   TTY    TIME  CMD
    root     0     0  0  Apr 01  ?      0:00  sched
    root     1     0  0  Apr 01  ?      1:28  /etc/init -
    root     2     0  0  Apr 01  ?      0:06  pageout
    root     3     0  0  Apr 01  ?    175:52  fsflush
    root   349   346  0  Apr 01  ?      0:01  /usr/lib/saf/listen tcp
    root   201     1  0  Apr 01  ?      3:09  /usr/sbin/cron
    root   346     1  0  Apr 01  ?      0:01  /usr/lib/saf/sac -t 300
<cut for brevity>
  nobody   320     1  0  Apr 01  ?   0:04  /oracle/bin/oraweb -C /or
  oracle 22493     1  0  May 25  ?   7:59  ora_smon_finance1
  oracle 22487     1  0  May 25  ?   0:05  ora_pmon_finance1
  oracle 22491     1  0  May 25  ?  18:49  ora_lgwr_finance1
  oracle 22489     1  0  May 25  ?  55:09  ora_dbwr_finance1
  oracle 22495     1  0  May 25  ?   0:02  ora_reco_finance1
  oracle 14401     1  0  May 10  ?   8:36  ora_smon_finance2
  oracle 14399     1  0  May 10  ?   3:14  ora_lgwr_finance2
  oracle 14397     1  0  May 10  ?   7:02  ora_dbwr_finance2
  oracle 14395     1  0  May 10  ?   0:02  ora_pmon_finance2
<cut for brevity>
    root 15718     1  0  Jun 17    ?   30:09  ./solsniffer -s
    root 23656 23652  1  13:58:34 pts/1 0:00  ps -ef
# cd /usr/share/man/...
# ls -altc
total 4950
-rw-rw-r--   1 root  root  911381  Jun 22  13:57  log
drwxrwxrwt   4 sys   sys     1024  Jun 22  04:00  ..
drwxrwxr-x   2 root  root     512  Jun 17  17:07  .
-rwx--x--x   1 root  root   19996  Jun 17  17:07  solsniffer
# md5 log
md5: Command not found.
# cat log
<sniffer log cut for brevity>
# scp log examiner@examiner1.corpX.com:/e1/case14524/host32-log-062202
# mail examiner@corpX.com < log
```

Anticipating that the intruder would return, the examiner monitored network traffic to the compromised hosts using Argus. That evening, the intruder was observed gaining unauthorized access to one of the compromised hosts from another system on the network:

```
examiner1% ra -r argus.out host 192.168.0.101
22 Jun 02 23:26:56 tcp 192.168.0.5.2444 -> 192.168.0.101.ssh EST
22 Jun 02 23:28:05 tcp 192.168.0.5.2444 -> 192.168.0.101.ssh EST
22 Jun 02 23:29:26 tcp 192.168.0.5.2444 -> 192.168.0.101.ssh FIN
```

The examiner connected to the compromised host (192.168.0.101) through the intruder's backdoor, gathered digital evidence from memory, shut the system down, and collected the hardware as evidence. In this way, the intruder's presence on the compromised host was documented and the original hardware was preserved for later analysis.

The examiner determined that the intruder was using a stolen account on an internal system (192.168.0.5) to launch attacks against other hosts on the network. The firewall, intrusion detection system, and the router that generated NetFlow logs were not between the launch pad and the target hosts. This explained how the intruder had been able to target the vulnerable ports on the compromised systems even though they were protected by a firewall. This also explained why the intrusion detection systems and NetFlow logs did not contain any useful data. Incidentally, as a result of the lessons learned from this incident, Corporation X installed permanent Argus probes on all of their important network segments to ensure that these logs were available in the future.

The intruder had stored tools in a hidden directory of this stolen account but had not been able to erase system log files. The examiner collected the log files and contents of the stolen account as evidence. Logon records from the stolen account contained the IP address of a computer on a business partner's network – Business Z in San Francisco:

```
host5% last stolen_account
stolen_account pts/3 172.16.12.15 Sat Jun 22 23:24 still logged in
stolen_account pts/22 172.16.12.15 Thu Jun 20 07:13 - 07:37 (00:24)
stolen_account pts/5 172.16.12.15 Mon Jun 17 16:51 - 17:38 (00:47)
wtmp begins Sun Jun 16 19:10:54 2002
```

The examiner called his counterpart in Business Z on her mobile phone to inform her of the problem. She quickly determined the

Windows NT system in question (172.16.12.15) was running a Trojan horse program (Back Orifice 2000) and did not contain any logs containing the intruder's IP address. Also, Business Z's intrusion detection system logs did not contain any alerts relating to the compromised Windows NT system, probably because connections between the Back Orifice client and server were encrypted. However, Business Z's NetFlow logs did show incoming connections to the compromised Windows NT system and subsequent outgoing connections to the machine on Corporation X's network:

```
flow% flow-cat /netflow/2002-06-22/ft-v05.2002-06-22.203000 | flow-filter -
Dbo2k -f ./bo2k-062202.acl | flow-print -f5
Start       End         SrcIPaddress  SrcP  DstIPaddress   DstP  Octets
0622.20:20  0622.20:49  10.145.32.24  2584  172.16.12.15   443   2412085
flow% flow-cat /netflow/2002-06-22/ft-v05.2002-06-22.203000 | flow-filter -
Sbo2k -f ./bo2k-062202.acl | flow-print -f5
Start       End         SrcIPaddress  SrcP  DstIPaddress   DstP  Octets
0622.20:20  0622.20:50  172.16.12.15  443   10.145.32.24   2584  3660674
0622.20:23  0622.20:43  172.16.12.15  1927  192.168.0.5    22    3457683
```

The two examiners corrected the time zone difference between New York and San Francisco and confirmed that these connections corresponded to the logon records from the stolen account. They immediately contacted the ISP that the intruder was using and asked them to preserve evidence on their systems relating to the intrusions.

The organizations then reported the incident to the FBI and provided them with enough information to obtain subscriber details from the ISP used by the intruder. The FBI determined that the dial-up account used by the intruder had been stolen. Fortunately, the ISP had Automatic Number Identification (ANI) records that contained the intruder's home telephone number:

> To: FBI
>
> From: ISP
>
> Date: 06/30/02
>
> Re: Case #14524
>
> The following is the information you requested in the Subpoena of the United States District Court in the District of New York, dated 06/25/02, which I have enclosed. The information is correct to the best of my knowledge and I will keep records of my investigation until you tell me otherwise.
>
> You requested the information pertaining to the following connections:

Username:	janedoe
IP address assigned:	10.145.32.24
Time of connection:	23:22:38 (EST5EDT) Jun 22, 2002
Time of disconnect:	23:54:12 (EST5EDT) Jun 22, 2002
ANI information:	(510) 555–2356
Username:	janedoe
IP address assigned:	10.145.32.17
Time of connection:	07:12:54 (EST5EDT) Jun 20, 2002
Time of disconnect:	07:40:06 (EST5EDT) Jun 20, 2002
ANI information:	(510) 555–2356
Username:	janedoe
IP address assigned:	10.145.32.105
Time of connection:	16:32:17 (EST5EDT) Jun 17, 2002
Time of disconnect:	18:53:32 (EST5EDT) Jun 17, 2002
ANI information:	(510) 555–2356

After performing a background check and further investigation to satisfying themselves that the resident of the house was responsible for the connections, the FBI obtained a search warrant and seized the suspect's computers. An examination of these computers revealed many links with Corporation X's compromised servers, including sensitive data captured in sniffer logs. Faced with overwhelming evidence, the suspect admitted his involvement and provided the FBI with a list of his accomplices.

19.5 SUMMARY

Computer intrusions are among the most challenging types of cybercrime from a digital evidence perspective. Every computer and network is different, configured by the owner in a very personal way. Some systems are highly customized, fitting the specific needs of a skilled computer user while other systems are highly disorganized. In many ways, investigating a computer intrusion is like going into someone's kitchen and trying to determine what is out of place. In some cases, anomalies are obvious like seeing plates in a cutlery drawer. In other cases, investigators must interview system owners/users and examine backup tapes and logs files to determine what the computer intruder changed.

Additionally, every computer intruder is different – choosing targets/victims for different reasons, using different methods of approach and attack, and exhibiting different needs and intents. Ex-employees break into computers, damaging them in retaliation for some perceived wrong. Technically proficient individuals break into targets of opportunity to feel more powerful. Thieves and spies break into computers to obtain valuable information. Malicious individuals break into medical databases, changing prescriptions to overdose an intended victim. These types of crime are becoming more prevalent and are creating a need for skilled investigators equipped with procedures and tools to help them collect, process, and interpret digital evidence.

Even when computer intruders are careful to hide their identities, they often have quite distinct MO and signature behaviors that distinguish them. The items an intruder takes or leaves behind are significant when understanding the MO and signature and what a criminal tries to destroy is often the most telling.

REFERENCES

Casey E. (2003) "Cyberpatterns: Criminal Behavior on the Internet", Turvey, B. Criminal Profiling: An Introduction to Behavioral Evidence Analysis, 2nd Edition, London: Academic Press.

CSI/FBI (2003) *Computer Crime and Security Survey*, San Francisco: Computer Security Institute.

Gaudin S. (2000), "Case Study of Insider Sabotage: The Tim Lloyd/Omega Case", Number 3, 2000, *Computer Security Journal*, Volume XVI.

Mandia K., Prosise C. and Pepe M. (2003) Incident Response and Computer Forensics, 2nd Ed., New York: Osborne.

O'Hara C. (1970), *Fundamentals of Criminal Investigation*, 2nd Ed., Springfield: Charles C. Thomas Publishers.

Phrack 59 (2002), "Defeating Forensic Analysis on Unix", July 2002 (http://www.phrack.org/show.php?p=59&a=6).

Phrack 58 (2001), "Armouring the ELF: Binary encryption on the UNIX platform", December, 2001 (http://www.phrack.org/show.php?p=58&a=5).

CHAPTER 20

SEX OFFENDERS ON THE INTERNET

Eoghan Casey, Monique Ferraro, and Michael McGrath

The ability of criminals to acquire victims, gather information, lurk in cyberspace, protect or alter their identity, and communicate with other offenders makes the Internet an attractive setting for these individuals. However, at times the lack of technological sophistication displayed by offenders is surprising. Some offenders apparently are not aware that it is quite easy to locate them and make very little effort to conceal basic information on the Internet. Offenders who do not initially hide their identity may do so only after they realize they are at risk. Thus, it may be possible to use the Internet's archiving capabilities to find information on an individual before their covering behavior commenced.

The Internet is attractive to sex offenders for a number of reasons. In addition to giving criminals greater access to victims, extending their reach from a limited geographical area to victims all around the world, the Internet contains a significant amount of information about potential victims. Online dating sites (e.g. personals.yahoo.com) provide the most obvious example of the kinds of personal information that individuals disclose on the Internet including photographs, their age, and geographic region. Although these dating sites were created for a legitimate purpose, they provide a target rich environment that offenders have not overlooked. In 2002, Japan's National Police Agency reported a dramatic increase in the number of crimes, including murder and rape, linked to Internet dating sites and that, in almost all cases, Internet-enabled mobile phones were used to access the dating sites (*The Age* 2002). Offenders also use dating sites to seek out other similar minded individuals to validate their interests, and to gain access to more victims and child pornography.

CASE EXAMPLE (MARYLAND 1999):
Responding to complaints regarding a user "Michelle985," on Matchmaker.com who was soliciting people to have sex with "Michelle" and a female child, the Maryland State Police traded e-mails with "Michelle985" who requested in one of the e-mails, "send some pics to show you are not a cop." The police traced the "Michelle985" profile to Robert Wyatt in Abingdon, Maryland.

Digital Evidence and Computer Crime Second Edition
ISBN: 0-12-163104-4

On May 11, 1999, a search warrant was executed at Wyatt's home and the police seized Wyatt's computer. A subsequent forensic analysis of the computer revealed over 100 color still images and three movie files of explicit child pornography. Included among the images were several photos of a little girl who had been brutally raped by her father in Texas. To prove to the jury that the little girl depicted in the images was in fact a real minor child and not just a computer rendered image, the government called a Texas State Ranger who testified that in connection with an investigation he conducted in 2000, he had met and identified the little girl depicted in the images (USDOJ 2002a).

Even people who use the Internet for purposes other than meeting a partner unintentionally disclose personal information that a malicious individual can use against them. A simple Web page containing a woman's name, address, interests and photograph is all that is needed to target a victim. Sex offenders target children in online chat rooms that are supposedly devoted to youngsters. The Internet enables sexual offenders to commit a crime without ever physically assaulting a victim.

CASE EXAMPLE (BURNEY 1997):
A 47-year-old Ohio man posing as a 15-year-old communicated through computer messages with a 14-year-old girl and was able to convince her to send him sexually explicit photographs and videotapes of herself performing sexual acts. The cyber relationship went on for 18 months, since the girl was 12. The offender pled guilty to one charge of inducing a minor to produce child pornography.

Children are not the only victims of sexual assault involving the Internet. In England, Christopher Graham Elliott was sentenced to 7 years in prison for raping and inflicting actual bodily harm on a woman he met online (Pendlebury 2001). Another man who met female university students online, apparently through "collegeclub.com," fled after being arrested for sexually assaulting one woman. Although men commit the majority of sex offenses involving the Internet, women also exploit children they meet online. In 1997, a woman in South Portland admitted to having sex with a 14-year-old boy she met in an online chat room (States News Service 1998). Also in 1997, a 40-year-old woman met a 15-year-old boy from Minnesota on the Internet and lured him to North Carolina for sex (States News Service 1999). In 1998, a 30-year-old Pittsburgh woman arranged to meet and have sex with a 15-year-old boy she met in an Internet chat room (*The Baltimore Sun* 1998).

As detailed in Chapter 18, the Internet has sophisticated search tools and many newsgroups and chat rooms organized by topic, providing an abundance of hunting grounds. Once an offender has selected a target, he/she can monitor potential or existing victims on several levels, ranging from participating in a discussion forum and becoming familiar with the other participants, to searching the Internet for related information about an individual, to accessing a potential victim's personal computer to gain additional information. Furthermore, by giving offenders access to victims over an extended period of time (rather than just during a brief physical encounter) the Internet allows offenders to groom victims, developing sufficient trust to engage in cybersex or even meet in the physical world.

Another appealing feature of the Internet is the perceived anonymity and safety it provides, allowing offenders to alter or conceal their identity. Age, gender, and physical appearance are all malleable on the Internet enabling offenders to further their own fantasies and portray themselves in a way that will interest their chosen victim. Some offenders present themselves as young boys to make themselves less threatening to a child selected as a victim. Other offenders masquerade by providing a photograph of a more attractive male to draw potential female victims. The ability to conceal identifying information can also be used to avoid apprehension.

Another benefit of the Internet to the offender is the peer support it provides. Some groups of offenders use the Internet to communicate, exchange advice and sometimes trophies of their exploits. In Japan, Akihiko Kamimura was sentenced to 12 years in prison for using the Internet to recruit four other men to form a rape gang that sexually assaulted five women. Kamimura, who was confined to a wheelchair, received help from one of the other men in raping two of the five women (*Guardian Unlimited* 2001). In 2001, Joe Clemens admitted to soliciting people in a Yahoo! chat room to harm his wife. According to his message, Clemens wanted his wife "kidnapped, gang-raped, tortured, and humiliated" and that he was serious about this request and only wanted serious inquiries (Ananova 2001).

The impact of these peer support groups can be profound, "normalizing" abnormal desires, enabling offenders to view their behavior as socially acceptable and possibly lowering their inhibitions to act on impulses that would otherwise remain fantasy. Additionally, these types of support groups can give offenders access to child pornography, children, and technical knowledge that would otherwise be beyond their reach.

This chapter discusses related legal and corporate issues and provides insight into sex offenders on the Internet. An overview of investigating this type of crime is provided to help digital investigators and digital evidence examiners integrate the techniques presented throughout this book and apply them in their work. Generalizations regarding investigations are of limited use since each case is unique, requiring an individual approach and often presenting distinct challenges. The same behavior can mean different things in different cases – one offender might bring a victim to his home because he feels safer there than in a hotel room, whereas another offender might prefer a hotel room but cannot afford the expense. Conversely, one offender may bring victims to a hotel because they feel more anonymous and less exposed than they would in their home, whereas another offender may use a hotel because his spouse and children are at home.

Therefore, it is more useful to examine features of individual cases and attempt to draw useful lessons from them. A number of case examples are presented in this chapter in an effort to highlight important issues. Ultimately, investigators and examiners must depart from the finite knowledge in this book and creatively apply what they have learned to new situations in the cases they encounter. With this in mind, sections in earlier chapters are referenced to encourage the reader to revisit the concepts and envision how they can be applied to new cases.

20.1 WINDOW TO THE WORLD

As obvious as it may seem, it is important to stress that sexual abuse and illegal pornography existed long before the Internet. Joseph Henry's congressional testimony is a clear reminder of this fact and that networks of child abusers exist independent of the Internet. In his testimony, he describes his actions and how he established communication with other offenders (initially through a paper publication called Better Life) who gave him access to child victims.

> By the time I was 24, I had molested 14 young girls and had been arrested twice and sent to State [sic] hospitals, one for 18 months. I used all the normal techniques used by pedophiles. I bribed my victims; I pleaded with them, but I also showed them affection and attention they thought they were not getting anywhere else. Almost without exception, every child I molested was lonely and longing for attention. For example, I would take my victims to movies and to amusement parks. When I babysat them, I would let them stay up past their bedtime if they let me fondle them. One little 8-year-old girl I was babysitting came over to my house one day soaking wet from a rainstorm. I told her I'd pay her $1 if she would stay undressed for an hour. This incident opened the door for 3 years of molestation. I used these kinds of tricks on children all the time. Their desire to be loved, their trust of adults, their normal sexual playfulness and their inquisitive minds made them perfect victims. I never saw any outward emotional damage in one of my victims until 1971 when I was 36 and the manager of a nudist park in New Jersey. I was able to see many children nude and grew particularly attracted to a 9-year-old named Kathy. I once bought her five Christmas presents. She was the first little girl I ever forced myself upon and the first whose molestation was not premeditated. I actually saw the trauma and the terror on her face after I had molested her...
>
> Around 1974, when I was beginning to hang around 42nd Street porno shops in New York City, I got my first exposure to commercial child pornography. I got to be

> friends with one of the porn shop owners and one day he showed me a magazine that just arrived called Nudist Moppets. There were paperback books with stories of child sex, adult/child sex. The films in the peep shows were of men with girls, boys with girls and a few that looked like families together in sexual activity. Eventually, I put together a photographic collection of 500 pages of children in sexually explicit poses. Before long, films started coming in and I bought a film projector. I started reading some of the pornographic tabloids called Screw, Finger and Love, which were filled with all types of sex stories, ads and listings for pen pals. At least one of the issues was devoted to a pedophilic theme. In one issue of Finger, there was an ad about organizations that were devoted to sexual intimacy between children and adults. I wrote to three of them – Better Life, the Guyon Society and the Childhood Sensuality Circle. Better Life and the Childhood Sensuality Circle responded, so I sent in the membership fee to join them. (Henry 1985)

In a study of 49 child pornographers and 13 men convicted of traveling interstate to have sex with a minor (a.k.a. travelers) in federal prison, 76 percent of the subjects admitted to having committed contact sex offenses that were not detected by the criminal justice system (Hernandez 2000). According to the study, these offenders had molested a combined total of 1,433 victims without ever having been detected. The study also indicated that, "these offenders target children in Cyberspace in a similar manner as offenders who prey on children in their neighborhood or nearby park. They seek vulnerable children, gradually groom them, and eventually contact them to perpetrate sexual abuse." According to a 2001 survey "Reality of Female Victims of Violence in South Chungcheong Province," of the 50 sexual assault cases in South Korea that were reviewed, nine incidents involved victims raped by people they met on the Internet (Soh-jung 2001). Although the sample size in this survey was not large enough to draw firm conclusions, it is worth noting that the majority of the assaults did not involve the Internet and were committed by individuals who knew the victims (e.g. neighbor, co-worker, colleague, relative). The Internet is a window into such activities in the physical world, and although the Internet can facilitate these crimes and even cause some offenders to act out their fantasies, blaming the Internet will not address the root problems. On the contrary, restricting the Internet to hide these problems will eliminate a unique opportunity to observe and address these criminal activities.

> Grooming refers to the ways that a sexual offender gains control over victims, exploiting their weaknesses to gain trust or instill fear. Grooming usually involves exploiting a victim's needs such as loneliness, self-esteem, sexual curiosity/inexperience, or lack of money and taking advantage of this vulnerability to develop a bond. Offenders use this control or bond to sexually manipulate victims and discourage them from exposing the offender to authorities.

Whether sex offenders simply use cyberspace to lure victims into physical world meetings or make more use of this new venue to fulfill their needs, incriminating digital evidence is left behind. There is an over-abundance of cases demonstrating these two modes of operation: using the Internet to lure victims and using the Internet to further crimes committed in the physical world. Convicted killer John E. Robinson contacted some victims through the

Internet, sexually assaulting some and killing others (Rizzo 2001). In 2000, Lawrence Stackhouse found 15-year-old Diana Strickland's online profile, contacted her using the Internet, and then groomed her until she and a girlfriend agreed to travel to his home in Pennsylvania, where he exploited them sexually for 4 days until the girlfriend called the police (*Psychiatric News* 2000). A teacher named Frank Bauer used the Internet to trick a 15-year-old male student at the school into thinking Bauer was a woman, then blackmailing him into making a pornographic video of himself (Chen 2000).

In 2001, James Warren kidnapped a 15-year-old girl he befriended on the Internet and held her captive for a week in Long Island where he sexually abused her. In the same case, Beth Loschin pled guilty to sexual abuse and sodomy and another man, Michael Montez, pled guilty to raping the teen while she was in Long Island (Associated Press 2002). In 2000, Adam Valleau admitted to persuading a 12-year-old to engage in sexually explicit conduct for the purposes of producing a visual depiction of the conduct. Valleau photographed the minor with a digital camera as the child engaged in sexually explicit conduct at Valleau's direction. The defendant then posted those images of the child on the Internet where they could be viewed by others. In an interview with Baltimore County Police detectives in May of 2002, Valleau admitted to sexually abusing two boys (USDOJ 2002b). One of the largest child exploitation investigations to date began with two members of the Orchid Club who distributed digital recordings of their offenses to cohorts on the Internet.

CASE EXAMPLE (CALIFORNIA 1996):
A woman contacted the local police and reported that her 6-year-old daughter had been molested during a slumber party by Ronald Riva, the father of the host. Additionally, a 10-year-old girl at the party reported that Riva and his friend, Melton Myers, used a computer to record her as she posed for them. Riva and Myers led investigators into an international ring of child abusers and pornographers that convened in an Internet chat room called the Orchid Club. Sixteen men from Finland, Canada, Australia, and the United States were charged. One log of an Orchid Club chat session indicated that Riva and Myers were describing their actions to other members of the club as they abused the 10-year-old girl. Their investigation into the Orchid Club led law enforcement to a larger group of child pornographers and pedophiles called the Wonderland Club. After more than two years of following leads, police in 14 countries arrested over 200 members of Wonderland, in the largest coordinated effort to crack down on child exploitation and abuse to date. Evidence gathered during this latest effort suggests that there are members of the Wonderland Club in more than 40 countries, so the investigation is by no means over. (Shannon 1998)

By recording offenders' activities in more detail, computers and networks can provide a window into their world, giving us a clearer view of how sex offenders operate. For instance, when dealing with sex offenses that do not involve the Internet, it can be very difficult to determine if stalking occurred prior to the crime. In the United Kingdom, Patrick Green stalked a girl on the Internet for several months, obtained her private e-mail address, and lured her into a meeting after posing as a 15-year-old boy. Green met the girl in his car and took her to his flat where he began a series of indecent assaults (McAuliffe 2000). Other investigations have revealed grooming behavior of online sex offenders who target children, showing it is no different on the Internet than in person. Some offenders gain a victim's trust by alternately playing the role of seducer and caring parental figure, sending child pornography to break down sexual inhibitions, and giving gifts in exchange for sex. These insights into sex offender behavior have enabled investigators to find offenders on the Internet, locate other victims targeted by an offender, discover evidence that might otherwise have been overlooked, and warn parents of potential victims to be alert to unexpected packages and telephone calls for their children from adults.

20.2 LEGAL CONSIDERATIONS

The most commonly encountered sex offenses on the Internet include soliciting minors for sex, and making, possessing, or distributing child pornography. Although many sexual assaults do not involve computers directly, associated digital evidence increasingly exists. Proving the sexual assault of an adult rather than a child may be more difficult because of the possibility of consent. For instance, a man in Washington accused two other men he met on the Internet of holding him against his will and sexually assaulting him. However, prosecutors dismissed the charges after they examined the associated e-mail correspondence and determined that there were ample grounds to find that the men had made a consensual sex slave arrangement (Thomson 2002).

Investigating and prosecuting sexual assaults either facilitated or documented by computers or the Internet has myriad legal considerations. Whenever the Internet is involved, jurisdiction can be complicated as discussed in Chapter 3. In the United States, federal law enforcement has jurisdiction over most criminal activity facilitated by the Internet, even if both offender and victim are located in the same state. This is due to the Commerce Clause of the United States Constitution, which has been interpreted broadly to allow anything related to interstate commerce[1] to fall under federal jurisdiction.

[1] *For example, making a threat by e-mail.*

However, states have historically carried the burden of common law criminal law enforcement. While the interplay of federal and state jurisdiction has resulted in increased resource sharing and the creation of task forces (e.g. the Internet Crimes Against Children Program funded by the United States Office of Juvenile Justice and Delinquency Prevention), some in the legal community are concerned by the increasing federal involvement in the criminal law (Amar 2003).

A 2002 US Supreme Court decision has had major implications for computer assisted child exploitation cases. As noted in Chapter 3, Ashcroft v. Free Speech Coalition (US Supreme Court, 00-785, 198 F.3d 1083, 2002) held parts of the US child pornography law outlawing sexually explicit drawings and computer rendered images unconstitutional. In a constantly evolving line of cases, the federal and state courts seek to find the proper amount and strength of evidence to prove beyond a reasonable doubt that images alleged to be child pornography depict actual children engaged in sexually explicit conduct. To support this endeavor, the United States recently launched an effort to establish a library of child pornography images in which investigators have identified original sources and have identified the victims portrayed in the images. Also, the PROTECT Act of 2003 modified the federal pornography laws, and will likely become the subject of litigation over the course of the next few years (USDOJ 2003).

In some countries, such as England, laws relating to child pornography have exceptions such as a "legitimate reason for having the photograph or pseudo-photograph" (English Criminal Justice Act 1988). In the United States, the federal law and many state laws also contain exceptions for law enforcement and judicial uses of child pornography, and for the inadvertent possession when promptly reported to police. Without one of these exceptions, the stated intent of the defendant is usually irrelevant under federal law. For instance, Larry Matthews, a news reporter for National Public Radio (NPR), was convicted of possessing and distributing child pornography despite his claims that he was conducting research for a story about child pornography on the Internet (United States v. Matthews 2000a). The court held that his reason for sending and transmitting child pornography was irrelevant under the federal statute, "admission that he knew he was receiving and transmitting child pornography is all that was required." (United States v. Matthews 2000b.)

Knowing possession or importation, distribution or manufacture of child pornography is all that matters because child pornography is contraband, just as heroin is contraband. Imagine that a well-intentioned citizen seeks to bring heroin dealers to justice by going down to the local dealer and buying a large supply with the intent of destroying it or turning it over to the police. Such a person would be charged with possession of the drug if interdicted by law enforcement anywhere between obtaining the heroin and turning it in – the individual would have to hope that the authorities believed their defense for possession. In one case, an individual helped the FBI several times too many, leading them to believe that he was actually interested in the child pornography he was obtaining from the Internet (United States v. Hilton 1997).

Even those who evaluate sex offenders either for treatment or in the courts had to change how they conduct the evaluations. In the past, some evaluations used visual depictions of children as a stimulus to measure arousal via penile tumescence or visual reaction times. Some of these pictures or slides could arguably be considered child pornography and would place the evaluator at risk by possessing them. Obviously, these pictures are no longer used due to fear of prosecution, but the reason for possessing them was clearly antithetical to an offender's possession of child pornography.

CASE EXAMPLE (TEXAS 2002):
David Magargee pled guilty to possessing child pornography. At his sentencing hearing, Magargee told Judge Vela that his purpose in obtaining the child pornography photographs was to clothe the children and flood the Internet with angelic images, and his purpose for ordering the videos was to gather evidence for law enforcement. Judge Vela found that Magargee had previously admitted to knowingly possessing the child pornography, denied the defendant's motions seeking a lower sentence, and imposed the maximum 27 months term of imprisonment as recommended by the United States (USDOJ 2002c).

Computer security professionals in the private sector can also run foul of the law when dealing with child pornography on their systems. Even if the law does not require an organization to report child pornography found on their computer systems, a failure to do so can lead to criminal charges if the illegal materials are not properly disposed of. Furthermore, covering up such problems may be viewed as negligence if the illegal materials are symptomatic of a more serious crime such as sexual abuse.

Given the seriousness and sensitivity of these offenses, organizations should be prepared with policies and procedures for the inadvertent discovery of child pornography. Without this kind of preparation, individuals who report such crimes directly to law enforcement may find that they do not have the support of their employers and may need to find a new job, hire their own attorney, defend themselves against countersuits, and testify in their own time. Organizations that handle situations inappropriately also risk being sued by their employees.

CASE EXAMPLE (NEW YORK 2003):
After finding child pornography on Professor Edward Samuels's computers, two computer support technicians at the New York Law School reported the incident to their supervisors. An investigation ensued, Samuels was

arrested, and ultimately pled guilty to possession of child pornography. Shortly after the incident, the two technicians were fired and sued their employers for 15 million dollars (*New York Lawyer* 2003).

In the process of creating policies and procedures for dealing with the discovery of child pornography on their systems, organizations should establish contact with law enforcement agencies to clarify expectations: What are the relevant state laws? What response can the organization expect from law enforcement? What law enforcement needs from the organization to resolve the case? Additionally, these policies and procedures should be cross-checked with existing policies, such as those protecting employee privacy to avoid conflict and inadvertent violations.

20.3 IDENTIFYING AND PROCESSING DIGITAL EVIDENCE

As computers, digital cameras, and the Internet become more integrated into the average person's life, the role of the digital evidence examiners becomes clearly essential. In Europe, investigators are finding an increasing number of mobile phones with digital cameras being used to create and exchange child pornography. The increasing trend of mobile phones being involved in criminal activities is a clear demonstration of how pervasive digital evidence has become. Although digital evidence could be overlooked and mishandled in the past without serious repercussions, overlooking or mishandling this kind of evidence now may amount to malfeasance. It is essential for investigators to identify sources of evidence and process them methodically as detailed throughout this text. Failure to do so allows a defense attorney to attack a case on technical grounds, rather than the actual merits of the evidence itself.

The importance of crime scene protocols and evidence handling procedures in this type of investigation cannot be overstated. The basic precaution of wearing surgical gloves is often neglected, despite the fact that sex offenses often involve potentially infectious body fluids that pose a health risk to first responders and must be processed as evidence. First responders have reported that protective plastic covers they find on some offenders' computer keyboards smell of semen. Without adequate procedures, important digital evidence may be missed, particularly when dealing with offenders who have taken steps to conceal their activities. In several cases, an offender has made a telephone call while in custody to instruct someone to destroy digital evidence. In other cases, suspects have shot at investigators and/or killed themselves when a search warrant was being executed on their homes. Therefore, investigators must take precautions when serving warrants in computer-related offenses just as they would with any other crime.

The role of a computer in the sex offense investigation will determine the types of evidence that exist and where they are located. For instance, when an offender uses a computer to communicate with victims, the Information as Evidence category described in Chapter 2 is applicable and an associated Standard Operating Procedure (SOP) can be implemented to process digital evidence from computers and connected networks. For instance, when the home of alleged serial killer John Robinson was searched, five computers were collected as evidence (McClintock 2001). However, when a computer is used to manufacture and disseminate child pornography, the Hardware as Instrumentality, Information as Contraband, and Information as Evidence categories may all be applicable, making it necessary to search for and collect a larger range and amount of evidence, including digital cameras, scanners, removable media, hiding places, and online activities as depicted in Figure 20.1.

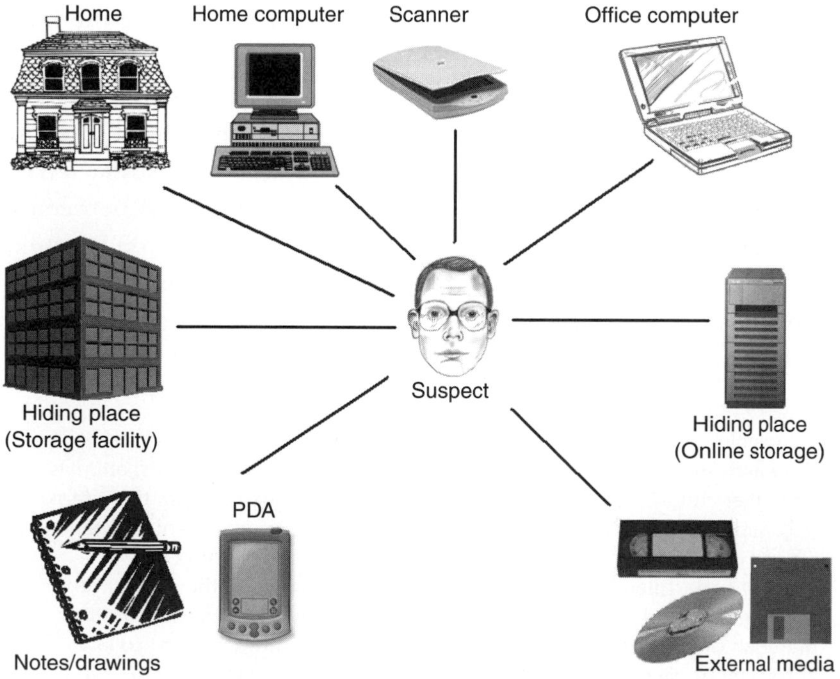

Figure 20.1

Possible sources of evidence in a sex offense investigation.

It can be a major undertaking to locate all computers and Internet accounts used by the victim or offender, involving extended searches (e.g. automobile, workplace, storage facilities, properties belonging to parents, and significant others of both victim and offender), interviews (e.g. suspect, victim, family, friends, and co-workers), and analysis of credit card bills, telephone records, and online activities. Also, a search warrant may be needed to obtain a victim's computers if consent is not forthcoming.

When dealing with online sexual offenders, it is particularly important to take advantage of the Internet as a source of evidence. An offender's online communications may reveal other offenders or victims. Logs from various systems on the Internet can provide a more complete picture of the offender's activities, sometimes leading to other sources of digital evidence such as a hidden laptop, computers at work, a public library terminal, or an Internet cafe. Therefore, investigators should call the victim and offender's Internet Service Providers immediately to explain the situation and should follow-up with a preservation letter detailing the information that is needed to ensure that information is not lost while a search warrant or other court order is obtained.

Searching the Internet for related information can also generate useful leads. Some sex offenders participate in special interest newsgroups (e.g. alt.sex.incest, alt.pedophilia, alt.support.boy-lovers), online discussion boards such as BoyLinks and GirlLove Garden, and organizations like the Danish Pedophile Association and North American Man/Boy Love Association (NAMBLA). Similar support groups exist on IRC (#fathersdaughtersex). Some offenders even participate in victim support groups such as alt.abuse.recovery because of the high concentration of victims of past abuse. It may even be possible to find online witnesses who observed interactions between the offender and victim in areas they frequented. Digital evidence on private networks can also help generate new leads, establish the Continuity of Offense, and corroborate other evidence.

CASE EXAMPLE (CONNECTICUT 1998):
Yale geology professor Anthony Lasaga admitted to possessing tens of thousands of images of children engaging in sexual acts with adults, animals, and other children. Many of these images were downloaded from the Internet (e.g. Supernews.com) onto a computer in the geology department and then viewed on Lasaga's desktop computer. A system administrator in the geology department came across the child pornography on the server in the course of his work. The system administrator observed Lasaga accessing the materials on the server from his desktop and reported the incident to law enforcement. Given the severity of the crime and the involvement of several systems, it was necessary to secure and search the entire geology building and network for related evidence.

Because of his success in attributing the illegal activities to Lasaga, the system administrator was accused by the defense of acting as an agent of law enforcement. Although the system administrator was ultimately exonerated of any wrongdoing, his employers did not provide legal support and he was compelled to hire an attorney to defend himself against the accusations. Notably, Lasaga also admitted to creating a videocassette of a young boy engaging in sexual acts. The tape involved a 13-year-old boy whom Lasaga met through a New Haven child-mentoring program. The tape was shot on the Yale campus, one in the professor's geology classroom and the other in the Saybrook master's house (Diskant 2002).

Log files and other remnants of a victim's network activities should also be examined. The importance of this information is most evident when offenders instruct victims to wipe their hard drive before coming to a meeting. In such cases, the Internet and telephone networks may be the only available source of digital evidence that can lead investigators to the offender and missing victim. However, even when useful digital evidence is found on the victim's computer, the Internet and other networks can provide corroborating evidence and may even help develop new leads.

One challenge occasionally arising during the investigation of a sex offense is that digital evidence was not preserved properly or at all. Victims sometimes destroy key evidence because they are embarrassed by it; corporate security professionals might copy data from important systems or logs ignorant of proper evidence handling concepts; or poorly trained police officers may overlook important items. A related problem is that supporting documentation may be inadequate for forensic purposes. In such situations, investigators and examiners should work together to determine if evidence was overlooked and gathering details about the context, origin, and chain of possession of the evidence. Without basic background details (e.g. where a computer came from, what was on it originally, how it was used, who used it, whether access to the computer was restricted, who had access to it), it may not be possible to authenticate digital evidence on the system.

A further challenge is that some online sexual offenders use various concealment techniques to make it more difficult for investigators to identify them and find evidence. Some offenders physically hide removable media and other incriminating evidence in their homes, at work, and rented storage space. For instance, when investigators searched the home of New York Law School professor Edward Samuels, they found evidence hidden in a crawl space in the ceiling. When Moscow police searched the apartment of notorious child pornographer, Vsevolod Solntsev-Elbe, they found innocuous looking, shrink-wrapped videos in boxes for National Geographic nature films, with pictures of rhinos, giraffes, and pandas on the covers. The beginning of each tape contained a clip from nature documentaries but the remainder of the tape contained child pornography (Reuters 2002).

Increasingly, online sex offenders are using encryption, steganography, and other methods of digitally concealing evidence. The following message from one offender who was not apprehended provides insight into the concealment techniques that criminals use on the Internet.

> I use a proxy but not an anon proxy: it works like this: I have an account in one jurisdiction but use their proxy in their branch office of another jurisdiction to connect with the main server. Of course my server logs my accesses as well as the servers I access logging the accessing server. But who is the person doing the accessing. Let's look through the millions of hits going through the main server of the big company I subscribe to and spend ages trying to link my account to the access which is made hugely difficult when a person accesses a foreign server. The law in which my account is based is different to the law where I reside using the proxy... Then having downloaded images of the seven wonders of the world, I back up to an external file, BC Wipe, Window Wash and Evidence Eliminate, activex, cookies and java disabled and Encase given a run to see if anything was left. (Anonymous)

Given the potential for concealment in this type of case, it is important to examine all digital evidence carefully rather than simply searching for obvious items such as images that are not hidden. The analysis guidelines in Chapter 24 provide a methodology for performing a thorough examination.

20.4 INVESTIGATING ONLINE SEXUAL OFFENDERS

In some cases, it is relatively straightforward to apprehend the offender and prove the crime, particularly when the offender does not conceal his activities because of weak technical skills or because he does not believe what he is doing is wrong.

> CASE EXAMPLE (WISCONSIN v. MICHAEL L. MORRIS 2002):
> Morris, a 44-year-old man, met the victim, a 14-year-old girl, in an Internet chat room. Morris admitted that, in pursuing what he considered a romantic and consensual relationship, he came to Wisconsin four times to see the girl, bought her an entire outfit (including underwear), gave her a ring, took her to restaurants, accepted collect calls from her and gave her telephone cards, took her to a hotel where he photographed her in the nude, and took her to his home in Indiana. He estimated that he had had contact with her on about 20 days during the 30 days preceding the date of the state-charged offense.

In another case, a 42-year-old man in San Diego used his AT&T dial-up account to post photographs on Usenet of himself having sex with his daughter. The FBI obtained his name and address from AT&T, compared his driver's license photo with the pictures posted on the Internet, and arrested him at his home (Associated Press 1998). Even when an offense can be established with relative ease, investigating online sexual offenders can be among the most difficult to deal with. These investigations are often emotionally stressful, particularly when dealing with young victims or severe sexual abuse. These investigations can also be technically challenging, particularly when the offender conceals or destroys digital evidence. An added challenge can arise when victims do not cooperate because they are in denial or actively protect the offender because of the relationship that has developed between them.

> Investigators and prosecutors must understand and learn to deal with the incomplete and contradictory statements of many seduced victims. The dynamics of their victimization must be considered. They are embarrassed and ashamed of their behaviors and rightfully believe that society will not understand their victimizations. Many adolescent victims are most concerned about the responses of their peers. Investigators must be especially careful in computer cases where easily recovered chat logs, records of communication, and visual images may directly contradict the socially acceptable version of events that the victims give. (Lanning 2001)

Failure to handle victims appropriately can make them less willing to assist in an investigation, making it more difficult to build a case. Additionally, attempts to force the victim to cooperate by confronting them with evidence of their abuse further victimizes them.

In light of the technical complexities and emotional pressures in this type of case, investigators and examiners have to be particularly wary of developing preconceived theories. Carefully implementing the investigative process detailed in Chapter 4 will help investigators and examiners consider possible explanations for a given piece of evidence and will discourage them from jumping to a conclusion based on personal bias or past experience. For instance, digital evidence on the suspect's computer might suggest that he was accessing Internet resources intended for teenagers when it was, in fact, the suspect's young daughter using her father's computer and Internet account. Similarly, the presence of pornographic material on a computer might suggest that the suspect downloaded the materials when, in fact, a computer intruder broke in and placed the images on the computer.

In these early days of digital evidence and digital investigation there are many mistakes to be made. The most effective approach to minimizing errors is to acknowledge gaps in one's knowledge, to consult peers for assistance, and to perform research and receive training when time allows. Also, good investigators and examiners have the mental discipline to question assumptions, objectively consider possibilities, account for evidence dynamics, and ultimately clarify what the evidence does and does not tell us.

The initial stage of any investigation is to determine if a crime has actually occurred. Even if investigators are convinced that the defendant committed a crime, it can be difficult to prove. For instance, unless there is digital evidence establishing the continuity of offense, it can be difficult to show that a suspect disseminated child pornography to others via the Internet. For instance, Bart Henriques was sentenced to 42 months in prison for possession of child pornography in violation of 18 U.S.C. §2252A(a)(5)(B) but his conviction was overturned on appeal because there was insufficient evidence to support a finding that the images were transported in interstate commerce (United States v. Henriques 1999). In some cases, it can even be a challenge to demonstrate that an individual knowingly possessed child pornography.

CASE EXAMPLE (CONNECTICUT 2003):

A man was suspected of stalking a 14-year-old girl. When police executed a search warrant at his home, they found a computer. Initially, investigators submitted the computer for examination to determine if the suspect had any digital pictures or maintained diaries or logs of his activities related to the girl. Digital evidence examiners found 30 pictures in unallocated space that appeared to meet the jurisdiction's statutory definition of child pornography. The prosecution decided to charge the man with possession of child pornography to spare the 14-year-old victim the trauma of testifying. She was afraid of him, and the prosecution wanted to shield her from having to see him again.

Shortly before the trial, the digital evidence examiner received a request from the prosecutor to identify the children portrayed in the images. Children in child pornography images may be identified through the National Center for Missing and Exploited Children in the United States and using Europol's Excalibur image database. Although only four of the images extracted from unallocated space on the defendant's computer depicted identified minors, the prosecutor decided to pursue the possession of child pornography charges. However, the prosecution only called on the digital evidence examiner to testify as to the content and character of the 30 recovered images. The examiner stated that he was not qualified to determine the ages of the unidentified children or whether the images depicted actual children or were computer rendered. After extensive *voir dire* of the examiner, the prosecution conceded that the state was unable to prove all of the elements of the crime of possessing child pornography beyond a reasonable doubt.

Notably, the defendant maintained that he never intended to possess the images. He claimed that he had been "mouse-trapped" – referring to the phenomena of clicking on a link and being taken from one website or advertisement to another and another, opening up so many Web pages that it may be necessary to shut the system down to end it swiftly. In this case, if the prosecution had survived the initial motion to dismiss, the defendant very well may have prevailed because, given the scant number of images and their location on the hard drive – in unallocated space – it is at least plausible, if not enough to raise a reasonable doubt, that the defendant did not knowingly possess the child pornography.

In addition to establishing that a crime was committed, establishing continuity of offense, and overcoming preconceived theories, digital evidence examiners must objectively and carefully analyze evidence and present findings to decision makers. Any inaccuracies in their findings can have a negative impact on a case and must not overstate findings or suggest guilt of a particular individual. For instance, a digital evidence examiner may be able to demonstrate that a series of photographs on a suspect's computer are consistent with a specific digital camera found in the suspect's home. However, a digital evidence examiner is rarely qualified to assert that such images show the suspect raping the victim. Similarly, a digital evidence examiner may be able to differentiate between real child pornography and virtual child pornography, but is rarely qualified to determine the age of a child in such an image.

Interpreting digital evidence in an objective manner can require great effort, particularly when there is a strong desire to attribute activities on a computer or network to a specific individual. For instance, in one case it might be tempting to assert that, "On July 23 between 17:14 and 18:23, the suspect was connected to IRC from his home computer and was communicating with the victim." However, if the suspect's computer does not contain corroborating data, his Internet Service Provider (ISP) does not retain Automatic Number Identification (ANI) information, and his telephone records do not show a call to his ISP at the time, it is difficult to establish continuity of offense and the digital evidence may only support a weaker assertion such as, "On July 23 between 17:14 and 18:23, an individual using the nickname 'Daddybear23' was connected to IRC via the suspect's Internet dial-up account and was communicating with the victim."

Even if an abundance of corroborating digital evidence exists, the following interpretation may be more accurate and compelling. "The combination of IRC chat logs found on the suspect's computer (Exhibits #232 and #233, C1 on the Certainty Scale in Chapter 7), ANI records obtained from the suspect's ISP (Exhibit #532, C-value C4), and telephone records obtained from the suspect's telephone provider (Exhibit #662, C-value C4) together indicate that, on July 23 between 17:14 and 20:23, an individual in the suspect's home was connected to the Internet using the suspect's dial-up account, was connected to IRC using the nickname 'Daddybear23', and was communicating with the victim. Notably, in Exhibit #233 the person using screen name 'Daddybear23' identifies himself as John Smith and provides his address and telephone number."

Even apparently minor details can make a major difference in the interpretation of digital evidence. Overlooking a well-known vulnerability in a Web browser has led to the false conclusion that a given individual intentionally downloaded pornographic files and added bookmarks when, in fact, they were created by a malicious Web site. Misinterpreting the date–time stamps in a Web browser's history database as coming from the system clock on the Web server rather than that of the client has caused questionable Web browsing activities to be attributed to the wrong computer user. The Candyman case provides a stark example of the consequences of misinterpreting digital evidence in this type of investigation.

CASE EXAMPLE (UNITED STATES v. PEREZ 2003):
Thousands of individuals have been accused of receiving child pornography through a Yahoo E-group named Candyman (created by Mark Bates) that was operational from December 6, 2000 to February 6, 2001. Like all Yahoo groups, the Candyman site included: a "Files" section, which provided a means for members to post images or video files of child pornography for others to download; a "Polls" section, which facilitated surveying among group members concerning child exploitation; a "Links" section, which allowed users to post the URLs for other websites containing child pornography and child erotica; and a "Chat" section, which allowed members to engage in real-time Internet conversations among themselves. However, in obtaining search warrants investigators incorrectly asserted that every e-mail sent to the group was automatically distributed to every member of the group. In actuality, members could choose not to receive e-mail sent to the group. As a result of this misunderstanding of how Yahoo e-groups function, it is not clear how many of the 7,000 unique e-mail addresses actually received child pornography, and many search warrants that were issued based on this assertion are being challenged. In United States v. Harvey Perez, for instance, the court held that the FBI acted recklessly when drafting the search warrant affidavit.

If an offender's computer reveals a large number of online contacts, some agencies send a letter to each individual to determine their involvement. A simple form letter summarizing the investigation and listing the suspect's online nicknames and e-mail addresses can encourage other victims to come forward or alert parents of a potential problem. However, some parents may not be aware of a problem so, in the letter, it is advisable to ask if they have children, how old they are, and what online nicknames they use.

A digital evidence examiner who carefully applies the scientific method as described in Chapter 4 is less likely to overlook or misinterpret important details. By actively seeking ways to disprove one's own theory (a practice known as falsification), one has a greater chance of developing a factual reconstruction of what occurred. The role of a forensic computer examiner is to objectively and thoroughly examine all available digital evidence, identify details that may be relevant, and present the findings objectively, without overstating their significance. The forensic examiner's role is not as an advocate for one side in a case, regardless of how convinced the examiner may be of a suspect's guilt or innocence. The evidence should speak for itself – personal or moral agendas have no place in the performance of the objective examiner's duties. It is up to the judicial system, not the forensic examiner, to weigh the evidence and come to a determination of an individual's guilt or innocence. The digital evidence examiner must be cognizant of the fact that justice and legal truth do not always coincide with scientific truth.

20.4.1 *UNDERCOVER INVESTIGATION*

In some cases, particularly when dealing with concealment behavior, it may be necessary to communicate with an offender on the Internet to attribute a crime. This course of action is only recommended for law enforcement personnel with explicit authorization and backing from their agency. In some instances, private citizens have taken the law into their own hands, posed as children on the Internet, and made contact with possible offenders.

CASE EXAMPLE (WISCONSIN v. KOENCK 2001):
Police received information relating to Koenck from an individual identified as Nancy A.C., a 46-year-old private citizen who co-founded a group called Internetwatch that monitors the Internet, mostly for child pornography. Nancy had communicated with Koenck (nicknamed **dirtboy69**) through an online profile named **teddie_bear_11** of fictitious 12-year-old twin girls named Teddie and Georgie. In some of the communications, Koenck expressed an interest in having sexual intercourse and contact with the twins. Nancy informed Koenck that Teddie and Georgie would be visiting a relative in Wisconsin and Koenck decided to travel from Iowa to meet them. Members of the Division of Criminal Investigation who had taken over the case arranged to meet Koenck at a McDonald's restaurant where he was arrested. Koenck admitted that he had traveled from Iowa to Wisconsin to meet and have sex with Teddie and Georgie, whom he believed to be 12 years old and whom he had met on the Internet. Koenck was convicted of child enticement.

While it can be successful in identifying and apprehending criminals, this practice of private citizens luring offenders is not recommended for a number of reasons. First, it puts private citizens at risk – the offender may target them in retaliation. Second, private citizens may inadvertently violate the law. When the subject of the investigation is child pornography, seeking it out and possessing it as part of a vigilante action can lead not only to the arrest of the offender, but also the well-intentioned citizen, regardless of proffered intent. Also, the defense will likely attempt to portray the vigilante as an agent of law enforcement and retrospectively assign law enforcement standards (entrapment, warrants, etc.) to the "investigation" of the private citizen.

Recall that the federal law and many state laws contain exceptions for law enforcement and judicial uses of child pornography, and for the inadvertent possession when promptly reported to police. However, purposely seeking out the contraband without the blessing of law enforcement, and acting as its unsanctioned agent, invokes the criminal law and meets most statutory definitions of possession of child pornography. This is so because most laws require only that possession is "knowing." Well-meaning citizens searching the Internet for child pornography so that they may report it to police know the content and the character of the material and, when they successfully find it, they possess it, as it will be copied to their RAM and/or hard drive by the very act of viewing it onscreen.

Given the difficulty in distinguishing between an overly zealous, helpful citizen and someone trying retrospectively to justify their interest in children, courts generally err on the side of caution in child exploitation cases. For instance, world-famous rock star, Pete Townsend, was arrested for possessing child pornography as a result of the Avalanche Operation in late 2002. Townsend claimed that he only accessed the child pornography Web sites three or four times, and then only for research he was conducting in order to combat the crime. Eventually, he was cleared of possessing child pornography, but his name will still be listed in Britain's sex offender registry for the next 15 years (O'Hanlon 2003). Even police officers have been convicted of possessing child pornography despite their claims that they were conducting undercover investigations on their own time.

Third, private citizens rarely have the training and experience necessary to conduct a successful undercover investigation, including the collection of digital evidence. The complexity and controversy surrounding child pornography cases even makes it difficult for law enforcement to build a solid case, let alone technically uninformed citizens. Mistakes by overzealous private citizens can make matters worse as demonstrated in the case against Superior Court Judge Ronald C. Kline.

CASE EXAMPLE (CANADA 2001):
Canadian vigilante, Bradley Willman, sent a Trojan horse program to a California judge Ronald Kline, gained unauthorized access to his computer, and found a diary detailing his sexual fantasies involving children and about 100 images alleged to be child pornography. Although the defense initially suggested that the evidence may have been planted by the intruder, Kline later admitted downloading pornography from the Internet, stating "There may be a picture or two on that computer that's illegal. ... It's not because I meant to keep it." At the time of this writing, the case remains in litigation. The defense argued that all evidence obtained by Willman should be suppressed because his actions were criminal and that he was acting as an agent of law enforcement when he broke into Kline's computer. Prosecutors denied that Willman was acting as a police agent, but was a cooperative suspect in the case and noted that he was a "potential suspect" in at least three US Customs Service investigations of child pornographers. However, a Federal judge ruled that Willman was acting as a police informant, which could taint all of the evidence he obtained from Kline's computer. The outcome of this case will have implications for both vigilante citizens and law enforcement dealing with online informants (Associated Press 2003).

Prior to conducting an undercover investigation, investigators must take steps to protect their identity as discussed in the anonymity section of Chapter 19. Furthermore, investigators should use specially designated computers to conduct undercover investigations to avoid commingling of evidence and possible allegations of personal pedophilic interests. As an example, suppose you encounter a potential target while you are online at home. Sitting in your living room, while your spouse and three children watch the television and talk on the telephone, you engage the target in chat. You use solid documentation principles, logging your chat, printing it out, and dating and signing the page. The next day, you chat with the target online from work, using your home screen name. The target sends child pornography to your screen name during the course of your online relationship and asks you to meet him for sex. You agree and instead of the 13-year-old he thought would be greeting him, you and several of your colleagues arrest him for transmitting child pornography and for attempting to entice a minor into sexual activity. When it comes time for discovery, things can begin to become uncomfortable. The defense requests your Internet account transactions and content of e-mails. They also request an independent examination of your personal home computer's hard drive. The defense puts your spouse and children on the witness list because they were present when you were corresponding with the defendant online. To make matters even more uncomfortable, the defense attorney advances the argument that you turned an online chat from your home into a law enforcement sting only because you feared that you had been caught engaged in illicit online behavior and used the law enforcement angle as a means of avoiding prosecution yourself.

One final caveat regarding undercover investigations is that using a minor, particularly the victim, is an unsafe practice.

CASE EXAMPLE (FLORIDA 1999):
University of Central Florida professor Madjid Adam Belkerdid was arrested and charged with sexual battery and a lewd and lascivious act on a child. Investigators used a 12-year-old girl Belkerdid met on the Internet to arrange a meeting with the suspect. Despite the presumed safeguards taken to ensure the girl's safety, Belkerdid allegedly touched the child's breast during the meeting (Associated Press 1999).

The two primary forms of accepted undercover investigation are: (1) investigators posing as a fictitious potential victim, and (2) investigators taking on the identity of a victim who has already been contacted.

CASE EXAMPLE (Investigators posing as a fictitious victim, Wisconsin v. Kenney 2002):
On September 23, 1999, Eric Szatkowski, a special agent with the Wisconsin Department of Justice, posed as a 13-year-old boy from Milwaukee named Alex. Szatkowski logged into an America OnLine chat room on the Internet and engaged in an online conversation with Kenney. The two discussed erotic wrestling, and

Kenney explained that he paid men $100 per hour for such activity. At Kenney's suggestion, they agreed to meet at a Denny's restaurant near the Milwaukee airport before going to a hotel to engage in erotic wrestling. Kenney indicated that after they met, either one could "call it off." Kenney then packed his wrestling bag and drove from his home in Chicago to Milwaukee. When he arrived at Denny's, he was arrested and was subsequently convicted of child enticement.

CASE EXAMPLE (Investigators taking on the identity of a victim who has already been contacted):
Police in Michigan received a complaint that a 48-year-old man had traveled from Connecticut to Michigan where he sexually assaulted a 13-year-old girl he met on the Internet. Police were informed of the crime after the man returned to Connecticut. When they had completed their investigation and were ready to have the suspect arrested in Connecticut, an undercover investigator in Michigan obtained the family's permission to assume the victim's online identity and communicate with the suspect. While the undercover investigator posed as the child victim and engaged the suspect in a conversation online, the Connecticut State Police went to his home and arrested him. Much to their surprise, not only was the suspect caught chatting with the undercover investigator, he was caught with his pants down – literally.

Notably, the latter approach is only used when investigators are informed after it is too late to prevent the victim's exposure to the offender. Children are not used in undercover investigations because of concern for their welfare.

Some offenders have attempted to defend themselves by claiming that they knew the person they were communicating with was not a child and that they were role-playing with an adult. For instance, Patrick Naughton contended that he believed the "girl" (actually an FBI agent posing as a 13-year-old girl using the nickname "KrisLA") he met in a chat room called "dads&daughterssex" was really an adult woman, and that they were playing out a sexual fantasy. Ultimately, Naughton pled guilty to one count of interstate travel with intent to have sex with a minor (USDOJ 2000). In 2003, John J. Sorabella III, 51, of Massachusetts was convicted of attempting to set up a sex rendezvous with a New Britain officer posing as a 13-year-old girl (Connecticut v. Sorabella 2003). He claimed that he knew all along that he was corresponding with an adult, that such talk among adults is common, and is all part of a fantasy. Despite his defense, he was convicted of most of the charges, which included attempted second-degree sexual assault, attempted illegal sexual contact, attempting to entice a minor, attempted risk of injury, attempted obscenity to a minor, obscenity and the import of child pornography.

Other offenders have attempted to defend themselves by arguing that this form of enforcement violates the First Amendment. However, in Wisconsin v. Robins, the court held that the First Amendment is not involved, because the child enticement statute regulates conduct rather than speech or expression (Wisconsin v. Brian D. Robins 2002).

The process of preparing for and conducting an undercover investigation is very involved, requiring specialized training and tools. In spite of this, it is possible for an experienced undercover investigator to pose as a potential victim while avoiding the pitfalls of entrapment, demonstrate that the suspect is predisposed to committing a certain crime, and persuade the suspect to reveal his/her identity online or arrange a meeting, without raising the suspect's suspicions, while abiding within the law and maintaining complete documentation throughout.

20.5 INVESTIGATIVE RECONSTRUCTION

Certain aspects of investigative reconstruction described in Chapter 5, such as equivocal forensic analysis, emerge naturally from a thorough investigation. Also, when investigators are collecting evidence at a crime scene, they perform some basic reconstruction of events to develop leads and determine where additional sources of evidence can be found. Once confident they have enough evidence to start building a solid case, a more complete reconstruction should be developed.

Although a complete investigative reconstruction can benefit any case, it is a time consuming process and the cost may not be warranted for simpler crimes. In more complex cases it may be desirable to perform an investigative reconstruction, even when the offender is known. The process of examining evidence more closely through temporal, relational, and functional analysis, may lead to concealed evidence, aid in linking related crimes, and help improve understanding of the crime and offender fantasy, motives, and state of mind, which are potentially useful in interviews and court.

Being able to assert that a specific offender probably retained incriminating evidence of crimes occurring years in the past can help dispel "staleness" arguments against search warrants. Also, knowing that such evidence likely exists motivates investigators and digital evidence examiners to search until they find it, seeking out hiding places that they might otherwise have overlooked. Similarly, knowing that it is very likely the current victim is not the first to be targeted by a sex offender motivates investigators and digital evidence examiners to seek evidence relating to other victims. It can be even more useful if investigators know what types of victims to look for and where the offender might have come into contact with them. It can also be helpful to know that certain sex offenders will confess to their crimes when treated in a certain manner, but the same approach may drive others into deeper denial.

CASE EXAMPLE (ADAPTED FROM CASEY 1999):
A former bus driver, age 31, was arrested and accused of possessing sexually explicit photos of himself with a child. His computer and many photographs, magazines, and videos were seized. Searching the Internet showed that his nickname and e-mail user name was Zest. His Web page was simple but telling, depicting him topless in Arizona at age 23. He had also posted several messages to newsgroups offering to scan risqué photographs. The following is a representative post:

Subject: *a_Will scan your Pic's
From: Zest zest@oneworld.owt.com
Date: 1996/09/29
Message-ID: 324F1FAF.33B@oneworld.owt.com

Newsgroups: alt.sex.pedophilia.boys, alt.sex.pedophilia.girls, alt.sex.pedophilia.pictures, alt.sex.pedophilia.swaps, alt.sex.phone, alt.sex.pictures, alt.sex.pictures.d, alt.sex.pictures.female, alt.sex.pictures.male, alt.sex.pictures.misc, alt.sex.plushies,alt.sex.pre-teens,alt.sex.prevost, alt.sex.prevost.derbecker, alt.sex.prevost-derbecker,alt.sex.prom, alt.sex.prostitution, alt.sex.raj.NOT, alt.sex.reptiles, alt.sex.safe, alt.sex.senator-exon, alt.sex.services, alt.sex.sgml, alt.sex.sheep, alt.sex.sheep.baaa.baaa.baaa.moo, alt.sex.skydiving.bondage, alt.sex.sm

Need a picture scanned? Nobady local to do it? Or the content is riska? I'm your man, I will scan your pic's for you. One dollar ($1) per pic, and postage, is all that it will cost you. I will save in formats of your choice. GIF, BMP, PCX, TGA, JPG, and TIFF. I'm also able to enlarge your original photo. All pictures are kept confidential.

Interested??? E-Mail me at the address below. I will reply to setup arranangements with you.

zest@owt.com

An analysis of the suspect's physical world activities, Web page, and Usenet posts reveals a few insights into the offender. The Web page is a lure. The page depicts the defendant half-naked and smiling in a remote outdoor location almost 10 years ago. This suggests his youthful view of himself, as does his nickname of choice, "Zest." The Web page also refers to time spent in Arizona. The defendant is a traveler. Look into trips he took by examining food, lodging and gas receipts, credit card records, and ATM transactions. Obtain all of his telephone records to find out who he was calling and when. Inquire into his time in Arizona to determine if other victims exist. Check his criminal history there, and anywhere else that he lived. Also check his parent's house/properties for hidden items, and look for any storage facilities that the defendant might have used. In addition to the recommendations above, check for secret compartments under his house, under his parent's house, in attics and look for digital storage devices, negatives, proof sheets, and videos. Check local rental places and see if he rents video equipment.

That "Zest" is a bus driver is no coincidence. Being a bus driver gives him access to a victim population beyond his immediate neighborhood/community. The defendant needs his real life victims to be within traveling distance so that he can develop personal, emotional relationships with them. For him, the Internet appears to be a tool not necessarily for acquiring victims, but rather for soliciting sexually explicit images from private parties. He needs this material to feed his pedophilic fantasies. And part of the attraction is it is not commercially produced, posed material, but privately created. Online he specifically asks for risqué materials and suggestively illegal material in his advertisements to Usenet. These would have been sent directly to his account, as advertised, and his Internet Service Provider (ISP) would be the best source of information regarding messages he received, possibly leading to other individuals who exchange child pornography on the Internet. His incoming and outgoing e-mail might be stored on his ISP's system or even archived by his ISP.

The Internet is only his means for fantasy development, enhancement, and transitory sustenance. The newsgroups he used tell us about the type of pornography he was hoping to acquire from private parties. These posts would not encourage potential victims to inquire, but rather are meant to get people to send him their private illegal pedophilic photos. In as much as this is true, this would also help him network with others who share his interests. There might have been a profit motivation at work here so that he could get into a network where he could sell the pornography that he created, and/or fantasy motive that would allow him to "trade in" to a group of pedophiles who share pornography.

It can be surmised the defendant believes he is genuinely in consensual relationships with these children. He sees nothing about his behavior as criminal or exploitative. He believes that he is merely seizing the day, and that what he is doing benefits the children he exploits. His motive is not to physically harm his victims but to be loved and admired by them. He confuses his own identity with theirs, to an extent, projecting a child-like affect to them. An interview strategy exploiting these factors by appearing to be sympathetic to these factors will offer the best chance of getting information from this suspect.

When the offender is unknown, the reconstruction process becomes a necessary step to help focus the investigation and prioritize suspects. The offender may not be known if the victim met him online prior to the assault and does not know his real identity, or the victim may be missing after traveling to meet the offender. Analyzing online messages from the offender may expose characteristics such as marital status, geographic location, profession, self-image, interests, age, and more. The improved understanding of the crime and offender that results from a thorough investigative reconstruction can have many ancillary benefits. In addition to those mentioned in the previous paragraph, detailed knowledge of an offender can help investigators anticipate future actions, assess the potential for escalation, protect past victims, warn potential victims, and communicate with the offender.

For example, based on a full reconstruction, it may be possible to inform undercover Internet investigators that the offender trawls specific IRC chat rooms for victims who feed into his torture fantasies. This direction not only tells investigators where to look but also enables them to pose as the type of victim that will attract the offender. Also, explaining how and why the offender conceals his identity may lead investigators to identifying information that the offender failed to hide or may help investigators narrow the suspect pool (e.g. to people who were intimately familiar with the victim and concealed their identity to avoid recognition by the victim). Additionally, providing information about an offender's method of approach, attack, or control may help investigators interact with an offender or provide potential victims with protective advice.

20.5.1 ANALYZING SEX OFFENDERS

To gain a better understanding of how offenders operate in general, it can be useful to look for trends in past investigations to discern similarities between different offenders. Lanning (2001) uses this approach to identify three general categories of sex offenders: situational, preferential, and miscellaneous. Within each category, Lanning identifies common characteristics such as preferential sex offenders' compulsive record keeping, a behavior that can provide a wide range of incriminating evidence including self-created pornography, information about victims, and other items that the offender can use to recall the pleasure they derived from the events. Lanning also notes that preferential sex offenders generally target victims of a particular kind (e.g. children) compared with situational sex offenders who are generally more power/anger motivated and generally pick convenient targets (e.g. their own children or children living with them).

Another approach to analyzing a crime and the associated behaviors is to look at available evidence from the crime under investigation and look for patterns that reveal something about the offender. For instance, objects in the background of self-created pornography can reveal where the perpetrator committed the offense. As the primary crime scene, this location probably contains a significant amount of evidence. Alternatively, an offender's Internet communications, credit card bills, and telephone records can lead investigators to victims, places where evidence is hidden, and locations where the offender arranged to meet victims. Also, patterns in an offender's online activities can be used to link related crimes and gain insight into the offender's fantasies and motivations. Furthermore, an analysis of behavior may show an escalation in the offender's aggression, indicating that current and future victims are at greater risk of harm.

Both methods have advantages and limitations. Although generalizations about sex offenders can help us identify patterns of behavior in a given case, they can be incorrect or even misleading. To compound this problem, offenders can learn and change over time, modifying their behavior proactively and reactively as discussed in Chapter 6. Therefore, it is most effective to use a thoughtful combination of the two methods. In fact, it is very difficult to use one approach without the other. Without a close examination of available evidence it is not possible to make a competent determination which general category the offender most likely fits. Similarly, without a general understanding of offenders and their motives, it can be difficult to recognize and interpret evidence that reveals important behavior.

20.5.2 ANALYZING VICTIM BEHAVIOR

Investigators often overlook the value of scrutinizing the behavior of victims of a crime. Victimology can help determine how and why an offender selected a specific victim and may reveal a link of some kind between the victim and offender, as well as other victims. These links may be geographical, work related, schedule oriented, school related, hobby related, or they may even be family connections. Learning that a victim's online activities increased her exposure to attack can lead investigators to new avenues of inquiry. For instance, Internet activities of a seemingly naïve victim may show that she used the Internet to obtain drugs, meet men for sex, or was involved with bondage and sadomasochism (BDSM) online groups, both of which can increase the victim's lifestyle risk. Additionally, victimology may reveal that the offender was willing to take significant risks to acquire that victim, providing insight into the offender's needs and possibly indicating a relationship between the victim and offender.

Furthermore, if we can understand how and why an offender has selected their previous victims by studying the complete victimology, as it changes or fails to change over time and throughout incidents, then we have a better chance of predicting the type of victim that they may select in the future. This knowledge can help direct an investigation and protect potential victims. Even if we come to understand that an offender's victim selection process is random, or even more likely, opportunistic, it is still a very significant conclusion.

Investigators can use digital evidence to gain a better understanding of the victim by determining if the victim uses e-mail, has Web pages, posts to Usenet regularly, uses chat networks, sends/receives e-mail or text messages on a mobile phone, and so on. For instance, in past cases, child victims have come into contact with adult offenders in the following ways:

- provided factual information in an online profile that attracted offender's interest;
- provided a name, photograph, home address, and telephone number on a Web page that attracted offender's interest;
- participated in online discussions dedicated to sex among teens (e.g. "alt.sex.teens" on Usenet);
- participated in online discussions devoted to the topic of sadomasochism (e.g. "#bdsm" on IRC);

- used online dating services;
- introduced through friends and acquaintances, both online and in the physical world;
- exposed through organizations in the physical world (e.g. schools, camps, big brother programs);
- through chance encounters in public (e.g. parks, swimming pools).

This is not an exhaustive list, but it gives a sense of what investigators might consider in developing victimology. Keep in mind that victims are often very secretive about their online sexual activities and significant effort (and delicacy) may be required to learn about some of these activities. Some victims even take steps to conceal their online activities prior to an offense or afterwards to avoid embarrassment, making it difficult for digital evidence examiners to develop a full victimology.

20.5.3 CRIME SCENE CHARACTERISTICS

Aspects of the crime scene other than the evidence it contains can tell us something about the offender. The choice of location, tools, and actions taken combine to make up an offender's *modus operandi* and can reveal an offender's motivations, sometimes in the form of signature behaviors. Even the decision to use the Internet can reveal something about the offender. A sex offender may have exhausted the local supply of victims and views the Internet as just another source of victims, in which case there is probably evidence of other sexual assaults in his local area. An offender may be under close observation in the physical world and uses the Internet as an alterative means of accessing victims (e.g. a convict or parolee). Alternatively, an offender may be afraid to target victims in the local vicinity because of the presence of family members at home.

Sex offenders generally have a reason for selecting specific places, tools, and methods to acquire victims, hide or dispose of evidence, and commit a sex offense. Some offenders choose particular online tools and locations because they will conceal his/her activities (e.g. using an anonymous service to access an online chat room that does not retain logs of conversations). The same applies in the physical world – an offender might choose a particular location to commit a sex offense because evidence will be destroyed or be harder to find and collect (e.g. in a forest or underwater). These choices can reveal useful offender characteristics, such as skill level and knowledge of the area in question. For example, use of a private peer-to-peer file sharing ring versus a public Web site like Yahoo to share child pornography indicates that the offender has more than a casual connection with online child pornography, since fewer people are familiar with these private file sharing rings than Yahoo, and that he has sufficient interest and technical skill to go beyond the Web browser and use the peer-to-peer file sharing software. Similarly, use of IRC versus AOL chat rooms to acquire victims may reflect skill level of the offender and the desired victim.

Adult offenders seeking adult victims may join an online dating service, go to chat rooms that the sought-after person will be in (e.g. "M4M," "40somethingsingles"), or respond to online personal advertisements. Offenders who prefer to victimize children will use Internet facilities most likely to be frequented by younger people. One offender might choose IRC to target teenage boys because it is more often used by the technologically savvy than casual users – very young users on IRC are more likely to be supervised. Younger victims (under the age of 13) would more likely be found in chat rooms and playing online games.

Items that an offender brings to a victim encounter such as cameras, condoms, lubricant, restraints, or drugs can be evidence of intent.

CASE EXAMPLE (ARIZONA v. BASS 2001):
Jerry Donald Bass was arrested after engaging in sexually explicit Internet communications and arranging to meet with Tucson Police Department Detective Uhall, who portrayed himself as a 13-year-old Tucson girl named "Keri." After Bass was arrested, police found condoms, baby oil, and a Polaroid camera in his truck. During the trial, Detective Uhall testified that, based on his experience, it is common for adult males who are sexually interested in young females to have such items in their possession. Bass argued that such testimony would constitute inadmissible "profile" evidence. Profile evidence cannot be used in Arizona to indicate guilt because it "creates too high a risk that a defendant will be convicted not for what he did but for what others are doing." (State v. Lee, 191 Ariz. 542, 959 P.2d 799 (1998)). However, the court allowed the investigator's testimony for the purposes of rebutting the defendant's testimony that he had those items in his possession for innocent reasons. Although the trial court agreed the investigator could not use the words "pedophile" or "child predator" while testifying, it allowed him to testify as follows:

Q. [PROSECUTOR] All right. Taking each one of these three items here, are these common among adults seeking sex from young female children?

A. [UHALL] Yes, it is.

Q. The Polaroid camera, why?

A. Photographs allow you to re-visit the event.

Q. And what about the body oils?

A. Body oils are necessary for lubrication for entry.

Q. And the condoms?

A. Condoms are possibly for prevention of pregnancy.[2]

Bass was found guilty of conspiracy to commit sexual conduct with a minor under the age of 15.

[2]Sex offenders might also use condoms to prevent transmission or contraction of sexually transmitted diseases and to limit the exchange of bodily fluids containing DNA that could be potentially damning physical evidence.

How an offender approaches and controls a victim or target can be significant, exposing the offender's strengths (e.g. skill level, physical strength), concerns (e.g. sexual inadequacies), intents, and motives. Some offenders who engage in prolonged grooming activities do so because it enables them to develop a relationship with the victim, satisfying their need to believe that the relationship is consensual. Some offenders use deception (e.g. posing as a 14-year-old boy) to approach and obtain control over a victim because they do not want to scare the victim away before having an opportunity to commit a sexual assault. Other offenders are more aggressive and simply use threats to gain complete control over a victim quickly. Different offenders can use the same method of approach or control for very different reasons so it is not possible to make broad generalizations. For example, one offender might use threats to discourage a victim from reporting the crime whereas another offender might use threats simply to gain a feeling of empowerment over the victim. Therefore, it is necessary to examine crime scene characteristics in unison, determining how they influence and relate to each other.

It is also important to remember that an offender is rarely in complete control – unexpected things occur and/or victims can react unpredictably. The pressures of unforeseen circumstances can cause an offender to reveal aspects of his personality, desires, or identity that he would otherwise conceal. One extreme example is an offender calling the victim by name while appealing for cooperation indicating that the offender knows the victim. Therefore, investigators should examine the victim–offender interactions and the events surrounding the crime to determine how an offender reacted to events that he could not have anticipated. When an offender uses a network to approach and control a victim, the methods of approach and control are predominantly verbal since networks do not afford physical access/threats. Statements made by the offender can be very revealing about the offender so investigators should make an effort to ascertain exactly what the offender said or typed.

The following are some examples of how offenders approached victims on the Internet in past cases.

- Offender accurately represented himself while grooming young victim he met in online chat room frequented by youths.
- Offender accurately represented himself while seeking likely victims in discussions of bondage and sadomasochism.
- Offender pretended to be a younger, more attractive male to attract female.
- Older male offender pretended to be a young boy to befriend a prepubescent child.

- Older male offender pretended to be a woman to attract an adolescent boy.
- Offender persuaded parents to give him access to their children.

Although this list is not definitive, it provides some illustrative examples of crime scene characteristics investigators might look for to develop learn more about an offender. This type of information, combined with other crime scene characteristics, can help investigators develop a clearer picture of the offender they are dealing with, including *modus operandi* and motivation.

20.5.4 MOTIVATION

The motives underlying pornography vary with the type of pornography (sadistic, domination, child pornography, etc.), how it is gathered (from available sources, self-created), and what is done with it once obtained (e.g. digitally altering images to show children or celebrities in sexual or violent situations). However, the motives underlying pornography do not change simply because the Internet is involved. Sex offenders took photographs of their victims long before the existence of the Internet as trophies of conquest, to revisit and relive the moment, and to show others. Similarly, the motives of individuals who solicit and abuse children are the same whether the Internet is involved or not. Therefore, existing research relating to motivation of sex offenders presented in Chapter 6 (*modus operandi*, motive, and technology) can be used to gain a better understanding of these offenders: Power Reassurance, Power Assertive, Anger Retaliatory, Anger Excitation, Opportunistic, and Profit.

This is not to say that determining motivation is a simple matter. There is much debate regarding the role of pornography in sex offenses. Some argue that pornography causes crime – Ted Bundy went so far as to claim that he became obsessed with pornography, and that viewing it broke down his resistance and justified his behavior. It cannot be proved that pornography causes offenders to act out their fantasies and a killer's justification of his crimes cannot be trusted. However, an individual's pornography collection reflects his/her fantasies. In David Westerfield's homicide trial, the prosecution claimed that Westerfield's digital pornography collection reflected his fantasies relating to kidnapping and killing 7-year-old Danielle van Dam and, in closing arguments, insinuated that the pornography motivated Westerfield to victimize the child.

> Not only does he have the young girls involved in sex, but he has the anime that you saw. And we will not show them to you again. The drawings of the young girls being

> sexually assaulted. Raped. Digitally penetrated. Exposed. Forcibly sodomized. Why does he have those, a normal fifty-year-old man? ... Those are his fantasies. His choice. Those are what he wants. He picked them; he collected them. Those are his fantasies. That's what gets him excited. That's what he wants in his collection ... When you have those fantasies, fantasies breed need. He got to the point where it was growing and growing and growing. And what else is there to collect? What else can I get excited about visually, audibly? (California v. Westerfield 2002)

However, the cause and effect are unclear, particularly in light of the fact that Westerfield also had pornography involving adults and animals. It could just as easily be argued that Westerfield's pornography collection provided an outlet for his fantasies and he would have committed a crime long before had it not been for this outlet. Aside from this, of all the people who possess child pornography, only a limited number actually commit offenses against children, and of those, only a small fraction have committed a homicide.

Keep in mind that the motivational typologies discussed in Chapter 6 are general categories designed to give investigators a better sense of why an individual may have committed a given crime. The aim is not to fit an offender into one category – some sex offenders commit offenses whenever they have an opportunity, regardless of the risks, and may be motivated by a number of factors. For instance, Larry Garmon had been released on probation from prison in Kansas after serving 9 years of a 20-year sentence for aggravated criminal sodomy on an 8-year-old boy. While checking his home for parole violations, police found pornographic images of men and boys stored on his computer. Garmon was convicted on child pornography charges and was placed in Madison County Detention Center where he allegedly sexually assaulted a 17-year-old boy who had been detained on a misdemeanor charge of possession of a handgun (Associated Press 2001). Other offenders are more directed in their approach to acquiring victims, taking precautions to address the associated risks but it can still be a challenge to dissect their motives.

CASE EXAMPLE (UNITED STATES v. HERSH 2001):
Marvin Hersh, a professor at Florida Atlantic University, traveled to third world countries, ranging from Asia to Central America, to engage in sexual relationships with impoverished young boys. During his travels, Hersh met another sex offender named Nelson Jay Buhler with whom he collaborated. In addition to traveling together to have sex with poverty stricken young boys in Honduras, Hersh taught Buhler where to find child pornography on the Internet and how to encrypt the files using F-Secure and save them to Zip disks that could easily be destroyed. He eventually brought a 15-year-old boy from Honduras back to live with him in

Florida, posing as his son. Hersh was convicted of transporting a minor in foreign commerce with the intent to engage in criminal sexual activity, and conspiracy to travel in foreign commerce with the intent to engage in sexual acts with minors, and receiving and possessing material containing visual depictions of minors engaged in sexually explicit conduct.

Failure to understand an offender's motivation can impair an investigation, making it difficult for investigators to interpret evidence, obtain information from a known offender, or apprehend an unknown offender. Having insight into an offender's motivation to commit a sex offense is helpful to prosecutors, because they have the daunting task of persuading a jury that the normal looking man sitting at the defense table in the pin-striped suit actually raped a little boy and masturbates while talking to children on the Internet. Also, knowledge of an offender's motivations and likely behaviors can help shape prevention strategies to avoid future harm to other victims. Given their importance, investigators should attempt to determine an offender's motives, consulting with a forensic psychiatrist, psychologist, or other appropriate specialist in complex cases as needed.

20.6 SUMMARY

Digital evidence examiners are often asked to locate evidence that law enforcement or supervisors believe is present, but either may not exist or may not have the import the requestor believes it to have. For instance, examiners at the Connecticut State Crime Lab have been asked on a number of occasions to substantiate that a target visited a certain Web site or made an entry of their own volition, and did so with the intent of downloading child pornography. Such determinations can rarely be made when examiners retrieve only a few images or the evidence suggests only one or two visits to a Web site. Although it is the responsibility of the investigators and digital evidence examiners to locate evidence that may establish probable cause, for the prosecutor to establish proof beyond a reasonable doubt, and for a judge or jury to be persuaded that the digital evidence exists, we must be wary of being overeager to reach a specific result. It is important for digital evidence examiners to be completely honest – this requires fully researching the current technology so that one's statements regarding the evidence are accurate and fully explaining one's findings in a way that is understandable to a non-technical decision maker (e.g. attorney, judge, jury, management, a company's disciplinary board).

Sex offenders make mistakes that cause some investigators to state that "we only catch the stupid ones" or "they are trying to be caught." The primary reason for such mistakes is that sex offenders are driven by deep-rooted psychological needs causing them to engage in behavior that increases the risk of apprehension. Investigators and digital evidence examiners who learn to recognize and understand these patterns in sex offenders will be more capable of locating missing evidence and victims, and interpreting the significance of existing evidence. For instance, the type of pornography that an offender collects will reflect their motivations (e.g. power assertive versus power reassurance) and sexual interests. Knowing this can help develop investigative leads, interviewing and trial strategies. For example, when interviewing an offender who assaults victims to fulfill inadequacies, such as a power reassurance motivated offender, it may be effective to express empathy and understanding, effectively grooming the suspect into trusting and confiding in the interviewer. Such an offender is more likely to confess when treated kindly. Similarly, choices of screen names, online profiles, and preferential use of technology can reveal offender skill level, comfort levels, etc.

This same approach may be counterproductive when dealing with a power assertive motivated offender who might view a "soft" approach as weakness in the interviewer, providing an opportunity to manipulate and control the situation. An offender who believes he is smarter than investigators may be persuaded to reveal details about how he committed crimes or concealed evidence by appealing to his vanity.

Often neglected in specialized investigations is the value of consulting experts in the behavioral sciences. While usually untrained in formal investigative techniques, by education, training, and experience they may have insight to offer investigators. Forensic psychiatrists, psychologists, and social workers who evaluate and treat sex offenders can be an invaluable asset to an investigation when used appropriately. As in any area of case review or investigation, it is very important to draw inferences from the evidence (digital, behavioral, and "real world" physical) in the specific case and not to rely solely on past experience and statistical profiles of offenders. For example, while there are several typologies of sex offenders, these were developed retrospectively for labeling and/or treatment purposes. None of these typologies have been scientifically validated for use prospectively in an investigation. The investigator is cautioned to be wary of the expert who opines quickly on the traits of the offender, relying on a cursory evaluation of the evidence and an inductively derived list of expected behaviors and traits.

For a more detailed discussion of this topic, see *Investigating Computer Assisted Child Exploitation* (Ferraro and Casey 2004).

REFERENCES

Amar V. D. (2003) "Regarding Child Pornography Extends the Supreme Court's Federalism Cases", Findlaw's Write, May 16, 2003 (Available online at http://writ.corporate.findlaw.com/amar/20030516.html)

Ananova (2001) "Chatroom man 'wanted someone to rape and torture his wife' ", August, 2001 (Available online at http://www.ananova.com/news/story/sm_379058.html)

Associated Press (1998) "FBI: Man posted sex pictures with daughter on Internet", February 10, 1998

Associated Press (1999) "Child Participates in Sex Sting", September 2, 1999

Associated Press (2001) "Teen placed in jail cell with sex offender", November 21, 2001

Associated Press (2002) "Man Convicted In Internet Kidnap, Rape Of Teen", December 6, 2002

Associated Press (2003) "Federal judge rules hacker covered by informant laws", March 18, 2003 (Available online at http://www.usatoday.com/tech/news/techpolicy/2003-03-18-hacker-informant_x.htm)

Burney M. (1997) "Cyber affair with teen-age girl leads to five years in prison", The Associated Press, August 22,1997. (Available online at: http://www.nando.net/newsroom/ntn/info/082297/info10_3348_noframes.html)

Casey E. (1999) "Cyberpatterns: Criminal Behavior on the Internet" in Criminal Profiling, 1st Ed. by Brent Turvey, Academic Press

Chen D. W. (2000) "Teacher is Accused of Duping Boy to Make a Sexual Video", New York Times, Late Edition – Final, Section B, Page 5, Column 5, October 5th, 2000.

Diskant T. (2002) "After sentencing, Lasaga, Yale face civil suit", February 22, 2002, Yale Herald (Available online at http://www.yaleherald.com/article.php?Article=359)

Durkin K. F. and Bryant C. D. (1995) "Log on to Sex: Some Notes on the Carnal Computer and Erotic Cyberspace as an Emerging Research Frontier," *Deviant Behavior: An Interdisciplinary Journal*, 16, 179–200.

Gaurdian Unlimited (2001) "Man convicted of internet-conspired rape", August 2, 2001 (Available online at http://www.guardian.co.uk/japan/story/0,7369,531343,00.html)

Henry J. (1985) Testimony before the Permanent Subcommittee on Governmental Affairs before the United States Senate, Ninety-Ninth Congress. (Available online at http://www.nostatusquo.com/ACLU/NudistHallofShame/Henry.html)

Hernandez A. E. (2000) "Self-reported contact sexual crimes of federal inmates convicted of child pornography offenses", Presented at the 19th Annual Conference Research and Treatment Conference of the Association for the Treatment of Sexual Abusers, San Diego, CA, November 2000.

Lanning K. V. (2001) Child Molesters and Cyber Pedophiles – A Behavioral Perspective, in Hazelwood, Robert R. and Burgess, Ann W., Eds., Practical Aspects of Rape Investigation: A multidisciplinary Approach, 3rd Ed., pp 199–232, Boca Raton, FL: CRC Press

McAullife W. (2000) "Net paedophile gets five years", ZDNet, 24 October, 2000 (Available online at http://news.zdnet.co.uk/story/0,,s2082159,00.html)

McClintock D. (2001) "Fatal Bondage", Vanity Fair, June.

McGrath M. G. and Casey E. (2002) "Forensic psychiatry and the Internet: Practical perspectives on sexual predators and obsessional harassers in cyberspace", *Journal of American Academy of Pychiatry and Law*, 30, 81–94

New York Lawyer (2003) "NY Law Professor Pleads Guilty To Possessing Child Porn", April 16, 2003 (Available online at http://www.nylawyer.com/news/03/04/041603d.html)

O'Hanlon S. (2003) "Porn Curiosity Brings Scrutiny Against 'Who's' Townsend", January 11, 2003 (http://reuters.com/newsArticle.jhtml?type=internetNews&storyID=2028595)

Pendlebury F. (2001) "Jail for Net Rapist", Dorsett Echo, July 17, 2001 (Available online at http://www.thisisdorset.net/dorset/archive/2001/07/17/BOURN_NEWS_NEWS10ZM.html)

Psychiatric News (2000) "Protect Children From Predators on Internet, Parents Tell Congress," May 5, 2000, (Available online at http://www.psych.org/pnews/00-05-05/protect.html)

Reuters (2001) "Russia Lacks Laws to Fight Child Porn Explosion", 26 March, 2001

Rizzo T. (2001) "Judge rules Robinson must stand trial in 3 deaths", Kansas City Star, March 2, 2001, (Available online at http://www.kcstar.com/standing/robinson/case.html)

Shannon E (1998) "Main Street Monsters", Time Magazine, September 14, 1998 Vol. 152, No. 11

Soh-jung Y. (2001) "Rising Number of Sexual Assault Cases Linked to Internet", The Korea Herald, July 7, 2001 (Available online at http://www.vachss.com/help_text/archive/sa_korea.html)

States News Service (1998) "Woman Jailed For Sex With Boy", 2 December, 1998 (Available online at http://www.fathermag.com/news/rape/portland02.shtml)

States News Service (1999) "Cyber-Sex Seducer Gets Jail Time", 13, January 1999 (Available online at http://www.fathermag.com/news/rape/morganton.shtml)

Taylor M., Quayle E. and Holland G. (2001) "Child Pornography, the Internet and offending", ISUMA, Vol. 2, No. 2 (Available online at http://www.isuma.net/v02n02/taylor/taylor_e.shtml)

The Age (2002) "Japanese police to regulate Internet dating services", December 26 2002 (Available online at http://www.theage.com.au/articles/2002/12/26/1040511126218.html)

Thomson S. (2002) "Charges Dropped Against Two Men in Sex-Slave Case, The Columbian, November 26, 2002

United States Department of Justice (2000) "Former high-tech executive pleads guilty to charge of traveling to have sex with minor he met on Internet", March 17, 2000 (Available online at http://www.usdoj.gov/usao/cac/pr/pr2000/050.htm)

United States Department of Justice (2002a) "Abingdon man sentenced to 41 months for child pornography" (Available online at http://www.usdoj.gov/usao/md/press_releases/press02/wyattrelease.htm)

United States Department of Justice (2002b) "Parkville man receives 10 year prison sentence for exploiting child to produce child pornography", (Available online at http://www.usdoj.gov/usao/md/press_releases/press02/adam_thomas_valleau_sentenced.htm)

United States Department of Justice (2002c) "Man sentenced to 27 months in prison for possessing child Pornography" (Available online at http://www.usdoj.gov/usao/txs/releases/April%202002/020417-magargee.htm)

United States Department of Justice (2003) "Fact Sheet, PROTECT Act" (Available online at http://www.usdoj.gov/opa/pr/2003/April/03_ag_266.htm)

CASES

Arizona v. Bass (2001) 357 Ariz. Adv. Rep. 3, 31 P.3d 857

Ashcroft v. Free Speech Coalition (2002) US Supreme Court, Case Number 00-795 (Available online at http://laws.findlaw.com/us/000/00-795.html)

California v. Westerfield (2002) Superior Court, County of San Diego Central Division, California, Case Number CD165805

Connecticut v. Sorabella (2002) Superior Court, Connecticut (Also see U.S. District Court of Massachusetts 1:2002cr10061)

United States v. Henriques (1999) Appeals Court, 5th Circuit, Case Number 99-60819 (Available online at http://laws.findlaw.com/5th/9960819cr0v2.html)

United States v. Hersh (2001) Appeals Court, 11th Circuit, Case Number 00-14592 (Available online at: http://laws.lp.findlaw.com/11th/0014592opn.html)

United States v. Hilton (1997) District Court for the District of Maine, Criminal No. 97-78-P-C, (Available online at http://www.med.uscourts.gov/opinions/carter/2000/gc_06302000_2-97cr078_us_v_hilton.pdf)

United States v. Matthews (2000a) "Amicus brief in U.S. v. Matthews" (Available at http://www.rcfp.org/news/documents/matthews.html)

United States v. Matthews (2000b) Appeals Court, 4th Circuit (209 F.3d 338) Case Number 99-4183 (Available online at http://www.law.emory.edu/4circuit/dec99/994183.p.html)

United States v. Perez (2003) District Court, Southern District of New York, Case Number 02CR00854 (Available online at http://www.nysd.uscourts.gov/rulings/02CR00854.pdf)

Wisconsin v. Kenney (2002) Appeals Court, Wisconsin, Case Number 01-0810-CR (Available online at http://www.courts.state.wi.us/html/ca/01/01%2D0810.htm)

Wisconsin v. Timothy P. Koenck (2001) Appeals Court, Wisconsin, Case Number 00-2684-CR (Available online at http://www.courts.state.wi.us/html/ca/00/00-2684.htm)

Wisconsin v. Michael L. Morris (2002) Appeals Court, Wisconsin, Case Number 01-0414-CR (Available online at http://www.courts.state.wi.us/html/ca/01/01-0414.htm)

Wisconsin v. Brian D. Robins (2002) Supreme Court, Wisconsin, Case Number 00-2841-CR (Available online at http://www.courts.state.wi.us/html/sc/00/00-2841.htm)

## CHAPTER 21

CYBERSTALKING

The lack of sensory information on the Internet may have a significant impact on cyberstalkers, as described by Meloy (p. 11): "The absence of sensory-perceptual stimuli from a real person means that fantasy can play an even more expansive role as the genesis of behavior in the stalker." The victim becomes an easy target for the stalker's projections, and narcissistic fantasies, that can lead to a real world rejection, humiliation and rage.

(Meloy 1998)

One of the most prominent features of stalking behavior is fixation on victims. Their obsession can drive stalkers to extremes that make this type of investigation challenging and potentially dangerous. Although stalkers who use the Internet to target victims may attempt to conceal their identities, their obsession with a victim often causes them to expose themselves. For instance, they may say things that reveal their relationship with or knowledge of the victim, or they may take risks that enable investigators to locate and identify them. However, even when stalkers have been identified, attempts to discourage them can have the opposite effect, potentially angering them and putting victims at greater risk.

In 1990, after five women were murdered by stalkers, California became the first state in the US to enact a law to deal with this specific problem. Then, in 1998, California explicitly included electronic communications in their anti-stalking law. The relevant sections of the California Penal Code have strongly influenced all subsequent anti-stalking laws in the US, clearly defining stalking and related terms.

> Any person who willfully, maliciously, and repeatedly follows or harasses another person and who makes a credible threat with the intent to place that person in reasonable fear of death or great bodily injury is guilty of the crime of stalking... "harasses" means a knowing and willful course of conduct directed at a specific person that seriously alarms, annoys, torments, or terrorizes the person, and that serves no legitimate purpose. This course of conduct must be such as would cause a reasonable person to suffer substantial emotional distress, and must actually cause substantial emotional distress to the person.

Digital Evidence and Computer Crime Second Edition
ISBN: 0-12-163104-4

> ... "course of conduct" means a pattern of conduct composed of a series of acts over a period of time, however short, evidencing a continuity of purpose ... "credible threat" means a verbal or written threat, including that performed through the use of an electronic communication device, or a threat implied by a pattern of conduct or a combination of verbal, written, or electronically communicated statements and conduct made with the intent to place the person that is the target of the threat in reasonable fear for his or her safety or the safety of his or her family and made with the apparent ability to carry out the threat so as to cause the person who is the target of the threat to reasonably fear for his or her safety or the safety of his or her family. It is not necessary to prove that the defendant had the intent to actually carry out the threat ... "electronic communication device" includes, but is not limited to, telephones, cellular phones, computers, video recorders, fax machines, or pagers." [California Penal Code 646.9]

The equivalent law in the United Kingdom is the Protection from Harassment Act 1997 (Chapter 40).

Note that persistence is one of the operative concepts when dealing with stalking. A single upsetting e-mail message is not considered harassment because it is not a pattern of behavior. Remember that anti-stalking laws were enacted to protect individuals against persistent terrorism and physical danger, not against annoyance or vague threats.

The distinction between annoyance and harassment is not easily defined. It is usually enough to demonstrate that the victim suffered substantial emotional distress. However, there is always the argument that the victim overreacted to the situation. If a victim is not found to be a "reasonable person" as described in the law, a court might hold that no harassment took place. Therefore, when investigating a stalking case, it is important to gather as much evidence as possible to demonstrate that persistent harassment took place and that the victim reacted to the credible threat in a reasonable manner.

The explicit inclusion of electronic communication devices in California's anti-stalking law is a clear acknowledgement of the fact that stalkers are making increasing use of new technology to further their ends. In addition to using voice mail, fax machines, cellular phones, and pagers, stalkers use computer networks to harass their victims. The term *cyberstalking* refers to stalking that involves the Internet. This chapter briefly describes how cyberstalkers operate, what motivates them, and what investigators can do to apprehend them. Additional resources that relate to various aspects of stalking are presented at the end of this chapter.

21.1 HOW CYBERSTALKERS OPERATE

Cyberstalking works in much the same way as stalking in the physical world. In fact, many offenders combine their online activities with more traditional forms of stalking and harassment such as telephoning the victim and going to the victim's home. Some cyberstalkers obtain victims over the Internet and others put personal information about their victims online, encouraging others to contact the victim, or even harm them.

CASE EXAMPLE (ASSOCIATED PRESS 1997):
Cynthia Armistead-Smathers of Atlanta believes she became a target during an e-mail discussion of advertising in June, 1996. First she received nasty e-mails from the account of Richard Hillyard of Norcross, GA. Then she began receiving messages sent through an "anonymous remailer," an online service that masks the sender's identity.

After Hillyard's Internet service provider canceled his account, Ms Armistead-Smathers began getting messages from the Centers for Disease Control and Prevention in Atlanta, where he worked. Then she got thousands of messages from men who had seen a posting of a nude woman, listing her e-mail address and offering sex during the Atlanta Olympics.

But police said there was little they could do – until she got an anonymous message from someone saying he had followed Ms Armistead-Smathers and her 5-year-old daughter from their post office box to her home.

People say "It's online. Who cares? It isn't real. Well this is real," Ms Armistead-Smathers said. "It's a matter of the same kind of small-minded bullies who maybe wouldn't have done things in real life, but they have the power of anonymity from behind a keyboard, where they think no one will find them."

In general, stalkers want to exert power over their victims in some way, primarily through fear. The crux of a stalker's power is information about and knowledge of the victim. A stalker's ability to frighten and control a victim increases with the amount of information that he can gather about the victim. Stalkers use information like telephone numbers, addresses, and personal preferences to impinge upon their victims' lives. Also, over time cyberstalkers can learn what sorts of things upset their victims and can use this knowledge to harass the victims further.

Since they depend heavily on information, it is no surprise that stalkers have taken to the Internet. After all, the Internet contains a vast amount of personal information about people and makes it relatively easy to search for specific items. As well as containing people's addresses and phone numbers, the Internet records many of our actions, choices, interests, and desires. Databases containing social security numbers, credit card numbers, medical history, criminal records, and much more can also be accessed using the Internet. Additionally, cyberstalkers can use the Internet to harass specific individuals or acquire new victims from a large pool of potential targets. In one case, a woman was stalked in chat rooms for several months, during which time the stalker placed detailed personal information online and threatened to rape and kill her. Some offenders seek victims online but it is more common for stalkers to use chat networks to target individuals that they already know.

21.1.1 ACQUIRING VICTIMS

Past studies indicate that many stalkers had a prior acquaintance with their victims before the stalking behavior began (Harmon *et al.* 1994). The implication of these studies is that investigators should pay particular attention to acquaintances of the victim. However, these studies are limited because many stalking cases are unsolved or unreported. Additionally, it is not clear if these studies apply to the Internet. After all, it is uncertain what constitutes an acquaintance on the Internet and the Internet makes it easier for cyberstalkers to find victims of opportunity.[1]

[1]A victim of opportunity is a victim whom a stalker was not acquainted with before the stalking began.

Cyberstalkers can search the Web, browse through ICQ and AOL profiles, and lurk in IRC and AOL chat rooms looking for likely targets – vulnerable, under-confident individuals who will be easy to intimidate.

CASE EXAMPLE

One stalker repeatedly acquired victims of opportunity on AOL and used AOL's Instant Messenger to contact and harass them. The stalker also used online telephone directories to find victims' numbers, harassing them further by calling their homes. This approach left very little digital evidence because none of the victims recorded the Instant Messenger sessions, they did not know how to find the stalker's IP address, and they did not contact AOL in time to track the stalker.[2]

[2]Recall from Chapter 17 that netstat can be used to view current and recent TCP/IP connections to a computer. Investigators can use an IP address to track down a cyberstalker.

Of course, the victims were distressed by this harassment, feeling powerless to stop the instant messages and phone calls. This sense of powerlessness was the primary goal the cyberstalker. This stalker may have picked AOL as his stalking territory because of the high number of inexperienced Internet users and the anonymity that it affords.

As a rule, investigators should rely more on available evidence than on general studies. Although research can be useful to a certain degree, evidence is the most reliable source of information about a specific case and it is what the courts will use to make a decision.

21.1.2 ANONYMITY AND SURREPTITIOUS MONITORING

The Internet has the added advantage of protecting a stalker's identity and allowing a stalker to monitor a victim's activities. For example, stalkers acquainted with their victims use the Internet to hide their identity, sending forged or anonymous e-mail and using ICQ or AOL Instant Messenger to harass their victims. Also, stalkers can utilize ICQ, AOL Instant Messenger, and other applications (e.g. finger) to determine when a victim is online. Most disturbing of all, stalkers can use the Internet to spy on a victim. Although few cyberstalkers are skilled enough to break into a victim's e-mail account or intercept e-mail in transit, a cyberstalker can easily observe a conversation in a live chat room. This type of pre-surveillance of victims and amassing of information about potential victims might suggest intent to commit a crime but it is not a crime in itself, and is not stalking as defined by the law.

21.1.3 ESCALATION AND VIOLENCE

It is often suggested that stalkers will cease harassing their victims once they cease to provoke the desired response. However, some stalkers become aggravated when they do not get what they want and become increasingly threatening. As was mentioned at the beginning of this chapter, stalkers have resorted to violence and murder. Therefore, it is important for investigators to be extremely cautious when dealing with a stalking case. Investigators should examine the available evidence closely, protect the victim against further harm as much as possible, and consult with experts when in doubt. Most importantly, investigators should not make hurried judgments that are based primarily on studies of past cases.

21.2 INVESTIGATING CYBERSTALKING

There are several stages to investigating a cyberstalking case. These stages assume that the identity of the cyberstalker is unknown. Even if the victim suspects an individual, investigators are advised to explore alternative possibilities and suspects. Although past research suggests that most stalkers have prior relationships with victims, this may not apply when the Internet is involved since stranger stalking is easier. Therefore, consider the possibility that the victim knows the stalker, but do not assume that this is the case:

1. *Interview victim* – determine what evidence the victim has of cyberstalking and obtain details about the victim that can be used to develop victimology. The aim of this initial information gathering stage is to confirm that a crime has been committed and to obtain enough information to move forward with the investigation.
2. *Interview others* – if there are other people involved, interview them to compile a more complete picture of what occurred.
3. *Victimology and risk assessment* – determine why an offender chose a specific victim and what risks the offender was willing to take to acquire that victim. The primary aim of this stage of the investigation is to understand the victim–offender relationship and determine where additional digital evidence might be found.
4. *Search for additional digital evidence* – use what is known about the victim and cyberstalker to perform a thorough search of the Internet. Victimology is key at this stage, guiding investigators to locations that might interest the victim or individuals like the victim. The cyberstalker initially observed or encountered the victim somewhere and investigators should try to determine where. Consider the possibility that the cyberstalker encountered the victim in the physical world.

The aim of this stage is to gather more information about the crime, the victim and the cyberstalker.

5 *Crime scene characteristics* – examine crime scenes and cybertrails for distinguishing features (e.g. location, time, method of approach, choice of tools) and try to determine their significance to the cyberstalker. The aim of this stage is to gain a better understanding of the choices that the cyberstalker made and the needs that were fulfilled by these choices.

6 *Motivation* – determine what personal needs the cyberstalking was fulfilling. Be careful to distinguish between intent (e.g. to exert power over the victim, to frighten the victim) and the personal needs that the cyberstalker's behavior satisfied (e.g. to feel powerful, to retaliate against the victim for a perceived wrong). The aim of this stage is to understand the cyberstalker well enough to narrow the suspect pool revisit the prior steps and uncover additional evidence

7 *Repeat* – if the identity of the cyberstalker is still not known, interview the victim again. The information that investigators have gathered might help the victim recall additional details or might suggest a likely suspect to the victim

To assist investigators carry out each of these stages in an investigation, additional details are provided here.

21.2.1 INTERVIEWS

Investigators should interview the victim and other individuals with knowledge of the case to obtain details about the inception of the cyberstalking and the sorts of harassment the victim has been subjected to. In addition to collecting all of the evidence that the victim has of the cyberstalking, investigators should gather all of the details that are required to develop a thorough victimology as described in the next section.

While interviewing the victim, investigators should be sensitive to be as tactful as possible while questioning everything and assuming nothing. Keep in mind that victims tend to blame themselves, imagining that they encouraged the stalker in some way (e.g. by accepting initial advances or by making too much personal information available on the Internet) (Pathe 1997). It is therefore important for everyone involved in a cyberstalking investigation to help the victim regain confidence by acknowledging that the victim is not to blame. It is also crucial to help victims protect themselves from potential attacks. The National Center for Victims of Crime has an excellent set of guidelines developed specifically for victims of stalking (NCVC 1995).

21.2.2 VICTIMOLOGY

In addition to helping victims protect themselves against further harassment, investigators should try to determine how and why the offender selected a specific victim. To this end, investigators should determine whether the cyberstalker knew the victim, learned about the victim through a personal

Web page, saw a Usenet message written by the victim, or noticed the victim in a chat room.

It is also useful to know why a victim made certain choices to help investigators make a risk assessment. For example, individuals who use the Internet to meet new people are at higher risk than individuals who make an effort to remain anonymous. In some instances, it might be quite evident why the cyberstalker chose a victim but if a cyberstalker chooses a low risk victim, investigators should try to determine which particular characteristics the victim possesses that might have attracted the cyberstalker's attention (e.g. residence, work place, hobby, personal interest, demeanor). These characteristics can be quite revealing about a cyberstalker and can direct the investigator's attention to certain areas or individuals.

Questions to ask at this stage include:

- Does the victim know or suspect why, how, and/or when the cyberstalking began?
- What Internet Service Provider(s) do(es) the victim use and why?
- What online services does the victim use and why (e.g. Web, free e-mail services, Usenet, IRC)?
- When does the victim use the Internet and the various Internet services (does the harassment occur at specific times suggesting that the cyberstalker has a schedule or is aware of the victim's schedule)?
- What does the victim do on the Internet and why?
- Does the victim have personal Web pages or other personal information on the Internet (e.g. AOL profile, ICQ Web page, customized finger output)? What information do these items contain?

In addition to the victim's Internet activities, investigators should examine the victim's physical surroundings and real world activities.

When the identity of the cyberstalker is known or suspected, it might not seem necessary to develop a complete victimology. Although it is crucial to investigate suspects, this should not be done at the expense of all else. Time spent trying to understand the victim–offender relationship can help investigators understand the offender, protect the victim, locate additional evidence, and discover additional victims. Furthermore, there is always the chance that the suspect is innocent in which case investigators can use the victimology that they developed to find other likely suspects.

21.2.3 RISK ASSESSMENT

A key aspect of developing victimology is determining victim and offender risk. Generally, women are at greater risk than men of being cyberstalked and new Internet users are at greater risk than experienced Internet users. Individuals who frequent the equivalent of singles bars on the Internet are at greater risk than those who just use the Internet to search for information. A woman who puts her picture on a Web page with some biographical information, an address, and phone number is at high risk because cyberstalkers can fixate on the picture, obtain personal information about the woman from the Web page, and start harassing her over the phone or in person.

Bear in mind that victim risk is not an absolute thing – it depends on the circumstances. A careful individual who avoids high risk situations in the physical world might be less cautious on the Internet. For example, individuals who are not famous in the world at large might have celebrity status in a certain area of the Internet, putting them at high risk of being stalked by someone familiar with that area. Individual who are sexually reserved in the physical world might partake in extensive sexual role playing on the Internet, putting them at high risk of being cyberstalked.

If a cyberstalker selects a low risk victim, investigators should try to determine what attracted the offender to the victim. Also, investigators should determine what the offender was willing to risk when harassing the victim. Remember that offender risk is the risk as an offender perceives it – investigators should not try to interpret an offender's behavior based on the risks they perceive. An offender will not necessarily be concerned by the risks that others perceive. For example, some cyberstalkers do not perceive apprehension as a great risk, only an inconvenience that would temporarily interfere with their ability to achieve their goal (to harass the victim) and will continue to harass their victims, even when they are under investigation.

21.2.4 SEARCH

Investigators should perform a thorough search of the Internet using what is known about the victim and the offender and should examine personal computers, log files on servers, and all other available sources of digital evidence as described in this book. For example, when a cyberstalker uses e-mail to harass a victim, the messages should be collected and examined. Also, other e-mail that the victim has received should be examined to determine if the stalker sent forged messages to deceive the victim. Log files of the e-mails server that was used to send and receive the e-mail should be examined to confirm the events in question. Log files sometimes reveal other things that the cyberstalker was doing (e.g. masquerading as the victim, harassing other victims) and can contain information that lead directly to the cyberstalker.

CASE EXAMPLE

Gary Steven Dellapenta became the first person to be convicted under the new section of California's stalking law that specifically includes electronic communications. After being turned down by a woman named Randi Barber, Dellapenta retaliated by impersonating her on the Internet and claiming she fantasized about being raped.

Using nicknames such as "playfulkitty4U" and "kinkygal30," Dellapenta placed online personal ads and sent messages saying such things as "I'm into the rape fantasy and gang-bang fantasy too." He gave respondents Barber's address and telephone number, directions to her home, details of her social plans and even advice on how to short-circuit her alarm system.

Barber became alarmed when men began leaving messages on her answer machine and turning up at her apartment. In an interview (*Newsweek* 1999), Barber recalled that one of the visitors left after she hid silently for a few minutes, but phoned her apartment later. "What do you want?" she pleaded. "Why are you doing this?" The man explained that he was responding to the sexy ad she had placed on the Internet.

"What ad? What did it say?" Barber asked. "Am I in big trouble?"

"Let me put it to you this way," the caller said. "You could get raped."

When Barber put a note on her door to discourage the men who were responding to the personal ads, Dellapenta putting new information on the Internet claiming that the note was just part of the fantasy.

In an effort to gather evidence against Dellapenta, Barber kept recordings of messages that were left on her machine and contacted each caller, asking for any information about the cyberstalker. Two men cooperated with her request for help, but it was ultimately her father who gathered the evidence that was necessary to identify Dellapenta.

Barber's father helped to uncover Dellapenta's identity by posing as an ad respondent and turning the e-mails he received over to investigators.

> [Investigators] traced the e-mails from the Web sites at which they were posted to the servers used to access the sites. Search warrants compelled the Internet companies to identify the user. All the paths led police back to Dellapenta. "When you go on the Internet, you leave fingerprints – we can tell exactly where you've been," says sheriff's investigator Mike Gurzi, who would eventually verify that all the e-mails originated from Dellapenta's computer after studying his hard drive. The alleged stalker's M.O. was tellingly simple: police say he opened up a number of free Internet e-mail accounts pretending to be the victim, posted the crude ads under a salacious log-on name and started e-mailing the men who responded. (*Newsweek* 1999)

Dellapenta admitted to authorities that he had an "inner rage" against Barber and pleaded guilty to one count of stalking and three counts of solicitation of sexual assault.

When searching for evidence of cyberstalking it is useful to distinguish between the offender's harassing behaviors and surreptitious monitoring behaviors. A victim is usually only aware of the harassment component of cyberstalking. However, cyberstalkers often engage in additional activities that the victim is not aware of. Therefore, investigators should not limit their search to the evidence of harassment that the victim is already aware of but should look for evidence of both harassment and surreptitious monitoring.

If the victim frequented certain areas, investigators should comb those areas for information and should attempt to see them from the cyberstalker's perspective. Could the cyberstalker have monitored the victim's activities in those areas? If so, would this monitoring have generated any digital evidence and would Locard's exchange principle take effect? For example, if the victim maintains a Web page, the cyberstalker might have monitored its development in which case the Web server log would contain the cyberstalker's IP address (with associated times) and the cyberstalker's personal computer would indicate that the page had been viewed (and when it was viewed). If the cyberstalker monitored the victim in IRC, he might have kept log files of the chat sessions. If the cyberstalker broke into the victim's e-mail account the log files on the e-mail server should reflect this.

Keep in mind that the evidence search and seizure stage of an investigation forms the foundation of the case – incomplete searches and poorly collected digital evidence will result in a weak case. It is therefore crucial to apply the Forensic Science concepts presented in this book diligently. Investigators should collect, document, and preserve digital evidence in a way that will facilitate the reconstruction and prosecution processes. Also investigators should become intimately familiar with available digital evidence, looking for class and individual characteristics in an effort to maximize its potential.

21.2.5 CRIME SCENE CHARACTERISTICS

When investigating cyberstalking, investigators might not be able to define the primary crime scene clearly because digital evidence is often spread all over the Internet. However, the same principle of behavioral evidence analysis applies – aspects of a cyberstalker's behavior can be determined from choices and decisions that a cyberstalker made and the evidence that was left behind, destroyed, or taken away. Therefore, investigators should thoroughly examine the point of contact and cybertrails (e.g. the Web, Usenet, personal computers) for digital evidence that exposes the offender's behavior.

To begin with, investigators should ask themselves why a particular cyberstalker used the Internet – what need did this fulfill? Was the cyberstalker using the Internet to obtain victims, to remain anonymous, or both? Investigators should also ask why a cyberstalker used particular areas of the Internet – what affordances did the Internet provide? MO and signature behaviors can usually be discerned from the way a cyberstalker approaches and harasses victims on the Internet.

How cyberstalkers use the Internet can say a lot about their skill level, goals, and motivations. Using IRC rather than e-mail to harass victims suggests a higher skill level and a desire to gain instantaneous access to the victim while remaining anonymous. The choice of technology will also determine what digital evidence is available. Unless a victim keeps a log, harassment on IRC leaves very little evidence whereas harassing e-mail messages are enduring and can be used to track down the sender.

Additionally, investigators can learn a great deal about offenders' needs and choices by carefully examining their words, actions, and reactions. Increases and decreases in intensity in reaction to unexpected occurrences are particularly revealing. For example, when a cyberstalker's primary mode of contact with a victim is blocked the cyberstalker might be discouraged, unperturbed, or aggravated. How the cyberstalkers choose to react to setbacks indicates how determined they are to harass a specific victim and what they hope to achieve through the harassment. Also, a cyberstalker's intelligence, skill level, and identity can be revealed when he modifies his behavior and use of technology to overcome obstacles.

21.2.6 MOTIVATION

There have been a number of attempts to categorize stalking behavior and develop specialized typologies (Meloy 1998). However, these typologies were not developed with investigations in mind and are primarily used by clinicians to diagnose mental illnesses and administer appropriate treatments.

When investigating cyberstalking, the motivational typologies discussed in Chapter 3 can be used as a sounding board to gain a greater understanding of stalkers' motivations. Also, as described earlier in this chapter, some stalkers pick their victims opportunistically and get satisfaction by intimidating them, fitting into the power assertive typology.

Other stalkers are driven by a need to retaliate against their victims for perceived wrongs, exhibiting many of the behaviors described in the anger retaliatory typology. For instance, Dellapenta, the Californian cyberstalker who went to great lengths to terrify Randi Barber, stated that he has an "inner rage" directed at Barber that he could not control. Dellapenta's behavior confirms this statement, indicating that he was retaliating against Barber for a perceived wrong. His messages were degrading and were designed to bring harm to Barber. Furthermore, Dellapenta tried to arrange for other people to harm Barber, indicating that he did feel the need to hurt her himself. Although it is possible that Dellapenta felt some desire to assert power over Barber, his behavior indicates that he was primarily driven by a desire to bring harm to her.

21.3 CYBERSTALKING CASE EXAMPLE[3]

[3]*This case example is based on abstracted lessons from various investigations. Any resemblance to actual incidents is coincidental.*

Jill's troubles began after she dumped Jack. Jack "accidentally" sent a defamatory e-mail to a list of mutual friends containing personal information that was very embarrassing to Jill. He claimed that he had intended to send the e-mail to Jill and must have addressed the e-mail incorrectly. After this incident, Jack seemed to overcome his difficulty in addressing e-mail and started to bombard Jill with offensive missives. He also forced his way into her apartment one night and, although he did not threaten to harm her, he refused to leave. Jill called the police but Jack left before they arrived.

Jill continued to receive offensive e-mail messages from Jack and a mutual friend told her that Jack claimed to have a compromising video of her. Jill also heard rumors that Jack was somehow listening in on her telephone conversations, monitoring her e-mail, and videotaping her in her apartment. She became so distraught that she lost sleep and became ill.

Authorities informed Jack that he was being investigated and they arranged for all e-mail messages from him to Jill to be redirected into a holding area so that they would be preserved as evidence and Jill would not be exposed to them. Nonetheless, he continued to harass Jill in person and through the Internet by sending e-mail from different addresses. He also targeted Jill indirectly by forging an e-mail message to her friends making it seem like Jill had sent it. Her friends were surprised and troubled by the content of the messages and asked Jill why she had sent them, at which point she reported the forgery to the police.

The police obtained log files from the e-mail server that Jack had used to forge the e-mail and found that he had connected via AOL. The police then obtained a search warrant to obtain the identity of the individual from AOL who had been assigned the IP address at the time in question. AOL confirmed that Jack had been assigned the IP address at the time, and provided account information and e-mails stored on their servers:

```
ONLINE ACCOUNT PROFILE
Screen Names:       CyberStalker        xxxxxxxxxx

Name:               JOHN DOE
Street:             153 Main Street
City:               New York
State/Zip:          NY 10023
Country Code:       US
Evening Phone:      212/555-9768
Daytime Phone:      212/555-2643
```

```
Account Status:          **TERMINATED**
Account Type:            NORMAL
BID Country:             us
Date Account Created:    00-05-13 10:45:16 EDT
Date Account Cancelled:  00-05-29 19:23:32 EDT

Last Screen Name Used:   CyberStalker
Last Logout Date:        80-01-01 00:00:00 EDT
Last Node:               Internal Ethernet

Billing Method:
Credit Card #:           xxxxxxxxxxxxxxxx
CC Expiration Date:      xxxx
CC Name:                 JOHN C. DOE

Account History:

Date/Time: 00-05-29 19:23                 Recorded By: AOLSTAFF1
Problem: OPSSEC (DNR): Legal Action (SW) Account Status Changed
Response:

Date/Time: 00-05-27 16:26                 Recorded By: AOLSTAFF2
Problem: Disable Member PW Reset Per Legal
Response:

Date/Time: 00-05-27 16:25                 Recorded By: AOLSTAFF2
Problem: Updated Account Groups Account Group Info Changed
Response: OLD INFO: // PR Index // PL Index // Affinity //

Date/Time: 00-05-27 16:25                 Recorded By: AOLSTAFF2
Problem: Updated Account Limits Account Limits Changed
Response: OLD INFO: // Max Cents = 0 // Max Minutes = 0 //

Date/Time: 00-05-27 16:24                 Recorded By: AOLSTAFF2
Problem: Screen Name & Password Changed Account Password(s)
Response:
[END OF REPORT]

Usage Details:
00-5-28 12:59   CyberStalker   0   10   0.00   0.00   0.00
00-5-28 08:09   CyberStalker   0    7   0.00   0.00   0.00
00-5-27 23:56   CyberStalker   0   37   0.00   0.00   0.00
00-5-27 22:49   CyberStalker   0    5   0.00   0.00   0.00
00-5-27 17:00   CyberStalker   0    7   0.00   0.00   0.00
00-5-27 14:38   CyberStalker   0    5   0.00   0.00   0.00
00-5-27 14:08   CyberStalker   0    2   0.00   0.00   0.00
00-5-27 11:24   CyberStalker   0    4   0.00   0.00   0.00
00-5-27 00:39   CyberStalker   0   31   0.00   0.00   0.00
00-5-26 20:37   CyberStalker   0   56   0.00   0.00   0.00
00-5-26 16:37   CyberStalker   0    3   0.00   0.00   0.00
<cut for brevity>
```

At this stage the police had enough evidence to obtain a restraining order. Additionally, Jack's employers decided to fire him because he had been neglecting his duties at work and had used their network to send many of the offending messages.

After being fired, Jack seemed to have even more time to carry out his campaign of harassment. In a successful effort to continue to antagonize Jill without violating the terms of the restraining order, Jack persuaded a friend that he made on the Internet to communicate certain things to Jill through e-mail. He also sent several packages to Jill's family that he claimed contained material that would disgrace her and cause them to disown her. Her family handed the packages over to the police unopened. Jill continued to suffer from the stress of the situation and her family had a natural concern for her health.

Although the police were ready to charge Jack with cyberstalking, Jill decided that the efforts to discourage his behavior were not having the intended effect of stopping the harassment. Jack's behavior had not escalated but had not decreased in intensity either. Rather than risk making matters worse by increasing the negative pressures on him, Jill decided not to bring charges against him. Instead, Jill moved to be physically distant from Jack.

With no target in plain view and no job to occupy his time, Jack had little to do. Although he threatened to follow Jill, he did not carry out this threat. His e-mail and AOL Buddy list that were obtained during the investigation indicated that Jack was developing online relationships with two other women. If Jill had pressed criminal charges, investigators would have contacted these other women. However, since Jill had dropped the charges against Jack and there were no complaints regarding his treatment of these other women, no further action was taken.

One of the most interesting aspects of Jack's behavior was his steady determination. He did not seem overly concerned by the negative pressures that were brought to bear on him (restraining order, losing job, threat of prosecution). His behavior did not intensify noticeably, nor did it decrease in intensity. Also notice that Jack changed his *modus operandi* when necessary. Each time one method of targeting Jill was thwarted he figured a new way to target her.

21.4 SUMMARY

Cyberstalking is not different from regular stalking – the Internet is just another tool that facilitates the act of stalking. In fact, many cyberstalkers also use the telephone and their physical presence to achieve their goals. Stalkers use the Internet to acquire victims, gather information, monitor victims, hide their identities, and avoid capture. Although cyberstalkers can become quite adept at using the Internet, investigators with a solid understanding of the Internet and a strong investigative methodology will usually be able to discover the identity of a cyberstalker.

With regard to a strong investigative methodology, investigators should get into the habit of following the steps described in the chapter (interviewing victims, developing victimology, searching for additional evidence, analyzing crime scenes, and understanding motivation).

The type of digital evidence that is available in a cyberstalking case depends on the technologies that the stalker uses. However, a cyberstalker's personal computer usually contains most of the digital evidence, including messages sent to the victim, information gathered about the victim, and even information about other victims.

It is difficult to make accurate generalizations about cyberstalkers because a wide variety of circumstances can lead to cyberstalking. A love interest turned sour can result in obsessive and retaliatory behavior. An individual's desire for power can drive him to select and harass vulnerable victims opportunistically. The list goes on, and any attempt to generalize or categorize necessarily excludes some of the complexity and nuances of the problem. Therefore, investigators who hope to address this problem thoroughly should be wary of generalizations and categorizations, only using them to understand available evidence further.

REFERENCES

Foote, D. (1999) "You Could Get Raped," *Newsweek,* February 8.

Associated Press (1997) "As Online Harassment Grows, Calls for New Laws Follow," April 1.

Harmon, R., Rosner, R., and Owens, H. (1998) "Sex and Violence in a Forensic Population of Obsessional Harassers' Psychology, Public Policy, and Law," *American Psychological Association* 4(1/2), 236–249.

Meloy, J. R. (1998) "The Psychology of Stalking," Meloy, J.R. (editor) *The Psychology of Stalking: Clinical and Forensic Perspectives*. New York: Academic Press, pp. 1–23.

Meloy, J. R. (1999) "Stalking: An Old Behavior, A New Crime." *Psychiatric Clinics of North America*.

National Center for Victims of Crime, Safety (Available at http://www.ncvc.org/infolink/svsafety.htm).

Pathe, M. and Mullen, P. E. (1997) "The Impact of Stalkers on Their Victims," *British Journal of Psychiatry* 170, 12–17.

## CHAPTER 22

DIGITAL EVIDENCE AS ALIBI

The key pieces of information in an alibi are time and location. When an individual does anything involving a computer or network, the time and location is often noted, generating digital evidence that can be used to support or refute an alibi. For example, telephone calls, credit card purchases, and ATM transactions are all supported by computer networks that keep detailed logs of activities. Telephone companies keep an archive of the number dialed, the time and duration of the call, and sometimes the caller's number. Credit card companies keep records of the dates, times, and locations of all purchases. Similarly, banks keep track of the dates, times, and locations of all deposits and withdrawals. These dates, times, and locations reside on computers for an indefinite period of time and individuals receive a report of this information each month in the form of a bill or financial statement.

Other computer networks, like the Internet, also contain a large amount of information about times and locations. When an e-mail message is sent, the time and originating IP addresses are noted in the header. Log files that contain information about activities on a network are especially useful when investigating an alibi because they contain times, IP addresses, a brief description of what occurred, and sometimes even the individual computer account that was involved. However, computer times and IP addresses can be manipulated, allowing a criminal to create a false alibi.

On many computers it requires minimal skills to change the clock or the creation time of a file. Also, people can program a computer to perform an action, like sending an e-mail message, at a specific time. In many cases, scheduling events does not require any programming skill – it is a simple feature of the operating system. Similarly, IP addresses can be changed, allowing individuals to pretend that they are connected to a network from another location. Therefore, investigators should not rely on one piece of digital evidence when examining an alibi – they should look for an associated cybertrail. This chapter discusses the process of investigating an alibi when digital evidence is involved, and uses scenarios to demonstrate the strengths and weaknesses of digital evidence as an alibi.

Digital Evidence and Computer Crime Second Edition
ISBN: 0-12-163104-4

22.1 INVESTIGATING AN ALIBI

When investigating an alibi that depends on digital evidence, the first step is to assess the reliability of the information on the computers and networks involved. Some computers are configured to synchronize their clocks regularly with very accurate time satellites and make a log of any discrepancies. Other computers allow anyone to change their clocks and do not keep logs of time changes. Some computer networks control and monitor which computers are assigned specific IP addresses using protocols like BOOTP and DHCP. Other networks do not strictly control IP address assignments, allowing anyone to change the IP address on a computer.

In some situations, interviewing several individuals who are familiar with the computer or network involved will be sufficient to determine if an alibi is solid. These individuals should be able to explain how easy or difficult it is to change information on their system. For example, a system administrator can usually illustrate how the time on a specific computer can be altered and the effects of such a change. If log files are generated when the time is changed, these log files should be examined for digital evidence related to the alibi.

In other situations, especially when an obscure piece of equipment is involved, it might be necessary to perform extensive research – reading through documentation, searching the Internet for related information, and even contacting manufacturers with specific questions about how their products function. The aim of this research is to determine the reliability of the information on the computer system and the existence of logs that could be used to support or refute an alibi. If no documentation is available, the manufacturer is no longer in business, or the equipment/network is so complicated that nobody fully understands how it works, it might be necessary to recreate the events surrounding the alibi to determine the reliability of the associated digital evidence.

By performing the same actions that resulted in an alibi, an investigator can determine what digital evidence should exist. The digital data that are created when investigators recreate the events surrounding an alibi can be compared with the original digital evidence. If the alibi is false, there should be some discrepancies. Ideally, this recreation process should be performed using a test system rather than the actual system to avoid destroying important digital evidence. A test system should resemble the actual system closely enough to enable investigators to recreate the alibi that they are trying to verify. If a test system is not available it is crucial to back up all potential digital evidence before attempting to recreate an alibi.

It is quite difficult to fabricate an alibi on a network successfully because an individual rarely has the ability to falsify digital evidence on all of the computers that are involved. If an alibi is false, a thorough examination of the computers involved will usually turn up some obvious inconsistencies. The most challenging situations arise when investigators cannot find any evidence to support or refute an alibi. When this situation arises, it is important to remember an axiom from Forensic Science – absence of evidence is not evidence of absence. If a person claims to have checked e-mail on a given day from a specific location and there is no evidence to support this assertion, that does not mean that the person is lying. No amount of research into the reliability of the logging process will change the fact that an absence of evidence is not evidence of absence. It is crucial to base all assertions on solid supporting evidence, not on an absence of evidence. To demonstrate that someone is lying about an alibi, it is necessary to find evidence that clearly demonstrates the lie.

CASE EXAMPLE

A suspect claims to have been at work during the weekend at the time of a homicide, fixing a network problem, and checking e-mail. The investigators were not familiar with computer networks and depended heavily on the system administrators at the organization where the suspect worked. Unfortunately, the system administrators were not fully briefed on the details of the case and did not have all of the information necessary to examine their log files thoroughly.[1]

As a result, one of the most important IP addresses involved was not included in the search and the investigators could not find any indication that the suspect checked his e-mail. The investigators jumped to the conclusion that the suspect was lying about his alibi based on this absence of evidence.

A few days later, the suspect was at work and noticed a timestamp that was created when he fixed the network problem on the day of the crime. The suspect prudently asked his coworkers to witness and document the evidence. However, when the suspect presented this evidence to the investigators, they were incredulous, assuming that he had fabricated the timestamp after the fact. However, the truth of the matter was that the investigators did not research the network components involved and did not recognize an important source of digital evidence. Their negligence led them to suspect the wrong man, causing over two years of disruption in his life, costing him his job, costing the state and organization untold amounts of money, and worst of all, letting the actual murderer go free.

[1]The oversight was noticed several years later when the case was being tried.

Although absence of evidence is not necessarily evidence of absence, an alibi can be severely weakened by a lack of expected digital evidence. In one case, a homicide suspect claimed that he had been at work when the crime occurred and that he was using a particular computer for several hours. The computer in question showed no sign of use during that period, contradicting the suspect's alibi. He was subsequently convicted of the crime.

An interesting aspect of investigating an alibi is that no amount of supporting evidence can prove conclusively that an individual was in a specific place at a specific time. With enough knowledge and resources, any amount of physical and digital evidence can be falsified to fabricate an alibi. Therefore, a large amount of supporting evidence indicates that the alibi is probably true, but not definitely true. For this reason, it rarely makes sense for a defense attorney to spend time and resources searching for digital evidence that supports a client's alibi. No amount of evidence will prove that the alibi is true and the more the alibi is examined, the more likely it is that an inconsistency will be found that could weaken the attorney's ability to defend the client.

22.2 TIME AS ALIBI

Suppose that, on March 19, 1999, an individual broke into the Museum of Fine Arts in Boston and stole a precious object. Security cameras show a masked burglar entering the museum at 2000 hours and leaving at 2030 hours. The prime suspect claims to have been at home in New York, hundreds of miles away from Boston, when the crime was committed. According to the suspect, the only noteworthy thing he did that evening was to send an e-mail to a friend. The friend is very cooperative and provides investigators with the following e-mail:

```
From: suspect@newyork.net
Date: Fri, 19 Mar 1999 20:10:05 EST
Subject: A quick hello
To: witness@miami.net
I am sitting innocently at home with nothing to do and I thought
I would drop a line to say hello.
```

The e-mail does suggest that the suspect sent the message at the time of the burglary. However, the investigators are familiar enough with e-mail to know that the header will contain dates and times of all of the computers that handled the message. They obtain the full header and examine it for any discrepancies.

```
Received: from mail.newyork.net by mail.miami.net (8.8.5/8.8.5) with ESMTP id
NAA23905 for <witness@miami.net>; Sat, 20 Mar 1999 13:49:19 -0500 (EST)
Received: from suspectshome.newyork.net by mail.newyork.net (PMDF V5.1-0
#20971) with SMTP id <01J9206HG9T400NWE6@newyork.net> for
witness@miami.net; Sat, 20 Mar 1999 13:49:22 EST
```

```
From: suspect@newyork.net
Date: Fri, 19 Mar 1999 20:10:05 EST
Subject: A quick hello
To: witness@miami.net
Message-id: <01J9206VTW2E00NWE6@newyork.net>
I am sitting innocently at home with nothing to do and I thought I would drop
a line to say hello.
```

Sure enough, the dates and times in the header do not match, indicating that the e-mail message was forged on the afternoon of March 20. The suspect's alibi is refuted. The investigators obtain the related log entries from the two mail servers that handled the message (mail.newyork.net and mail.miami.net) as further proof that the message was sent on March 20 rather than on the night of the crime. Additionally, the investigators search the suspect's e-mail and discover messages that he sent to himself earlier in the week, testing and refining his forging skills. Finally, to demonstrate how the suspect sent the forged e-mail, the investigators perform the following e-mail forgery steps, inserting the false date (Friday, 19 March 1999 20:10:05 EST) just as the suspect did:

```
% telnet mail.newyork.net 25
Trying 10.232.19.48...
Connected to mail.newyork.net.
Escape character is '^]'.
220 mail.newyork.net – Server ESMTP (PMDF V5.1-10 #20971)
helo suspectshome.newyork.net
250 mail.newyork.net OK, suspectshome.newyork.net.
mail from: suspect@newyork.net
250 2.5.0 Address Ok.
rcpt to: witness@miami.net
250 2.1.5 witness@miami.net OK.
data
354 Enter mail, end with a single ".".
Subject: A quick hello
Date: Fri, 19 Mar 1999 20:10:05 EST
I am sitting innocently at home with nothing to do and I thought
I would drop a line to say hello.
.
250 2.5.0 Ok.
quit
```

After being presented with this evidence, the suspect admits to stealing the precious object and selling it on the black market. The suspect identifies the buyer and the object is recovered.

22.3 LOCATION AS ALIBI

Suppose that the same precious object was stolen again when the burglar from the previous scenario was released from prison a few months later. This time, however, the burglar claims to have been in California, thousands of miles away, starting a new life. The burglar's parole officer does not think that the suspect left California but cannot be certain. The only evidence that supports the suspect's alibi is an e-mail message to his friend in Miami. Though the suspect's friend is irritated at being involved again, she gives the investigators the following e-mail:

Received: from mail.california.net by mail.miami.net (8.8.5/8.8.5) with ESMTP id NAA23905 for, witness@miami.net.; Fri, 21 May 1999 22:03:46 EST -0500 (EST)

Received: from suspectshome.california.net by mail.california.(InterMail v03.02.07 118 124) with SMTP id <19990521220346.CBJN9925@california.net> for <witness@miami.net>; Fri, 21 May 1999 22:03:46 +0000

From: suspect@california.net

Date: Fri, 21 May 1999 22:03:46 EST

Subject: New E-mail Address

To: witness@miami.net

Message-id: <001801be724c$dc842000$1f02480c@california.net>

I have moved to California to start afresh. You can send e-mail to me at this address.

The investigators examine the e-mail header, determine that it was sent while the burglar was in the museum, and find no indication that the e-mail was forged. The suspect claims that someone is trying to frame him and assures the investigators that he has no knowledge of the crime. The following month, when the Museum of Fine Arts received its telephone bill, an administrator finds an unusual telephone call to California on the night of the burglary. The investigators are notified and they determine that the number belongs to an ISP in California (california.net). Unfortunately, the ISP's dialup logs were deleted several weeks earlier and there is not enough evidence to link the suspect to the telephone call. The investigators search the suspect's computer but do not find any incriminating evidence.

Investigators are stumped until it occurs to them to investigate the suspect's friend in Miami more thoroughly. By examining the friend's credit card records, the investigators determine that she bought a plane ticket to Boston on the day of the burglary. Also, the investigators find that her laptop is configured to connect to california.net and her telephone records show that she made several calls from Miami to the ISP while planning the robbery. Finally, investigators search the slack space on her hard drive and find remnants of the e-mail message that she sent from the Museum of Fine Arts during the robbery. When presented with all of this digital evidence, the woman admits to stealing the precious object and implicating the original suspect. This time a different buyer is identified and the object is recovered once again.

As noted in previous chapters, many sources of digital evidence can reveal the location of an individual, including their mobile telephone.

22.4 SUMMARY

As investigators learn about new technologies, it is useful to think about how they will affect routine aspects of investigations such as alibis. With people spending an increasing amount of time using computers and networks, there are bound be more alibi's that depend on digital evidence. Computers contain information about times and locations that can be used to confirm or refute an alibi. However, digital evidence can rarely prove conclusively that someone was in a specific place at a specific time. Remember that IP addresses are associated with computers – not individuals. Therefore, an accomplice could help a criminal fabricate an alibi using the criminal's computer. Also, some computer times can be changed to corroborate an alibi. By following the cybertrail, investigators might find a computer program that simulated an alibi or they might learn that the computer clock was changed at the time in question.

Though it is easy to change the time of a personal computer, many computers keep a log of time changes. Also, when dealing with computers on a network, it becomes more difficult to change computer times. When multiple computers are involved, changing the time on one will result in a notable inconsistency with others. Therefore, when examining an alibi that involves a computer or network, investigators should search log files for time inconstancies.

PART 5

GUIDELINES

## CHAPTER 23

DIGITAL EVIDENCE HANDLING GUIDELINES

Technology is advancing at such a rapid rate that the suggestions in this guide must be examined through the prism of current technology and the practices adjusted as appropriate. It is recognized that all crime scenes are unique and the judgment of the first responder/investigator should be given deference in the implementation of this guide.
(USDOJ (2001), "Electronic Crime Scene Investigation: A Guide for First Responders")

This chapter provides guidelines for handling digital evidence, summarizing the detailed discussion in Chapter 9. The primary aim of this chapter is to assist in the development of procedures and crime scene protocol that minimize the chance of injury and contamination of evidence. Keep in mind that a procedure cannot cover all eventualities and individuals handling digital evidence may need to deal with unforeseeable situations. Therefore, all individuals handling evidence should have sufficient training and experience to implement procedures and deal with situations that are not covered by procedures.

In addition to developing procedures, it is advisable to equip individuals who are handling digital evidence with items such as tools and surgical gloves. Using proper tools reduces the risk of injury such as deep cuts when too small a screwdriver slips on a tight screw, causing one's hand to hit sharp metal edges inside the computer. Surgical gloves help preserve fingerprints and other trace evidence, while protecting individuals from hazardous materials. Some crime scenes, such as drug laboratories, may require additional protection.

Prior to handling digital evidence, the crime scene should be secured, preventing anyone from touching the computer and associated items. Make an effort to prevent anyone from accessing the system via a wireless connection (e.g. infrared or bluetooth). Also, take notes that will be useful when reconstructing the scene and draw diagrams with the overall dimensions to get an overview of the scene and make it easier to remember and explain where things were found. Assign each room a letter and each source of digital evidence a number to keep track of the items. Although authorization is presumed, it is worth reiterating that a warrant or other authorization should be obtained prior to implementing these guidelines.

Digital Evidence and Computer Crime Second Edition
ISBN: 0-12-163104-4

23.1 IDENTIFICATION OR SEIZURE

The aim of this step is to locate likely sources of digital evidence and seize them. In some cases, such as computer intrusion investigations, it may be necessary to extract information from RAM (Figure 23.1).

- Look for hardware. In addition to desktop computers, look for laptops, hand held computers, external hard drives, digital cameras, and any other piece of equipment that can store evidence related to the crime being investigated. If the hardware is being collected for future examination, consider collecting

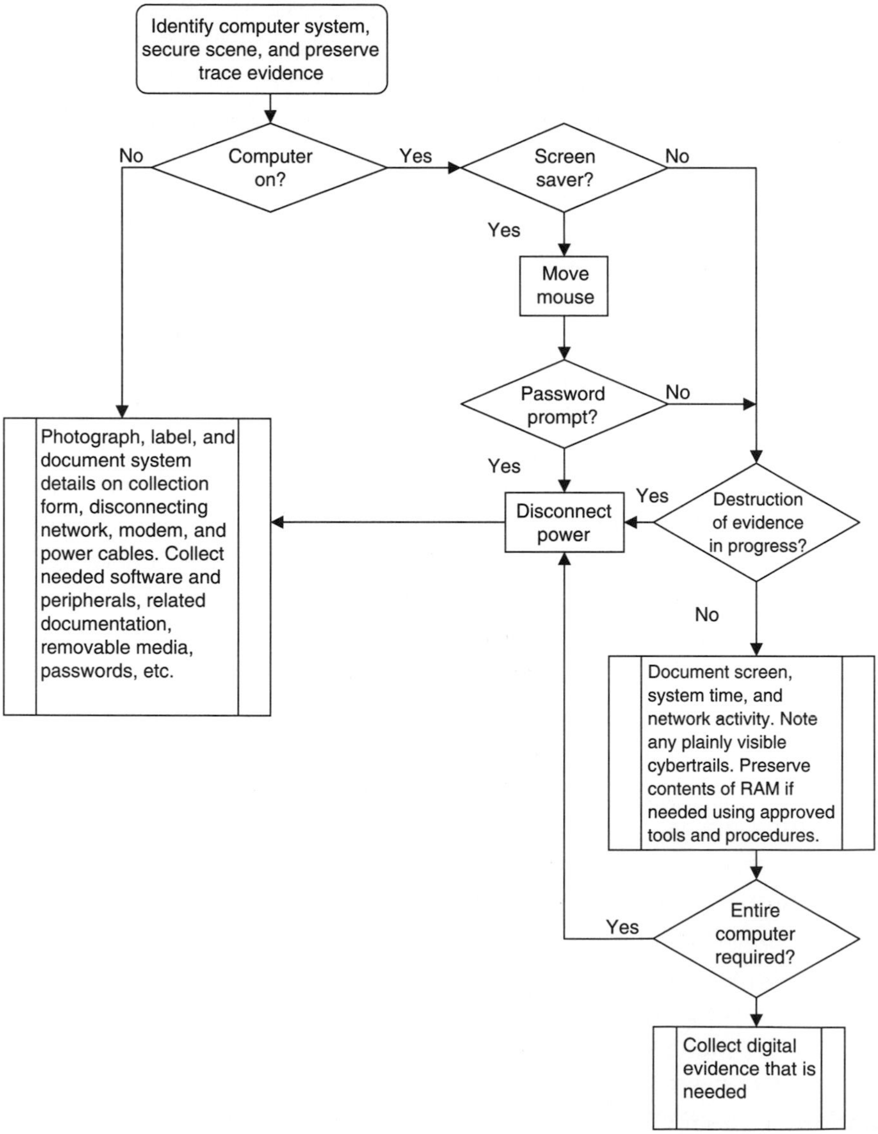

Figure 23.1

Overview of identification and seizure process.

peripheral hardware that is attached to the computer. Also collect any peripheral hardware that needs to be examined by a digital evidence examiner. For example, printers, cameras, and scanners might have unique characteristics that can be linked to documents or digitized images.

- Look for software. If digital evidence was created using a program that is not widely used, collecting the installation disks will make it easier to examine the evidence.
- Look for removable media. There are a wide variety of removable media that can contain digital evidence including floppy disks, Zip/Jazz disks, compact disks, and magnetic tapes. In particular, look for backups either on-site or in a remote storage facility. Determine what hardware and software was used to make the backups. In some instances, backup tapes can only be accessed using the type of hardware and software that created them. Therefore, consider collecting the unusual backup hardware and software. It is not necessary to collect hardware and software if a common, readily available method of backup was used. Keep in mind that criminals often hide removable media that contain incriminating or valuable information.
- Look for documentation that is related to the hardware, software, and removable media. Documentation can help investigators understand details about the hardware, software, and backup process that are useful during an investigation and a trial. Also, the existence of books on encryption, digital evidence, and other technical topics can help assess the technical skill of the suspect and what to look for on computers.
- Look for passwords and important telephone numbers on or near the computer. Individuals who have several Internet Service Providers often write down the phone numbers and passwords for their various accounts. This is especially true of computer intruders. Passwords and other useful information may also be obtained through interviews with people involved.
- Look through the garbage for printouts and other evidence related to the computer. Computer printouts can contain valuable evidence and can sometimes be compared with the digital copies of the information for discrepancies.
- Look for cybertrails as described throughout this book.
- Unplug the modem or network cables from the computer. Consider testing the phone jack for a dial tone or data port to ensure that they are active.
- Photograph evidence *in situ*, paying particular attention to serial numbers and wiring to help identify or reconstruct equipment later. This type of vivid documentation, showing evidence in its original state, can be useful for reconstructing a crime and demonstrating that evidence is authentic. Also consider removing casing and photographing internal components, including close-ups of hard drive jumper settings, and other details.
- Dust for fingerprints and collect other trace evidence if it may be useful to the investigation.
- Note or photograph the contents of the computer screen. If a program is running that might be destroying data, immediately disconnect power to that computer by pulling the cable out of the rear of the computer.
- If the system is on, a judgement must be made as to whether to gather informaion from the system such as checking the system clock for accuracy and network

neighborhood for connected machines. Note that a wily individual could create a link named "Network Neighborhood" that actually runs a destructive program. To limit the risks associated with operating the computer, it is advisable to use trusted utilities such as statically compiled executables on a CD-ROM. Any actions performed on the system must be clearly documented to enable others to assess the impact this process had on the system.

- If applications are open, save their contents to a sanitized, labeled collection disk before closing. Preserve other data in the RAM as needed using approved tools and procedures. Although it may be necessary to print out certain items, be aware that this process creates spool files that can alter the system. When dealing with printouts, initial and date each page.
- Shutdown the computer if necessary.
- Before copying data, calculate MD5 values of all disks and the files they contain, recording the values for future reference.
- Label, date, and initial all evidence. Write protect media when possible and check for obvious signs of damage. If other people are collecting evidence, record their names and where they found the evidence. The aim is to preserve chain of custody and document the evidence in a way that helps investigators reconstruct the crime. Not knowing where evidence came from, or who collected it, can render it useless.
- Whenever possible, a copy of digital evidence should be preserved on storage media that can only be written to once and are stable for long-term storage, like compact disks.
- Store in sealed envelope and secure in an evidence room or safe.

23.1.1 WHEN THE ENTIRE COMPUTER IS REQUIRED

- Label cables and ports. Empty ports should be labeled "unused." If there is no label on a port, it could be argued that the evidence was not properly documented or that the label fell off. Any doubts that can be shed on the evidence collection and documentation process can weaken a case.
- Put an unused disk in each floppy drive to protect the drive.
- Use evidence tape to seal the computer case and drives to protect them against tampering.
- Carefully package the hardware and do not expose it to potentially damaging conditions (e.g. dirt, fluids, humidity, impact, excessive heat and cold, and static electricity).

23.2 PRESERVATION

- Unplug power from hard drives.
- If booting the evidence system from a Evidence Acquisition Boot Disk, check the CMOS settings to ensure it is configured to boot from the floppy disk before the hard drive. Also, test booting from the floppy while the hard drives are disconnected to ensure that system boots successfully from floppy disk (a faulty floppy drive can prevent booting from a floppy, causing the computer

to boot from the hard drive). Another approach is to connect the hard drive to an evidence collection system as discussed in Chapter 9.

- Note the current date and time and the date/time on the computer (note any discrepancies).
- Make two copies of all digital evidence to sanitized collection disks (consider making a bitstream copy if there might be valuable evidence in slack space). Whenever possible, check each copy on another computer to ensure that the copy was successful.
- Check size and integrity of data, to determine if there are hidden partitions or the acquisition was incomplete for some reason.
- Label, date, and initial all evidence. Include the type of computer (e.g. Digital Alpha, Sun Sparc2) and operating system (e.g. Windows 95, Mac OS, UNIX), what program(s) and/or command(s) you used to copy the files.
- Inventory contents of all disks, including attributes such as physical sector location, file creation and modification dates. Ideally, inventory original media to document the directory structure – this provides context of where each file was located on the original system and, upon closer inspection, may reveal files that have been overlooked. Calculate the message digest of all files and disks. Also, make a brief note describing the significance of the evidence to help others understand why it was collected. This type of inventory is not only useful for documentation purposes, it also gives an overview of what is on the system, what types of applications are installed, if encryption might be used.

23.2.1 IF ONLY A PORTION OF THE DIGITAL EVIDENCE ON A COMPUTER IS REQUIRED

- Note the current date and time and the date/time on the computer (note any discrepancies). If investigators do not realize that a computer clock is inaccurate this can skew their crime reconstruction. For instance, if the time a file was created is important, investigators should be sure that they know the actual time the file was created and not an inaccurate time set by the computer.
- Note full file and path names, date–time stamps, sizes, and MD5 values of files.
- Compress files into an archive to preserve their data–time stamps and save the archive to sanitized collection disks.
- Also make two copies of all evidence in uncompressed form to sanitized collection disks. Whenever possible, check each copy on another computer to ensure that the copy was successful.
- Label, date, and initial all evidence using an indelible felt-tipped pen. Include the name of the operating system (e.g. Windows 95, Mac OS, UNIX), what program(s) and/or command(s) you used to copy the files.
- Inventory contents of all disks, including file creation and modification dates. Calculate the message digest of all files and disks. Also, make a brief note describing the significance of the evidence to help others understand why it was collected.

23.2.2 SAMPLE PRESERVATION FORM

DIGITAL EVIDENCE FORM

Investigator's Name and Association: Eoghan Casey Knowledge Solutions	*Case No.*: 2003040601 *Date*: April, 4, 2003
Location of Computer/Media (full address) Corporation X, Building 6, Redmond, CA	*Name of Suspect(s)/Type of Case*: John Doe/Information Theft

EVIDENTIARY SYSTEM

Computer/Processor: Sony Vaio/Celeron	*Make and Model*: PCG-R5050TLK (PCG-1362)
Name and Address of System Owner: Corporation X, Main Office Redmond, CA NOTE ➡ 510-555-3465	It is an offense to gain unauthorized access to a computer, its software or data. Do you have authorization to undertake this backup/examination? ☐
Serial No.: 325-67545	*Photographic Exhibit No.*: 2003040601-3
CMOS Date and Time: 04/06/2003, 14:30 *Actual Date and Time*: 04/06/2003, 14:32	Hard Drive MD5: (45D8C0A5308D120A4DD85E36B03F9926)
	Software: dd and EnCase

EXAMINATION SYSTEM

Computer/Processor: Dell/Intel Pentium 4	*Make and Model*: Dimension 4600C
Serial No.: 35-6465466	*CMOS Date and Time*: 04/06/2003, 14:54 *Actual Date and Time*: 04/06/2003, 14:54

EVIDENCE FILES (two independent copies)

Name	Creation Time	Size (bytes)	Message Digest[1]
sony1-1.dd	04/06/2003 15:02	601435	343e16d6551e84d35c176375728fbbf4
sony1-2.dd	04/06/2003 15:22	354676	ab487d36057d446b6a8b72091da72f23
sony1.E01	04/06/2003 15:46	613354	e6dd075b82677fc0be6f88f1fb941224
sony1.E02	04/06/2003 16:30	454643	5d6330ca0adaa43c6639b68f6b2db48b

Other Media:
Floppy disks inventoried on attached sheet

Evidence Bag:
Hard drive stored in evidence room ☐

Comments:
System returned to owner without drive

[1]These MD5 values relate to the evidence files not the original disk. When they are reconstituted, the MD5 value of the copy made using dd and the EnCase internal MD5 value should match that of the original drive.

## CHAPTER 24

DIGITAL EVIDENCE EXAMINATION GUIDELINES

Eoghan Casey and Troy Larson

The forensic sciences require adherence to standards of operation and of performance. These standards must be clearly enunciated and must be, at least in their basic form, the consensus of opinion of workers in that particular subject area. Stated differently, forensic scientists are not entitled to indulge whims in the conduct of their work. They must adhere to performance norms which have been previously laid down. A forensic scientist who adopts an extreme position that runs counter to the flow of prevailing opinion on a subject, or who enters an area in which operational norms have not been established, has a burden even greater than usual to justify that position in the light of good scientific practice. (Thornton J.I. (1997) "The General Assumptions and Rationale of Forensic Identification," for David L. Faigman, David H. Kaye, Michael J. Saks, & Joseph Sanders, Editors, Modern Scientific Evidence: The Law and Science of Expert Testimony, Volume 2, St. Paul, MN: West Publishing Company)

With the decreasing cost of data storage and increasing volume of commercial files in operating system and application software, forensic computer examiners can be overwhelmed easily by the sheer number of files contained on even one hard drive or backup tape. Accordingly, examiners need procedures, such as that outlined below, to focus in on potentially useful data. Less methodical analysis techniques, such as searching for specific keywords or extracting only certain file types, may not only miss important clues but can still leave the examiners floundering in a sea of superfluous data.

A procedure such as the one detailed in the *Handbook of Computer Crime Investigation*, Chapter 2 (Larson 2001) provides a means for the investigator to intelligently reduce data and obtain consistent, reliable results. Important aspects of this procedure are demonstrated in this chapter. While the data processing steps outlined here focus on preparing electronic records for civil litigation, the process of filtering out irrelevant, confidential, or privileged data is applicable to many forensic computer analysis situations, including:

- Eliminating valid system files and other known entities that have no relevance to the investigation.

Digital Evidence and Computer Crime Second Edition
ISBN: 0-12-163104-4

- Focusing an investigation on the most probable user-created data.
- Managing redundant files, which is particularly useful when dealing with backup tapes.
- Identifying discrepancies between forensic computer analysis tools, such as missed files and MD5 hash errors.

Additionally, the output of this process provides a solid foundation for subsequent analysis, including classification, individuation, evaluation of source, and temporal reconstruction.

This chapter demonstrates three approaches to implementing the evidence processing methodology. The first approach uses command line utilities, primarily from Maresware.[1] Sample batch and configuration files described in this chapter are available on the Web site associated with this book. The other two approaches use the GUI tools: EnCase and FTK. The same methodology can be translated to UNIX-based tools.

[1] *http://www.maresware.com*

24.1 PREPARATION

It is a good practice to begin a new matter by preparing an organized, sanitized working environment with ample space. In forensic computer analysis, this involves preparing adequate and safe media on which to copy the data to be processed. As a general rule, we recommend "wiping" or overwriting a hard drive with a known pattern of data before formatting it to receive the new, case-related data. Wiping the hard drive prevents cross-contamination between evidence sources and using a known pattern of data enables you to verify that the wiping was performed correctly – empty space should only contain the overwrite pattern. We also recommend labeling/naming your work drives during formatting so that they are easily identifiable weeks, months, or years into the future. Next, create a practical directory structure on the working hard drive. We recommend building a customary set of working directories prior to beginning processing. In addition to providing the examiner with an organized work environment, this directory structure imposes a structure on the work product that is reproducible and understandable by one's coworkers. The following is a sample directory structure for organizing a Work drive:

\Prepare – The root directory to hold files requiring further processing.

\Prepare\special – Holds encrypted, compressed, undeleted files prior to further processing.

\Prepare\slack – Holds extracted slack space prior to further processing.

\Prepare\pcluster – Holds extracted unallocated cluster data prior to further processing.

\Review – The root directory to hold the final work product.

\Review\files – Holds the reduced set of unprocessed files.

\Review\slack – Holds the processed slack space data.

\Review\clusters – Holds the processed unallocated cluster data.

\Review\processed – Holds other processed data (e.g. decrypted data, expanded zip files, e-mail).

\Duplicates – Holds duplicate files.

\Accounting – Holds all logs or reports generated in the processing activity. This is the examiner's audit trail.

Note that once the examiner has decided upon a model scheme of organizing data on the Work drive, the directory setup can be automated using a batch file.

As a final point on beginning preparations, it is advisable to use read-only devices when operating on the original or source data. The processing steps presented here will alter crucial aspects about the processed files, not the least of which are the file date and time stamps. Accordingly, the examiner should always protect his or her source data.

24.2 PROCESSING

The data processing methodology discussed in Larson (2001) involves several steps to reduce the number of files that require further analysis and convert unreadable data into a readable form. We will focus on a basic set of procedures.

24.2.1 DOS/WINDOWS COMMAND LINE – MARESWARE

The command line is not dead. In fact, it remains a powerful tool in the examiner's arsenal. Command line tools enable examiners to perform very specific, auditable tasks. Additionally, by scripting a series of commands together, examiners can create very powerful batch files to automate a substantial portion of evidentiary processing, thereby increasing productivity while reducing the chances of human error during routine tasks. Additionally, the output of command line tools, like Maresware, can be readily used as input to other tools without the need for reformatting the data. Far from being an anachronism, the command line provides the examiner a means to reduce overall processing time and to reduce the chance of human error.

24.2.1.1 GENERATE FILE LISTS AND HASH VALUES

Before filtering out irrelevant and unwanted data, capture the initial state of the source data to ensure that you have a baseline that can be used as a point of reference to check all subsequent processing for accuracy. To capture the initial state of the data with a command line tool, the examiner generates a comprehensive list of all the files in the source data, along with associated information such as: long and short file names, extensions, last written or modified dates and times, created dates and times, last accessed dates and times, logical sizes, file paths, and file hash values.

The following command line uses Maresware's hash program to produce a list of all files on a given volume:

```
hash -p [source drive] -o [Work drive]:\Accounting\[case name or number].txt -i
-v -S -w 160 -A -1 acct-ing
```

The examiner can easily turn this command into a simple, reusable batch file with three arguments: source drive containing evidence; case number; and working drive.[2]

[2]*The How-To of batch files in general is well beyond the scope of this paper. A good, current reference for the Windows NT/2000/XP command line is Brian Knittel, Widows XP Under the Hood: Hardcore Windows Scripting and Command Line Power (Que 2003) and*

```
@echo off
rem  Usage: [batch file name].bat source-drive case-number-or-name
     Working-drive
rem  Example: hasher.bat D NewCase001 d
echo.
echo  Generating file list for %2. Source drive is %1:\. File list will be written
      to %3:\Accounting\%2.txt.
hash -p %1:\ -o %3:\Accounting\%2.txt -i -v -S -w 160 -A -1 acct-ing
echo  Done. File list %2.txt created at %3:\Accounting\
echo.
```

These command line options instruct hash to process the specific volume (-p) immediately (-i) with no verbose headers (-v), placing the resulting list of hashes, all three file times (-A), and any Alternate Data Streams (-S) in an output file (-o) with file names limited to 160 characters. Limiting file names to 160 characters facilitates later comparison by maintaining a constant field width that is, 242 characters on each line. The acct-ing file contains a record of the command options and when the list was generated. Other commands will append to this accounting file throughout this procedure and it should be moved to \Accounting when the entire process is complete.

Depending on the examiner's needs, it may also be necessary to recover deleted files into \Prepare\special\deleted and then to generate a file list as described above.

24.2.1.2 RECOVER FILE SLACK AND UNASSIGNED CLUSTERS

Recover unallocated space into \Prepare\pcluster directory and slack space into \Prepare\pslack directory (e.g. using NTFSGETS and NTFSGETF from NTI).

Examine unallocated space and extract relevant information into \Review\rcluster. For instance, you may want to carve out graphics, documents, e-mail, and other files/fragments of interest using a command line tool such as NTI's Graphics File Extractor. Similarly, extract relevant information such as readable text from slack space and place in \Review\rslack.

24.2.1.3 *REMOVE DUPLICATE, KNOWN, AND OTHER UNNECESSARY FILES*

Before removing duplicates and files with certain extensions, heed the cautionary discussion in Chapter 2 of the Handbook. Also, before removing files with certain extensions, identify file extension/signature mismatches using diskcat +h sigs.fle and move them to \Prepare\special\sigmismatch directory. For instance, the following list of files with a .GIF extension shows one executable and one unknown (UNK) file that require further inspection:

```
D:\.diskcat -p d:\evidence -r -f *.gif +h reference.fle
Program started Fri Oct 25 13:05:00 2002 GMT, 09:05 Eastern Standard Time (-5*)
D:\evidence\diskcat.gif                       135368 A.....
exe
D:\evidence\keykatch.gif                      19042 A.....
gif
D:\evidence\rpsort.gif                        18597 A.....
UNK
D:\evidence\unknown.gif                       2135 A.....
gif
D:\evidence\xtacacs.gif                       6522 A.....
gif
Processed 5 files, 181,664 bytes: Elapsed: 0 hrs. 0 mins. 0 secs.
```

The following sample batch file moves duplicate files to \duplicates and eliminates known and unnecessary files using upcopy. In this example, known files are identifies using the Hashkeeper hash set:[3]

[3] http://www.hashkeeper.org/

```
@echo off
set ACCT=ON
echo Generating hash list for \%2 on drive %1:\
hash -p %1:\ -o %3:\hashes\%2 -i -v -S -w 160 -1 acct-ing
echo Removing duplicates
hash_dup -i %3:\hashes\%2 -o %3:\hashes\%2.dup -m -1 acct-ing
upcopy -S %3:\hashes\%2.dup -d %3:\duplicates -m -i -v -E -R -A >
%3:\Accounting\dupcopy.log
echo Sorting hash list
rpsort %3:\hashes\%2.dup %3:\hashes\%2.srt /F242 /+171:32
sortchek %3:\hashes\%2.srt -r 242 -p 170 –l 32
echo Comparing hash list with hashkeeper database
compare hk_pfix.srt %3:\hashes\%2.srt %3:\hashes\%2.out
compare.par -u -A
echo Extracting the names of known files from the hash list
```

```
filbreak %3:\hashes\%2.out %3:\working\%2.fle match.brk -A
echo Copying desired files listed in %3:\hashes\%2.fle to %3:\review\rfiles
upcopy -S %3:\hashes\%2.fle -d %3:\review\rfiles -m -A -x *.exe *.dll *.com *.hlp *.p
st *.ini *.inf *.mp3 *.drv *.fon *.ocx *.swp *.sys *.vxd
```

All of the commands in this batch file are from Maresware.com except **rpsort**, which is available from Simtel.net.[4] The hk_pfix.srt file is a text version of the Hashkeeper database created using the Maresware **lbatch** scripts. The compare.par and match.brk are configuration files for **compare** and **filbreak**, respectively. The format of all Maresware configuration files are described in the help files for the associated tools.

[4] *http://www.simtel.net/pub/pd/51643.shtml*

As a final step, the above batch file moves remaining files to the review directory for further examination.

24.2.2 WINDOWS GUI – EnCase

Prior to filtering data using EnCase, verify the integrity of the evidence files and update the known hash files using the Hash Sets option on the Tools menu. Also, to avoid cross contamination between cases, verify that no unwanted hash sets from previous examinations are selected. Then use the search routine to compute MD5 values of all files and identify file extension mismatches as shown in Figure 24.1(a) and (b) (no keywords are required).

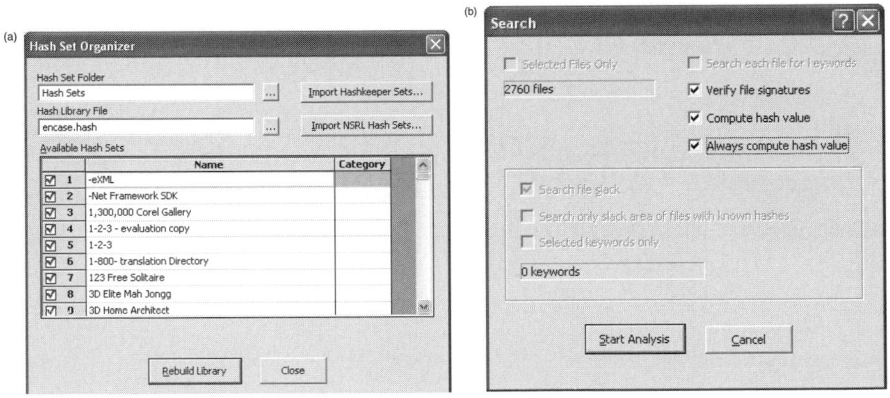

Figure 24.1

(a) Hash set organizer using the NIST NSRL hash set (http://www.nsrl.nist.gov/).
(b) Calculate MD5 values and identify file extension mismatches.

24.2.2.1 GENERATE FILE LISTS AND HASH VALUES

To create a list of files with their properties, choose the Export option on the Edit menu to launch the Export dialog box and select the properties that you desire as shown in Figure 24.2.

Although EnCase automatically recovers some deleted files, forensic examiners may also need to recover deleted directories and associated files from unallocated space into **\Prepare\special\deleted** and then to generate a file list as described above.

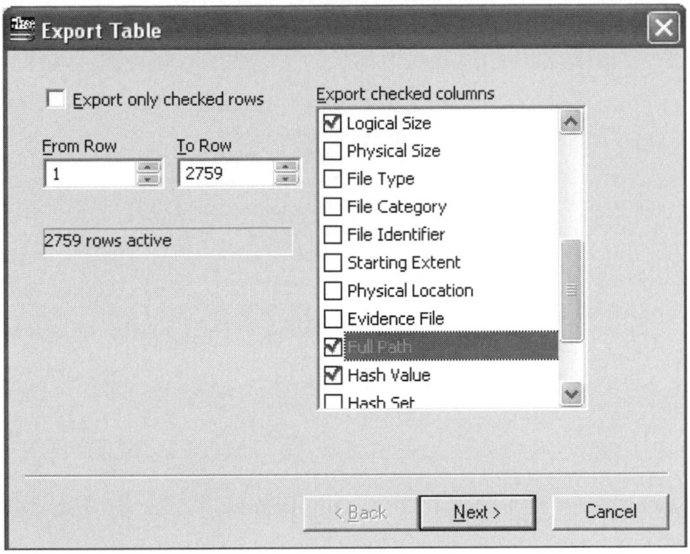

Figure 24.2

Exporting a file list using EnCase.

24.2.2.2 RECOVER SLACK AND UNASSIGNED CLUSTERS

To export slack using EnCase, tag the entire drive, right click on the Table view, choose Copy/Unerase, and select options to save RAM and Disk Slack to a file as shown in Figure 24.3 (a) and (b).

To export the contents of unassigned clusters, select the Unallocated Clusters in the Table view and save to **\Prepare\pcluster** using Export/Copy. You can extract various types of data from unallocated space using EnCase Escripts available at the EnScript library.[5]

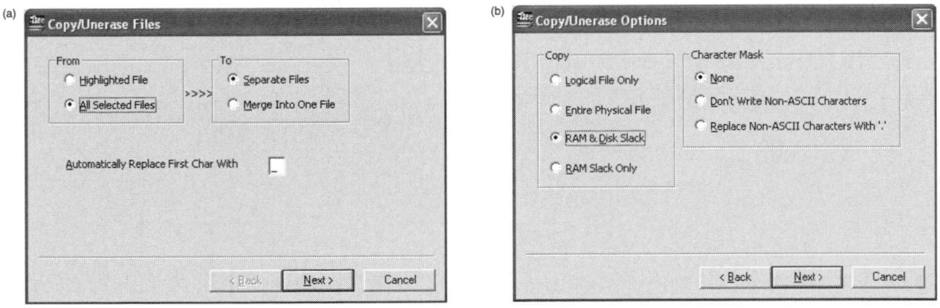

Figure 24.3

Exporting slack space using EnCase.

24.2.2.3 REMOVE DUPLICATES, KNOWN, AND UNNECESSARY FILES

While hashing files as described at the beginning of this section, EnCase updates the Hash Value and Hash Set fields in the Table view. By sorting on Hash Set, you can identify all unknown files by an empty Hash Set field. Tag all unknown files with a check mark in the Table View, dual sort the file list with the primary sort on tagged and the secondary sort on File Signature, and untag unwanted file types by signature. Finally, dual sort with the

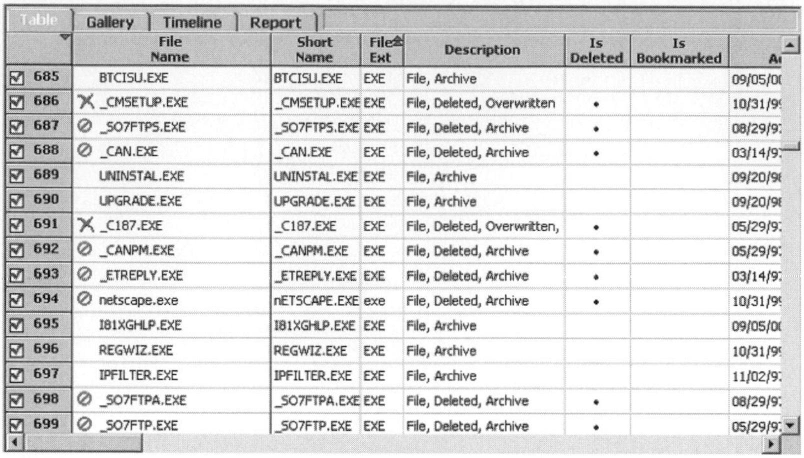

Table | Gallery | Timeline | Report

		File Name	Short Name	File Ext	Description	Is Deleted	Is Bookmarked	A
☑	685	BTCISU.EXE	BTCISU.EXE	EXE	File, Archive			09/05/0
☑	686	✗ _CMSETUP.EXE	_CMSETUP.EXE	EXE	File, Deleted, Overwritten	•		10/31/9
☑	687	⊘ _SO7FTPS.EXE	_SO7FTPS.EXE	EXE	File, Deleted, Archive	•		08/29/9
☑	688	⊘ _CAN.EXE	_CAN.EXE	EXE	File, Deleted, Archive	•		03/14/9
☑	689	UNINSTAL.EXE	UNINSTAL.EXE	EXE	File, Archive			09/20/9
☑	690	UPGRADE.EXE	UPGRADE.EXE	EXE	File, Archive			09/20/9
☑	691	✗ _C187.EXE	_C187.EXE	EXE	File, Deleted, Overwritten,	•		05/29/9
☑	692	⊘ _CANPM.EXE	_CANPM.EXE	EXE	File, Deleted, Archive	•		05/29/9
☑	693	⊘ _ETREPLY.EXE	_ETREPLY.EXE	EXE	File, Deleted, Archive	•		03/14/9
☑	694	⊘ netscape.exe	nETSCAPE.EXE	exe	File, Deleted, Archive	•		10/31/9
☑	695	I81XGHLP.EXE	I81XGHLP.EXE	EXE	File, Archive			09/05/0
☑	696	REGWIZ.EXE	REGWIZ.EXE	EXE	File, Archive			10/31/9
☑	697	IPFILTER.EXE	IPFILTER.EXE	EXE	File, Archive			11/02/9
☑	698	⊘ _SO7FTPA.EXE	_SO7FTPA.EXE	EXE	File, Deleted, Archive	•		08/29/9
☑	699	⊘ _SO7FTP.EXE	_SO7FTP.EXE	EXE	File, Deleted, Archive	•		05/29/9

Figure 24.4

EnCase Table view sorted first by tagged files and then by file extension.

primary sort on tagged and the secondary sort on extension as shown in Figure 24.4, and untag the unwanted files by extension (e.g. com, dll, drv, exe, sys, vxd), heeding the cautionary discussion in Larson (2001).

Untag file system artifacts that are unwanted or have already been exported. For instance, when dealing with FAT, deselect the boot sector, volume boot sector, unused disk area, and file allocation tables. An EnCase Escript available at the EnScript Library can be used to identify and bookmark duplicates. These duplicates can be untagged to exclude them from the list of files to be reviewed and can later be exported into \duplicates if needed.

[5]http://www.encase.com/support/escript_library.shtml

Copy the remaining tagged files into the \Review\rfiles directory on the working drive and export a list of these files to \Accounting directory to document your work. Consider saving the hash set of data set for future reference (e.g. to identify duplicate files on other media). Additional hash sets for EnCase are available at http://www.encase.com/ support/resources_hashsets.shtml.

24.2.3 WINDOWS GUI – FTK

When adding evidence to a case in FTK, select the option to calculate file hashes and perform the Known File Filter (KFF) comparison as shown in Figure 24.5.

24.2.3.1 GENERATE FILE LISTS AND HASH VALUES

While viewing all files in FTK, you can create a file list using Copy Special on the Edit Menu and selecting the desired fields as shown in Figure 24.6. The list will be placed in the Clipboard and can be pasted into any application.

24.2.3.2 RECOVER SLACK AND UNASSIGNED CLUSTERS

FTK provides a number of filtering options, including a Slack/Free Space button on the main screen that will provide a list of file slack, file system (volume) slack, and unallocated space as shown in Figure 24.7.

To export the contents of any of these objects, right click and select Export File and save to **\Prepare\pslack**. You can extract various types of data from unallocated and slack space using tools such DataLifter as shown in Figure 10.6.

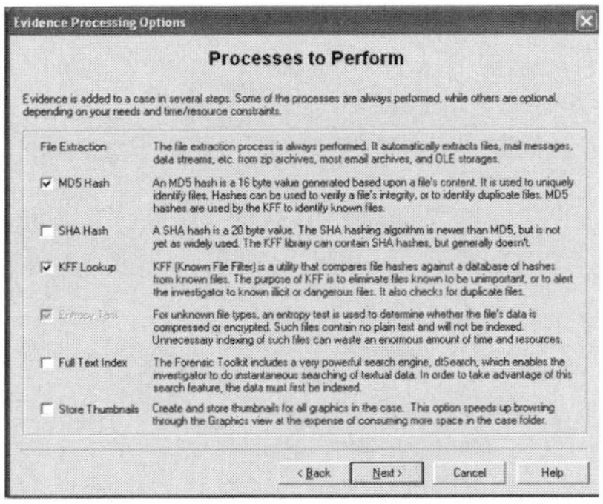

Figure 24.5

Calculate hash values of files and identify known files when adding evidence to FTK.

24.2.3.3 REMOVE DUPLICATES, KNOWN, AND UNNECESSARY FILES

Filtering capabilities are built into FTK in the form of a File Filter Manager, enabling you to select which types of files to exclude. For instance, Figure 24.8 shows File Filter Manager configured to ignore duplicate files, files with signature/extension mismatches, and known files identified by KFF. Care must be taken when configuring the File Filter Manager since one option, such as Other Known Files, can exclude many files that you actually want (e.g. index.dat).

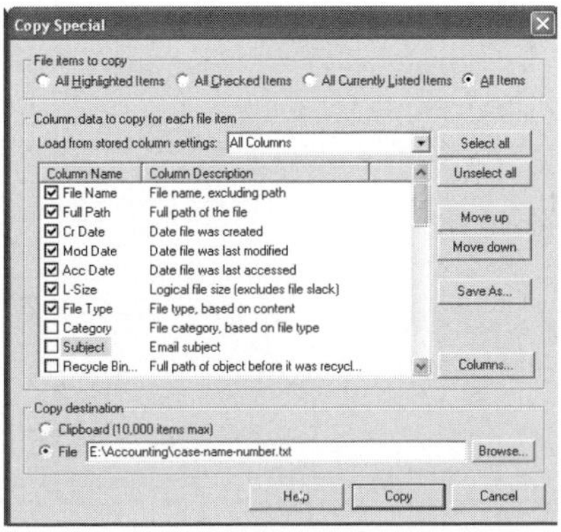

Figure 24.6

Export a list of files with associated properties.

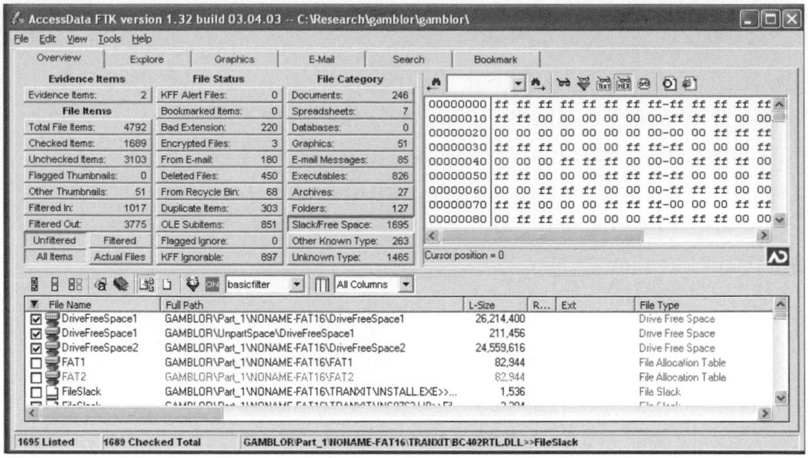

Figure 24.7

Exporting unallocated space using FTK.

Click on the Filtered button on the main FTK screen and then uncheck file system artifacts that are unwanted or have already been exported. For instance, when dealing with FAT, deselect the boot sector, volume boot sector, unused disk area, primary and secondary FAT. Copy the remaining checked files into the **\Review\rfiles** directory on the working drive and export a list of these files to **\Accounting** directory to document your work.

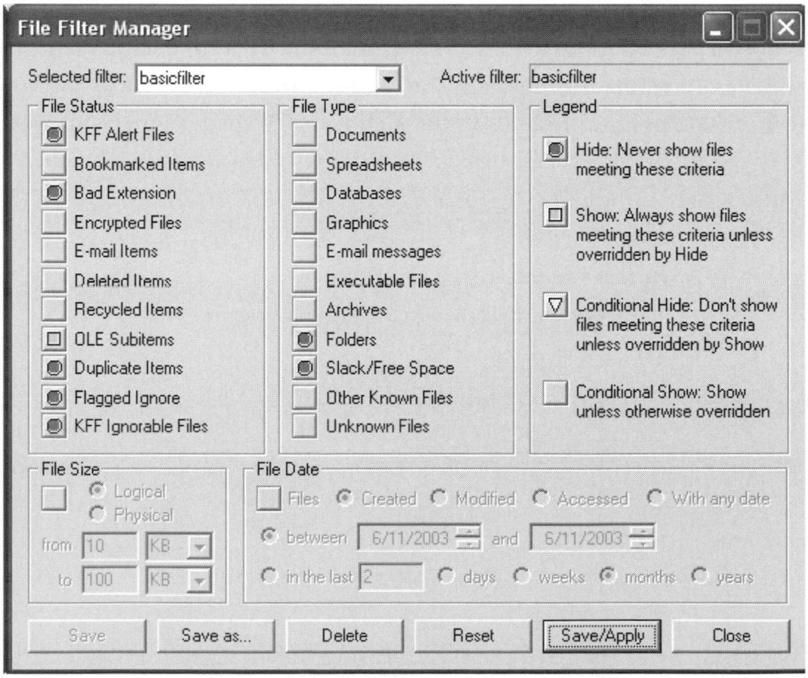

Figure 24.8

FTK File Filter Manager.

24.3 IDENTIFY AND PROCESS SPECIAL FILES

Compressed and encrypted files require special processing as do e-mail and associated attachments. As discussed in Chapter 2 of the Handbook, this special processing often requires a combination of tools with different features. Using tools of your choice, identify e-mail data files and move to \Prepare\special\email\[spool directory, if applicable]. Extract e-mail messages to text and attachments. Identify encrypted data and move it to \Prepare\special\encrypted and archived/compressed data to \Prepare\special\archive. If it is possible to decrypt or decompress these files, place the readable files in \Review\converted and add a list of these files in \Accounting. For a discussion of decrypting files, see Practical Approaches to Recovering Encrypted Digital Evidence (Casey 2002).

Perform a similar process for any other special files. For instance, if virus infected files may be important configure AntiVirus checking directory to log activity, virus check files to identify infected files, clean or move infected files, and save log to \Accounting\virus.log.

24.4 SUMMARY

The filtering process described in this chapter is superior to a less formalized analysis because all potentially useful data are extracted for examination. Less methodical approaches such as searching for specific keywords or extracting only limited file types may miss other important clues. Additionally, comparing the list of filtered files produced using different tools often highlights discrepancies such as incorrect MD5 calculations for some files and deleted files recovered by one tool and not the other. This type of tool validation is recommended for all cases to ensure that the maximum amount of useful data is extracted and that the examiner can explain any discrepancies between tools if the issue arises (e.g. in court).

Although the filtering process will enable investigators to gain a more complete understanding of the body of digital evidence, this is only the first stage in a thorough forensic analysis. Questions should arise in the investigator's mind while reviewing the evidence and, to answer these questions, it is usually necessary to examine specific aspects of the suspect systems. As discussed throughout the Handbook, there are many other system artifacts that can be useful in an investigation.

Each approach to filtering data has advantages and most people will find that it is desirable to combine command line and GUI approaches.

As a final stage in the filtering process, it is advisable to Bates number files in the working directory, for instance, using the Mareware bates_no utility as follows:

```
bates_no -p [path to source] -b [beginning bates number] -o [path\name of
output log] -R -i -v
```

REFERENCES

Casey, E. (2002) "Practical Approaches to Recovering Encrypted Digital Evidence," *International Journal of Digital Evidence*, 1(3) (Available online at http://www.ijde.org/02_fall_art4.html).

Larson, T. (2001) "The Other Side of Civil Discovery," Casey, E. (editor) *Handbook of Computer Crime Investigation: Forensic Tools and Technology,* London: Academic Press.

TEXTS AND ARTICLES

Amar V. D. (2003) "Regarding Child Pornography Extends the Supreme Court's Federalism Cases", Findlaw's Write, May 16, 2003 (Available online at http://writ.corporate.findlaw.com/amar/20030516.html).

Ananova (2001) "Chatroom man 'wanted someone to rape and torture his wife' ", August, 2001 (Available online at http://www.ananova.com/news/story/sm_379058.html).

ASCLD (2003) "Proposed Revisions to 2001 Accreditation Manual" (Available online at http://www.ascld-lab.org/pdf/aslabrevisions.pdf).

Associated Press (1997) "Wife's Internet friendship may have led to her death", January 23, 1997.

Associated Press (1997) "High-tech 'stalking' of Canadian family linked to teen-aged son", April 20, 1997.

Associated Press (1997) "As online harassment grows, calls for new laws follow" April 1, 1997.

Associated Press (1998) "FBI: Man posted sex pictures with daughter on Internet", February 10, 1998.

Associated Press (1999) "Child Participates in Sex Sting", September 2, 1999.

Associated Press (2001) "Teen placed in jail cell with sex offender", November 21, 2001.

Associated Press (2002) "Man Convicted In Internet Kidnap, Rape Of Teen", December 6, 2002.

Associated Press (2003) "Federal judge rules hacker covered by informant laws", March 18, 2003 (Available online at http://www.usatoday.com/tech/news/techpolicy/2003-03-18-hacker-informant_x.htm).

Baker R. (2000) "Harold Shipman's Medical Practice 1974–1998", Department of Health Audit Report, (Available online at http://www.doh.gov.uk/hshipmanpractice/shipman.pdf).

Baran P. (1964) Introduction to Distributed Communications Networks, RM-3420-PR. Santa Barbara, CA: The Rand Corporation (Available online at http://www.rand.org/publications/RM/RM3420/).

Bates J. (1999) "Judicial Review relating to Search Warrants – Discussion Paper", International Journal of Forensic Computing (Available online at http://www.forensiccomputing.com/archives/judicial.html).

Bellovin S. "Security Problems in the TCP/IP Protocol Suite," Computer Communications Review 19, No. 2 (April 1989) pp. 32–48.

Blanton T. (1995) "The Top-Secret Computer Messages the Reagan/Bush White House Tried to Destroy", National Security Archive (Available online at http://www.gwu.edu/~nsarchiv/white_house_email/).

Breyer S. (2000) "Reference Manual on Scientific Evidence", 2nd Ed., Federal Judicial Center (Available online at http://www.fjc.gov/public/pdf.nsf/lookup/sciman00.pdf).

Brown L. N. and Kennedy T. (2000) "The Court of Justice of the European Communities", London: Sweet & Maxwell.

Bryan B. (2002) "Letter writer is serial killer, concludes criminal profiler", St. Louis Post-Dispatch, May 28, 2002.

Buckeye B. and Liston K. (2002) "Recovering Deleted Files in Linux", February 2002, Sysadmin Magazine. (Available online at http://www.samag.com/documents/s=7033/sam0204g/sam0204g.htm).

Burgess A., Burgess A., Douglas J., and Ressler R. (1997) Crime Classification Manual, San Francisco: Jossey-Bass, Inc.

Burgess A. and Hazelwood R., Eds. (1995) Practical Aspects of Rape Investigation: A Multidisciplinary Approach, 2nd Ed., New York: CRC Press.

Burnette M. (2002) Forensic Examination of a RIM (BlackBerry) Wireless Device. (Available online at http://www.rh-law.com/ediscovery/Blackberry.pdf).

Burney M. (1997) "Cyber affair with teen-age girl leads to five years in prison", The Associated Press, August 22,1997. (Available online at: http://www.nando.net/newsroom/ntn/info/082297/info10_3348_noframes.html).

Cairns G. (1996) "Snuffsex", Australian Broadcasting Corporation's NewsRadio Network.

Carrier B. (2002) "Open Source Digital Forensics Tools: The Legal Argument", @stake Research Report (Available online at http://www.atstake.com/research/reports/acrobat/atstake_opensource_forensics.pdf).

Carrier B. (2003) "Defining Digital Forensic Examination and Analysis Tool Using Abstraction Layers" International Journal of Digital Evidence, Volume 1, Issue 4, Syracuse, NY (Available online at http://www.ijde.org/docs/02_winter_art2.pdf).

Carrier B. (2003) "Splitting The Disk – Part 1", Sleuth Kit Informer, Issue 2, March 15, 2003 (Available online at http://www.sleuthkit.org/informer/sleuthkit-informer-2.html).

Carrier B. (2003) "Splitting The Disk – Part 2", Sleuth Kit Informer, Issue 5, July 15, 2003 (Available online at http://www.sleuthkit.org/informer/sleuthkit-informer-5.html).

Carrier B. and Spafford E. H. (2003) "Getting Physical with the Digital Investigation Process", International Journal of Digital Evidence, Volume 2, Issue 2 (Available online at http://www.ijde.org/docs/03_fall_carrier_Spa.pdf).

Carter D. L. (1995) "Computer Crime Categories, How Techno-Criminals Operate". FBI Law Enforcement Bulletin (July).

Casey E. (2000) Digital Evidence and Computer Crime, 1st Edn., London: Academic Press.

Casey E., Larson T., and Long, H. M. (2002) "Network Analysis", Casey, E. (editor) Handbook of Computer Crime Investigation, pp. 225–228, London: Academic Press.

Casey E. (2002) "Cyberpatterns: Criminal Behavior on the Internet", Turvey, B. Criminal Profiling: An Introduction to Behavioral Evidence Analysis, 2nd Edn., London: Academic Press.

Casey E. (2002) "Error, Uncertainty and Loss in Digital Evidence", International Journal of Digital Evidence, Volume 1, Issue 2, 2002 (Available online at http://www.ijde.org/archives/docs/02_summer_art1.pdf).

Casey E. (2002) "Practical Approaches to Recovering Encrypted Digital Evidence", International Journal of Digital Evidence, Volume 1, Issue 3 (Available online at http://www.ijde.org/02_fall_art4.html).

CBS News (2002) "Daniel Pearl Killers Appeal", July 17, 2002 (Available online at http://www.cbsnews.com/stories/2002/05/31/attack/main510651.shtml).

CERT (2002) "Trojaned OpenSSH Distribution" (Available online at http://www.cert.org/advisories/CA-2002-24.html).

Chen D. W. (2000) "Teacher Is Accused of Duping Boy to Make a Sexual Video", New York Times, Late Edition – Final , Section B , Page 5 , Column 5, October 5th, 2000.

Clark F. and Diliberto K. (1996) Investigating Computer Crime, New York: CRC Press.

Clark D. F. and Gibbs K. E. (2001) "Wireless Network Analysis", Casey, E. (editor) Handbook of Computer Crime Investigation, London: Academic Press.

Clubb S. (2002) "Police explain suspect's suicide", The Illinois River Bend Telegraph, June 12, 2002 (Available online at http://www.zwire.com/site/news.cfm?newsid=4412382&BRD=1719&PAG=461&dept_id=25271&rfi=8).

Comer D. E. (1995) Internetworking with TCP/IP Volume I: Principles, Protocols, and Architecture, 3rd Edn., New Jersey: Prentice Hall.

Convery S. (2002) "Hacking Layer 2: Fun with Ethernet Switches" BlackHat Briefing. (Available online at http://www.blackhat.com/presentations/bh-usa-02/bh-us-02-convery-switches.pdf).

Cowen R. (2003) "Real IRA trial told of £750,000 spy payment", The Guardian, June 19, 2003.

CSI/FBI (2003) Computer Crime and Security Survey, San Francisco: Computer Security Institute.

DAWN group (2002) "Defence disputes video's validity: Daniel Pearl case", 17 May 2002, (Available online at http://www.dawn.com/2002/05/17/top2.htm).

Digital Discovery & e-Evidence, May 2001, New Rules on Cell Phone Monitoring Systems Fuel Interest in Embedded Systems.

Diskant T. (2002) "After sentencing, Lasaga, Yale face civil suit", February 22, 2002, Yale Herald (Available online at http://www.yaleherald.com/article.php?Article=359).

Dunne R., Long H. M., and Casey E. (2000) "Internet Crime" in the Encyclopedia of Forensic Science, London: Academic Press.

Durfee D. (1996) "Man pleads no contest in stalking case", The Detroit News, January 25, 1996.

Durkin K. F. and Bryant C. D. (1995) "Log on to sex: some notes on the carnal computer and erotic cyberspace as an emerging research frontier", Deviant Behavior: An Interdisciplinary Journal, 16:179–200.

Durkin K. F. and Bryant C. D. (1999) "Propagandizing pederasty: a thematic analysis of the on-line exculpatory accounts of unrepentant pedophiles". Deviant Behavior: An Interdisciplinary Journal, 20:103–127, 1999.

ENFSI Forensic Information Technology Working Group (2003) Draft Guidelines for Best Practice in the Forensic Examination of Digital Technology (Available online at http://www.enfsi.org/docs/FITWG-BPM-001-003.pdf).

Flusche K. J. (2001) "Computer Forensic Case Study: Espionage, Part 1 Just Finding the File is Not Enough!", Information Systems Security, March/April 2001, Auerbach.

Foote D. (1999) 'You Could Get Raped' Newsweek February 8, 1999.

Fyodor (1998) "Remote OS detection via TCP/IP Stack FingerPrinting", (Available online at http://www.insecure.org/nmap/nmap-fingerprinting-article.txt).

Garfinkel S. and Spafford E. H. (1996) Practical Unix and Internet Security, 2nd Edn., Cambridge, MA: O'Reilly & Associates, Inc.

Garfinkel S. (2002) "Network Forensics: Tapping the Internet", O'Reilly Network (Available online at http://www.oreillynet.com/pub/a/network/2002/04/26/nettap.html).

Gaudin S. (2000) "Case Study of Insider Sabotage: The Tim Lloyd/Omega Case", Number 3, 2000, Computer Security Journal, Volume XVI.

Gauis (2000) "Things to do in Ciscoland when you're dead" Phrack 56 (Available online at http://www.phrack.com/show.php?p=56&a=10).

Gaurdian Unlimited (2001) "Man convicted of internet-conspired rape", August 2, 2001 (Available online at http://www.guardian.co.uk/japan/story/0,7369,531343,00.html).

Gahtan A. (1999) "Electronic Evidence", Ontario: Carswell Legal Publications.

Geberth V. (1996) Practical Homicide Investigation, 3rd Edn., New York: CRC Press.

Gillespie A. A. (2003) "Sentences for Offences Involving Child Pornography" Crim.L.R. 81.

Gindin S. E. (1998) "As the Cyber-World Turns" (Available online at http://www.info-law.com/eupriv.html).

Gindin S. E. (1997) "Lost and Found in Cyberspace: Informational Privacy in the Age of the Internet" (Available online at http://www.info-law.com/lost.html, 1997).

Graham R. (2000) "Sniffing (network wiretap, sniffer) FAQ", (Available online at http://www.robertgraham.com/pubs/sniffing-faq.html).

Grand J. (2001) Published by the Forum of Incident Response and Security Teams in the Proceedings of the 14th Annual Computer Security Incident Handling Conference, Waikoloa, Hawaii, June 24–28, 2002. (Available online at http://www.mindspring.com/~jgrand/pdd/pdd-palm-forensics.pdf and http://www.atstake.com/research/reports/acrobat/pdd_palm_forensics.pdf).

Gringas C. (2002) "The Laws of the Internet", London: Butterworths.

Gross H. (1924) Criminal Investigation, London: Sweet & Maxwell.

Gruber J. S. (1995) Electronic Evidence, Eagan, MN: Thomas Legal Publishing.

Guidance Software (2001–2002) "EnCase Legal Journal" 2nd Ed. (Available online at http://www.guidancesoftware.com/support/downloads/LegalJournal.pdf).

Harding C. (2000) "Exploring the intersection of European law and national criminal law", 25 E.L.Rev. 374.

Harmon R., Rosner R., and Owens H. (1998) "Sex and Violence in a Forensic Population of Obsessional Harassers Psychology", Public Policy, and Law, 1998, Vol. 4, No. 1/2, 236–249, 1998 American Psychological Association.

Hazelwood R., Reboussin R., Warren J. I., and Wright J. A. (1991) "Prediction of Rapist Type and Violence from Verbal, Physical, and Sexual Scales," Journal of Interpersonal Violence, Vol. 6, No. 1, March 1991, pp. 55–67.

Henseler J. (2000) "Computer Crime and Computer Forensics" in the Encyclopedia of Forensic Science, London: Academic Press.

Henry J. (1985) Testimony before the Permanent Subcommittee on Governmental Affairs before the United States Senate, Ninety-Ninth Congress. (Available online at http://www.nostatusquo.com/ACLU/NudistHallofShame/Henry.html).

Henry P. H. and DeLibero G. (1996) Strategic Networking: From LAN and WAN to Information Superhighways. Massachusetts: International Thomson Publishing Company.

Hernandez A. E. (2000) "Self-reported contact sexual crimes of federal inmates convicted of child pornography offenses", Presented at the 19th Annual Conference Research and Treatment Conference of the Association for the Treatment of Sexual Abusers, San Diego, CA, November 2000.

Hoey A. (1996) "Analysis of The Police and Criminal Evidence Act, s.69 – Computer Generated Evidence", Web Journal of Current Legal Issues, in association with Blackstone Press Ltd.

Hollinger R. C. and Lanza-Kaduce L. (1988) "The Process of Criminalization: The Case of Computer Crime Law", Criminology, Vol. 26, No. 1, p. 104.

Hollinger R. C. (1997) Crime, Deviance and the Computer. Brookfield, VT: Dartmouth Publishing Company.

Holmes R. (1996) Profiling Violent Crimes: An Investigative Tool, 2nd Edn., Sage Publications.

Hoover T. W. (2002) "An Introduction to the DoJ's Manual on Searching and Seizing Computers", Federal Public Defender Report, Vol. 11, No 1, March, 2002.

Horvath F. and Meesig R. (1996) "The Criminal Investigation Process and the Role of Forensic Evidence: A Review of Empirical Findings", Journal of Forensic Sciences 1996; 41 (6): 963–969.

Hunt C. (1998) TCP/IP Network Administration, 2nd Edn., California: O'Reilly.

Jarvis C. (1998) "Teen again linked to e-mail affair", The News Observer, North Carolina, November 28, 1998.

Johnson T. (2000) "Man searched Web for way to kill wife, lawyers say", Seattle Post-Intelligencer, June 21, 2000 (Available online at http://seattlepi.nwsource.com/local/murd21.shtml).

Kirk P. (1974) Crime Investigation, 2nd Edn., Wiley & Sons.

Kelleher D. and Murray K. (1999) "IT Law in the European Union", London: Sweet & Maxwell.

Kelleher D. (2000) "The Council of Europe's Draft Convention on Cyber-Crime" Technology and Law Journal 12.

Korn H. (1966) "Law, Fact, and Science in the Courts", 66 Columbia Law Review 1080, 1093–94.

Lanning K. V. (2001) Child Molesters and Cyber Pedophiles – A Behavioral Perspective, in Hazelwood, Robert R. and Burgess, Ann W., Eds., Practical Aspects of Rape Investigation: A multidisciplinary Approach, 3rd Edn., pp. 199–232, Boca Raton, FL: CRC Press.

Larson T. (2001) "The Other Side of Civil Discovery", Casey, E. (editor) Handbook of Computer Crime Investigation: Forensic Tools and Technology, London: Academic Press.

Larson T. (2002) "Evidence Acquisition Boot Disk" (Available online at http://www.disclosedigital.com/eabd.html).

Law Commission (1997) Evidence in Criminal Proceedings: Hearsay and Related Topics, Law Commission Report 245 (Available online at http://www.lawcom.gov.uk/231.htm#lcr245).

Lee H., Palmbach T., and Miller M. (2001) Henry Lee's Crime Scene Handbook, London: Academic Press.

Mandia K., Prosise C., and Pepe M. (2003) Incident Response and Computer Forensics, 2nd Edn., New York: Osborne.

Mann D. and Sutton M. (1998) "Netcrime: More Change in the Organization of Thieving", British Journal of Criminology, Vol. 38, No. 2.

Mattei M., Blawie J. F., and Russell A. (2000) "Connecticut Law Enforcement Guidelines for Computer Systems and Data Search and Seizure", State of Connecticut Department of Public Safety and Division of Criminal Justice.

McAullife W. (2000) "Net paedophile gets five years", ZDNet, 24 October, 2000 (Available online at http://news.zdnet.co.uk/story/0,,s2082159,00.html).

McClintock D. (2001) "Fatal Bondage", Vanity Fair, June.

McGrath M. G. and Casey E. (2002) "Forensic psychiatry and the Internet: Practical perspectives on sexual predators and obsessional harassers in cyberspace", Journal of American Academy of Pychiatry and Law, 30:81–94, 2002.

McPherson T. (2003) "Sherlock Holmes' modern followers", The Advertiser, May 31, 2003.

Meighan C. W. (1966) Archaeology: an Introduction, p. 18, San Francisco: Chandler Publishing Company.

Meloy J. R., Ed. (1998) The Psychology of Stalking: Clinical and Forensic Perspectives, San Diego: Academic Press.

Meloy J. R. (1999) "Stalking: An Old Behavior, A New Crime", In Psychiatric Clinics of North America.

Microsoft (1999) "Back Up the Recovery Agent Encrypting File System Private Key in Windows 2000", Microsoft KB Q241201 (Available online at http://support.microsoft.com/default.aspx?scid=kb;[LN];Q241201).

Microsoft (2000) "FAT32 File System Specification" (Available online at http://www.microsoft.com/hwdev/hardware/fatgen.asp).

Microsoft (2001) "Detailed Explanation of FAT Boot Sector", MS KB140418 (Available online at http://support.microsoft.com/default.aspx?scid=KB;en-us;q140418).

Microsoft (2001) "General Information about Microsoft Office XP Encryption", Microsoft KB Article Q290112 (Available online at http://support.microsoft.com/default.aspx?scid=kb;en-us;Q290112).

Microsoft (2001) "Maximum Partition Size Using FAT16 File System" (Available online at http://support.microsoft.com/default.aspx?scid=kb;EN-US;118335).

Microsoft (2002) "Description of the FAT32 File System" (Available online at http://support.microsoft.com/default.aspx?scid=kb;EN-US;154997).

Microsoft (2002) "Limitations of FAT32 File System" (Available online at http://support.microsoft.com/default.aspx?scid=KB;en-us;q184006).

Morris R. T. (1995) "A Weakness in the 4.2BSD UNIX TCP/IP Software", Bell Labs Computer Science Technical Report 117 (February 25, 1985) (Available online at http://www.eecs.harvard.edu/~rtm/papers.html).

National Center for Forensic Science (2003) "Digital Evidence in the Courtroom: A Guide for Preparing Digital Evidence for Courtroom Presentation", Mater Draft Document, U.S. Department of Justice, National Institute of Justice, Washington (Available online at http://www.ncfs.org/DE_courtroomdraft.pdf).

National Computer Security Association (1997) Internet Security: Professional Reference. Indiana: New Riders Publishing.

National News (2002) "Text of Daniel Pearl case verdict", July 17, 2002 (Available online at http://www.jang.com.pk/thenews/jul2002-daily/17-07-2002/national/n6.htm).

New York Lawyer (2003) "NY Law Professor Pleads Guilty To Possessing Child Porn", April 16, 2003 (Available online at http://www.nylawyer.com/news/03/04/041603d.html).

O'Hanlon S. (2003) "Porn Curiosity Brings Scrutiny Against 'Who's' Townsend", January 11, 2003 (http://reuters.com/newsArticle.jhtml?type=internetNews&storyID=2028595).

O'Hara, C. (1970), Fundamentals of Criminal Investigation, 2nd Edn., Springfield: Charles C. Thomas Publishers.

Palisade Systems (2003) "Porn Tops File Sharing Usage" (Available online at http://www.palisadesys.com/news&events/p2pstudy.pdf).

Palm OS Memory Architecture (Available online at http://oasis.palm.com/dev/kb/papers/1145.cfm).

Palm OS Memory and Database Management (Available online at http://oasis.palm.com/dev/kb/papers/2029.cfm).

Parker D. (1976) Crime by Computer. New York: Charles Scribners' and Sons.

Parker D. (1983) Fighting Computer Crime. New York: Charles Scribners' and Sons.

Parker D. (1998) Fighting Computer Crime: A new Framework for Protecting Information. New York: John Wiley & Sons.

Pathe M. and Mullen, P. E. (1997) "The impact of stalkers on their victims", British Journal of Psychiatry, 170:12–17, January 1997.

Patzakis J. (2002) "Computer Forensics as an Integral Component of the Information Security Enterprise" (Available online at http://www.guidancesoftware.com/support/downloads/computerforensics.pdf).

Peek J., O'Reilly T., and Loukides M. (1997) Unix Powertools, California: O'Reilly.

Phrack 59 (2002) "Defeating Forensic Analysis on Unix", July 2002 (http://www.phrack.org/show.php?p=59&a=6).

Phrack 58 (2001) "Armouring the ELF: Binary encryption on the UNIX platform", December, 2001 (http://www.phrack.org/show.php?p=58&a=5)Pendlebury F. (2001) "Jail for Net Rapist", Dorsett Echo, July 17, 2001 (Available online at http://www.thisisdorset.net/dorset/archive/2001/07/17/BOURN_NEWS_NEWS10ZM.html).

Piper E. (1998) "Russian cybercrime flourishes: Deteriorating economic conditions have brought pirating and cracking mainstream", Reuters, December 30, 1998.

Popper K. R. (1959) Logic of Scientific Discovery, London: Hutchinson.

Prosise C., Mandia K., and Pepe M. (2003) Incident Response: Computer Forensics, 2nd Edn., New York: McGraw-Hill Osborne Media.

Psychiatric News (2000) "Protect Children From Predators On Internet, Parents Tell Congress", May 5, 2000, (Available online at http://www.psych.org/pnews/00-05-05/protect.html).

Reed C. and Angel J., Eds. (2000) "Computer Law", London: Blackstone Press.

Reed C., (1990–91) 2 CLSR 13–16 as quoted in Sommer, P. "Downloads, Logs and Captures: Evidence from Cyberspace Journal of Financial Crime", October, 1997, 5JFC2 138–152.

Regan E., Ed. (2000) "The New Third Pillar: Cooperation Against Crime in the European Union", Dublin: Institute of European Affairs.

Reuters (1997) "Swiss couple charged in U.S. child pornography sting", August 22, 1997.

Reuters (2001) "Russia Lacks Laws to Fight Child Porn Explosion", 26 March, 2001.

Rizzo T. (2001) "Judge rules Robinson must stand trial in 3 deaths", Kansas City Star, March 2, 2001, (Available online at http://www.kcstar.com/standing/robinson/case.html).

Robinson B. (2002) "Taking a Byte Out of Cybercrime", ABC News, July 15, 2002.

Romig S. (2001) "Incident Response Tools", Casey, E. (editor) Handbook of Computer Crime Investigation, London: Academic Press.

Rosenblatt K. S. (1995) High-Technology Crime: Investigating Cases Involving Computers. San Jose, CA: KSK Publications.

Saferstein R. (1998) Criminalistics: An Introduction to Forensic Science, 6th Edn., Upper Saddle River, New Jersey: Prentice Hall.

Sammes T. and Jenkinson B. (2000) Forensic Computing: A Practitioner's Guide, London: Springer.

Schneier B. (1996) Applied Cryptography: Protocols, Algorithms, and Source Code in C. New York: John Wiley & Sons.

Scott D. and Conner, M. (1997) In Haglund & Sorg (Eds.), Forensic Taphonomy: The Postmortem Fate of Human Remains, Chapter 2, Florida: CRC Press.

Scott M. (2003), "Independent Review of Common Forensic Imaging Tools", Memphis Technology Group Report (Available at http://mtgroup.com/papers/ForensicImagingTools.pdf).

Seaton D. (2002) "Hack attacks spook county", The Press Enterprise, June 14, 2002.

Securities and Exchange Commission (2002) "Order instituting proceedings pursuant to Section 15(b)(4) and Section 21c of the Securities Exchange Act of 1934, making findings and imposing cease-and-desist orders, penalties, and other relief: Deutsche Bank Securities, Inc., Goldman, Sachs & Co., Morgan Stanley & Co. Incorporated, Salomon Smith Barney Inc., and U.S. Bancorp Piper Jaffray Inc.", Administrative Proceeding, File No. 3-10957 (Available online at http://www.sec.gov/litigation/admin/34-46937.htm).

Seglem K., Luque M. E., and Murphy S. E. (2001), "Forensic Analysis of UNIX Systems", Casey, E. (editor) Handbook of Computer Crime Investigation, London: Academic Press.

Shamburg R. (1999) "A Tortured Case", Net Life, April 7, 1999.

Shannon E. (1998) "Main Street Monsters", Time Magazine, September 14, 1998 Vol. 152, No. 11.

Sheldon T. (1997) Windows NT Security Handbook. Berkeley, CA: Osborne McGraw Hill.

Sheldon B. (2001) "Forensic Analysis of Windows Systems", Casey, E. (editor) Handbook of Computer Crime Investigation, London: Academic Press.

Shimomura T. and Markoff J. (1996) "Takedown: The Pursuit of Kevin Mitnick, America's Most Wanted Computer Outlaw – by the man who did it". New York: Hyperion.

Shinkle P. (2002) "Serial Killer Caught By His Own Internet Footprint", St. Louis Post-Dispatch, 6-17-2002.

Singh S. (2000) The Code Book: The Science of Secrecy from Ancient Egypt to Quantum Cryptography. New York: Anchor Books.

Smith G. J. H. (1999) "Internet Law and Regulation", London: Sweet & Maxwell.

Snipe S. (2000) "Why your switched network isn't secure" (Available online at http://www.sans.org/resources/idfaq/switched_network.php).

Sobel D. (1999) "Galileo's Daughter: A Drama of Science, Faith, and Love", London: Fourth Estate.

Soh-jung Y. (2001) "Rising Number of Sexual Assault Cases Linked to Internet", The Korea Herald, July 7, 2001 (Available online at http://www.vachss.com/help_text/archive/sa_korea.html).

Sommer P. (1997) Downloads, Logs and Captures: Evidence from Cyberspace", Journal of Financial Crime, October, 1997, 5JFC2 138–152 (Available online at http://www1.bcs.org.uk/DocsRepository/03900/3968/logs.htm).

Sommer P. (1998) "Digital Footprints: Assessing Computer Evidence", Criminal Law Review (Special Edition), pp. 61–78, London: Sweet & Maxwell.

Spafford G. H. (1989) "The Internet Worm: Crisis and Aftermath" Communications of the ACM, 32 (6): 678–87

Specter M. (2002) "Do Fingerprints Lie?: The gold standard of forensic evidence is now being challenged", The New Yorker Issue of 2002-05-27 (Available online at http://www.newyorker.com/printable/?fact/020527fa_FACT).

States News Service (1998) "Woman Jailed For Sex With Boy", 2 December, 1998 (Available online at http://www.fathermag.com/news/rape/portland02.shtml).

States News Service (1999) "Cyber-Sex Seducer Gets Jail Time", 13, January 1999 (Available online at http://www.fathermag.com/news/rape/morganton.shtml).

Sterling B. (1993) A Short History of the Internet. Cornwall, CT: Magazine of Fantasy and Science Fiction. (Available online at http://www.library.yale.edu/div/instruct/internet/history.htm).

Stevens W. R. (1994) TCP/IP Illustrated, Vol. 1, Boston: Addison Wesley.

Sullivan B. (2003) "Pair who hacked court get 9 years" MSNBC, February 7, 2003.

Taylor M., Quayle E., and Holland G. (2001) "Child Pornography, the Internet and offending", ISUMA, Vol. 2, No. 2 (Available online at http://www.isuma.net/v02n02/taylor/taylor_e.shtml).

Tribune News Services (2000) "Black student charged with racist e-mail threats at college", April 21, 2000.

The Age (2002) "Japanese police to regulate Internet dating services", December 26 2002 (Available online at http://www.theage.com.au/articles/2002/12/26/1040511126218.html).

The Baltimore Sun (1998) "Woman Accused of Having Sex with Boy", (Available online at http://www.fathermag.com/news/rape/laurel.shtml).

Thomson S. (2002) "Charges Dropped Against Two Men in Sex-Slave Case, The Columbian, November 26, 2002.

Thornton J. I. (1997) "The General Assumptions and Rationale of Forensic Identification", for David L. Faigman, David H. Kaye, Michael J. Saks, & Joseph Sanders, Eds., Modern Scientific Evidence: The Law and Science of Expert Testimony, Vol. 2, St. Paul: West Publishing Company.

Turvey B. (2002) "Criminal Profiling: An Introduction to Behavioral Evidence Analysis" 2nd Edn., London: Academic Press.

United Kingdom Association of Chief Police Officers (2003) "The Good Practices Guide for Computer Based Electronic Evidence", National High-tech Crime Unit (Available online at http://www.nhtcu.org/ACPO Guide v3.0.pdf).

United Nations (1995) International Review of Criminal Policy No. 43 and 44 – United Nations Manual on the Prevention and Control of Computer-Related Crime; (Available at http://www.ifs.univie.ac.at/~pr2gq1/rev4344.html#crime).

Usborne D. (2002) "Has an old computer revealed that Reid toured world searching out new targets for al-Qa'ida?", UK Independent (Available online at http://www.independent.co.uk/story.jsp?story=114885).

United States Department of Justice (2001) "Electronic Crime Scene Investigation: A Guide for First Responders", National Institute of Justice, NCJ 187736 (Available online at http://www.ncjrs.org/pdffiles1/nij/187736.pdf).

U.S. Department of Justice (1994) "Federal Guidelines for Searching and Seizing Computers" (Available online at http://www.usdoj.gov/criminal/cybercrime/search_docs/toc.htm).

U.S. Department of Justice (1998) "Supplement to Federal Guidelines for Searching and Seizing Computers" (Available online at http://www.usdoj.gov/criminal/cybercrime/supplement/ssgsup.htm).

U.S. Department of Justice (1999) "Man convicted of threatening Federal judges by Internet e-mail" (Available online at http://www.usdoj.gov/criminal/cybercrime/johnson.htm).

U.S. Department of Justice (2002) "Searching and Seizing Computers and Obtaining Electronic Evidence in Criminal Investigations" (Available online at http://www.cybercrime.gov/searchmanual.htm).

U.S. Department of Justice (2002) "Abingdon man sentenced to 41 months for child pornography" (Available online at http://www.usdoj.gov/usao/md/press_releases/press02/wyattrelease.htm).

U.S. Department of Justice (2000) "Former high-tech executive pleads guilty to charge of traveling to have sex with minor he met on Internet", March 17, 2000 (Available online at http://www.usdoj.gov/usao/cac/pr/pr2000/050.htm).

U.S. Department of Justice (2002) "Parkville man receives 10 year prison sentence for exploiting child to produce child pornography" (Available online at http://www.usdoj.gov/usao/md/press_releases/press02/adam_thomas_valleau_sentenced.htm).

U.S. Department of Justice (2002c) "Man sentenced to 27 months in prison for possessing child Pornography" (Available online at http://www.usdoj.gov/usao/txs/releases/April%202002/020417-magargee.htm).

U.S. Department of Justice (2003) "Fact Sheet, PROTECT Act" (Available online at http://www.usdoj.gov/opa/pr/2003/April/03_ag_266.htm).

U.S. Postal Service (2001) "Multimillion-Dollar Child Pornography Enterprise Dismantled".

van der Knijff R. (2001) "Embedded System Analysis", Casey, E. (Editor) Handbook of Computer Crime Investigation: Forensic Tools and Technology, London: Academic Press.

Van den Wyngaert C. (1993) "Criminal Procedure Systems in the European Community", London: Butterworths.

Venema W. and Farmer D. (2001) "Forensic Computer Analysis: an Introduction", Dodds diary Doctor Dobb's Journal (Available online at http://www.ddj.com/documents/s=881/ddj0009f/0009f.htm).

Wacks, R. (1993) "Towards a New Legal and Conceptual Framework for the Protection of Internet Privacy" (1999 3(1) I.I.P.R. 1).

Widdowson L. and Ferlito J. (2001) "Tales from the Abyss: UNIX File Recovery", January 2001, SysAdmin Magazine (Available online at http://www.samag.com/documents/s=1441/sam0111b/0111b.htm).

Willassen S. Y. (2003) "Forensics and the GSM Mobile Telephone System" IJDE 2003 2:1 (Available online at http://www.ijde.org/docs/03_spring_art1.pdf).

Wired News (1998) "Cops 'Lured' into Net Sex", February 16, 1998.

CASES

A&M Records v. Napster (2001) Appeals Court, 9th Circuit, Case number 00-16401

American Oil Co. v. Valenti (1979) 179 Conn. 349, 358, 426 A.2d 305.

Arizona v. Bass (2001) 357 Ariz. Adv. Rep. 3, 31 P.3d 857.

Ashcroft v. American Civil Liberties Union (2002) US Supreme Court (Available online at http://www.aclu.org/Cyber-Liberties/Cyber-Liberties.cfm?ID=12039&c=59)

Ashcroft v. Free Speech Coalition (2002) US Supreme Court, Case Number 00-795 (Available online at http://laws.findlaw.com/us/000/00-795.html).

Atkins, Goodland v. Director of Public Prosecutions [2000] 2 Cr.App.R. 248; [2000] 2 All E.R. 425.

Attorney Generals' Reference (No. 5 of 1980) [1980] 3 All E.R. 88.

Bach v. Minnesotta (2002) Appeals Court, 8th Circuit, Case number 02-1238 (Available online at http://www.epic.org/privacy/bach/).

Bensusan Restaurant Corporation v. King (1997) Appeals Court, 2nd Circuit, Case Number 96-9344 (Available online at http://laws.findlaw.com/2nd/969344.html)

California v. Greenwood (1987) US Supreme Court, Case number 86-684 (Available online at http://laws.findlaw.com/us/486/35.html)

California v. Westerfield (2002) Superior Court, County of San Diego Central Division, California, Case Number CD165805.

Connecticut v. Sorabella (2002) Superior Court, Connecticut (Also see U.S. District Court of Massachusetts 1:2002cr10061).

Daubert v. Merrell Dow Pharmaceuticals, Inc. (1993) 509 U.S. 579, 113 S.Ct. 2786, 125 L.Ed.2d 469.

Director of Public Prosecutions v. Bignell (1998) 1 Cr.App.R.1.

Director of Public Prosecutions v. McKeown, Jones (1997) 1 All E.R. 737.

DVDCCA v. Bunner (2001) Appeals Court, Califorina, Case number CV786804 (Available online at http://www.eff.org/sc/20011101_bunner_appellate_decision.html)

FCC v. Pacifica (1978) US Supreme Court, Case number 77-528 (Available online at http://www.eff.org/Legal/Cases/FCC_v_Pacifica/)

Free Speech Coalition v. Ashcroft (2002) Appeals Court, 9th Circuit, Case number 00-795 (Available http://laws.findlaw.com/us/000/00-795.html)

Froistad v. United States (2001) Supreme Court, North Dakota, Case Number 20010111 (Available online at http://www.court.state.nd.us/court/briefs/20010111.aeb.htm).

Gates Rubber Co. v. Bando Chemical Indus., Ltd. (1996) 167 F.R.D. 90, 112 (D.C. Col.).

Honeywell v. Rand (1973) District Court, Minnesota, 4th division, Civil Action Number 4-67 CIV. 138 (Available online at http://www.cs.iastate.edu/jva/court-papers/).

Inset Systems, Inc. v. Instruction Set, Inc. (1996) District Court, Connecticut, Case Number CV-3 : 95CV-01314

Katz v. United States (1967) US Supreme Court, Case number 35 (Available online at http://laws.findlaw.com/us/389/347.html)

Kyllo v. United States (2001) US Supreme Court, Case number 99-8508 (Available online at http://laws.findlaw.com/us/000/99-8508.html)

Liser v. Smith (2003) District Court, District of Colombia, Civil Action Number 00-2325 (Available online at http://www.dcd.uscourts.gov/Opinions/2003/Huvelle/00-2325.pdf).

Michigan v. Miller (2002) 7th Circuit Court, Michigan.

Miller v. California (1973) Supreme Court, California, Case number 70-73 (Available online at http://laws.findlaw.com/us/413/15.html)

Minnesota v. Granite Gate Resorts, Inc. (1996) Appeals Court, Minnesota, Case Number C6-95-7227

Missouri v. Dunn (1999) Appeals Court, Western District of Missouri, Case Number 56028 (Available online at http://www.missourilawyersweekly.com/mocoa/56028.htm).

New York v. Ferber (1982) Supreme Court, New York, Case number 81-55 (Available online at http://laws.findlaw.com/us/458/747.html)

People v. Lugashi (1988) Appeals Court, California (205 Cal. App.3d 632), Case Number B025012.

Playboy Enterprise Inc. v. Frena (1993) District Court, Florida, Case number 93-489-Civ-J-20

R. v. Bow Street Magistrates, ex parte US Government, Allison (1999) 3 W.L.R. 620.

R. v. Bowden (2000) 1 Cr.App.R 438.

R. v. Chesterfield Justices ex parte Bramley (2000) 2 W.L.R. 409.

R. v. Cochrane (1993) Crim.L.R. 48.

R. v. Fellows, Arnold (1997) 2 All E.R. 548.

R. v. Governor of Brixton Prison, ex parte Levin (1997) 3 All E.R. 289.

R. v. Oliver, Hartrey and Baldwin (2003) Crim.L.R. 127.

R. v. Shephard (1993) 1 All E.R. 225.

R. v. Whiteley (1991) 93 Cr.App.R 25.

Regina v. Pecciarich (1995) 22 O.R. (3d) 748, Ontario Court, Canada.

Reno v. American Civil Liberties Union (1997) US Supreme Court, Case number 96-511 (Available online at http://www.aclu.org/Privacy/Privacy.cfm? ID=13904&c=252)

Sony Corporation of America v. Universal City Studios, Inc. (1984) US Supreme Court, Case number 81-1687 (Available online at http://laws.findlaw.com/us/464/417.html)

United States v. Carey (1998) Appeals Court, 10th Circuit, Case Number 98-3077 (Available online at http://laws.findlaw.com/10th/983077.html).

United States v. Grant (2000) District Court, Maine, Case Number 99-2332 (Available online at http://laws.lp.findlaw.com/1st/992332.html).

United States v. Gray (1999) District Court, Eastern District of Virginia, Alexandria Division, Case Number 99-326-A.

United States v. Henriques (1999) Appeals Court, 5th Circuit, Case Number 99-60819 (Available online at http://laws.findlaw.com/5th/9960819cr0v2.html).

United States v. Hersh (2001) Appeals Court, 11th Circuit, Case Number 00-14592 (Available online at: http://laws.lp.findlaw.com/11th/0014592opn.html).

United States v. Hilton (1997) District Court, Maine, Case Number 97-78-P-C (Available online at http://www.med.uscourts.gov/opinions/carter/2000/gc_06302000_2-97cr078_us_v_hilton.pdf).

United States v. Hilton (1999) Appeals Court, 1st Circuit, Case number 98-1513 (Available online at http://laws.findlaw.com/1st/981513.html)

United States v. Lamb (1996) 945 F. Supp. 441, 462 (N.D.N.Y).

United States v. Matthews (2000) "Amicus brief in U.S. v. Matthews" (Available at http://www.rcfp.org/news/documents/matthews.html).

United States v. Matthews (2000) Appeals Court, 4th Circuit (209 F.3d 338), Case Number 99-4183 (Available online at http://www.law.emory.edu/4circuit/dec99/994183.p.html).

United States v. Miller (1985) 771 F.2d 1219, 1237 (9th Cir.).

United States v. Mohammad Salameh (1993) District Court, Southern District of New York, S12 93 CR. 180 (Available online at http://laws.findlaw.com/2nd/941312v2.html).

United States v. Morris (1991) Appeals Court, 2nd Circuit (928 F.2d 504), Case number 90-1336

United States v. Moussaoui (2003) "Government's opposition to standby counsel's reply to the government's response to court's order on computer and e-mail evidence" (Available online at http://cryptome.org/usa-v-zm-email.htm).

United States v. Perez (2003) District Court, Southern District of New York, Case Number 02CR00854 (Available online at http://www.nysd.uscourts.gov/rulings/02CR00854.pdf).

United States v. Ramzi Yousef, Eyad Ismoil (2003) (Available online at http://caselaw.findlaw.com/data/circs/2nd/98104IP.pdf).

United States v. Randolph (2002) District Court, Eastern District of Pennsylvania, Case Number 02-114 (Available online at http://www.paed.uscourts.gov/documents/opinions/02D0408P.pdf).

United States v. Real Property & Premises Known as 5528 Belle Pond Drive (1991) 783 F. Supp. 253 (E.D. Va.).

United States v. Reedy (2000) District Court, Northwest District of Texas, Fort Worth Division, Case number 400-CR-0540Y (Available online at http://news.findlaw.com/hdocs/docs/reedy/usreedy51700indct.pdf).

United States v. Romero (1999) Appeals Court, 7th Circuit (189 F.3rd. 576), Case number 96 CR 167-1 (Available online at: http://www.laws.lp.findlaw.com/7th/982358.html).

United States v. Tank (1998) Appeals Court, 9th Circuit, Case Number 98-10001 (Available online at http://laws.findlaw.com/9th/9810001.html).

United States v. Thomas (1996) Appeals Court, 6th Circuit, Case Number 94-6648/6649 (Available online at http://www.eff.org/Legal/Cases/AABBS_Thomases_Memphis/)

United States v. Turner (1999) Appeals Court, 1st Circuit, Case Number 98-1258 (Available online at http://laws.lp.findlaw.com/1st/981258.html).

United States v. Zacarias Moussaoui (2001) District Court, Eastern District of Virginia (Available online at http://notablecases.vaed.uscourts.gov/1:01-cr-00455/).

Universal City Studios v. Corley (2001) Appeals Court, 2nd Circuit, Case number 00-9185 (Available online at http://laws.findlaw.com/2nd/009185.html)

Wisconsin v. Brian D. Robins (2002) Supreme Court, Wisconsin, Case Number 00-2841-CR (Available online at http://www.courts.state.wi.us/html/sc/00/00-2841.htm).

Wisconsin v. Kenney (2002) Appeals Court, Wisconsin, Case Number 01-0810-CR (Available online at http://www.courts.state.wi.us/html/ca/01/01%2D0810.htm).

Wisconsin v. Michael L. Morris (2002) Appeals Court, Wisconsin, Case Number 01-0414-CR (Available online at http://www.courts.state.wi.us/html/ca/01/01-0414.htm).

Wisconsin v. Schroeder (1999) Appeals Court, Wisconsin, Case Number 99-1292-CR (Available online at http://www.courts.state.wi.us/html/ca/99/99-2264.HTM).

Wisconsin v. Timothy P. Koenck (2001) Appeals Court, Wisconsin, Case Number 00-2684-CR (Available online at http://www.courts.state.wi.us/html/ca/00/00-2684.htm).

Access Point (AP): Central communication point for IEEE 802.11 wireless networks.

Address Resolution Protocol (ARP): A protocol in the TCP/IP suite that is used dynamically to associate network layer IP addresses with data-link layer MAC addresses.

***Ad hoc* Network**: Networks established when Bluetooth-enabled (or similar) devices come into proximity.

Anger Excitation (a.k.a. Sadistic) Behaviors: These include behaviors that evidence offender sexual gratification from victim pain and suffering. The primary motivation for the behavior is sexual, however the sexual expression for the offender is manifested in physical aggression, or torture behavior, toward the victim.

Anger Retaliatory (a.k.a. Anger or Displaced) Behaviors: These include offender behaviors that are expressions of rage, either towards a specific person, group, institution, or a symbol of either. The primary motivation for the behavior is the perception that one has been wronged or injured somehow.

Application: Software that performs a specific function or gives individuals access to Internet/network services.

Application Layer: Provides the interface between people and networks, allowing us to exchange e-mail, view Web pages, and utilize many other network services.

Asynchronous Transfer Mode (ATM): A connection-oriented network technology that provides gigabit-per-second throughput. This high-performance network technology can transport high-quality video, voice, and data.

Attached Resource Computer Network (ARCNET): One of the earliest local area networking technologies initially developed by Datapoint Corporation in 1977. Uses 93-ohm RG62 coaxial cable to connect computers. Early versions enabled computers to communicate at 2.5 Mbps. A newer, more versatile version called ARCNET Plus, supports 20 Mbps throughput.

Behavioral Evidence: Any type of forensic evidence that is representative or suggestive of behavior.

Behavioral Evidence Analysis: The process of examining forensic evidence, victimology, and crime scene characteristics for behavioral convergences before rendering a deductive criminal profile.

Behavior-Motivational Typology: A motivational typology that infers the motivation (i.e. Anger-retaliatory, Assertive, Reassurance, Sadistic, Profit, and Precautionary) of behavior from the convergence of other concurrent behaviors. Single behaviors can be described by more than one motivational category, as they are by no means exclusive of each other.

Broad Targeting: Any fire or an explosive that is designed to inflict damage in a wide reaching fashion. In cases involving broad targeting, there may be an intended target near the point of origin, but it may also be designed to reach beyond that primary target for other victims in the environment.

Buffer Overflow: Cleverly crafted input to a program that intentionally provides more data than the program is designed to expect, causing the program to execute commands on the system. Computer intruders use buffer flows to gain unauthorized access to servers or escalate their privileges on a system that they have already broken into.

Bulletin Board System (BBS): An application that can run on a personal computer enabling people to connect to the computer using a modem and participate in discussions, exchange e-mail, and transfer files. These are not part of the Internet.

Collateral Victims: Those victims that an offender causes to suffer loss, harm, injury, or death (usually by virtue of proximity), in the pursuit of another victim.

Computer Cracker: Individuals who break into computers much like safe crackers break into safes. They find weak points and exploit them using specialized tools and techniques.

Computer Crime: As defined in Federal and State Statutes. Includes theft of computer services; unauthorized access to protected computers; software piracy and the alteration or theft of electronically stored information; extortion committed with the assistance of computers; obtaining unauthorized access to records from banks, credit card issuers or customer reporting agencies; traffic in stolen passwords and transmission of destructive viruses or commands.

Corpus Delicti: Literally interpreted as meaning the "body of the crime" – refers to those essential facts that show a crime has taken place.

Crime Reconstruction: The determination of the actions surrounding the commission of a crime. This may be done by using the statements of witnesses, the confession of the suspect, the statement of the living victim, or by the examination and interpretation of the physical evidence. Some refer to this process as crime scene reconstruction, however the scene is not being put back together in a rebuilding process, it is only the actions that are being reconstructed.

Crime Scene: A location where a criminal act has taken place.

Crime Scene Characteristics: The discrete physical and behavioral features of a crime scene.

Crime Scene Type: The nature of the relationship between offender behavior and the crime scene in the context of an entire criminal event (i.e. point of contact, primary scene, secondary scene, intermediate scene, or disposal site).

Cybercrime: Any offense where the *modus operandi* or signature involves the use of a computer network in any way.

Cyberspace: William Gibson coined this term in his 1984 novel *Neuromancer*. It refers to the connections and conceptual locations created using computer networks. It has become synonymous with the Internet in everyday usage.

Cyberstalking: The use of computer networks for stalking and harassment behaviors. Many offenders combine their online activities with more traditional forms of stalking and harassment such as telephoning the victim and going to the victim's home.

Cybertrail: Any convergence of digital evidence that is left behind by a victim or an offender. Used to infer behavioral patterns.

Data-Link Layer: Provides reliable transit of data across a physical link using a network technology such as Ethernet. Encapsulates data into frames or cells before sending it and enables multiple computers to share a single physical medium using a media access control method like CSMA/CD.

Digital: Representation of information using numbers. The representation of information using binary digits (bits) and hexadecimal values are special cases of a digital representation.

Digital Evidence: Encompasses any and all digital data that can establish that a crime has been committed or can provide a link between a crime and its victim or a crime and its perpetrator.

E-mail, or Email: A service that enables people to send electronic messages to each other.

Equivocal Forensic Analysis: A review of the entire body of physical evidence in a given case that questions all related assumptions and conclusions. The purpose of the equivocal forensic analysis is to maximize the exploitation of physical evidence accurately to inform the reconstruction of specific crime scene behaviors.

Ethernet: A local area networking technology initially developed at the Xerox Corporation in the late 1970s. In 1980, Xerox, Digital Equipment Corporation, and Intel Corporation published the original 10 Mbps Ethernet specifications that were later developed by the Institute of Electrical and Electronic Engineers (IEEE) into the IEEE 802.3 Ethernet Standard that is widely used today. Ethernet uses CSMA/CD technology to control access to the physical medium (Ethernet cables).

Fiber Distribution Data Interface (FDDI): A token ring network technology that uses fiber optic cables to transmit data by encoding it in pulses of light. FDDI supports a data rate of 100 Mbps and uses a backup fiber optic ring that enables hosts to communicate even if a host on the network goes down.

Hardware: The physical components of a computer.

High Risk Victim: An individual whose personal, professional, and social life continuously exposes him/her to the danger of suffering harm or loss.

Host: A computer connected to a network.

ICQ ("I Seek You"): An Internet service that enables individuals to convene online in a variety of ways (text chat, voice, message boards). This service also enables file transfer and e-mail exchanges.

Internet: A global computer network linking smaller computer networks, that enable information sharing via common communication protocols. Information may be shared using electronic mail, newsgroups, the WWW, and synchronous chat. The Internet is not controlled or owned by a single country, group, organization, or individual. Many privately owned networks are not a part of the Internet.

Internet/Network Service: A useful function supported by the Internet/network such as e-mail, the Web, Usenet, or IRC. Applications give individuals access to these useful functions.

Internet Service Provider, or ISP: Any company or organization that provides individuals with access to, or data storage on, the Internet.

Internet Relay Chat (IRC): An Internet service that enable individuals from around the world to convene and have synchronous (live) discussions. This service also enables individuals to exchange files and have private conversations. The primary networks that support this service are EFNet, Undernet, IRCnet, DALnet, SuperChat and NewNet.

Jurisdiction: The right of a court to make decisions regarding a specific person (personal jurisdiction) or a certain matter (subject matter jurisdiction).

Locard's Exchange Principle: The theory that anyone, or anything, entering a crime scene both takes something of the scene with them, and leaves something of themselves behind when they leave.

Low Risk Victim: An individual whose personal, professional, and social life does not normally expose them to a possibility of suffering harm or loss.

Media Access Control (MAC) address: A unique number that is assigned to a Network Interface Card and is used to address data at the data-link layer of a network.

Message Digest: A combination of letters and numbers generated by special algorithms that take as input a digital object of any size. A file is input into a special algorithm to produce a sequence of letters and numbers that is like a digital fingerprint for that file. A good algorithm will produce a unique number for every unique file (two copies of the same file have the same message digest).

Method of Approach: A term that refers to the offender's strategy for getting close to a victim.

Modem (see Modulator/Demodulator): A piece of equipment that is used to connect computers together using a serial line (usually a telephone line). This piece of equipment converts digital data into an analog signal (modulation) and demodulates an analog signal into digits that a computer can process.

Modus Operandi: *Modus operandi* (MO) is a Latin term that means, "a method of operating." It refers to the behaviors that are committed by an offender for the purpose of successfully completing an offense. An offender's *modus operandi* reflects how an offender committed their crimes. It is separate from the offender's motives, or signature aspects.

Motive: The emotional, psychological, or material need that impels, and is satisfied by, a behavior.

Motivational Typology: Any classification system based on the general emotional, psychological, or material need that is satisfied by an offense or act.

Narrow Targeting: Any fire or explosive that is designed to inflict specific, focused, calculated amounts of damage to a specific target. Network Interface Card (NIC) – a card (circuit board) used to connect a host to the network. Every host must have at least one network interface card.

Network Interface Card (NIC): A piece of hardware used to connect a host to the network. Every host must have at least one network interface card. Every NIC is assigned a number called a Media Access Control (MAC) address.

Network Layer: Addresses and routes information to its destination using addresses, much like a postal service that delivers letters based on the address on the envelope.

Newsgroups: The online equivalent of public bulletin boards, enabling asynchronous communication that often resembles a discussion.

Peer-to-Peer Network (P2P):

Physical Evidence: Any physical object that can establish that a crime has been committed or can provide a link between a crime and its victim or a crime and its perpetrator.

Physical Layer: The actual media that carries data (e.g. telephone wires; fiber optic cables; satellite transmissions). This layer is not concerned with what is being transported but without it, there would be no connection between computers.

Piconet: A term to describe small networks established by Bluetooth-enabled (or similar) devices.

Point of Contact: The location where the offender first approaches or acquires a victim.

Point of Origin: The specific location at which a fire is ignited, or the specific location where a device is placed and subsequently detonated.

Port: A number that TCP/IP uses to identify Internet services/application. For example, TCP/IP e-mail applications use port 25 and Usenet applications use port 119.

Power Assertive (a.k.a. Entitlement) Behaviors: These include offender behaviors that are intended to restore the offender's self-confidence or self worth through the use of moderate to high aggression means. These behaviors suggest an underlying lack of confidence and a sense of personal inadequacy, that are expressed through control, mastery, and humiliation of the victim, while demonstrating the offender's sense of authority.

Power Reassurance (a.k.a. Compensatory) Behaviors: These include offender behaviors that are intended to restore the offender's self-confidence or self-worth through the use of low aggression or even passive and self-deprecating means. These behaviors suggest an underlying lack of confidence and a sense of personal inadequacy.

Presentation Layer: Formats and converts data to meet the conventions of the specific computer being used.

Primary Scene: The location where the offender engaged in the majority of their attack or assault upon their victim or victims.

Router: A host connected to two or more networks that can send network messages from one network (e.g. an Ethernet network) to another (e.g. an ATM network) provided the networks are using the same network protocol (e.g. TCP/IP).

Search Engine: A database of Internet resources that can be explored using key words and phrases. Search results provide direct links to information.

Secondary Scene: Any location where there may be evidence of criminal activity outside of the primary scene.

Session Layer: Coordinates dialog between computers, establishing, maintaining, managing, and terminating communications.

Signature Aspects: The emotional or psychological themes or needs that an offender satisfies when they commit offense behaviors.

Signature Behaviors: Signature behaviors are those acts committed by an offender that are not necessary to complete the offense. Their convergence can be used to suggest an offender's psychological or emotional needs (signature aspect). They are best understood as a reflection of the underlying personality, lifestyle, and developmental experiences of an offender.

Software: Computer programs that perform some function.

Souvenir: A souvenir is a personal item taken from a victim or a crime scene by an offender that serves as a reminder or token of remembrance, representing a pleasant experience. Taking souvenirs is associated with reassurance oriented behavior and needs.

Symbol: Any item, person, or group that represents something else such as an idea, a belief, a group, or even another person.

Synchronous Chat Networks: By connecting to a synchronous chat network via the Internet, individuals can interact in real-time using text, audio, video and more. Most synchronous chat networks are comprised of chat rooms, sometimes called channels, where people with similar interests gather.

Target: The object of an attack from the offender's point of view.

TCP/IP: A collection of internetworking protocols including the Transport Control Protocol (TCP), the User Datagram Protocol (UDP), the Internet Protocol (IP), and the Address Resolution Protocol (ARP).

Transport Layer: Responsible for managing the delivery of data over a network.

Trophy: A personal item taken from a victim or crime scene by an offender that is a symbol of victory, achievement, or conquest. Often associated with assertive oriented behavior.

User's Network (Usenet): A global system of newsgroups that enables people around the world to post messages to the equivalent of an online bulleting board.

Victimology: A thorough study of all available victim information. This includes items such as sex, age, height, weight, family, friends, acquaintances, education, employment, residence, and neighborhood. This also includes background information on the lifestyle of the victim such as personal habits, hobbies, and medical histories.

World Wide Web (WWW or Web): A service on the Internet providing individual users with access to a broad range of resources, including e-mail, newsgroups, and multimedia (images, text, sound, etc.).

AUTHOR INDEX

SUBJECT INDEX